Evolution on Planet Earth

The Impact of the Physical Environment

Evolution on Planet Earth

The Impact of the Physical Environment

Edited by

Lynn J. Rothschild
NASA Ames Research Center, USA

and

Adrian M. Lister
University College London, UK

ACADEMIC PRESS
An imprint of Elsevier
Amsterdam • Boston • Heidelberg • London • New York
Oxford • Paris • San Diego • San Francisco
Singapore • Sydney • Tokyo

CENTRE FOR ECOLOGY & EVOLUTION

THE LINNEAN SOCIETY OF LONDON

Academic Press
An Imprint of Elsevier
84 Theobald's Road, London WC1X 8RR, UK
http://www.academicpress.com

Academic Press
An Imprint of Elsevier
525 B Street, Suite 1900 San Diego, California 92101-4495, USA
http://www.academicpress.com

ISBN 0-12-598655-6

Library of Congress Catalog Number: 2002115394

British Library Cataloguing in Publication Data
Evolution on planet Earth: the impact of the physical environment
1. Evolution (Biology)
I. Rothschild, Lynn J.   II. Lister, Adrian
576.8

Designed and Typeset by Keyword Typesetting Services, Wallington, Surrey
Printed and bound in Italy

Contents

Contributors

Chaloner, Bill G., Department of Geology, Royal Holloway, University of London, Egham, TW20 0EX, UK

Clarke, Andrew, Biological Sciences, British Antarctic Survey, High Cross, Madingley Road, Cambridge CB3 0ET, UK

Forster, Catherine A., Department of Anatomical Sciences, Health Sciences Center, State University of New York at Stony Brook, Stony Brook, NY 11794, USA

Grady, Monica M., Department of Mineralogy, The Natural History Museum, Cromwell Road, London SW7 5BD, UK

Harris, David R., Institute of Archaeology, University College London, 31-34 Gordon Square, London WC1H 0PY, UK

Hewitt, Godfrey, Biological Sciences, University of East Anglia, Norwich NR4 7TJ, UK

Horneck, Gerda, German Aerospace Center (DLR), Institute of Aerospace, Medicine, Radiation Biology, D-51170 Köln, Germany

Jablonski, David, Department of Geophysical Sciences, University of Chicago, 5734 South Ellis Avenue, Chicago IL 60637, USA

Jain, Ravi, Molecular Biology Institute and Department of MCD Biology, UCLA, Los Angeles CA 90095, USA

Janis, Christine, Department of Ecology and Evolutionary Biology, Brown University, Providence RI 02912, USA

Lake, James A., Molecular Biology Institute, UCLA, Los Angeles CA 90024, USA

Lenton, Timothy M., Centre for Ecology and Hydrology – Edinburgh, Bush Estate, Penicuik, Midlothian EH26 0QB, UK

Lister, Adrian M., Department of Biology, University College London WC1E 6BT, UK

MacLeod, Norman, Department of Palaeontology, The Natural History Museum, Cromwell Road, London SW7 5BD, UK

Mancinelli, Rocco L., SETI Institute, Mail Stop 239-4, NASA Ames Research Center, Moffett Field, CA 94035-1000, USA

Moore, Jonathan, Molecular Biology Institute and Department of MCD Biology, UCLA, Los Angeles CA 90095, USA

Morey-Holton, Emily R., NASA Ames Research Center, Moffett Field, CA 94035-1000, USA

Nisbet, Euan G., Department of Geology, Royal Holloway, University of London, Egham TW20 0EW, UK

Potts, Richard, Human Origins Program, National Museum of Natural History, Smithsonian Institution, Washington DC 25060, USA

Raven, John A., Department of Biological Sciences, University of Dundee, Dundee DD1 4HN, UK

Rawson, Peter, Department of Geological Sciences, University College London WC1E 6BT, UK

Rayner, Jeremy M.V., School of Biology, University of Leeds, Clarendon Way, Leeds LS2 9JT, UK

Rivera, Maria, Molecular Biology Institute and Department of MCD Biology, UCLA, Los Angeles CA 90095, USA

Rothschild, Lynn J., Mail Stop 239-20, NASA Ames Research Center, Moffett Field, CA 94035-1000, USA

Skene, Keith R., Department of Biological Sciences, University of Dundee, Dundee DD1 4HN, UK

Sleep, N.H., Department of Geophysics, Stanford University, Stanford, CA 94305, USA

Vermeij, Geerat J., Department of Geology, University of California at Davis, One Shields Avenue, Davis, CA 95616, USA

Introduction

Lynn Rothschild and Adrian Lister

Long before Darwin placed the study of evolution on a firm footing, a relationship between organisms and their physical environment was apparent. Polar bears do not sunbathe in the Bahamas, and trees do not grow on the South Pole. Even a parasite living in the gut of an unsuspecting mammal – apparently insulated from the physical environment – must know which way is up (gravity), and is usually exposed to the outside world at some stage in its life history.

In this book we seek to explore how physical parameters have constrained, moulded and triggered the evolution of life on our planet. Such parameters can be considered under two rough headings:

1. Fundamental physical and chemical parameters, such as gravity or the nature of key chemical elements.
2. Major physical forces that are contingent on the nature of the Earth itself, such as sea level change or volcanic activity.

Clearly these two categories interact: the gravitational force on Earth is determined by the size and composition of the planet, and atmospheric composition has been strongly influenced by Earth processes such as vulcanism and photosynthesis.

The sense in which we mean 'physical factors' is in contradistinction to biological or 'biotic' factors. 'Physical' is thus synonymous with 'abiotic' and encompasses not only phenomena conventionally under the purview of physics, but also chemistry, geology, cosmology, and so

on. Even this dichotomy is not always clear. When largely determined by a biological process (photosynthesis), is the oxygen tension of the atmosphere, so crucial to animal life, a physical or a biotic factor?

The originally intuitive connection between organisms and their environment was ultimately translated into evolutionary theory. Evolutionary ideas were spawned first by the desire to explain the diversity of living forms, to which subsequently was added the realization that species have changed through geological time. The two principal attempts to understand the mechanisms behind change focused on use and disuse plus the internal 'will' of the organism (Lamarckian evolution), or selection of natural variation among organisms (the Darwin–Wallace theory of natural selection).

Under either evolutionary model, organisms somehow become better suited to their environment. But the fundamental break made by Darwin and Wallace from earlier evolutionary thinkers was that they introduced external conditions as a major force, as opposed to the 'internally driven' evolution of Lamarckism or orthogenesis. For Darwin, the primary force of natural selection on an organism came from its 'conditions of life', under which he explicitly included both biotic and physical factors. The latter included 'disaster caused by earthquakes, deluges and heavenly bodies' (Darwin, 1859).

With the intellectual connection between Darwinism and Mendelism in the first half of the twentieth century, the focus shifted to

population genetics and mechanisms of speciation. Evolutionary biologists such as Ernst
Mayr and Theodosius Dobzhansky developed
new concepts of speciation as part of what is
now known as the Modern Synthesis. This time
the physical environment acted not only in
selecting for adaptations, but also to physically
isolate portions of a previously cohesive interbreeding species in the process of allopatric
speciation.

During the 1940s and 1950s, research on patterns of change through geological time was
initiated by G.G. Simpson (1944, 1953), triggering a new wave of palaeontological investigation (e.g. Eldredge and Gould, 1972; Gingerich,
1976; Stanley, 1979). Patterns of phenotypic
change within lineages (microevolution), and
of clade diversity through time (macroevolution), invited hypotheses as to the role of physical factors in evolutionary change. According to
one model, species stay stable in constant environments, or because they tend to track environmental changes by adjusting their geographic
range. Major environmental changes would
be required to trigger evolutionary change,
perhaps as concerted speciation and extinction
events across a broad phylogenetic spectrum
(Vrba *et al.*, 1995). On the other hand, Sheldon
(1996) has suggested a model whereby stasis
prevails in moderately varying environments,
and rapid evolution in constant ones, and this
has received some support (Gould, 2002).

Since the 1980s, patterns of diversification
on a broader, cladal level (macroevolution)
have increasingly been a focus of study, and
the implication of physical factors is a major
area of investigation (e.g. Benton, 1996;
Jablonski, Chapter 13). These focus on major
global and regional environmental changes
through the geological record, and attempt to
find correlations with major biotic innovations
and shifts of relative diversity, including 'mass
extinction' events (Hallam and Wignell, 1997;
MacLeod, Chapter 14).

Whatever its macroevolutionary deploy-
ment, the evolution of individual species is
still fundamental, entailing local genetic adaptation to environment or environmental
change. In many cases, the biotic or abiotic
(physical) nature of the selective force seems
evident. Mimicry and other forms of protective

coloration are well known among butterflies,
fish and so on, and are clearly biotically driven.
The recent burgeoning of fields such as behavioural ecology and population biology stress
largely biotic interactions in shaping adaptation, in areas such as sexual selection and the
setting of life-history parameters. Yet adaptations to the physical environment are of equal
importance and remain the subject of much
research. Antarctic fish survive in frigid temperatures by producing so-called 'antifreeze' in
their blood (Clarke, Chapter 11); organisms living in other extreme environments such as
exceptionally low pH evolve mechanisms to
allow survival in otherwise inhospitable habitats (Rothschild and Mancinelli, 2001). Even
today new examples are being uncovered.
Mosquitoes in pitcher plants have shifted
photoperiodism in response to global warming
(Bradshaw and Holzapfel, 2001).

Many of these examples, moreover, illustrate the complex interaction of physical and
biotic factors. Polar bears are white because
they live in a white environment, driven by
climate and ultimately by higher-level physical
actors. Yet the proximate cause is biotic: there
would be no need to be white if feeding did not
depend on creeping up on prey unseen. In this
volume we stress the abiotic component, but it
is clear that a full understanding of evolution
must treat and synthesize both aspects. Indeed,
the interaction between the two is a central
question. In current research, the 'Red Queen'
hypothesis (van Valen, 1973) has come to
represent evolution driven by biotic interactions. Barnosky (2001) has suggested that the
Red Queen is dominant over shorter timescales
and in restricted ecological or geographical
areas, while physical factors are most effective
over longer timescales and wider areas.

It is clear that the physical environment is as
important for evolutionary biology today as it
was to our nineteenth century forebears.
Today, our knowledge of the physical history
of the Earth and other planets, and the physical
limits for life, is increasing rapidly. The new
meta-discipline of astrobiology recognizes the
importance of understanding the role of the
physical environment in the origin and evolution of life if we are to ask the questions 'how
did life on Earth arise?', 'what is its future?',

and 'are we alone in the universe?' At the same time, the field has become complex, and has fragmented along disciplinary lines, so the time is ripe to gather together the threads as a starting-point for a new synthesis.

This book grew out of a meeting organized at the Linnean Society of London in 1999. The idea behind the meeting was born when Lynn Rothschild, a specialist in environmental influences on microbial evolution, read Jared Diamond's (1996) *Guns, Germs and Steel*. The thesis of his work is that the history of human societies is influenced by physical factors in the environment, from climate to geography. From the importance of the environment to the question of life on Mars, to the origin of life on Earth, to the evolution of our own species, the meeting would take in the breadth of the subject. Adrian Lister, a specialist in Quaternary mammal evolution, agreed to become the meeting's co-organizer, and from that point on we have been jointly responsible for the project.

What follows are chapters devoted to exploring the influence of many key factors in the physical environment on a wide range of evolutionary phenomena. Because of the multidisciplinary nature of the work, each author was encouraged to present basic information followed by the latest advances in the field including the author's own thinking. In cases the assignment itself spurred a reinterpretation of existing data in light of the theme. In covering such a large canvas in one book, we had to invite one person to represent huge areas of scientific endeavour. Important contributions have been made by many other colleagues, and we trust they will find their work represented here; complementary volumes include Vrba *et al.* (1995) and Culver and Rawson (2000).

The span of the book is vast, from before the origin of life to the birth of human civilization. Thus, our use of the term 'evolution' is mostly in its conventional biological sense, though at the two extremes also touches on chemical evolution and cultural evolution respectively. The organization is not deliberately chronological, but takes on a broad time-directional stamp because the discussion of fundamental forces centres more around early events in life's history, while Earth forces (such as plate tectonics or ice ages) become important once life has

diversified and developed strong biogeographic patterns. However, there are important exceptions, such as the importance of gravity for vertebrate locomotion, or of vulcanism in the early Earth.

Nor have we limited ourselves to purely terrestrial phenomena but have stepped away from the Earth to encompass extraterrestrial influences. Grady's chapter (Chapter 21) looks at the possible origin of life on Mars as a sort of 'control' for the Earth, and also considers Martian influences on terrestrial evolution. Horneck (Chapter 7) discusses interplanetary movement of life, while the Sun is considered by Rothschild (Chapter 6). Nisbet and Sleep (Chapter 1) discuss impact history on early Earth, while MacLeod's (Chapter 14) treatment of extinction includes later impacts.

We recognize, however, that other aspects of this area are not covered. We have not given voice to a proponent of bolide impacts as a major cause of extinction. The possible role of cometary input as a source of water, organics or even life is absent. The Moon is mentioned, but its potential importance both for the Earth and for the habitability of other planets could be developed further. These areas provide an excellent starting-point for future work.

In the process of putting the meeting together, we, together with Jim Lake, realized how little molecular biologists and biochemists, let alone virologists, have thought about the effect of the physical environment on evolution at that level: another area ripe for study. We are particularly excited that the present volume has already provided impetus to research in this area (Jain *et al.*, 2003).

Conspicuous by its absence is a chapter on the effect of high temperature on evolution. Regrettably, John Baross (University of Washington) was unable to attend the meeting for health reasons. The interested reader could start by consulting Rothschild and Mancinelli (2001).

The layout of the book broadly follows that of the meeting. The chapters are grouped into eight areas: The Earth's atmosphere and the building blocks of life, Radiation, Molecular evolution, Gravity, Temperature, The dynamic Earth, Climate, and Implications for life elsewhere. Many of the major physical factors are indicated by the chapter headings. Others can be located via the Index (e.g. volcanism – see chapters by MacLeod, Vermeij, and Forster; fire – see chapter by Clarke; magnetism – see chapter by Rothschild). Overlaps and crossovers are rife, which is as it should be: this aspect is explored by many authors, and is discussed in the Epilogue.

Many people helped in the organization of the meeting and the production of this volume. The immediate and sustained enthusiasm of the Linnean Society of London, particularly that of John Marsden, in producing the meeting was critical. The financial assistance of NASA for the US participants, and the Centre for Ecology and Evolution (University College London) allowed the idea to take flight; to all three bodies we remain profoundly grateful. Bruce Runnegar gave an excellent lecture on the effect of oxygen on evolution, but was unable to write the chapter. Tim Lenton kindly filled in the void and promptly contributed a different, but equally stimulating chapter considering this topic. To both we are grateful.

We would also like to offer our sincere thanks to Academic Press, and especially to Andrew Richford, for their sustained interest and patience. All chapters were reviewed by two independent referees: our sincere thanks to all these colleagues for their invaluable contribution. Others who deserve mention are Peter Sheldon, Samantha Fallon, Marquetia Baird and Jackie Holding.

We believe that the unprecedented variety of chapters in this volume forms an intellectual feast. In stimulating thought and future avenues of research, may it at least serve as an hors d'oeuvre.

References

Barnosky, A.D. (2001). Distinguishing the effects of the Red Queen and Court Jester on Miocene mammal evolution in the northern Rocky Mountains. *Journal of Vertebrate Paleontology*, **21**: 172–185.

Benton, M.J. (1996). Testing the roles of competition and expansion in tetrapod evolution. *Proceedings of the Royal Society of London, Series B*, **263**: 641–646.

Bowring, S.A., Erwin, D.H., Jin, Y.G. et al. (1998). U/Pb zircon geochronology and tempo of the end-Permian mass extinction. *Science*, **280**: 1039–1045.

Bradshaw, W.E. and Holzapfel C.M. (2001). Genetic shift in photoperiodic response correlated with global warming. *Proc. Natl. Acad. Sci., USA*, **98**: 14509–14511.

Culver, S.J. and Rawson, P.F. (eds) (2000). *Biotic Response to Global Change: the Last 145 Million Years*. Cambridge: Cambridge University Press.

Darwin, C. (1859). *On the Origin of Species by Means of Natural Selection*. London: John Murray.

Diamond, J. (1996). *Guns, Germs and Steel: the Fate of Human Societies*. New York: Norton.

Eldredge, N. and Gould, S.J. (1972). Punctuated equilibria: an alternative to phyletic gradualism. In T.J.M. Schopf (ed.) *Models in Paleobiology*. San Francisco: Freeman, Cooper, pp. 82–115.

Gingerich, P.D. (1976). Paleontology and phylogeny: patterns of evolution at the species level. *Am. J. Sci.*, **276**: 1–28.

Gould, S.J. (2002). *The Structure of Evolutionary Theory*. Cambridge, Mass.: Belknap Press.

Hallam, A. and Wignell, P.B. (1997). *Mass Extinctions and their Aftermath*. Oxford: Oxford University Press.

Jain, R., Rivera, M.C., Moore, J.E. and Lake, J.A. (2003). Horizontal gene transfer accelerates genome innovation and evolution. *Molecular Biology and Evolution*, in press.

Rothschild, L.J. and Mancinelli, R.L. (2001). Life in extreme environments. *Nature (London)*, **409**: 1092–1101.

Sheldon, P.R. (1996). Plus ça change – a model for stasis and evolution in different environments. *Palaeogeog. Palaeoclimatol. Palaeoecol.*, **127**: 209–227.

Simpson, G.G. (1944). *Tempo and Mode in Evolution*. New York: Columbia University Press.

Simpson, G.G. (1953). *The Major Features of Evolution*. New York: Columbia University Press.

Stanley, S.M. (1979). *Macroevolution: Pattern and Process*. San Francisco: W.H. Freeman.

Van Valen, L. (1973). A new evolutionary law. *Evol. Theory*, **1**: 1–30.

Vrba, E.S., Denton, G.H., Partridge, T.C. and Burckle, L.H. (eds) (1995). *Paleoclimate and Evolution, with Emphasis on Human Origins*. Yale University Press.

PART 1

The Earth's atmosphere and the building blocks of life

1

The physical setting for early life

E.G. Nisbet and N.H. Sleep

ABSTRACT

See Plates 1 and 2

The solar system is nearly 4.6 Ga (1 Ga $= 10^9$ years) old. The oldest strong evidence for life on Earth is from rocks about 3.5 Ga old, formed 1 Ga after the Earth's accretion, by which time complex oxygenic photosynthesis had already evolved. There is possible evidence for photosynthetic life in rocks from Isua, Greenland (*c*. 3.8 Ga), and life may have existed on Earth as early as 4 Ga ago. Much prior to that, heavy meteorite bombardment is likely to have boiled the oceans; it is unlikely that any life could have continuously survived prior to about 4.2–4.0 Ga ago, with the possible exception of hyperthermophile organisms living deep in hydrothermal systems. Around 4.5 Ga ago, a major impact could have melted the entire Earth. Other planets offer alternative early sites for the first life: Venus had oceans, but early Mars (with fewer large impacts than Earth) is an attractive candidate, with transfer of life to Earth (or even exchange), via meteorite, around 4 Ga ago.

Earth around 4 Ga ago, at the close of the Hadean, probably had deep oceans. Loss of hydrogen to space may have supplied minor oxidation power to the air/water system, which was probably slightly oxidizing, enough to supply sulphate for reduction by bacteria; water may have been rich in bicarbonate. The last common ancestor (LCA) of all cells may date from around this time. It may have been a hyperthermophile (an organism living at temperatures over 75°C). Whether the first common ancestor (FCA) lived in high temperature water is unclear, however. Frequent large meteorite impacts may have caused brief but catastrophic events of raised seawater temperature, and it is possible that only a hyperthermophile LCA survived.

Since the accretion, oceans have been stable on the planet's surface. Partitioning of water between the Earth's interior and the large surface reservoir that is the ocean may have been determined by the thermal budget of the mantle, with degassing (loss of water from the interior) and engassing (take up of water by the interior) controlled by the evolving geotherm. Photosynthesis may have evolved in stages, possibly as an adaptation of thermotaxis around hydrothermal vents. By 3.5 Ga ago, after the onset of the modern photosynthetic carbon cycle mediated by rubisco (ribulose bisphosphate carboxylase/oxygenase – the cause of the strong C isotope fractionation), the oxidation state was adequate to deposit sulphate in shallow water, and life had occupied mesothermophile habitats: the Earth's surface temperature was probably broadly similar (±0–30°C) to today.

1.1 Introduction

The Hadean aeon (from about 4.56 to 4.0 Ga ago: 1 Ga $= 1 \times 10^9$ years) and the succeeding Archaean aeon (from about 4.0 to 2.5 Ga ago) saw the start of life, and its early occupation of the planet. The Earth's physical environment was very different from today: this early environment must have had a crucial role in shaping the early evolution of life. The purpose of this review is to set out some of what is known

Evolution on Planet Earth
ISBN 0-12-598655-6

about the early stages of the planet's history, and to speculate on some of the deeper controls on the early history of life. In addition, the intent is to explore some of the possible consequences to the early evolution of life of the events that took place in the Hadean and early Archaean on the Earth and neighbouring planets. How did the early physical environment control the development of life?

1.2 Synopsis of accretion

The solar system is roughly 4.6 Ga old: evidence from meteorites gives ages of about 4.56 Ga (Tilton, 1988) for the initial condensation of material from a dust cloud that gave rise to the sun and planets. The start of the accretion of the Earth dates from this time. The sequence of events began with a dust cloud from one or more supernovae explosions, followed by the formation of small bodies (planetesimals) from particles in the dust cloud. These planetesimals grew by collision, and the largest became the nuclei of the planets, which swept up the smaller bodies (see review by Ahrens, 1990). In this phase the proto-Earth would have been uninhabitable, but the events that took place would have had a profound impact on the later history of the planet.

At a time during the early accretion of the Earth, when the radius was less than 1400 km (comparable to the present-day Moon), accreted material retained a full complement of volatiles such as water, carbon dioxide, sulphur dioxide, ammonia and methane. As the Earth grew, the impact shocks of bigger infalling planetesimals were greater, becoming capable of vaporizing incoming volatiles so that they reacted with metallic iron. Incoming water reacted with iron and Mg silicate in the early Earth to release hydrogen: the escape of this hydrogen to space could have left a weakly oxidizing environment in the interior of the small proto-planet. Later, during the bulk of the accretion, as the planet grew further, a massive water (steam) atmosphere may have developed. This acted like a thermal blanket over the accreting planet, blocking the escape

of heat. This opacity to outgoing infrared radiation may have induced a magma ocean on the Earth. Fe was in excess of available oxidants, and under these conditions, complete oxidation of iron did not occur, producing a metallic core and magnesium-rich mantle. Physical removal of Fe to the core removed it from reaction and allowed more oxidizing conditions elsewhere. Carbon gases in the atmosphere would have been oxidized to carbon dioxide, while the material in the interior was correspondingly reduced. In the atmosphere, light may have split (photolysed) water into H and OH, with subsequent loss of H to space. Any loss of hydrogen to space by photolysis would also have contributed to the oxidation power of the atmosphere. The surface temperature would have been buffered around 1200°C by the melting of wet silicates, and the gas content of the atmosphere would have been controlled by its solubility in silicate melt.

In the penultimate stages of accretion, late giant impacts would have occurred, as the largest competing planetesimals crossing the Earth's orbit finally impacted. Very large impacts would have been capable of driving off much of the early proto-atmosphere; the biggest impact could have reworked the entire planet. In the final stage, as impacts became less frequent and mostly smaller (tens to hundreds of km), the impactors would have gardened the surface of the Earth on a grand scale. In a sense, accretion is still unfinished because the Earth still collects meteorites, but only rarely are these on a 100 m scale or larger.

The accretion of the Earth would have been at the same time as accretion of Mars and Venus. Mars is much smaller that Earth (only about one-ninth of its mass) and would thus have had a rather different accretion history; on Mars, individual accretionary events may have been large in comparison to the size of the planet, but the subsequent re-equilibration of the planet may have been quicker. Even at a late stage (say 4–3.5 Ga ago) impacts capable of ejecting material from the Martian surface would have been relatively frequent. In contrast to Mars, Venus is closer to Earth in gravitational attraction (82% of its mass) and would have had a broadly comparable early accretion history. However, the history of the two planets diverged when Earth acquired a large Moon.

1.3 Impactor history on Earth

Some of the history of the Earth's accretion can be worked out, by considering the character of the planet and solar system we have inherited. For example, though a heavy rain of bodies of varying sizes probably continued for hundreds of millions of years, the size of the biggest impacts probably varied over time, and the frequency of impacts declined. Earth is an active planet, and there is no direct evidence of Hadean impacts left on its surface, but for the later impacts (4 to 3.5 Ga ago) there is some comparable record on the Moon. Some deductions are possible, from modelling and from comparison with the Moon's record.

1.3.1 The late great impact

The first event that made Earth radically different from Venus was probably a single late great impact (see reviews in Newsom and Jones, 1990; Halliday, 2000) (Figure 1.1). Late impactors on the Earth had profound consequences. As the planet grew, so did other proto-planetary bodies in orbits near the Earth. The late stages of accretion must have involved one or more giant impacts, as these bodies collided with the Earth. Subsequently, the enlarged body, with its somewhat larger gravitational attraction, would have continued to sweep out any smaller bodies in Earth-crossing orbits, with the size of impacts decreasing exponentially. Large impacts would have continued until the early Archaean (until roughly 3.6 Ga ago), but the mean size of impacts would have become progressively less, and large impacts would have been much less common by the mid- or late Archaean. By this stage, the rate of impact would have been far less than in the great Hadean events. On the modern Earth, impacts of life-threatening meteorites are now rare and – though perhaps unpleasant to dinosaurs – not immediately significant to the daily

processes of life. However, like the risk of large earthquakes, the near certainty of future impact is present though the time period may be long.

The most significant of the impacts was the putative event that is thought to have formed the Moon, which took place around 4.5 Ga ago. This was roughly 50 million years after the main initial accretion of the Earth. In this event, a proto-planet comparable to modern Mars, or bigger (i.e. 10% of the Earth's mass or more), hit the early proto-Earth at a stage when it was around 95% or less of its present mass. To put it into context, the impact of a body of 1000 km diameter is capable of so shocking a 100 bar H_2O-rich atmosphere (equivalent to 1 km of water) that the gases contained in the atmosphere would have reached a velocity over 11 km/sec, enough to escape the Earth. A larger *c.* 4000 km diameter body (e.g. the Moon-forming impactor) would have driven off a substantial amount of the proto-atmosphere.

A massive impact between two planets of a scale large enough to eject enough debris to form the Moon must have been rare in the formation of the solar system, and the Earth's impact would have taken place as the number of lesser impacts was declining, and most of the matter had already been swept up into planet-sized bodies. If a large impact of this magnitude took place fairly late in the accretion, at a time when the infall rate of other bodies had substantially declined, it would have caused a major change in the thermal history of the surface. The early proto-atmosphere, which had been thick enough to trap outgoing heat and to blanket a magma ocean, would not reform after the impact. The succeeding atmosphere would be thinner and allow more infrared heat loss from the surface to space, and consequently the surface of the Earth would then cool rapidly. The crust of the Earth would rapidly solidify (Ahrens, 1990), creating a place for life.

The Moon-forming event would have had immense long-term effects on the Earth (Ahrens, 1990). It tilted the planet and the momentum of the impact added to whatever spin the Earth had already acquired. The tilt of the Earth gave us the seasons, and facilitates the equator-to-pole transfer of heat that makes the Earth much more habitable. The angular momentum in the rotation has decayed, as the Moon has retreated and as tides have dissipated energy, to the present 24-hour day.

The impact probably melted the entire planet, even if it was not already molten, and thus imparted some of the heat that has driven the Earth's later geological evolution. To this day, a significant fraction of the heat dissipated by the plate system, and by plume volcanoes such as those in Hawaii, is primordial heat from the accretion events. Volcanic heat sources drive hydrothermal systems, providing habitat for hyperthermophile organisms, and continually renew the surface of the planet, supplying cations and anions to the oceans. Thus the heat that drives modern tectonic activity and plume volcanic activity comes in part from the lunar-forming impact.

The aggregation of the core of the impactor to the core of the pre-existing Earth may also have had important biological consequences, by creating a large enough liquid zone in the core to sustain, over the aeons, a planetary magnetic field. This has acted to protect the surface of the Earth and the atmosphere from the solar wind, and may be an important contributory factor in the long-term habitability of the Earth.

There is also a small but interesting possibility that the Moon-forming impactor, which was a planet in its own right, the size of Mars, could have hosted life. If so, there would be the faint chance that a cell on a piece of the surface ejected on impact could have survived, frozen in space, and then much later fallen back to Earth or to a briefly habitable Moon.

1.3.2 Impacts of bodies >250 km (from beginning to 4.4–3.8 Ga ago)

In the later stages of accretion, and up to as late as around 4.4 to 3.8 Ga ago, the Earth would have sustained a few impacts from bodies greater than 250–300 km diameter (Figure 1.2). These were capable of boiling much of the ocean and heating the water above 100°C. As the surface water boils, the atmosphere is added to by steam and pressure rises: an ocean 3.5 km deep must be heated to about 350°C to boil completely. Any living organisms in the ocean would have been sterilized. It is just possible that hyperthermophile organisms living in hydrothermal systems in the rock may have survived briefly. For example, an organism capable of surviving 115°C but living in a hydrothermal setting that was 'normally' at 75°C, might survive if deep enough in the vol-

Late Moon-forming impact (4.5 Ga)

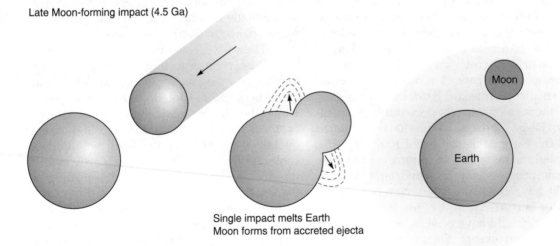

Single impact melts Earth
Moon forms from accreted ejecta

Figure 1.1 The Moon-forming event, about 4.5 Ga ago. Note that Figures 1.1–1.5 are cartoons, which are not properly to scale.

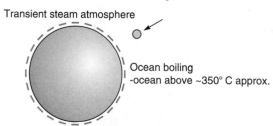

Impacts of bodies >~250 km (ocean boiling); likely prior to ~4.2–3.8 Ga ago

Transient steam atmosphere

Ocean boiling -ocean above ~350° C approx.

Ejecta blanket – >1 km basalt debris from large impact; reacts with CO_2 in air/sea

Figure 1.2 Bombardment by large (ocean-boiling) objects (>~250 km) (not to scale). These would have occurred with declining frequency in the period prior to ~4 Ga ago or later.

canic edifice (e.g. 1 km below surface). However, for larger ocean-boiling impacts this would be improbable – the heat would soon enter the hydrothermal circulation system (possibly reversing it) and sharply changing chemical conditions. Perhaps such an organism could survive being blown into space within a rock, and then falling back to the Earth thousands of years later after as the steam condensed.

Simultaneously, as the ocean boiled, massive blankets of basalt ejecta, thrown out by the impact, would react with the hot steam/CO_2 vapour. The basalt debris would have buffered the CO_2 content in the air and managed pH, such that pH may not have been far from that in modern seawater, and the ratio of CO_2 in water to atmosphere may have been around 60:1 in the settled periods between impacts, as it is now. The reaction between atmosphere and ejecta blankets may have been very important in capturing carbon dioxide as carbonate (later subducted) and hence reducing the greenhouse warming of the atmosphere. Indeed, among the early perils may have been the danger that in time intervals between impacts the oceans would have become too cool to sustain life. The early Sun was faint, and between heating impacts the loss of heat to space may have large enough to freeze the water. However, even in such events, lanes and leads in the global ice pack would have

remained, and hot to cool hydrothermal systems would have been habitable. A comparison can be drawn from the modern Earth, on which 5°C hydrothermal vents exist in the high Arctic, and Antarctic lakes do not freeze because ablation and latent heat maintain finite ice thickness if there is some sunlight (McKay *et al.*, 1985).

1.3.3 *Impacts of bodies in the size range to 100 km diameter (frequent until c. 3.5 Ga)*

Impacts in the size range up to 100 km diameter (Figure 1.3) would have been 'frequent' (in the geological sense) as late as 3.5 Ga ago: the marks of such impacts can still be seen with the naked eye on the face of the Moon. The larger impacts in this range would have been capable of heating the surface waters of the oceans >100°C (Figure 1.4), not enough to boil the oceans (see above). Any free-living or bottom-living oceanic life would probably have been sterilized. For most such impacts, the impact heat would dissipate to space relatively quickly, within 100 years. Hyperthermophile life capable of living up to 115°C and living in the subsurface would probably have survived such impacts. Thus whatever the nature of the first common ancestor, the last common ancestor is likely to have been hyperthermophile (Gogarten-Boekels *et al.*, 1995).

1.4 Atmospheres and water

1.4.1 *Late accretion*

Life exists by exploiting contrasting oxidation states. For example, in the breathing process reduced matter is, in effect, combusted against oxygen from the air. In photosynthesis, life uses light energy to regain reduction power, managing the oxidation state of carbon compounds. Reactions between the atmosphere and the rock during accretion, core precipitation, and loss of hydrogen to space, would have had major long-term consequences for the distribution of oxidized and reduced ma-

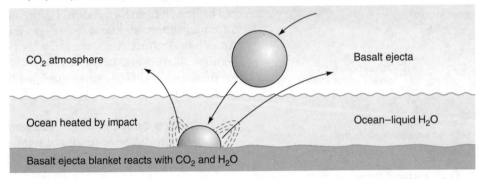

Impacts of 1–250 km (mid-sized)
(frequent prior to ~3.5 Ga or later)

CO$_2$ atmosphere

Basalt ejecta

Ocean heated by impact

Ocean–liquid H$_2$O

Basalt ejecta blanket reacts with CO$_2$ and H$_2$O

Figure 1.3 Bombardment by 1–250 km size objects (ocean warming) (not to scale). These would have been frequent prior to ~3.5 Ga ago.

terial in the planet. All these would have had profound influence in making the planet habitable. In the earliest stages of accretion, the accretion of large amounts of volatiles (e.g. CO$_2$ and H$_2$O) would have had a general oxidizing effect on the Earth's bulk. Then, with larger impacts, and reduction of MgFe silicate by CO for instance, the reduced core would have precipitated and CO$_2$ would have been given off: the results would have been a huge mass of reduced iron in the interior, and a more oxidized exterior. In the waning phases of accretion, as late impacts brought in less

energy and a solid crust and lithosphere was established, the overall regime would have had the effect of oxidizing the mantle. At the same time, if significant amounts of steam had been present in the upper atmosphere, photolysis of water would have occurred when light split H$_2$O to H and OH, with subsequent accumulation of oxidation power (OH) as the H was lost to space.

The rare giant impacts would briefly have returned the interior to reducing conditions, in consequence oxidizing the 'exosphere' (outer part of the Earth). From this sequence

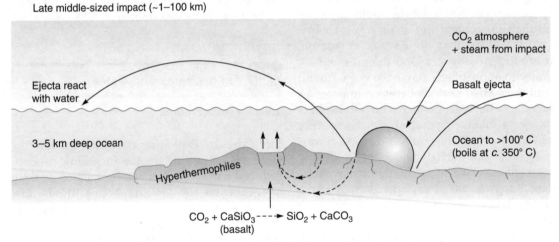

Late middle-sized impact (~1–100 km)

CO$_2$ atmosphere
+ steam from impact

Ejecta react
with water

Basalt ejecta

3–5 km deep ocean

Ocean to >100° C
(boils at *c*. 350° C)

Hyperthermophiles

CO$_2$ + CaSiO$_3$ ---▸ SiO$_2$ + CaCO$_3$
(basalt)

Basalt reactions buffer CO$_2$ and Na from basalt controls Cl. Ocean pH close to modern values.
Any life in oceans is killed by 50–100 km impact but hyperthermophile life in rock hydrothermal
systems may survive

Figure 1.4 Detail of bombardment by late middle-sized impacts (~3.4–100 km) (not to scale).

the late Hadean/early Archaean Earth that emerged probably had a relatively thin, relatively oxidized CO_2/H_2O-rich exterior ocean/atmosphere system, overlying a more reduced mantle and deeper interior. This oxidation contrast between a more oxidized ocean/atmosphere and a more reduced mantle may have been crucial to early life.

During the period after a massive Mars-size impact, the Earth's surface would have been very hot. As it cooled, there is the possibility of a brief interval of habitability, during which the surface temperature was around 100°C, capable of sustaining hyperthermophiles. Sleep *et al.* (2001) showed that the available internal heat could have sustained the surface temperature in a habitable condition for at most 2 million years. However, a massive CO_2 atmosphere, if present, could have maintained the surface for up to 20 million years, until the gas was consumed by reactions with rock to form carbonates. While these time intervals are short compared to the vastness of geological time, it is not inconceivable that life could have begun in such an interval, perhaps later to become extinct on Earth but ejected in the bombardment to another planet and surviving there, before crossing back again.

1.4.2 Why did the atmosphere allow free water to exist on the surface?

On Mars today, because of the surface temperature, liquid water does not exist on the surface. There is water on the surface, but it is ice or vapour, and there is evidence for an aquifer of brine below the surface (Carr, 1996). On Venus, although the early inventory of water may have been roughly comparable to Earth's, there is now virtually no water. Hydrogen is now very scarce, and probably most easily found in sulphuric acid in the clouds. It is likely that both planets had oceans in their early days. Venus lost its water by photolytic production of H and subsequent upwards loss of hydrogen to space, while Mars has lost its water downwards, sinking in to the cold crust or remaining in the cool interior. Only Earth has oceans that remain stable.

The first question, discussed in this section, is why the water on Earth was not lost by H loss upwards. The second question is why the water was not lost downwards into the interior (see section 1.4.3 below).

The Earth's earliest rock vapour atmosphere would have been very short-lived, succeeded by an early steam and carbon gas atmosphere. Sources of energy were meteorite impact, energy from the interior, and energy from the Sun. As the impacts became less frequent, impact energy would have been delivered episodically with, over time, longer gaps between impacts, and in smaller quantities of energy. The much smaller input of energy to the atmosphere/ocean system from the interior of the Earth would also have declined; today virtually all the energy delivered into the atmosphere comes from the Sun.

The very long-term stability of a steam or methane-rich atmosphere is limited. In steam/atmosphere conditions the so-called 'cold-trap' of the atmosphere is not well developed. A cold trap exists today on the modern Earth because sunlight, passing through the air, warms the ground. Air heated by the ground temperature moves upwards and cools as it expands, so the top of the lower air (or troposphere) at altitudes of 10 km (near the poles) to 30 km (near the equator) is cooler than air on the ground. Water condenses in this cool area and falls as rain back to the ground: it cannot pass up into the stratosphere. This 'cold trap' defines the top of the troposphere. Above this level in the modern Earth's lower stratosphere, ozone absorbs ultraviolet energy (and hence is warmer); above that the air is cooler upwards, but at the very top of the atmosphere absorption of 'harder' short-wave ultraviolet and X-ray radiation means that temperatures are much higher, up to thousands of degrees centigrade. At these temperatures, and bombarded by hard radiation and cosmic rays, any water molecules would be photolysed, and the hydrogen atoms released would be likely to be knocked out into space. But, because of the cold trap, there is very little water present at these levels, and the Earth loses little: its hydrogen is safely locked below the cold trap.

In a steam atmosphere, however, water vapour is present in the high air and can be split by ultraviolet light, with loss of the loose H atoms to space, leaving behind residual oxygen. Moreover, in any atmosphere containing large amounts of methane (not subject to the cold trap) H loss can also occur.

On Venus, the early development of the atmosphere remains unclear (e.g. see discussion in Lewis and Prinn, 1984). One possible model of the early history of the atmosphere is that an early steam atmosphere developed. Steam is a strong greenhouse gas. The cold trap was absent or weak, and water vapour reached the top of the atmosphere. Light energy split water molecules to H and OH, and H was ejected to space by radiation and impact. The planet lost water; more water partitioned into the high air; more was lost; volcanism delivered water to the surface and this in turn was lost; the melting temperature of the mantle would have risen and the mantle would have been further dehydrated, until it became dry and much more oxidized by exchange with residual atmospheric oxygen. Any residual H on Venus is highly enriched in the heavier D isotope, less likely to be ejected. Simultaneously, the carbon dioxide delivered by the volcanism would have collected in the air, increasing the temperature in a runaway greenhouse, until after a few hundred million years stability was reached with carbon dioxide in equilibrium with the rock surface. Sulphur gases were also emitted, forming the sulphur gas-rich clouds.

On Earth, a different process may have taken place. An early steam atmosphere is likely, as on Venus. But, further from the Sun and bigger, the Earth may have had slightly more water volume, by proportion to surface area, degassed onto the surface, and would have been possibly somewhat cooler, with a lower loss rate of H from the top of the air. A liquid water ocean would have collected. The Earth's surface would have to be above 350°C to evaporate an ocean equal to the present inventory of water. In a liquid ocean, dissolved carbon dioxide flushed through hydrothermal systems around volcanoes would have precipitated carbonate, which would have been returned to the interior via any subduction tak-

ing place. This would deplete the air of carbon dioxide and reduce the greenhouse, cooling the system. In effect, a race would take place between the loss of H to space from the top of the air, and the removal of CO_2 from the base of the ocean. On Venus, the H was lost and the CO_2 then built up again in the air as subduction was difficult in the absence of the cool surface. On Earth, the opposite may have happened. It is possible that substantial loss of H did occur by this mechanism (Yung *et al.*, 1989), but any hot-ocean event was not prolonged enough to dry the planet (i.e. not for hundreds of millions of years: the Earth would not sustain 800 Ma of hot oceans). If this model is correct, it is conceivable that the planet lost H equivalent to 1–2 km of ocean (see Yung *et al.*, 1989: the D/H ratio on Earth is somewhat enriched), but the planet's oceans retained the bulk of their water. CO_2 was probably partitioned down into the mantle.

What is left as the residue after 1–2 km of ocean has lost its water is oxygen (or OH). This excess oxygen would have provided a small but useful source of atmospheric oxidation power, capable, for instance, of converting sulphur and sulphur gases in the air to sulphur dioxide, or thence dissolved sulphate. This would have been a resource available to prospective life in the end of the Hadean (we define the Hadean/Archaean boundary as the date of the last common ancestor on Earth 4.2–3.8 Ga ago).

Plate tectonics can probably only be sustained on planets that have liquid oceans. This is because plate tectonics is driven by the falling in of cold slabs of surface plate into warmer mantle (subduction), and a cool surface is essential to make dense cool slabs. Moreover, the O surplus from water photolysis, input into the mantle via hydrothermal systems, would have oxidized the mantle somewhat. In sum, the late Hadean Earth would have retained its oceans, under a cool, lower CO_2 air, and a mantle that contained reintroduced CO_2 and was relatively oxidized. This process, and its obverse on Venus, would have happened before about 4 Ga ago.

Over the billions of years since then, hydrogen-containing water 'snow' has been added to the planet from space in micrometeorites, and some H has been lost, either from water vapour or methane rising into the upper air, but overall the oceans appear to have been stable. On Earth since plate tectonics began, carbon dioxide has been degassed in volcanoes and reintroduced to the mantle via hydrothermal systems in crust destined for subducted slab. If too much carbon dioxide was consumed, weathering would slow, decreasing the Ca supply. In extreme, the surface would cool and ice caps form, eventually (if this process went too far) dropping sea levels below the ridges and slowing hydrothermal reintroduction of carbon dioxide. Carbon dioxide would build up again. The system would be somewhat stable.

1.4.3 Carbon dioxide cycling and the carbon dioxide content of the early air

Carbon dioxide in the air is buffered by chemical reactions with the crust. One example, which is analogous to processes on Venus, is:

$$CO_2 + CaSiO_3 \longleftrightarrow SiO_2 + CaCO_3$$

In the reaction above, $CaSiO_3$ represents rock silicate minerals (such as pyroxenes) that react with carbon dioxide in air. They produce quartz and calcite. On Venus, the surface conditions are not far from the equilibrium of this reaction. On Earth, water is involved but in the end silicate weathering produces quartz sand and calcite limestones in the sea. The reaction is reversed by metamorphism in the crust. Sleep and Zahnle (2001) point out that in this cycle, warmer temperatures speed silicate weathering and carbonate formation, reducing atmospheric carbon dioxide and hence promoting global cooling. Over long periods of time, cycling of carbon dioxide into and out of the mantle also dynamically buffers carbon dioxide in the air. In this mantle cycle, CO_2 is outgassed at ridge axes and island arcs, while subduction returns CO_2 to the mantle. Negative feedback is provided because the amount of basalt carbonatization depends on the availability of CO_2 in seawater, and, hence, on CO_2 in air. Any large impacts would produce massive deposits of reactive

impact ejecta, which would provide an effective carbon dioxide sink. Flux calculations suggest that in the Hadean, with CO_2 cycling by tectonic activity, and in addition with a high rate of reaction of atmosphere/ocean carbon dioxide with basaltic ejecta from impacts, and in the light of a Sun with lower luminosity, the Earth should have been cool. The likely state would have been with atmospheric and oceanic CO_2 levels sufficiently low that the climate was cold unless another greenhouse gas was present (e.g. methane). On such an Earth the climate would be warm enough to produce some fresh snow but dry enough that significant areas of land would be free of high albedo snow and ice. Sea lanes of open water would exist.

1.4.4 Is the Earth self-fluxing? How deep were Archaean oceans?

The presence of water on the Earth's surface is not only dependent on the control of H loss from the top of the atmosphere: it also depends on the relationship between volcanism and subduction. Why does the water not simply sink into the Earth, the way water sinks into the ground after rain? This is, after all, what has happened on Mars. On Mars, there appears to have been an original inventory of several hundred metres of water: a puddle ocean (Carr, 1996). Today, this is absent. The water now near the surface is present either as ice cap, within hydrated minerals, or as a probable brine layer in the brecciated crust, where the low geothermal gradient is warm enough to allow brine to remain liquid.

There is evidence that Mars may still have occasional breakout phases of brine flows (Carr, 1996). Much earlier, on the young Mars, plate tectonics appears to have operated, and there may have been a shallow Martian swamp-ocean. This swamp-ocean has either been subducted, or has left its remnants in the postulated brine layer. Possibly, as the planet cooled, more and more water was accepted back into the mantle via subduction, fluxing the melting, until eventually even the wettest pockets of mantle were too cool to generate melt except very rarely in the giant plumes such as Olympus. Movement stopped, and the residual surface water simply soaked into the crust, or ablated as vapour, to condense at the poles.

Earth avoided this. Why? To have free water on the Earth's surface means either (1) that the mantle does not 'want it' (cannot accept it), or (2) that the mantle has not had time to accept it. The first control is equilibrium – if it applies, then the present mantle is effectively saturated with water, and the excess resides on top. Any overwet part of the mantle will melt, and the water will be expelled with the melt. This happens locally beneath most island arcs. But if the mantle cools, it will be able to accommodate more water before melting where the geotherm intersects the solidus (adding water to hot rock under pressure helps it melt).

The second control is kinetic. If it applies, the mantle is capable of accepting more water, but convection has not had time to deliver it yet. The Earth convects slowly (^{40}Ar has not been degassed), so this is likely. Moreover, if the notion of kinetic control is correct, two further deductions are possible. Either, at some time, the mantle was indeed effectively saturated, or alternatively, the Earth was

assembled in parts, with water outer, and the present distribution still reflects that primeval assembly history. However, sustaining any primeval heterogeneity of water content is unlikely given the evidence for planetary melting after a Mars-size impact. Perhaps the Earth was once effectively saturated so that volcanism expelled excess water to the surface, but the planet may have become progressively undersaturated as it cooled, all the while accepting more water back into the mantle.

On the modern Earth, with plate tectonics, there are nested water cycles (Figure 1.5). Water is expelled from the mantle at mid-ocean ridges and at volcanoes above plumes (such as Hawaii). Water is incorporated into the crust at mid-ocean ridges, and in sediments. This crustal water is subducted. In subduction zones nearly all the water is expelled again. Much of the expelled water helps to promote melting in the mantle wedge above the subduction zone and this water travels upwards with the magma. As water remains in the melt, some is eventually vented at volcanoes, either in the melt or as gas. Other melt solidifies in mid-crust, and induces metamorphism in surrounding rocks. The heat drives circulation of fluids

Water in atmosphere and ocean

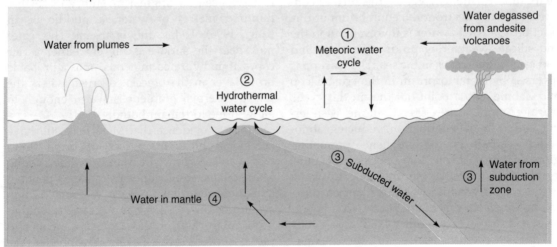

Figure 1.5 Nested water cycles on the modern Earth. The meteoric water cycle (1) (rain and snow) has very large fluxes. The hydrothermal systems (2) flux volumes of water equal to the volume of the oceans in times of the order of 10^6–10^7 years. A small amount of this water remains in the oceanic crust and is subducted (3). Most of this is returned via the volcanoes above subduction zones (3). The slowest cycle is the long-term flux into the mantle (4), and direct degassing out of the mantle at plumes and ridges.

in the metamorphic pile and the solidifying magma. Overall, the continental crust can probably broadly be described as 'saturated' in water for its thermal state if above a subduction zone. This is not to say that any particular rock is saturated, but that in a sustained subduction setting, well after inception of subduction, the introduction of water and heat from below leads to the efflux of a nearly equivalent amount of water out of the top of the continental pile. Of course, this is in broad terms: new subduction zones may add fluid to the overlying continental wedge for some time before volcanic and metamorphic efflux matches input.

Consider a subduction zone that is in steady operation, with a constant temperature in the overlying mantle wedge, and allowing the normal inwards mantle convection, consequent on subduction, around the mantle wedge above the subduction zone, to maintain temperature and supply fresh overlying mantle to the wedge region. Imagine for some reason adding more water down the subduction zone. The extra water will induce not a wetter melt, but more melt. Now extend the logic to consider Archaean subduction with a hot mantle. The same amount of subducted water as in the modern example will in these circumstances generate far more melt, and probably dryer melt (the temperature of the Archaean mantle wedge being higher). Next, consider processes at mid-ocean ridges and in plumes. Today, the small amount of water in the mantle and potential temperature of *c.* 1280°C allows melting to create roughly 8 km of oceanic crust and an olivine-rich residue. Add more water to the source region, and more melt would form (and hence a thicker crust). Then think about the early Archaean, and postulate a higher potential temperature. With the same amount of water as today, far more melt would form, giving a thicker crust and soon extracting the excess water from the mantle: in other words a hot stable mantle then would be less hydrous than today.

This leads to the deduction that the Earth may be self-fluxing. Initially, if the mantle adiabat were hotter than today, the accretionary water content of the mantle would lead to massive volcanism and much of the water would be expelled from the hot mantle, via magma, to condense as oceans. Komatiite volcanism (komatiites are lavas with more than 18% MgO, and erupted as very hot magmas), from a hot and nearly dry mantle, would have dominated. Towards the end of the Hadean and in the early Archaean, consider a mantle adiabat say 200–300°C hotter than today: if so, the dominant volcanism may have become komatiitic basalt or basalt, with komatiites confined to plume eruptions fed by hot mantle. Subducted water would have been almost completely expelled, fluxing voluminous tonalite magmatism. As the planet's mantle slowly cooled, it gradually accepted more subducted water, and the melting continued to be possible at the ridges, though the thickness of the oceanic crust decreased.

If this is correct, the implication is that the total volume of the oceans has declined slowly over time, with water slowly reaccepted into the mantle as it very slowly cools. Deep oceans mean thick continents (from isostasy and erosional controls: see Nisbet, 1987); moreover, continents may have grown in bulk over time. Assuming then that the area of the continents was less in the early Archaean, the net effect of more voluminous oceans, somewhat offset by smaller continents, may have meant that Archaean oceans were at least as deep as they are today, or deeper, and probably 3 to 5 km deep on average.

1.5 Is the depth of the modern ocean controlled by the critical point of water?

There are alternative opinions about the controls on the depth of the oceans. One possibility is that the depth is simply accident: this opinion is valid, given our lack of knowledge, but can only be noted rather than tested. There is, however, an alternative that *can* be tested: this is that the depth of the modern ocean is controlled by the critical properties of water. Kasting and Holm (1992) noted that fast-spreading mid-ocean ridges are typically submerged to a depth at which the pressure is close to the critical pressure for seawater. This may be chance, but they put forward arguments to show that this depth ensures optimal convective heat transfer and hence maximizes the penetration of hydrothermal fluids into the ridge axis. By doing so, the ocean depth optimizes the operation of plate tectonics, and also the circulation by the plate system of water into the subduction zones and back out of andesite volcanoes. Primary degassing to the air/ocean system probably only occurs at plumes on the modern Earth, as ridge outgassing is immediately reincorporated.

Kasting and Holm (1992) argued that ocean depth is balanced by this process, which ensures that plate ingassing rate (taking up water by subducting crust that has been hydrated via ridge hydrothermal systems) is balanced with plate outgassing rate (from andesite volcanoes fed by water from subduction zones). There is direct outgassing from plume volcanoes, balanced by a small input to the mantle of water returned by subduction. The stable state is set by the critical pressure of water at the ridges, which optimizes the hydration of the crust: hence the stable state of the ocean, in this model, is when the fast spreading ridges that feed subduction are at a pressure of about 250 bars, giving a pressure in the hydrothermal systems (1–2 km into the rock) of roughly 300 bars and exiting water at about 350°C. The optimal depth is set within a range of about 100 bars pressure (1 km of water) and is controlled by the turbulence of hydroconvection (Kasting and Holm, 1992).

This argument is persuasive: but what if the Earth had early oceans that were shallower or deeper? If the early oceans were shallower, rates of inward intake of volatiles would be slow, and the boiling of vent fluids at low pressure would produce saline fluids at the base of the systems that would limit further circulation (N. Sleep, in Kasting and Holm, 1992). Moreover, if the water were not in the ocean but still in the hot early mantle, the mantle would melt profusely, with abundant Hawaii-like plumes. The result would be rapid degassing and a deeper ocean. If on the other hand the early ocean were deeper than the critical depth, uptake of water by hydrothermal systems would be less, while degassing continued, and the deep oceans would tend to get deeper. Intake of water at subduction zones would help melt the hot mantle and prompt profuse plume volcanism, which would rapidly degas the mantle.

The two arguments – that the Earth is self-fluxing; and that the depth of the ocean is controlled by the critical depth – can be reconciled on this Earth (though not necessarily on Venus), given the 1 km range of the optimal depth of water over the ridge. In this synthesis, assume that the early Earth degassed profusely, so that even after the end of major H loss to space a deepish ocean (say 5 km) remained, limiting the hydrothermal circulation. But, slowly as the now-dry mantle cooled, the plume activity would be suppressed. The mantle would accept water from subduction. Thus water would be engassed until the ocean depth came into the stable range of pressure over the ridge, when an efficient cycling of water through the plate system was achieved. This would allow plate tectonics to be an efficient process. Continental growth would reduce ocean area slightly. As the Earth slowly cooled further, maintenance of 3.5 km deep oceans, by the critical pressure feedback, would sustain a comparatively dry mantle and further reduce plume formation, reducing primary outgassing. But, in turn, this reduction in primary outgassing would tend to increase the mantle's water content (as subduction added water). The whole system may be semi-buffered, capable of maintaining deep oceans (near critical-depth) over ridges for a long period, but slowly engassing as the Earth ages and cools. Thus both the 'self-fluxing' and the 'constant pressure' models can be partially correct.

The other possible starting position, that early oceans were shallow, is easy to dispose of: in this case both the constant pressure and self-fluxing controls would act in the same way, to encourage degassing until the ocean was at least 3–4 km deep. An Earth with abundant komatiite plumes would probably rapidly degas deep oceans.

The shallow or deep ocean puzzle, in principle, is testable, as continental freeboard depends on ocean depth over the ridges (see Nisbet, 1987). A deep early ocean would mean thick elevated continents of small area; a shallow ocean would imply wide thin continents. But the test cannot yet be applied: there is not enough agreed evidence to distinguish between these (although what evidence there is tends to point to early thick small-area continents).

What is clear is that oceans have been present on Earth from very early on, and that after the first accretion the normal state of the oceans was as liquid water, though they may on occasion have been converted to steam or have been capped by ice floes. It is probably unlikely that they became wholly ice, given the amount of carbon dioxide likely to have been in the atmosphere, though sustained ice events cannot be excluded (a snowball or 'cat's-cradle world'). An ice ocean reflects sunlight and so intrinsically tends to be self-stabilizing, as it cools the planet. However, stagnant ice, with an ablation/precipitation rate greater than dust infall rate except near sources, has a relatively low albedo. Since it takes very little open water to take care of global CO_2 exchange, the likely open lanes and leads in a global ice ocean would allow the exchange of greenhouse gases between water and air (e.g. see Mckay *et al.*, 1985). On a snowball planet the result of these factors is that the existence of an ice-covered ocean would kill land glaciation and hence create positive albedo/ice feedback. Moreover, if the oceans did ice over very early, on the active Earth ejecta from large impacts or dust from huge volcanic eruptions would have probably been adequate to change the albedo enough to melt the oceans. Perhaps this is why, though 'snowball' episodes may have occurred, Earth has not been a 'snowball' for long-sustained periods.

1.6 Earliest life – the geological setting

1.6.1 Location

There is reasonable geological ground for suspecting that life dates back to somewhere around 4 ± 0.3 Ga ago (summarized by Nisbet and Fowler, 1996a) (Figure 1.6). There are various claims of the first evidence for life in rocks from West Greenland that date 3.8–3.7 Ga ago or perhaps earlier. Among the more persuasive claims is the evidence from ^{13}C depleted carbon microparticles in > 3.7 Ga old rocks in Isua, West Greenland (Rosing, 1999). Biological organisms fractionate carbon, preferentially fixing the lighter ^{12}C isotope from the ambient C in CO_2 into organic carbon. Consequently, the biogenic carbon is isotopically 'light' in its $^{13}C/^{12}C$ ratio, while the remaining carbon in the inorganic reservoir is correspondingly 'heavier'. The discovery of isotopically 'light' carbon microparticles in Isua may be interpreted as evidence for life: it is a reasonable explanation that the particles are derived from biogenic detritus. If so, the debris may have come from plankton living freely in the ocean, or sea-bottom organisms living on the water/mud interface (Rosing, 1999).

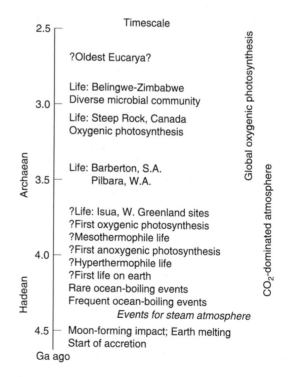

Figure 1.6 Approximate timescale of events on early Earth (from various sources).

The evidence set out above clearly implies that there were significant oceans on Earth 4 Ga ago. On occasion, all or part of the oceans would have been heated by impacts. Below the oceans, and also on flanks of volcanoes protruding from these oceans, hydrothermal systems would have been very common, in various geological settings. Submarine hydrothermal systems at mid-ocean ridges would have been not dissimilar from modern mid-ocean ridge systems. The reason is that they would have operated in a new crust not greatly dissimilar to today, made of either basalt or (possibly at 4 Ga ago) komatiitic basalt, and that the same chemical processes as today would have occurred. This would also imply that very broadly the ocean chemistry would be similar to today. Moreover, subaerial hydrothermal systems, fed by rainwater, would have been similar to those on Iceland today.

However, it should be noted that the composition of some lavas would have been more magnesian than today: these were produced by plume volcanoes erupting komatiites, which may have formed large shields (mountains like Mauna Loa in Hawaii, but much larger, wider and flatter). Komatiites are ultramafic (Mg-, Fe-rich) rocks and hydrothermal systems in such rocks tend to be alkaline (e.g. alkaline hot springs). Komatiites also can host large nickel sulphide deposits. In some cases, hydrothermal penetration of nickel sulphide deposits in komatiite flows may have provided local Ni-rich fluids. Many key proteins are Ni-based: for example, urease, a key part of the nitrogen cycle, is built around Ni. Given that early cells were probably not highly skilled at extracting metals from the environment, it is possible that Ni-enzymes used in basic cellular housekeeping (and thus likely to be of great antiquity) evolved in a setting where Ni sulphides were abundant and obtruded into early cells, perhaps serving as sites for the first Ni-metal proteins. It is plausible to imagine that such settings would have been around komatiite hydrothermal systems that penetrated Ni sulphide deposits in komatiite flows. More generally, the crucial roles of metal proteins and heat-shock proteins in the basic housekeeping operations of living organisms may record an early hydrothermal heritage (Nisbet and Fowler, 1996b).

Hyperthermophiles

Figure 1.7 summarizes the 'standard' model of microbial evolution. At the core are hyperthermophile organisms. If it is accepted that life began early in the Earth's history, the likelihood is thus that a large impact (> 300 km diameter) would have heated the oceans *after* the first appearance of life – one or more ocean-boiling impacts capable of sterilizing both the oceans and also any organisms living in hydrothermal systems at up to 115°C would have been likely prior to 4.2–3.8 Ga ago. If so, any mesothermophile organisms living in the water or on or near the seafloor prior to the impact would have been eliminated by heating; only hyperthermophiles in the deep subsurface would have survived. It is thus probable that life passed through a hyperthermophile 'bottleneck' (Gogarten-Boekels et al., 1995): only hyperthermophiles survived. The first life may or may not have been hyperthermophilic, but an event causing ocean water to warm > 100°C would have been highly likely. Thus on geological grounds it should be expected that the last common ancestor of all

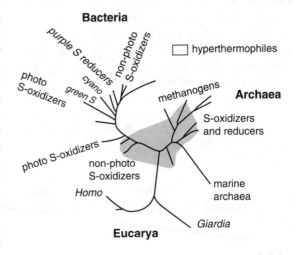

Figure 1.7 'Standard rRNA-based tree of life (Woese, 1987; Pace, 1997). Note position of hyperthermophiles (central shaded area). Note also that other 'trees' are possible (Doolittle, 1999).

extant life was hyperthermophilic (but see also Galtier *et al.*, 1999, for a possibly contrary opinion). Note, however, that the first common ancestor, whether it lived on Earth or Mars, or elsewhere, need not have been hyperthermophilic – this is because the interpretation of the molecular history of modern extant organisms, however it is derived, can only give information about the genes contained in the last common ancestor (though gene families give information before the last common ancestor with this limitation). Possibly a wide variety of life (and viruses) existed before that, only to be wiped out by a large impact, with only a few, or even a single hyperthermophile survivor. Note also that if one possible alternative scenario is correct, that life on Mars seeded the Earth after about 3.7 Ga and extinguished native life by competition, then the logic would also fail. Perhaps even multiple exchanges took place, in the heavy early flux of meteorites between the planets.

1.7 Testing the 'standard' model of microbial phylogeny: does it fit the geological evidence?

The 'standard model' (Woese, 1987; Pace, 1997; but see also Doolittle, 1999 for a contrary opinion) of the phylogeny of microbial life (Figure 1.7) has several distinctive features that bear on this problem, and may be subject to testing by geological constraints.

The model, most simply interpreted, implies that the last common ancestor was a hyperthermophile (Figure 1.7). Given it is unlikely that the whole ocean could stay at $>75°C$ for a sustained time without dehydrogenating the planet, the habitat of an early hyperthermophile may have been in a rock-hosted hydrothermal system. By inference, the last common ancestor lived in a hydrothermal system.

Here the possible evidence from Greenland (Rosing, 1999) for very early planktonic life is interesting. The oceans by 3.7 Ga were probably generally cool ($< 75°C$); if not, the H loss to space may have been so large that the water would have been lost. If so, any plankton

would have been mesothermophilic, not hyperthermophilic. To continue the reasoning, if the possible evidence for planktonic life at 3.7 Ga (Rosing, 1999) is accepted, then the early hyperthermophile stage of the standard model must have been complete by this time. By implication, the hyperthermophile-only biosphere must have ended prior to 3.7 Ga: a time range from the start of life of anywhere between as much as 500 Ma and 100 Ma or less.

Geologically, the assumption that our early ancestors were hyperthermophile is very reasonable, whether born in hydrothermal systems, or the survivors after impact heating of the oceans (though note that life itself is unreasonable, thus one should not expect life's history to be retro-predictable). The voluminous volcanism on the early Earth would have hosted innumerable hydrothermal systems. These would have occurred in deep oceans at mid-ocean ridges and around plume volcanoes like Hawaii today; in meteoric hydrothermal systems on land, if land existed (it probably did); and around subduction zones if subduction took place in some form or other. Thus a diverse array of hydrothermal settings would have been present, capable of sustaining local environments of redox contrast in S and C compounds and thus of sustaining life. During this time, opportunistic organisms evolved from thermophiles may have repeatedly colonized low-temperature environments only to be destroyed by impacts.

The standard model implies that early metabolisms were sulphur based and non-photosynthetic. In the model, anoxygenic photosynthesis comes next, then later comes oxygenic photosynthesis. This model-derived phylogenetic history is subject to test, especially if Rosing's (1999) possible evidence for planktonic photosynthesis at >3.7 Ga is considered. First: is there enough time? The model implies that the last common ancestor existed 4.2–3.8 Ga ago; by about 3.5 Ga ago, there is probable evidence for oxygenic photosynthesis (Schidlowski and Aharon, 1992). Could bacterial evolution have proceeded so far in 300–700 Ma? One approach to answering this question is to attempt to calibrate the pace of other aspects of bacterial evolution (e.g. in the S cycle). Generally, however, the question may be unanswerable: rates may vary, and even 300 Ma (the lower time range) is an enormous length of time.

Whether the specific line of descent predicted by the standard model can be found in the geological record is a matter of dispute. The model would imply that sulphur-processing organisms evolved first, in a hyperthermophile setting, then anoxygenic bacterial photosynthesis began, as in purple and green S bacteria, followed by anoxygenic photosynthesis as carried out by green bacteria (very oxygen intolerant); lastly in this chain, oxygenic photosynthesis would have appeared. Is this testable? Probably not yet, though in principle it may be accessible to isotopic constraint. For example, S isotope fractionation may be very slight in biogenic material in Archaean rocks (Canfield and Teske, 1996): if this is correct, in rocks that postdate the onset of the light (biological fractionation) carbon isotopic signature in the geological record, then the standard model is likely to be

wrong as it is difficult to reconcile with the record. However, our own S isotope results from the late Archean (Grassineau et al., 2001, 2002) show a wide spread in S isotopic fractionation: if our results are correct, the standard model is not proven right, but it is also not proven wrong.

Atmospheric models can help constrain the discussion. Kasting et al., 1989 considered the sulphur species that would have been present in the earliest air. If volcanic degassing of SO_2 were significant, then photochemical processes would act so that S was present in the atmosphere/ocean system in the full range of oxidation states ($+4$ to -2) and enough sulphur vapour may have been produced photochemically in the air from the volcanic emissions to act rather as the ozone layer today, to protect the surface from UV radiation. Lovelock (1988) suggested a methane-based smog as an alternative (Figure 1.8): a possible option if the $CH_4:CO_2$ ratio were high, but if so, there would have been considerable H loss to space.

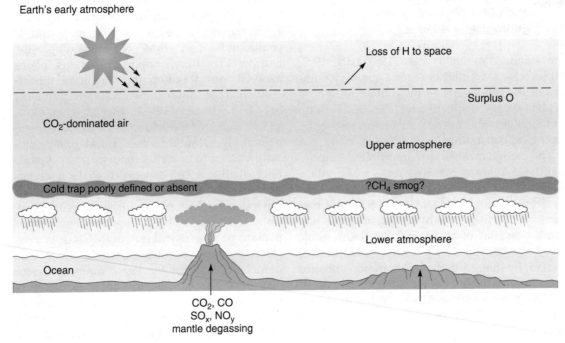

Earth's early atmosphere

Loss of H to space

Surplus O

CO_2-dominated air

Upper atmosphere

Cold trap poorly defined or absent

?CH_4 smog?

Lower atmosphere

Ocean

CO_2, CO
SO_x, NO_y
mantle degassing

Figure 1.8 Possible model of Earth's early atmosphere. Note that there are many other possible hypotheses, and this diagram is only illustrative. In particular, at times methane may have been much more important than shown. See also discussion in Lovelock (1988).

Even if sulphur were degassed in a less oxidized state, production of sulphate in the air would occur. An atmosphere/ocean system with abundant S in varied oxidation states is consistent with the demands of the 'standard' phylogenetic model, as it would have been capable of sustaining S-dependent organisms. Traditionally the eukaryotes have been thought of as arising relatively recently, in post-Archaean time. In contrast, recent evidence suggests that the eukarya are of great antiquity (Brocks *et al.*, 1999) which is also not inconsistent with the standard model, though again not proving it. In general, the geological record can provide many tests of the standard model: the model cannot be proven correct, but it can be falsified. To date, it has passed the tests.

1.8   Onset of photosynthesis: geological consequences

1.8.1 *Isotopic evidence for the start of photosynthesis*

In the 'standard' phylogenetic model of microbial life (Woese, 1987), prior to photosynthesis, life would have depended on the primary production in the vicinity of the hydrothermal vents. When photosynthesis began, life had access to energy that did not depend on the vagaries of the supply of reduction power from fluids circulating through the rock. Thus the evolution of photosynthesis allowed biology to escape from the hydrothermal ghetto (Nisbet and Sleep, 2001; Nisbet, 2002). According to the implications of the 'standard model', anoxygenic photosynthesis would have developed before oxygenic photosynthesis.

Much of the interpretation of the first possible geological evidence for widespread organic processes, from >3.7 Ga old rocks in west Greenland (see above) is very controversial. The material is difficult to interpret as it is hosted in high-grade metamorphic rocks in which post-formational processes may have rewritten the evidence. Rosing's (1999) interpretation of carbon microparticles in rocks, if correct, is interesting. The particles were in sediment interbedded in ironstones, and the

carbon was isotopically moderately fractionated. Assuming that metamorphic isotopic resetting has not been strong, the carbon may possibly record anoxygenic photosynthesis in plankton.

In 3.6–3.5 Ga old rocks, the first reasonably firm evidence of photosynthetic life is present. C isotopes in carbonates in rocks from the Pilbara (Western Australia) and Barberton (South Africa) have $\delta^{13}C$ around 0‰ (this ‰, or 'per mil' notation signifies deviance, in parts per thousand, of the $^{13}C/^{12}C$ ratio from an arbitrary fossil standard, roughly equivalent to modern limestones). By capturing carbon rich in ^{12}C, biological organisms become 'light' at around $\delta^{13}C$ −10 to −30‰ (and recycled carbon much lighter), while the residual carbonate precipitates 'heavier' carbon with $\delta^{13}C$ around 0‰. Thus, the record of 0‰ carbonates, even 3.5 Ga ago in the 3.5 Ga Barberton and Pilbara rocks, implies strongly that the photosynthetic carbon cycle was in full operation, controlling the carbon budget as today (Schidlowski and Aharon, 1992). Given that $\delta^{13}C$ in 'primitive' carbon gases emitted from the mantle was probably −5 to −7‰ even in the Archaean, the existence 3.5 Ga ago of inorganic carbonate with $\delta^{13}C$ at 0‰ implies the corresponding existence of a light, biological reservoir.

The earliest living community, which diversified after the last common ancestor, may have included of an array of differing specialists. It may have been based on hyperthermophile chemolithotrophs living 4.4–3.8 Ga ago (see reviews in Nisbet and Fowler, 1996a, b and Nisbet 1995), supplemented by early specialist organisms recycling redox power. The evolution of *Chloroflexus*-like anoxygenic photosynthesis (with a photosynthetic reaction centre similar to Reaction Centre II) then allowed colonization of anaerobic shallow water niches. The arrival of rubisco in a microaerobic or aerobic ancestor of purple bacteria presumably conferred a small selective advantage on its host. Possibly it was used to dispose of surplus oxidation power, or it may have been used as a defence in toxic warfare between cells, in which some cells evolved oxygen as a toxin, while others were faced with the problem of oxygen disposal.

The evolutionary step that created the cyanobacteria (Schidlowski and Aharon, 1992; Blankenship, 1992; Pierson, 1994; Blankenship and Hartman, 1998) would, by releasing oxygen, have immediately dominated the planet's ecology. This step, which may have been a corporate merger between rubisco-using purple bacteria (Reaction Centre II) and green S bacteria (Reaction Centre I), was possibly taken before 3.5 Ga ago. Once cyanobacteria were present, their oxygen output would have created abundant local aerobic conditions, in which rubisco-mediated carbon management would have held a monopoly. The dual role of rubisco (Lorimer and Andrews, 1973; Lorimer, 1981), managing O_2 as well as CO_2, probably allowed users of the Calvin–Benson cycle, particularly cyanobacteria, to occupy the best-lit top level of microbial mats, and permitted cyanobacteria to live as oceanic plankton (Nisbet and Fowler, 1999). Oxygenic cyanobacteria would then have maintained their topmost position by sustaining an upper environment which was aerobic and hostile to anaerobic competitors. Rubisco (Hartman and Harpel, 1994) allowed life to escape the early hyperthermophile ghetto, but only oxygen-managers would have been allowed to participate in the best-lit newly aerobic superstructure. They would have spread worldwide in a geological instant. With the dominance of rubisco, the normal processes of evolutionary competition would have fine-tuned all other cellular processes in the cyanobacteria to maximize their compatibility with rubisco, and would have modified the rubisco itself to the maximum efficiency possible in the enzyme's role and setting.

Any prospective later photosynthetic competitor to rubisco, which attempted to emerge from the lower anaerobic layers of microbial mats, would have been unable to establish itself, as every available transitional niche would already have been rapidly occupied by organisms adapted to a rubisco-dominated world. Thus the double jump from anaerobic to aerobic conditions (imposed by the cyanobacteria), and from anoxygenic to oxygenic photosynthesis, would have been extremely difficult for a later prospective non-rubisco competitor. The hopeful competitor would have had to evolve through the many intermediate stages from other anoxygenic photosynthesizers, in the face of the established oxygenic cyanobacteria already using all available nutrient.

Rubisco is 'inefficient' as an enzyme, capable of working in reverse, and the photosynthetic process in general is also inefficient (e.g. green light is reflected unused, which gives us aesthetically pleasant green leaves). Since evolution is not teleological, evolutionary processes may be unable to invest the capital needed (many generations of capital investment without return) to produce a competitor to the rubisco system. In evolution, each small developmental stage must be immediately advantageous ('profitable'), and it is thus presumably very difficult to produce an alternative to an established monopolistic system as immensely complex as oxygenic photosynthesis. Thus the qwerty keyboard may survive pro-tem as the main means of communication with the internet. The growth of monopolies in the market economy provides many parallels, for example in the aircraft industry and the computer industry. Once an established monopoly or near-monopoly has been built up, only long-term subsidy or regulatory intervention allows prospective competitors to create viable alternatives to the dominant corporations. In the natural world such sustained subsidy would be unlikely to happen. Monopolistic qwerty enzymes, having the advantage that they already exist and that other cell systems are designed around them, become unchallengeable when once deeply enough embedded: rubisco and perhaps nitrogenase may thus persist simply because so much capital is invested in them, the entry barriers are high, and no natural subsidy is available to develop a competitor.

There is risk in this too: biology has now, by producing genetic engineers, become capable of teleology. There is now a motive for subsidy. Attempts to modify rubisco itself may likely fail (Tolbert, 1994), as evolutionary accident has probably fine-tuned rubisco as much as possible. However, a thorough redesign to create a wholly new photosynthetic process, though presently beyond our computing ability, and certainly far past our present manipulative skill to embed in a cell, is potentially within our power. We may be able, some years hence, to create and insert a wholly different and much more efficient enzyme. This would, however, be dangerous (see Tolbert, 1994) in the extreme. If more efficient than rubisco, it would inevitably find its way into the natural world, displacing rubisco users. The prokaryotes into which it went would rapidly outcompete not only cyanobacterial mat formers and plankton, but possibly all plants: ecocide.

1.8.2 *Did rubisco stabilize an oxygen-rich atmosphere?*

Tolbert (1994) made the interesting point that the balance between carbon dioxide and oxygen in the atmosphere will be set by the competing possibilities of rubisco, provided all other nutrients are available. Moreover, atmospheric ammonia and nitrogen are also managed within the photosynthetic and photorespiratory cycles. In addition, modern marine organisms (and probably Archean stromatolite-building cyanobacteria) manage calcium budgets within the photosynthetic cycle: thus the biogenic white cliffs of Dover.

The CO_2 compensation point is a feature of rubisco's schizophrenia (Tolbert, 1994). This is the CO_2 equilibrium level in air at which gross CO_2 released by photorespiration is all being refixed by photosynthesis: there is no net fixation. This is the minimum lower level for the C3 plants without other sources of CO_2: with 21% O_2 it is around 40–65 ppm CO_2. If atmospheric CO_2 falls below this level, C3 plants will cease fixing carbon: all other things being equal (in the absence of C4 plants), volcanic emissions would then build up carbon dioxide until the system started up again. At higher atmospheric levels of CO_2, the opposite happens: photorespiration is decreased, more CO_2 is fixed (given adequate supply of other essential nutrients), and carbon dioxide is reduced until balance is attained. For example, if CO_2 is 1000 ppm at modern O_2 levels, photorespiration is reduced by about 90% (Tolbert, 1994).

On the modern Earth feedbacks such as the greenhouse effect also obviously play roles in the cybernetics of this system (Lovelock, 1988). Moreover, the compensation point is linked to oxygen level – O_2 inhibition of photosynthetic CO_2 fixation occurs (Tolbert, 1994). At 2–5% O_2 there is little inhibition of CO_2 fixation; at 21% O_2 (modern air) there is 25% inhibition of CO_2 fixation, while at 40–60% O_2 the carbon fixation is inhibited by 50%. On a world populated by eukaryotes, the key role of the mitochondria, in partnership with the chloroplast and nitrogen-fixation, in managing oxygen is thus central to the management of the O_2:CO_2 balance.

Whether, over the aeons, rubisco and the mitochondrion/chloroplast switch (Joshi and Tabita, 1996) can be the primary control on CO_2:O_2 ratio in the air is disputable. The slow, long-term chemical controls relating cations (such as Ca^{++}) to carbon fixation, and subduction and erosion to nutrient supply, must weigh in the equation. So too must storage of reduced carbon in sediment, in clays, and in granitoids. Basalt weathering controls oxidation. The redox balance of the surface can be influenced over the aeons by organic carbon sequestration, but ultimately the larger scale of the inorganic controls must also be powerful (Walker, 1994). However, the rate at which the biosphere pumps carbon between reservoirs may always have overwhelmed the inorganic processes, except when massive impacts or plume eruptions damaged the biosphere, so it is possible that the inorganic controls always been fine-tuned by biology (Nisbet, 2002).

1.9 Did photosynthetic life stabilize the presence of water on the Earth?

One of the most interesting currently unanswerable questions is the Gaian proposition that life has stabilized the presence of water. The argument is as follows. The Sun is progressively delivering more and more heat flux to the top of the atmosphere, as the star ages. If by accident there was a liquid ocean 4 Ga ago, it should now be much warmer, and the air much richer in water vapour: if so the water-induced moist greenhouse might long ago have dehydrated the planet by H loss from the top of the atmosphere. The hypothesis is that life manages the greenhouse, and has progressively reduced the atmospheric greenhouse increment by reducing CO_2 in the air. By switching half-way through the planet's history to an O_2 atmosphere, managed by rubisco, life has maintained a cool surface (c. 15–30°C) and hence kept the planet away from a moist greenhouse. If the surface heats substantially (e.g after CO_2 emission by plume volcanism), life flourishes in the warm wet world, carbon is captured, equilibrium is restored. If too much carbon is sequestered (e.g. in limestones or down subduction zones), the planet cools, even to a snowball. Life becomes less abundant, and volcanic return soon builds up the carbon dioxide. Hence, in this view, life has sustained liquid water on the planet throughout its history. Recently (in the past few tens of millions of years), as the Sun has heated, it can be proposed as a hypothesis that life has reduced the greenhouse by reducing CO_2 levels, while developing C4 plants capable of operating in very low CO_2 levels: life can thus be sustained for many years to come (Lovelock, 1988).

The opposite argument (Walker, 1994 , 1977) is easily put: carbon dioxide sequestration is determined by Ca availability (e.g. from weathering, or uptake in hydrothermal systems). If carbon dioxide rises for some reason, then consequently so does Ca availability from weathering under the acid rain: the stability of the cycling is predetermined. But which control is operative – Gaian or abiological – is not clear.

1.10   Late Archaean: environmental reconstruction

Fluxes of carbon, oxygen, sulphur and nitrogen in the Archaean may have been less than today (assuming that the land surface was barren except for cyanobacteria and swamp bacteria), but the isotopic ratios in the carbonates show

that organic cycling of carbon was already on a scale to partner inorganic management of carbon isotopes. This also implies that oxygen production, though less than today, was nevertheless of an order comparable to that on the modern Earth (though the standing crop of O_2 was low), and that the sulphur cycle was also established, as production of sulphate by oxidation would have occurred. The carbon cycle depends on the supply of fixed nitrogen: whether the inorganic supply of fixed nitrogen (e.g. from lightning) was adequate to allow a biosphere productive enough to set carbonate $\delta^{13}C$ at 0‰ is debatable. Probably, as early as 2.7 Ga ago, the biosphere was sustained with usable N by bacterial fixation (e.g. by cyanobacterial heterocysts). If so, the atmosphere may have supplied N_2 to bacteria and, the modern cycling of N between nitrogen fixers and denitrifyers may have been in operation. (See Chapter 2 for an expanded discussion of the evolution of the nitrogen cycle.) By implication then, N, O and C fluxes in the biosphere were all biologically managed, and excepting the noble gases, the late Archaean atmosphere, as today, was a biological construct.

Thus it is possible that by the late Archaean, biological management of carbon (and perhaps sulphur) was on a sufficient scale that the modern carbon cycle was already long established. Globally, carbon was being degassed from the mantle and partitioned in the atmosphere/ocean system as on the modern Earth between organic and inorganic capture (Schidlowski and Aharon, 1992). The absence in other studies (Canfield and Teske, 1996) of evidence to identify the sulphuretum cycle may be in part a consequence of metamorphic isotopic homogenization, or of the difficulty caused by mechanical differences between an Archaean microbial ecology, with thin but efficient mats, and latest Proterozoic and Phanerozoic bioturbated muds with eukaryotes, including metazoa, producing thicker zones of seabed biological redox management. It may be that only multisample high-resolution isotopic studies can identify relicts of Archaean S mats (Grassineau *et al.*, 2002): homogenization in analysing conventional large samples may make it difficult to see more highly fractionated clumps.

Liquid water is the crucial difference between Earth and the other planets. Liquid water enables the plate system to operate. Liquid water is needed to make subduction zones operate, to create granitoid intrusions, and andesite volcanoes, and to maintain the continents. If the water were lost, CO_2 would degas, the plate system would probably be replaced by a plume-led system, and the world would be very different, either Venus-like or, ultimately, Mars-like. If the Gaian explanation is correct, and life has managed the surface temperature, then life is the architect of the planet. If, on the contrary, the inorganic explanation is valid, then life is simply an accidental veneer, important to us, but of little impact on the long-term geological evolution of the planet.

Acknowledgements

Thanks to Lynn Rothschild for being so persistent and so tolerant, and for her many comments, and to staff at RHUL for help with diagrams. Kevin Zahnle and Mary Fowler sifted ideas, but no one bears responsibility for error: in a book such as this the intent was to stimulate discussion. Work by EGN was supported by Leverhulme and NERC.

References

Ahrens, T.J. (1990). Earth accretion. In: H.E. Newsom and J.H. Jones (eds) *Origin of the Earth*. Oxford: Oxford University Press, pp. 211–227.

Blankenship, R.E. (1992). Origin and early evolution of photosynthesis. *Photosynthesis Research*, **33**: 91–111.

Blankenship, R.E. and Hartman, H. (1998). The origin and evolution of oxygenic photosynthesis. *TIBS*, **23**: 94–97.

Brocks, J.J., Logan, G.A., Buick, R. and Summons, R.E. (1999). Archaean molecular fossils and the rise of eukaryotes. *Science*, **285**: 1033–1036.

Canfield, D.E. and Teske, A. (1996). Late Proterozoic rise in atmospheric oxygen concentration inferred from phylogenetic and sulphur-isotope studies. *Nature*, **382**: 127–132.

Carr, M. (1996). *Water on Mars*. Oxford: Oxford University Press. 229pp.

Doolittle, W.F. (1999). Phylogenetic classification and the universal tree. *Science*, **284**: 2124–2128.

Galtier, N., Tourasse, N. and Gouy M. (1999). A non-hyperthermophile common ancestor to extant life forms. *Science*, **283**: 220–221.

Gogarten-Boekels, M., Hilario, E. and Gogarten J.P. (1995). The effects of heavy meteorite bombardment on the early evolution – the emergence of the three domains of Life. *Origins of Life and Evolution of the Biosphere*, **25**: 251–264.

Grassineau, N.V., Nisbet, E.G., Bickle, M.J. *et al.* (2001). Antiquity of the biological sulphur cycle: evidence from S and C isotopes in the 2.7 Ga rocks of the Belingwe greenstone belt, Zimbabwe. *Proceedings of the Royal Society of London B*, **268**: 113–119.

Grassineau, N.V., Nisbet, F.G., Fowler, C.M.R. *et al.* (2002). Stable isotopes in the Archaean Belingwe belt, Zimbabwe: evidence for a diverse microbial mat ecology. In Fowler, C.M.R., Ebinger, C.J. and Hawkesworth, C.J. (eds), *The Early Earth: Physical, Chemical and Biological Development*. Geological Society of London (Special Publication), 199, pp. 309–328.

Halliday, A.N. (2000). Terrestrial accretion rates and the origin of the Moon. *Earth and Planetary Science Letters*, **176**: 17–30.

Hartman, F.C. and Harpel, M.R. (1994). RuBisCo. *Annual Review of Biochemistry*, **63**: 197–234.

Joshi, H.M. and Tabita F.R. (1996). A global two-component signal transduction system that integrates the control of photosynthesis, carbon dioxide assimilation, and nitrogen fixation. *Proceedings National Academy of Sciences*, **93**: 14515–14520.

Kasting, J.F., Holm, N.G. (1992). What determines the volume of the oceans? *Earth and Planetary Science Letters*, **109**: 507–515.

Kasting, J.F., Zahnle, K.J., Pinto, J.P. and Young, A.T. (1989). Sulphur ultraviolet radiation and the early evolution of life. *Origins of Life and Evolution of the Biosphere*, **19**: 95–108.

Lewis, J.S. and Prinn, R.G. (1984). *Planets and their Atmospheres*. Orlando: Academic Press.

Lorimer, G.H. (1981). The carboxylation and oxygenation of ribulose 1,5-bisphosphate: the primary events in photosynthesis and photorespiration. *Annual Review of Plant Physiology*, **32**: 349–383.

Lorimer, G.H. and Andrews, T.J. (1973). Plant photorespiration – an inevitable consequence of the existence of atmospheric oxygen. *Nature*, **243**: 359.

Lovelock, J. (1988). *The Ages of Gaia*. Oxford: Oxford University Press.

McKay, C.P., Clow, G.D., Wharton, R.A. Jr. and Squyres S. (1985). Thickness of ice on perenially frozen lakes, *Nature*, **313**: 561–562.

Newsom, H.E. and Jones J.H. (1990). *Origin of the Earth*. New York: Oxford University Press.

Nisbet, E.G. (1987). *The Young Earth*. London: G. Allen and Unwin, pp. 147–8.

Nisbet, E.G. (1995). Archaean ecology: a review of evidence for the early development of bacterial biomes, and speculations on the development of the early Earth. In: M.P. Coward and A.C. Ries (eds) *Early Precambrian Processes*. Geological Society of London, Special Publication 95: 27–51.

Nisbet, E.G. (2002) Fermor lecture: the influence of life on the face of earth: garnets and moving continents. In Fowler, C.M.R., Ebinger, C.J. and Hawkesworth, C.J. (eds), *The Early Earth: Physical, Chemical and Biological Development*. Geological Society of London (Special Publication), 199, pp. 275–307.

Nisbet, E.G. and Fowler, C.M.R. (1996a). Some like it hot. *ature*, **382**: 404–405.

Nisbet, E.G. and Fowler, C.M.R. (1996b). The hydrothermal imprint on life: did heat shock proteins, metalloproteins and photosynthesis begin around hydrothermal vents? In: C.J. MacLeod, P.A. Tyler and C.L. Walker (eds) *Tectonic, Magmatic, Hydrothermal and Biological Segmentation of Mid-Ocean Ridges*. Geological Society of London, Special Publication 118: 239–251.

Nisbet, E.G. and Fowler, C.M.R. (1999). Archaean metabolic evolution of microbial mats. *Proc. Royal Society London B*, **266**: 2375–2382.

Nisbet, E.G. and Sleep, N.H. (2001). The habitat and nature of early life. *Nature*, **409**: 1083–1091.

Pace, N.R. (1997). A molecular view of microbial diversity and the biosphere. *Science*, **276**: 734–740.

Pierson, B.K. (1994). In: S. Bengtson (ed.) *Early Life on Earth*. Nobel Symposium 84, Columbia University Press, New York, pp. 161–180.

Rosing, M.T. (1999). ^{13}C depleted carbon microparticles in >3700-Ma sea-floor sedimentary rocks from West Greenland. *Science*, **283**: 674–676.

Schidlowski, M. and Aharon, P. (1992). Carbon cycle and carbon isotope record: geochemical impact of life over 3.8 Ga of Earth history. In: M. Schidlowski *et al.* (eds) *Early Organic Evolution: Implications for Mineral and Energy Resources*. Berlin: Springer-Verlag, pp. 147–175.

Sleep, N.H. and Zahnle, K. (2001). Carbon dioxide cycling and implications for climate on ancient Earth. *Journal of Geophysical Research*, **106**: 1373–1399.

Sleep, N.H., Zahnle, K. and Neuhoff, P.S. (2001). Initiation of clement surface conditions on the earliest Earth. *Proceedings National Academy of Sciences, USA*, **98**: 2001.

Tilton, G.W. (1988). Age of the solar system. In: J.F. Kerridge and M.S. Matthews (eds) *Meteorites and the Early Solar System*. University of Arizona, Tucson, pp. 899–923.

Tolbert, N.E. (1994). Role of photosynthesis and photorespiration in regulating atmospheric CO_2 and O_2. In: N.E. Tolbert and J. Preiss (eds) *Regulation of Atmospheric CO_2 and O_2 by Photosynthetic Carbon Metabolism*. Oxford: Oxford University Press, pp. 8–33.

Walker, J.C.G. (1977). *Evolution of the Atmosphere*. Macmillan: New York.

Walker, J.C.G. (1994). Global geochemical cycles of carbon. In: N.E. Tolbert and J. Preiss (eds) *Regulation of Atmospheric CO_2 and O_2 by Photosynthetic Carbon Metabolism*. Oxford: Oxford University Press, pp. 75–89.

Woese, C.R. (1987). Bacterial evolution. *Microbiological Reviews*, **51**: 221–271.

Yung, Y.L., Wen, J.S., Moses, J.I. *et al.* (1989). Hydrogen and deuterium loss from the terrestrial atmosphere: a quantitative assessment of nonthermal escape fluxes. *Journal of Geophysical Research*, **94**: 14971–14989.

2

What good is nitrogen: an evolutionary perspective

Rocco L. Mancinelli

ABSTRACT

See Plate 3

Nitrogen is an essential element for life. It is present in all living systems, occurring in several important molecules including proteins and nucleic acids. As a consequence, the origin of the nitrogen cycle and the development and evolution of organisms with the ability to transform nitrogen into compounds associated with the nitrogen cycle are essential to understanding the origin and evolution of life. Available nitrogen is presently, and has through Earth's history, been plentiful as dinitrogen (N_2) in the atmosphere. Dinitrogen, however, is only useful to most organisms when it is 'fixed' either into NH_3, NO_x, or N that is chemically bound to either inorganic or organic molecules, and releasable by hydrolysis to NH_3, or NH_4^+. Of these forms, NH_3, organic-N, and some oxides of nitrogen are biologically useful. Because life requires fixed nitrogen its origin depended on the availability of fixed nitrogen on the early Earth. Fixed nitrogen on the early Earth resulted from in-fall from comets as organic-N, and was produced abiotically from the thermal shock waves of lightning and meteorites. Models of shock production by meteor impact on the early Earth suggest that impact-produced fixed nitrogen could have resulted in a large portion of the reservoir of earth's N_2 being converted into fixed nitrogen, primarily as NO_2^- and NO_3^- through HNO, at the end of accretion. This scenario suggests that the first forms of fixed nitrogen on the early Earth were organic-N, NO_2^-, and NO_3^-. Based upon this scenario and what is current knowledge about microbe/environment interactions, as well as microbial physiology and ecology, I hypothesize the following evolutionary sequence for the biological nitrogen cycle: ammonification followed by denitrification, followed by nitrification. Nitrogen fixation would have evolved whenever the demand of fixed nitrogen exceeded the supply.

2.1 Introduction: the chemistry of nitrogen

Life as we know it is equally dependent on nitrogen and carbon. Without nitrogen, there would be no proteins, no nucleic acids, no amino sugars, and no amino lipids – only carbohydrates; a world of 'candy'. Yet, nitrogen is often regarded as carbon's 'poor cousin', and is somewhat neglected. Nitrogen, however, is not carbon's 'poor cousin'. Granted, carbon is extremely important, so if carbon is 'king', nitrogen is 'queen', and H, O, P, and S are members of the 'Royal Court'.

Nitrogen (N) has an atomic number equal to seven and is part of the non-metal group V elements. The oxidation states of N range from -3 to $+5$. The five most common oxidation states of N are presented in Table 2.1.

In the oxidized state ($+$), the outer electrons of N serve to complete the electron shell of other atoms. In the reduced state ($-$), the electrons required for the outer shell of N are supplied by other atoms. In the oxidized state, nitrogen acts as an oxidizing agent by oxidiz-

Evolution on Planet Earth
ISBN 0-12-598655-6

Table 2.1 **The five most common oxidation states of N and their forms.**

Form of N	Oxidation state	Use by organism
NH_3	–3	N-source/energy source
NH_4^+	–3	N-source/energy source
Organic-NH_2	–3	N-source
N_2	0	N-source
N_2O	+2	Energy source
NO_2^-	+3	N-source/energy source
NO_3^-	+5	N-source/energy source

ing other compounds and becoming more reduced. These nitrogen oxidation states also represent the common pools of nitrogen present in the biosphere. The exceptions to this fact are organic NH_2, and the NH_3/NH_4^+ pools, which have the same valence, but are two separate pools. With the exception of the NH_3/NH_4^+ and organic-NH_2 pools, changes in the oxidation state of a nitrogen species brought about by oxidation/reduction reactions result in the transfer of nitrogen from one nitrogen pool to another.

2.2 Nitrogen cycling

2.2.1 *Nitrogen reservoirs*

Reservoirs of nitrogen include the lithosphere (the Earth's crust and upper mantle), atmosphere, hydrosphere, and biosphere. Most of the nitrogen on Earth exists in the lithosphere, and most of this nitrogen is locked up in igneous rocks that make up the crust and mantle, making it unavailable for use in the biosphere. Most of the remainder of the nitrogen exists as N_2 in the atmosphere, with very small amounts of NO_x (oxides of nitrogen) and

organic-N. In the hydrosphere, nitrogen exists as dissolved N_2, NH_4^+, NO_3^-, NO_2^-, and organic-N. The forms of nitrogen that occur in the biosphere vary from complex organic-N compounds to nitrogen oxides and NH_4^+. Because nitrogen is in a constant state of flux due to cycling through the biosphere in various forms (i.e. N_2, NO_x, NH_4^+, organic-N) it is difficult to quantify each nitrogen species, but most of it occurs as organic-N (reviewed by Mancinelli, 1992).

Nitrogen cycling results from abiotic and biotic transformation reactions. The biological and non-biological transfer of N among the various reservoirs constitutes the N-cycle at the reservoir level and is usually referred to as the global nitrogen cycle (Figure 2.1). The total amount of N in a given reservoir at any one time represents a balance between gains and losses. For example, N can be added to the hydrosphere from the lithosphere through soil erosion and runoff into rivers, and from the atmosphere through fixation of N_2. A reservoir may contain many ecosystems. For example, a site in the ocean containing a hydrothermal vent and associated biota constitutes an ecosystem within the hydrosphere reservoir. The nitrogen within each ecosystem is present in nitrogen pools. The transfer of nitrogen from one pool to another within an ecosystem represents a loss from one pool and a gain to another, with no net change in either the total nitrogen in the ecosystem, or the reservoir. These nitrogen transformation reactions within an ecosystem constitute the nitrogen cycle at the ecosystem level.

2.3 Abiotic N-cycle

Nitrogen cycling on the pre-biotic earth was significantly altered by the origin and evolution of life. The abiotic nitrogen cycle is outlined in Figure 2.2. The elemental abundance of nitrogen is plentiful now and always has been. The amount of nitrogen in Earth's early atmosphere ($\sim$4.0 Gyr) has been estimated to have been $\sim$3.9 $\times$ 10^{18} kg (Walker, 1977; Borucki and Chameides, 1984; Mancinelli and McKay, 1988). Dinitrogen (N_2) in the atmo-

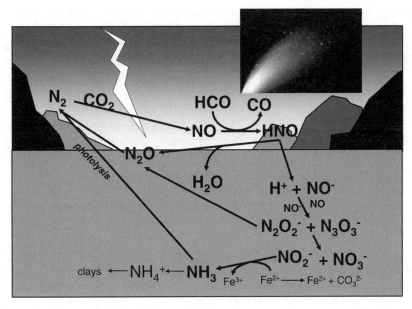

Figure 2.1 Schematic diagram of the abiotic and biotic nitrogen cycle on present Earth.

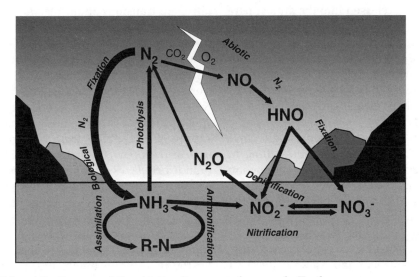

Figure 2.2 Schematic diagram of the abiotic nitrogen cycle on early Earth.

sphere was the most available form of nitrogen on the early Earth, but N_2 is not useful to most organisms. Dinitrogen must be 'fixed' in order to be useful for life (NH_3, NH_4^+, NO_x, or N bound to inorganic or organic compounds that are hydrolyzable to NH_3 or NH_4^+). Because life requires fixed nitrogen, the origin of life depended on the availability of fixed nitrogen.

The Earth's early atmosphere was primarily made of N_2, CO_2, and water vapor (e.g. Hunten, 1993). Thermal shock waves resulting from lightning and meteoritic in-fall released sufficient energy into the atmosphere to break the N_2 triple bond. When this occurs in the presence of CO_2 and O_2 (from the breakdown of water vapor) NO is produced (Borucki and Chameides, 1984; Walker, 1977). The NO readily forms nitrosyl hydride (HNO) (Kasting and Walker, 1981) (Figure 2.2). When these reactions occur today, however, sufficient O_2 is readily available and does not depend on the breakdown of water.

The fate of the HNO produced today and in the pre-biotic atmosphere of early Earth was described by Mancinelli and McKay (1988). In summary, HNO is a reactive N^{1+} species (Grätzel *et al.*, 1970). In the atmosphere it dimerizes and dehydrates to form N_2O and H_2O at a rate of $4 \times 10^{-15}\,cm^3\,s^{-1}$ (Hampson and Garvin, 1977). This slow rate of dimerization combined with the high solubility of HNO in water allows nearly all of the HNO formed to go into solution in the ocean. In the ocean, the HNO dissociates into H^+ and NO^-. Once in solution, two molecules of NO^- combine to form $N_2O_2^-$ and $N_3O_3^-$ and their conjugate acids. These species rapidly decay into N_2O, N_2O^- and N_3O^- (Doyle and Mahapatro, 1984). The N_2O is released into the atmosphere and is readily photolyzed to N_2. Any HNO_2 formed as a side product falls into the ocean and forms NO and HNO_3. The HNO_3 dissociates into H^+ and NO_3^-. These reactions led to increases in the NO_2^- and NO_3^- nitrogen pools of the hydrosphere on the early Earth (Figure 2.2).

The sources of fixed nitrogen on the early Earth included comets and meteors falling to earth from space, as well as production through thermal shock waves produced by lightning and meteors falling through the atmosphere. In fact, it has been calculated that $\sim 10^9\,kg\text{-}N\,yr^{-1}$ could have been fixed as HNO through this process on the early pre-biotic Earth (Mancinelli and McKay, 1988), of which nearly all ended up in the ocean. Because these processes occurred not only during the origin and early evolution of life, but also prior to its origin throughout much of the early meteor and cometary bombardment period, that is for at least 5×10^8 years, the process built up fixed N on the pre-biotic Earth. These events could have resulted in substantial amounts of available fixed N ($\sim 5 \times 10^{17}\,kg\text{-}N$ representing $\sim 13\%$ of the N in the atmosphere), allowing for an abundant supply of fixed nitrogen on the early Earth in the form of NO_3^- and NO_2^-. This amount is substantial when considering that the current rate of fixation on the Earth by biological and non-biological sources totals only $\sim 1.7 \times 10^{11}\,kg\text{-}N\,yr^{-1}$ (Schlesinger, 1991) with nearly all of it resulting from biological fixation and relatively little from thermal shock in the atmosphere. Thus, based on the current consumption rate (determined by the fact that the fixation rate roughly equals the rate at which N is returned to the atmosphere) the supply of fixed N on the early Earth could, in theory, supply today's biosphere for 2.9×10^6 years.

2.4 Supply and demand: the biological N-cycle

The first organisms to inhabit the Earth were microbes, and as a result the first biological nitrogen transformation reactions were performed by microbes. Microbes will use what is the most available and most efficient source of nitrogen. The most abundant supply of fixed nitrogen on the early Earth included organic-N from comets and carbonaceous chondrites, NO_3^- and NO_2^- pools in the ocean from atmospheric thermal shocks, and the small available supply of NH_4^+/NH_3 from NO_x reduction by iron in the ocean (Summers and Chang, 1993; Brandes *et al.*, 1999). These organisms need to produce proteins and nucleic acids. Because organisms ultimately incorporate NH_3 and not NO_3^- or NO_2^-, it would have been more efficient (i.e. requiring fewer enzymes, less complex cellular machinery, and less energy) for the first organisms to utilize the relatively small available supply of NH_4^+/NH_3 rather than the organic-N, or the NO_3^- or NO_2^-. This line of reasoning is reinforced by the fact that today many algae prefer NH_4^+ to NO_3^-, presumably for reasons of cellular efficiency and energy (e.g. Goldman and Glibert, 1983; Wheeler, 1983).

2.4.1 N-assimilation

Nitrogen assimilation is the conversion of NH_3 to organic-N and biomass. The NH_3 is assimilated in a series of enzymatically catalyzed reactions via two basic pathways. In one pathway, α-keto carboxylic acid is aminated to form an amino acid. In the other pathway, the enzyme glutamine synthetase catalyzes production of glutamine from glutamate, NH_4^+ or NH_3 and ATP. Then, through a series of reactions amino acids are produced that are used subsequently in the production of other compounds for the cell (reviewed in Merrick and Edwards, 1995). Virtually all organisms have the ability to assimilate NH_4^+ at some level. Assimilation of NH_4^+/NH_3 would have been the first step to evolve in the biogeochemical nitrogen cycle.

$$NH_3 \rightarrow \text{Organic-N}$$

2.4.2 Ammonification

Once NH_4^+/NH_3 became limiting in an ecosystem on the early Earth, organic-N, NO_3^- and NO_2^- were left as the primary available sources of fixed nitrogen. The simplest pathway to follow from the nitrogen assimilation pathway is the deamination of organics to yield NH_4^+. This enzymatic process of organic-N conversion to NH_4^+ is called ammonification. Because there is a wide array of N-containing organic compounds belonging to different chemical classes, a wide array of enzymes is required that break them down to produce NH_4^+. These enzymes include hydrolases, oxidases, aminases, ureases, proteases, nucleases, and lyases.

Ammonification occurs in a phylogenetically diverse group of organisms ranging from microbes to plants and animals. This great diversity has led several researchers to suggest that this metabolic process arose early in the evolution (e.g. Walker, 1977; Mancinelli and McKay, 1988). Further, from a physiological standpoint, it seems more efficient to evolve from assimilation of NH_4^+ to cleaving an NH group from an organic compound and assimilating it through an existing metabolic pathway rather than evolving an entirely new pathway. This process allows for ammonification via simple fermentation reactions as the next probable biological step to evolve in the nitrogen biogeochemical cycle.

2.4.3 NO_x reduction and assimilation

When the demand for NH_4^+/NH_3 and organic-N exceeded the supply, organisms that evolved the ability to use NO_3^- and NO_2^- as a nitrogen source would have had a selective advantage over those that were unable to use these sources. A phylogenetically diverse group of organisms including microbes and plants possesses the nitrate and nitrite reductases that catalyze the reduction of NO_3^- and NO_2^- to NH_3 for assimilation. In these organisms, the NO_3^- is reduced in a single step to NO_2^- by an assimilatory nitrate reductase. This reaction is followed by the reduction of NO_2^- to NH_3 through hydroxylamine (an enzyme bound intermediate) by an assimilatory nitrite reductase requiring six electrons per molecule of NH_3 produced (Merrick and Edwards, 1995). The NH_3 that is formed is then assimilated as previously described. The availability of NO_3^- and NO_2^- suggests that this step was the next to evolve in the biological N-cycle.

2.4.4 Denitrification

The availability of NO_3^- and NO_2^- in the environment would give organisms that have the ability to respire NO_3^- and NO_2^-, that is, perform denitrification, a selective advantage over fermentative microbes. The selective advantage arises from the fact that respiration produces cellular energy (i.e. more ATPs) more efficiently than fermentation reactions.

Denitrification is the dissimilatory reduction of NO_3^- to N_2O, or N_2 by denitrifying bacteria. Because it is coupled to the production of ATP and electron transfer occurs via the cytochrome system it is a form of anaerobic respiration. The process usually occurs under anaerobic conditions, although denitrification by *Thiosphaera* sp. has been reported to occur under aerobic conditions (Robertson *et al.*, 1995). Because denitrifiers are facultative anaerobes, with few exceptions they preferentially use O_2 as their terminal electron acceptor over the nitrogen oxides because it is a more efficient energy producing process. The organism generates cellular energy (ATP) by the transport of electrons via the cytochrome system from an organic or inorganic source to NO_3^-, or to a more reduced nitrogen oxide (e.g. NO_2^-, NO, N_2O) derived from NO_3^-. The nitrogen oxides that form during the process serve as electron acceptors during denitrification and proceed along the following pathway:

$$2NO_3^- \rightarrow 2NO_2^- \rightarrow 2[NO] \rightarrow N_2O \rightarrow N_2$$

Each step of the pathway is catalyzed by an enzyme, a nitrogen oxide reductase that transfers electrons from the electron transport chain to the particular intermediate of the denitrification pathway (Zumft, 1997 and references therein).

Denitrification is a widespread phenomenon in nature occurring in aquatic and terrestrial environments. It has been identified among a phylogenetically diverse group of microbes ranging from the Bacteria to the Archaea. The phylogenetic diversity of this capability argues for its antiquity. Further, until biological denitrification evolved no efficient mechanism existed that would have put N_2 back into the atmosphere. Without the early evolution of biological denitrification prior to evolution of oxygenic photosynthesis (~ 2.5 Gya or possibly earlier) approximately 51% of the atmospheric N_2 could have been locked up in biomass and in the hydrosphere as NO_3^- and NO_2^- salts (based on a total endowment of 3.9×10^{18} kg N_2 and a fixation rate of 10^9 kg N yr^{-1} for 2×10^9 yr, as discussed previously). This process would have resulted not only in a change in the early atmosphere, but also in a reduction in the early Earth's atmospheric pressure.

2.4.5 *Biological nitrogen fixation*

When the demand for fixed nitrogen exceeds the supply, organisms that have evolved the ability to fix N_2 will have a selective advantage. Biological N-fixation refers to the ability of an organism to transform N_2 from an atmospheric gas into NH_3. Only prokaryotes are capable of nitrogen fixation. Nitrogenase, the enzyme responsible for nitrogen fixation, is a complex metaloenzyme coded for by the *nif* genes that initiate, regulate, and direct production of the enzyme when it is induced. The overall reaction catalyzed by nitrogenase is:

$$N_2 + 8e^- + 8H^+ + 16ATP \rightarrow 2NH_3 + H_2 + 16ADP + 16P_i$$

Sixteen molecules of ATP are required to break the nitrogen-to-nitrogen triple bond in N_2. The requirement for such a large number of ATP molecules makes biological nitrogen fixation a very energy-expensive process. Because of the high energy cost, organisms preferentially use fixed nitrogen when it is available and fix nitrogen only when the demand exceeds the supply. The NH_3 produced is assimilated as previously described.

Nitrogenase is inhibited by O_2. As a result, microorganisms have developed a variety of strategies to handle the sensitivity of nitrogenase to O_2. For example, in heterotrophic aerobic bacteria, such as *Azotobacter*, nitrogen fixation occurs when the respiration rate of the cell equals or exceeds the rate of diffusion of O_2 into the cell, creating a low O_2 environment (reviewed by Mancinelli, 1992). Certain cyanobacteria (e.g. *Nostoc* and *Anabaena*) house nitrogenase in heterocysts, which are specialized cells that do not have Photosystem II and therefore do not produce O_2. Some filamentous non-heterocystous cyanobacteria (e.g. *Trichodesmium* and *Microcoleus*) form aggregates that produce micro-aerobic conditions in the center of the aggregate where N_2 fixation is thought to occur.

Nitrogen fixers represent a diverse array of physiological types of prokaryotes. The diazotrophs can be divided into autotrophs and heterotrophs, depending on their source of carbon. They can be further subdivided and designated as free-living or symbiotic depending on their habitat. Examples of free-living heterotrophic diazotrophs include members of the genera *Azotobacter*, *Clostridium* and *Klebsiella*; whereas members of the heterotrophic symbiotic type include members of the genera *Rhizobium* and *Frankia*. Common examples of autotrophic diazotrophs include the photosynthetic bacterium *Rhodospirillum rubrum* and the cyanobacteria (reviewed by Mancinelli, 1992). The *nif* genes occur on a plasmid that can be transferred laterally among the prokaryotic community and this process may account for its dispersal among the wide variety of phylogenetic types of microbes.

2.4.6 *Nitrification*

Nitrification is the oxidation of NH_4^+ to NO_3^-. Organisms that are capable of oxidizing NH_4^+ to NO_2^- and NO_2^- to NO_3^- are nitrifiers. They belong to two distinct types of chemoautotrophic bacteria: the ammonia oxidizers and the nitrite oxidizers. The metabolism of each type corresponds to the two steps involved in nitrification, that is, step 1: $NH_4^+ \rightarrow NO_2^-$ and step 2: $NO_2^- \rightarrow NO_3^-$. These nitrifiers synthesize all of their cellular constituents from CO_2 via the Calvin cycle and an incomplete tricarboxylic acid (TCA) cycle. Consequently, the reactions of the TCA cycle that they possess can only be used for biosynthesis, and not for the production of cellular energy (ATP) (reviewed by Mancinelli, 1992; Ward, 1996).

Organisms that oxidize NH_4^+ to NO_2^- belong to genera whose names begin with the prefix 'nitroso', which refers to the ability of the organism to oxidize NH_4^+ to nitrous acid (HNO_2). Nitrous acid is completely ionized under physiological conditions leaving free NO_2^-. These genera include *Nitrosomonas*, *Nitrosolobus*, *Nitrosospira*, *Nitrosococcus*, and *Nitrosovibrio*. The nitrosobacteria oxidize ammonia *via* the following series of enzymatically catalyzed reactions:

$$NH_3 + \tfrac{1}{2}O_2 \rightarrow [NH_2OH] \rightarrow [HNO] + 2H^+$$

$$[HNO] + \tfrac{1}{2}O_2 \rightarrow H^+ + NO_2^-$$

The intermediates, hydroxylamine (NH_2OH) and nitroxyl (HNO) are depicted in brackets because they are thought to be enzyme bound and do not exist free in the cell (Mancinelli 1992; Ward, 1996). Nitrous oxide (N_2O) and nitric oxide (NO) are produced as by-products of ammonia oxidation by these organisms.

Those organisms that oxidize NO_2^- to NO_3^- belong to genera whose names begin with 'nitro', which refers to the ability of the organism to oxidize NO_2^- to nitric acid (HNO_3). As is true for nitrous acid, under physiological con-

ditions, the nitric acid is ionized leaving free NO_3^-. The genera capable of NO_2^- oxidation include *Nitrobacter*, *Nitrospina*, and *Nitrococcus*.

The oxidation of NO_2^- to NO_3^- by the nitrobacteria is carried out by a nitrite oxidase, with two electrons transferred to O_2 via cytochromes. The reaction is coupled to the generation of ATP, and is as follows:

$$NO_2^- + \tfrac{1}{2}O_2 \rightarrow NO_3^-$$

The transfer of two electrons results in a small change in the N oxidation state, from +3 to +5. This change in the N oxidation state is small enough to obviate the need for any intermediates.

Although nitrification in nature is principally carried out by these two types of chemoautotrophic bacteria, a variety of heterotrophic bacteria (e.g. species of *Arthrobacter*) and fungi (e.g. species of *Aspergillus*) are also capable of nitrification. These heterotrophic organisms, unlike the chemoautotrophs, fail to couple the oxidation reactions to cellular energy production, or any other proven useful metabolic process (reviewed in Mancinelli, 1992; Ward, 1996). Since nitrification requires O_2, it most likely evolved after oxygenic photosynthesis, and was probably the last step in the biological N-cycle to evolve.

2.5 **Present-day N-cycle**

Although nitrogen is transformed non-biologically, most of it is transformed biologically. All organisms need nitrogen and thus must participate in the nitrogen cycle at some level, but microbes are the primary mediators of N-transformation reactions on present-day earth. Figure 2.1 depicts a generalized scheme of the present nitrogen cycle. The largest available pool of nitrogen exists as N_2 in the atmosphere, with very small amounts of NO_x and organic-N, totaling $\sim 3.86 \times 10^{18}$ kg. In fact, one square meter column of the atmosphere rising from the Earth's surface to the top of the troposphere contains about 24 300 kg of N, more than any other element. The amount of nitrogen in the hydrosphere (N_2, NH_4^+, NO_3^-, NO_2^- and organic-N) is two orders of magnitude lower than the atmosphere, that is $\sim 2.3 \times 10^{16}$ kg. The nitrogen pools in the biosphere (i.e. N_2, NO_x, NH_4^+, organic-N) total $\sim 2.8 \times 10^{14}$ kg with organic-N being the most abundant. Nitrogen is currently being fixed at the rate of about 92 million metric tons per year, predominantly (>90%) by biological nitrogen fixation. The total amount that is denitrified and returned to the atmosphere is about 83 million metric tons per year. The difference (9 million metric tons) represents the rate at which fixed N is building up in the biosphere (Mancinelli, 1992).

2.6   Nitrogen and the eukaryotes

Organisms other than prokaryotes contribute to nitrogen transformation reactions. The primary role of plants is the assimilation of NO_3^- and NH_4^+ into plant biomass. Except for the release of trace amounts of gaseous NO_x during over fertilization by humans, the nitrogen in plants is locked up in biomass until the plant dies. Fungi and protists take up and incorporate organic nitrogen into their biomass. They do not actively excrete nitrogen compounds. Animals take up and incorporate organic nitrogen compounds and excrete organic nitrogen compounds. The primary excretory product is ammonia, derived from the digestion of proteins.

Because ammonia is highly toxic to most animals, it must be eliminated quickly and efficiently. This has led to the evolution of diverse and sometimes complex physiological systems to regulate nitrogen release from the organism, ranging from simple diffusion mechanisms in protozoans and small aquatic animals to complex mechanisms in many terrestrial animals. In terrestrial animals, as well as large aquatic animals, ammonia must be converted to less toxic substances before being excreted, such as urea in mammals and the insoluble uric acid in insects, birds, and reptiles.

Invertebrates have evolved a complex system whereby nitrogenous waste is excreted through nephridia (paired tubules). One end of the nephridium opens into the body cavity,

and the other to the outside environment. The excretory organ of mollusks is the renal gland, a wide tube opening at one end into the sac surrounding the heart and at the other end to the exterior environment. The urine is formed through filtration of the blood, and its composition altered by reabsorption and secretion. The paired coxal glands of aquatic arthropods pass from the coelomic sac where blood filtration occurs to external openings at the base of limbs, notably the antennae.

Insects have evolved a very different type of complex excretory system. Malpighian tubules (varying in number from 2 to 100 depending upon species) open at one end into the blood space (body cavity) and at the other into the rectum, part of the alimentary canal. The primary urine is formed not by filtration, but secretion of ions and water from the blood. In the rectum the composition of the urine is changed radically; soluble urate is converted to insoluble uric acid, and water and the soluble products of digestion are reabsorbed. Birds, like reptiles and insects, excrete uric acid into an extension of the alimentary canal.

In most fishes, amphibians, and mammals, nitrogen is detoxified in the liver and excreted as urea, a readily soluble and harmless product. Nitrogenous wastes in freshwater fish are essentially converted into ammonia that diffuses out through the skin.

The human excretory or urinary system is a highly evolved complex regulatory system characteristic of that of all mammals. It consists of two kidneys where urine is produced by filtration, secretion, and reabsorption; the ureters, tubes that transport the urine; the bladder where the urine is stored; and the urethra through which the urine is voided. Malfunction of the excretory system can lead to dehydration or edema, and the dangerous build-up of waste and toxic substances that can kill the organism.

2.7   Conclusion

The extent to which one reaction proceeds in favor of another is determined by environmental conditions (e.g. aerobic vs anaerobic) and nutrient availability, that is, supply and demand. Biological nitrogen fixation occurs only in the absence of fixed nitrogen. Denitrification occurs in the presence of NO_3^- or NO_2^-, in anaerobic environments. Nitrification occurs under aerobic conditions when NH_4^+ is readily available.

The available N on early Earth was N_2 in the atmosphere, organic-N from comets and carbonaceous chondrites. The abiotic N-cycle of early Earth formed NO_3^- and NO_2^- pools in the early abiotic ocean. When life arose, the available nitrogen sources were N_2, organic-N, NO_3^-, and NO_2^-. From the above discussion regarding environmental conditions, physiological simplicity, and phylogenetic diversity, it appears that assimilation of available NH_4^+/NH_3 was the first step to evolve in the biological N-cycle, followed by ammonification of organic-N. When organic-N became limiting, those organisms capable of reducing NO_3^-/NO_2^- to NH_3 had the selective advantage.

The debate whether denitrification evolved prior to or after aerobic respiration began in the early 1970s (e.g. Broda, 1975; Egami, 1973). Broda argued from an atmospheric chemistry standpoint that the evolution of denitrification followed that of aerobic respiration because there was no available NO_3^- on the early Earth (Broda, 1975). Egami proposed a sequence of events where denitrification preceded aerobic respiration based on physiological considerations (Egami, 1973). Broda's atmospheric chemistry arguments were based on the outdated assumption that the early Earth's atmosphere consisted of CH_4, H_2O, and NH_3. The lack of N-oxides in the early Earth's atmosphere has always been the primary argument against the evolution of denitrification prior to aerobic respiration. This argument lost its validity by the demonstration of N-oxide formation due to lightning and meteorite impacts (e.g. Kasting, 1990; Mancinelli and McKay, 1988; Yung and McElroy, 1979). Formation of NO is predicted for both CO_2-rich and CO-rich atmospheres, with the former also generating a significant amount of N_2O (Fegley *et al.*, 1986). Removal of this principal obstacle led Mancinelli and

McKay (1988) to demonstrate how NO_3^- and NO_2^- pools could accumulate in the early ocean and provide the substrates necessary for the evolution of denitrification. This model allowed them to conclude that denitrification arose prior to aerobic respiration (Mancinelli and McKay, 1988).

It is clear that biological N_2 fixation arose when the supply of fixed N was exceeded by the demand (e.g. Navarro-Gonzalez *et al.*, 2001). When this event occurred on the geological timescale is uncertain. The demand is a function of biomass, that is, the greater the biomass the greater the demand for fixed nitrogen. So, another way to address the issue is to determine when the biomass was large enough that abiotic fixation mechanisms became an insufficient source. When photosynthesis evolved, the opportunity arose for large quantities of CO_2 to be fixed into biomass. The evolution of oxygenic photosynthesis led to a significantly larger increase in biomass, increased biodiversity, and a much greater demand for fixed nitrogen, such that the abiotic fixation rate probably was insufficient. This scenario suggests that if biological N-fixation had not evolved prior to oxygenic photosynthesis it would have evolved immediately subsequent to or concomitantly with it. The debate regarding its antiquity based on phylogenetic grounds is untenable primarily due to the fact that the *nif* genes occur on a plasmid that can be transferred laterally among different microbial genera.

Biological nitrification requires the presence of O_2 and a plentiful supply of NH_4^+/NH_3. This fact indicates that biological nitrification arose after oxygenic photosynthesis. Prior to oxygenic photosynthesis, O_2 and NH_4^+/NH_3 would not have been available for nitrification to occur. These facts combined with the previous discussion suggest that nitrification was probably the last step of the biological nitrogen cycle to evolve.

Obviously, the N-cycle is a well-orchestrated system with all of the different transformation reactions balancing each other on a planetary scale. For example, for the Earth as a whole, the amount of N fixed is equal to the amount of N returned to the atmosphere. If this were not true, then the Earth's atmosphere would eventually become depleted of N. It appears that there is a balanced interplay between life and its environment, as illustrated through the biogeochemical cycle of nitrogen.

Clearly, because nitrogen is so critical to life, its various forms in the environment have had an impact on the evolution of all life, ranging from microbes to humans. Specifically, all life needs fixed nitrogen, and because it is so physiologically important, organisms have evolved mechanisms for regulating its uptake and excretion. The mechanisms that evolved to excrete nitrogenous wastes differ considerably in various organisms and environments. This diversity ranges from simple diffusion in aquatic animals and protozoa to complex regulatory systems, such as the kidneys in mammals and malphigian tubules in insects.

References

Borucki, W.J. and Chameides, W.L. (1984). Lightning: estimates of the rates of energy dissipation and nitrogen fixation. *Rev. Geophys.*, **22**: 363–372.

Brandes, J.A., Boctor, N.Z., Cody, G.D. *et al.* (1999). Abiotic nitrogen reduction on the early Earth. *Nature*, **395**: 365–367.

Broda, E. (1975). *The Evolution of the Bioenergetic Processes.* Oxford: Pergamon Press.

Doyle, M.P. and Mahapatro, S.N. (1984). Nitric oxide dissociation from trioxodinitrate(II) in aqueous solution. *J. Am. Chem. Soc.*, **106**: 3678–3679.

Egami, F. (1973). A comment to the concept on the role of nitrate fermentation and nitrate respiration in an evolutionary pathway of energy metabolism. *Z. Allg. Mikrobiol.*, **13**: 177–181.

Fegley, B. Jr, Prinn, R.G., Hartman, H. and Watkins, G.H. (1986). Chemical effects of large impacts on the earth's primitive atmosphere. *Nature*, **319**: 305–308.

Goldman, J.C. and Glibert, P.M. (1983). Kinetics of inorganic nitrogen uptake by phytoplankton. In: E.J. Carpenter and D.G. Capone (eds) *Nitrogen in the Marine Environment.* London: Academic Press, pp. 233–274.

Grätzel, V.M., Taniguchi, S. and Henglein, A. (1970). Pulsradiolytische untersuchung zwischenpro-

dukte der NO-reduktion in wäbriger lösung. *Ber. Bunsen-Ges. Phys. Chem.*, **74**: 1003–1010.

Hampson, R.F. and Garvin, D. (1977). Reaction rate and photochemical data for atmospheric chemistry. *Nat. Bur. Stand. Spec. publ.*, **513**: 210–222.

Hunten, D.M. (1993). Atmospheric evolution of the terrestrial planets. *Science*, **259**: 915–920.

Kasting, J.F. (1990). Bolide impacts and the oxidation state of carbon in the earth's early atmosphere. *Origins Life Evol. Biosphere*, **20**: 199–231.

Kasting, J.F and Walker, J.C.G. (1981). Limits on oxygen concentration in the prebiological atmosphere and the rate of abiotic fixation of nitrogen. *J. Geophys. Res.*, **86**: 1147–1158.

Mancinelli, R.L. (1992). Nitrogen cycle. *Encyclopedia of Microbiology*, **3**, 229–237. Academic Press, NY.

Mancinelli, R.L. and McKay C.P. (1988). Evolution of nitrogen cycling. *Orig. Life Evol. Biosph.*, **18**: 311–325.

Merrick, M.J. and Edwards, R.A. (1995). Nitrogen control in bacteria. *Microbiol Rev.*, **59**: 604–622.

Navarro-Gonzalez, R., McKay, C.P. and Mvondo, D.N. (2001). A possible crisis for Archaean life due to reduced lightning fixation by lightning. *Science*, **412**: 61–64.

Robertson, L.A., Dalsgaard, T., Revsbech, N.-P. and Kuenen, J.G. (1995). Confirmation of aerobic denitrification in batch cultures, using gas chromatography and ^{15}N mass spectrometry. *FEMS Microbiol Ecol.*, **18**: 113–120.

Schlesinger, W.H. (1991). *Biogeochemistry and Analysis of Global Change.* Academic Press, NY.

Summers, D. and Chang S. (1993). Prebiotic ammonia from iron(II) reduction of nitrite on the early Earth. *Nature*, **365**: 630–633.

Walker, J.C.G. (1977). *Evolution of the Atmosphere*, New York: Macmillan Publishing Co., Inc., pp. 179–273.

Ward, B.B. (1996). Nitrification and ammonification in aquatic systems. *Life Support Biosphere Sci.*, **3**: 25–29.

Wheeler, P.A. (1983). Kinetics of inorganic nitrogen uptake by phytoplankton. In: E.J. Carpenter and D.G. Capone (eds) *Nitrogen in the Marine Environment.* London: Academic Press, pp. 309–346.

Yung, Y.L. and McElroy, M.B. (1979). Fixation of nitrogen in the prebiotic atmosphere. *Science*, **203**: 1002–1004.

Zumft, W. (1997). Cell biology and molecular basis of denitrification. *Molecular and Biological Reviews*, **61**: 533–661.

3

The coupled evolution of life and atmospheric oxygen

Timothy M. Lenton

ABSTRACT

See Plates 4 and 5

This review highlights the reciprocal influences of evolution on oxygen and of oxygen on evolution. Without life there would be very little free oxygen at the Earth's surface. Over Earth history, oxygen has risen from negligible concentrations to ~0.21 atm (~21% of the present atmosphere), in a series of steps. Atmospheric oxygen is the product of oxygenic photosynthesis, which may have evolved over 3.5 Gyr ago (billion years ago), and was definitely occurring in cyanobacteria by about 2.7 Gyr ago. The capacity to detoxify reactive oxygen species and the potential for aerobic respiration arose before oxygenic photosynthesis. Oxygen remained below 0.0008 atm until 2.2 Gyr ago, but probably reached significant dissolved concentrations in localized oxygen oases, including microbial mats. Localized accumulation of oxygen triggered an increase in defences against oxygen toxicity, the development of metabolic pathways utilizing oxygen, and facilitated the appearance of eukaryotic cells with mitochondria. Between 2.2 and 2.0 Gyr ago the Earth underwent a great oxidation event in which atmospheric oxygen rose above 0.002 atm and possibly above 0.03 atm. Exhaustion of some oxygen sinks and increased organic carbon burial (a major oxygen source) both contributed. The rise of oxygen generated an ozone layer providing effective UV protection, increased the availability of nitrogen in the ocean, but reduced the availability of many bio-essential metals. Eukaryotes with chloroplasts appeared at this time, possibly as a result of increased nitrogen availability. From 2.0 to 0.8 Gyr ago oxygen concentrations remained above 0.002 atm and were sufficient for the evolutionary radiation of multi-cellular eukaryotes, but fossil evidence for this does not appear until 1.0 Gyr ago. Oxygen rose again between 0.8 and 0.57 Gyr ago, exceeding 0.03 atm. Episodes of enhanced organic carbon burial in this interval have been attributed to favourable geologic conditions and the evolution of metazoans that produced fast-sinking faecal pellets. The deep ocean may not have been oxygenated until this time. Ventilation of the ocean opened an evolutionary window for the diversification of large metazoans seen in the Cambrian explosion 0.54 Gyr ago. By 0.37 Gyr ago oxygen had exceeded 0.14 atm, and by 0.3 Gyr ago it had reached at least 0.21 atm, driven upwards by vascular plants colonizing the land surface and increasing organic carbon burial. Since then, the atmospheric fraction of oxygen has been stabilized close to 21% by feedback mechanisms involving terrestrial and marine biota. These have prevented oxygen rising to dangerously flammable concentrations for forests (~25%), or falling to suffocating concentrations for large animals (~15%). Previous estimates that oxygen reached ~35% of the atmosphere in the late Carboniferous (0.3 Gyr ago) need to be reconsidered.

Evolution on Planet Earth
ISBN 0-12-598655-6

3.1 Introduction

An appreciation that air, fire and life are linked can be found throughout much of recorded history, but it was not until the Enlightenment that the connecting substance was isolated (Gilbert, 1981). Joseph Priestley, a liberal minister in England, is usually credited with the discovery of oxygen in 1774. Priestley realized that since even a small candle uses a large amount of pure air, there must be a provision in nature to replace it, and he found that a sprig of mint would revive the foul air left after combustion, such that it could support a flame once more. When heating mercury(II) oxide, Priestley discovered that it gave off large amounts of a gas in which a candle would burn with an enlarged flame, and in which mice lived longer than in a similar volume of normal air. Priestley named the gas 'dephologisticated air' because he firmly believed the erroneous theory that combustion releases a substance called phlogiston into the air. Karl Wilhelm Scheele, a pharmacist in Sweden, independently isolated oxygen about two years before Priestley but was slower to publish his finding. He named the gas 'fire air' because it supports combustion. Antoine Lavoisier discovered the mechanism of oxidation in 1775 and was the first to appreciate the chemical significance of the action of oxygen. He repeated Priestley's experiments and by 1778 was convinced that the gas given off by heating mercury(II) oxide is the same one that is present in the air and combines with substances during combustion. Lavoisier called it 'oxygine' (from the Greek 'acid producing').

Oxygen is a colourless, odourless, gaseous element. Its common form is the diatomic oxygen molecule, O_2 (dioxygen). Oxygen reacts with most other elements to form oxides. It is the most abundant element in the Earth's crust (49.2% by weight, 58% of the atoms), and the dioxygen molecule, O_2, comprises 20.946% of the Earth's atmosphere by volume (Duursma and Boisson, 1994).

O_2 is released during oxygenic photosynthesis. There are two principal reactions. In the light-dependent reaction, energy from sunlight is absorbed by photosynthetic pigments (principally chlorophyll) and used for the photolysis of water. The electrons released pass along an electron transfer chain, losing their energy, some of which is used to convert ADP to ATP (photophosphorylation). Electrons and protons released in the photolysis of water are also used to reduce NADP to NADPH. The ATP and NADPH produced in the light reaction provide energy and reducing power, respectively, for the ensuing light-independent (or 'dark') reaction. In this reaction carbon dioxide is reduced to carbohydrate in a metabolic pathway known as the Calvin cycle. Oxygenic photosynthesis can be summarized by the equation:

$$CO_2 + H_2O \rightarrow CH_2O + O_2 \qquad (3.1)$$

O_2 is the strongest molecular oxidant widely available to life, meaning that in oxidation reactions it gives the greatest energy yield, and hence aerobic respiration is widespread:

$$CH_2O + O_2 \rightarrow CO_2 + H_2O \qquad (3.2)$$

All eukaryotes use aerobic respiration with organic carbon as fuel. In contrast, prokaryotes show great metabolic diversity, using many different oxidants. As well as being essential to aerobes, O_2 is very toxic. Obligate anaerobes cannot use free O_2 for metabolism, and are typically killed by exposure to it. Facultative anaerobes are normally aerobic but can respire anaerobically during periods of O_2 shortage.

There is a triatomic form of oxygen, O_3, commonly known as ozone. Ozone (O_3) is a colourless gas, produced by the action of high-energy ultraviolet radiation on O_2:

$$O_2 + hv \rightarrow O + O \qquad (3.3)$$

$$O + O_2 \rightarrow O_3 \qquad (3.4)$$

Its presence in the stratosphere acts as a protective screen for ultraviolet radiation called the ozone layer. However, ozone is also a potent greenhouse gas and a powerful oxidant. Human and natural sources of NO_x and volatile organic carbon compounds (VOC) lead to photochemical production of O_3 in the troposphere, and can cause ozone to reach sufficient concentrations in the boundary layer to cause oxidative damage to plants and animals (Mudd, 1996).

This chapter reviews the contemporary oxygen cycle (section 3.2), the history of atmospheric oxygen (section 3.3), its toxicity (section 3.4), its influences on evolution (section 3.5), and the mechanisms that have controlled its concentration (section 3.6).

3.2 The oxygen cycle

Figure 3.1 summarizes the global oxygen cycle, which is currently close to steady state. O_2 is liberated in net primary production (which is the balance of oxygenic photosynthesis and aerobic respiration by the primary producers) and a corresponding amount of organic matter is generated. Almost all of this O_2 and organic carbon is consumed in aerobic respiration by heterotrophs, or in methane production (by methanogens) and its subsequent oxidation (by methanotrophs or by chemical reaction). A small remainder (currently $\sim 0.1\%$) of the O_2 produced in oxygenic photosynthesis provides a net source of O_2 to the atmosphere. This source flux corresponds to the small amount of organic carbon that is buried in new sediments,

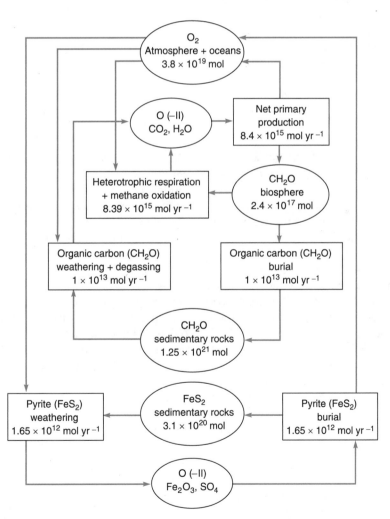

Figure 3.1　The oxygen cycle. Ovals indicate reservoirs in moles of O_2 or reducing equivalents. Boxes indicate fluxes in moles of O_2 per year. Estimated values are from Schlesinger (1997), Betts and Holland (1991), Berner and Canfield (1989) and Van Cappellen and Ingall (1996).

having escaped aerobic respiration or conversion to methane and subsequent oxidation. The effect of organic carbon burial is described by the same chemical equation as photosynthesis (equation 3.1) (with 'CH$_2$O' referring to the organic carbon being buried in rocks). The corresponding sink flux of O$_2$ is the oxidation of organic carbon exposed in rocks on the continents (oxidative weathering) or degassed by volcanic and metamorphic activity. Oxidative weathering of organic carbon is sometimes called the 'respiration' of the rocks, because the overall chemical reaction is identical to aerobic respiration (equation 3.2).

There are also significant burial and weathering fluxes of the other main redox elements – iron and sulphur. Iron pyrite (FeS$_2$), which contains both iron and sulphur in reduced form, is produced by bacteria that use sulphate and Fe(III) as oxidants. For each mole of sulphur buried 15/8 moles of O$_2$ are liberated. The reaction is reversed when pyrite undergoes oxidative weathering:

$$2Fe_2O_3 + 16Ca^{2+} + 16HCO_3^- + 8SO_4^{2-}$$
$$\leftrightarrow 4FeS_2 + 16CaCO_3 + 8H_2O + 15O_2 \quad (3.5)$$

Currently $\sim 1.65 \times 10^{12}$ mol O$_2$ yr^{-1} are exchanged due to pyrite burial and weathering (Berner and Canfield, 1989) and $\sim 1 \times 10^{13}$ mol O$_2$ yr^{-1} due to organic carbon burial and weathering (Betts and Holland, 1991). Hydrogen escape to space provides a further source of oxygen and oxidation of reduced volcanic gases such as H$_2$ a further sink, but these are minor components of the current O$_2$ budget. 3.75×10^{19} mol O$_2$ is present in the atmosphere and 3.1×10^{17} mol O$_2$ is dissolved in the ocean (Duursma and Boisson, 1994). Dividing the total atmosphere/ocean oxygen reservoir by the net source (or sink) flux gives a 'residence time' for O$_2$ of ~ 3.25 million years. This is the average amount of time an oxygen atom spends in the atmosphere/ocean system.

Far more O$_2$ has been produced since the origin of oxygenic photosynthesis than is currently present in the atmosphere and ocean. The reservoir of organic carbon in the crust is $\sim 1.25 \times 10^{21}$ moles (Berner and Canfield, 1989; Berner, 1991) and assuming that oxygenic photosynthesis has been the dominant pathway of carbon fixation since the origin of life, a corresponding number of moles of O$_2$ have been produced. A further $\sim 3.1 \times 10^{20}$ moles of O$_2$ have been liberated due to FeS$_2$ burial in the crust. The atmospheric and oceanic O$_2$ reservoirs are $\sim 2.4\%$ and $\sim 0.02\%$ of the total, respectively. The remaining ~ 40 times as much oxygen must have formed oxidized compounds, which are incorporated in the rocks of the crust and the mantle. These include ferric oxide in the form of banded iron formations and redbeds and sulphate minerals such as gypsum.

3.3 History of atmospheric oxygen

Over the history of life on Earth, atmospheric oxygen has risen intermittently from essentially zero to its present concentration and the oxygen cycle has undergone major reorganizations. This section reviews the constraints on the oxygen trajectory that have been derived from models of atmospheric chemistry, geochemical evidence, and the O$_2$ requirements of past organisms. The discussion spans the four geologic eons of Earth history: Hadean, ~ 4.5–4 Gyr ago (billion years ago); Archean, 4–2.5 Gyr ago; Proterozoic, 2.5–0.57 Gyr ago; and Phanerozoic, 0.57–0 Gyr ago. Past oxygen concentrations are generally expressed as partial pressures (pO$_2$) in atmospheres (1 atm $= 101\,325$ pascals $= 760$ mm Hg) or as fractions or percentages of present atmospheric level (PAL), which is ~ 0.21 atm. These measures are independent of the total atmospheric pressure. However, Phanerozoic oxygen levels are often expressed as fractions or percentages of the atmosphere, because the proportion of oxygen to unreactive gas determines flammability and this in turn sets narrow constraints on O$_2$ variation. Usually it is assumed that the reservoir of inert gas, mostly N$_2$, was constant (~ 0.79 atm) over Phanerozoic time (Holland, 1978). Oxygen partial pressure and atmospheric fraction (f$_{O2}$) can then be easily interconverted: f$_{O2}$ = pO$_2$/(pO$_2$ + 0.79), pO$_2$ = 0.79f$_{O2}$/(1 − f$_{O2}$).

After the formation of the Earth but before the origin of life, in the Hadean, there can only have been negligible concentrations of atmospheric O_2. O_2 would have been produced by photo-dissociation of H_2O and CO_2 at high altitudes, reaching $\sim 10^{-3}$ atm at 60 km. However, there was an excess of H_2 supply from volcanoes over O_2 supply at low altitudes, and the by-products of water vapour photolysis catalyses the reaction between these two gases. This kept $O_2 < 10^{-12}$ PAL ($pO_2 < 10^{-13}$ atm) near the Earth's surface (Kasting, 1993).

There is isotopic evidence for life on Earth early in the Archean, 3.85 Gyr ago, in the oldest sedimentary rocks (Mojzsis *et al.*, 1996). Reduced carbon microparticles in rocks from >3.7 Gyr ago are of biogenic origin and may have been derived from planktonic photoautotrophs (Rosing, 1999). Oxygenic photosynthesis evolved early in the history of life, but how early is still poorly constrained. Photosystem II, which liberates oxygen, evolved after photosystem I. Hence the first photosynthesizers with just photosystem I were anaerobic and may have used H_2S as a reductant to produce free sulphur (as in green sulphur bacteria) or sulphate. These sulphur bacteria would have drained the limited supplies of H_2S in the ocean, thus inadvertently selecting for oxygenic photosynthesis using abundant H_2O as a reductant. Molecular fossils (biomarkers) are indicative of cyanobacteria, the first oxygenic photosynthesizers, 2.7 Gyr ago (Brocks *et al.*, 1999) and 2.5 Gyr ago (Summons *et al.*, 1999). Recent genomic sequence analyses date their origin at 2.6 Gyr ago (Hedges *et al.*, 2001). Indirect evidence suggests cyanobacteria may have originated long before this, probably in ancient microbial mats (DesMarais, 1991; Schopf, 1993). Bacterial microfossils from 3.5 Gyr ago bear a striking resemblance to modern cyanobacteria (Schopf, 1993). Organic nitrogen isotopes are consistent with cyanobacteria 3.4–2.7 Gyr ago (Beaumont and Robert, 1999). Cyanobacteria were almost certainly part of microbial mats that formed stromatolites in lakes 2.7 Gyr ago (Buick, 1992), and the stromatolite record extends back to 3.5 Gyr ago (DesMarais, 1991). The banded iron formations (BIFs) present from 3.5 Gyr ago may have been formed by reaction of ferrous iron with O_2 produced by oxygenic photosynthesis, but formation by anoxygenic photosynthesis is also possible (Widdel *et al.*, 1993).

Figure 3.2 summarizes the constraints on atmospheric oxygen that have been estimated for the past 3 billion years. The unlabelled points in Figure 3.2 are all paleosols (ancient soils that were exposed to the atmosphere), summarized by Rye and Holland (1998) who determined the pO_2 limits for each. In order of decreasing age: upper limit paleosols are Mt Roe (2.765 ± 0.01 Gyr), Pronto/NAN (2.46 ± 0.02 Gyr), Hokkalampi (2.32 ± 0.12 Gyr), Hekpoort (2.15 ± 0.1 Gyr); lower limit paleosols are Drakenstein (2.08 ± 0.16 Gyr), Flin Flon (1.85 ± 0.05 Gyr), Sturgeon Falls ($\sim$1.1 Gyr), Arisaig ($\sim$0.45 Gyr). Wolhaarkop (2.08 ± 0.16 Gyr) is also a paleosol, for which the pO_2 lower limit was determined by a different technique (Holland and Beukes, 1990). Uraninites upper limit is from Holland (1984). Biological lower limits on pO_2 are updated from Runnegar (1991) with earlier records of sterols (Brocks *et al.*, 1999) and *Grypania* (Han and Runnegar, 1992), and additional estimates for *Beggiatoa* (Canfield and Teske, 1996) and the Cambrian fauna (Holland, 1984). These biological constraints may reflect localized dissolved O_2 concentrations rather than the global atmospheric pO_2 concentrations suggested. The appearance of charcoal (Rowe and Jones, 2000) is the first convincing evidence for pO_2 nearing the present level.

Despite the existence of oxygenic photosynthesis, atmospheric oxygen remained at very low concentrations during the late Archean, because the source of O_2 was insufficient to overwhelm the flux of reduced gases from volcanoes and from photo-oxidation of iron and other reduced elements in the surface ocean. Late Archean $pO_2 \sim 10^{-11}$ atm has been estimated from a photochemical model (Kasting, 1991), while weathering profiles suggest higher concentrations, but still <0.0005 atm before 2.44 Gyr ago (Rye and Holland, 1998). Methane was an important component of the late Archean atmosphere ($\sim$2.8 Gyr ago) that contributed significantly to the greenhouse effect (Lovelock, 1995),

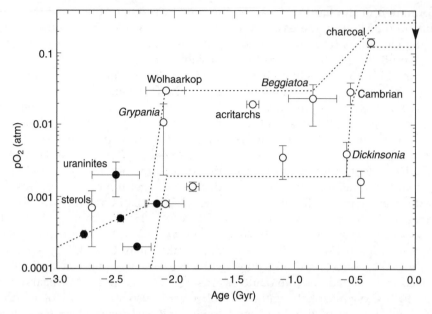

Figure 3.2 Geochemical and biological constraints on atmospheric oxygen over the past 3 billion years. Filled circles are estimated upper limits on oxygen. Empty circles are estimated lower limits on oxygen. The inverted triangle indicates the present partial pressure of oxygen. Age error bars indicate uncertainty in dating or the range of ages over which sediments were deposited. Partial pressure error bars indicate uncertainties in the estimates. The dashed lines are approximate upper and lower bounds on atmospheric oxygen. Unlabelled points are all paleosols (ancient soils that were exposed to the atmosphere) – see text for details.

with pCH_4 >0.00002 atm (Hayes, 1994; Rye and Holland, 2000) and probably pCH_4 >0.0001 atm (Pavlov *et al.*, 2000). Local 'oxygen oases' in the surface ocean where photosynthetic activity was high, could have had up to ~10% of the current dissolved oxygen concentration (Kasting, 1991), while in the active parts of cyanobacterial mats, oxygen could have reached supersaturating concentrations (L.J. Rothschild, personal communication). Evidence of oxygen demanding metabolic processes sets independent lower limits on local, dissolved oxygen concentrations. Isotopic evidence suggests that methanotrophs were globally abundant ~2.8 Gyr ago and these require pO_2 ~0.0005 atm to oxidize methane (Hayes, 1994). The production of sterols unique to eukaryote biochemistry was occurring 2.7 Gyr ago (Brocks *et al.*, 1999) and requires pO_2 >0.0002 atm (0.001 PAL) (Runnegar, 1991). pO_2 >0.002 atm (0.01 PAL) is required for aerobic respiration (Runnegar, 1991), and this probably also evolved in the

Archean (Knoll, 1992). However, in each case it is likely that the oxygen consumers thrived in close proximity to oxygenic photosynthesizers, e.g. in microbial mats (DesMarais, 1991) or ponds (Rye and Holland, 2000), while the bulk of the environment remained essentially oxygen free and chemically reducing.

Between about 2.2 and 2.0 Gyr ago in the early Proterozoic, a global oxidation event occurred in which atmospheric pO_2 rose from <0.0008 atm to >0.002 atm (Holland *et al.*, 1989; Rye and Holland, 1998) and possibly as high as >0.03 atm (Holland and Beukes, 1990). This was the first major step in the oxidation of the Earth's surface environment (Holland, 1999). The bounds on O_2 are inferred from the weathering profiles of paleosols – ancient soils exposed to the atmosphere (Rye and Holland, 1998). These set limits on the O_2/CO_2 ratio of the atmosphere. When combined with an independent estimate of pCO_2 <0.02 atm (Rye *et al.*, 1995) (obtained for 2.75–2.25 Gyr ago and assumed to hold 2.25–2.0 Gyr ago), a limit on the O_2 content of the atmosphere is obtained. Reduced iron is mobile while oxidized iron is not. If Fe was mobile during weathering, as it is in paleosols older than 2.2 Gyr, this indicates that O_2 must have been below a certain concentration. If Fe was immobile, as it is in later paleosols, this indicates that O_2 must have been above a critical concentration. Red beds (oxidized layers) deposited above paleosols aged ~ 2.2 Gyr are indicative of more oxidizing conditions after this time (Rye and Holland, 1998), and the disappearance of large ore deposits of uraninite by 2.0 Gyr ago is also consistent with a major increase in O_2 (Holland, 1984). Organelle-bearing eukaryote fossils from 2.1 Gyr ago resemble the photosynthetic algae *Grypania* (Han and Runnegar, 1992). This organism respired aerobically and probably required $pO_2 \sim 0.002–0.02$ atm, but if it was photosynthetic it could have generated its own oxygen supply locally (Runnegar, 1991). The inference that pO_2 exceeded 0.03 atm in the interval 2.2–2.0 Gyr ago comes from a high degree of Fe retention in the Wolhaarkop paleosol (Holland and Beukes, 1990). If this reflects the global situation, O_2 may have peaked before relaxing to somewhat lower concentrations consistent with later constraints. A large positive carbon isotope shift of carbonates 2.2–2.0 Gyr ago suggests a massive increase in organic carbon burial, releasing an estimated 12–22 times the current atmospheric O_2 inventory, more than sufficient to generate the inferred rise in atmospheric O_2 (Karhu and Holland, 1996). Hypotheses for the ultimate cause of the oxidation event are discussed in section 3.6.

After the first major rise in oxygen, there was an interval of relative stability during the middle of the Proterozoic, 2.0–1.0 Gyr ago (although the precise oxygen concentrations are poorly constrained). Paleosols from 1.9–1.8 Gyr ago and ~ 1.1 Gyr indicate pO_2 >0.002 atm. Spherical acritarch cells of 0.6 mm diameter from 1.4–1.3 Gyr ago can be used to derive higher minimum oxygen concentrations, assuming diffusion-limited respiration, but this is not robust because respiration may not have occurred in the centre of the cells (Runnegar, 1991). The last of the major banded iron formations were deposited ~ 1.8 Gyr ago

and this has usually been attributed to a rise in atmospheric oxygen ventilating the deep sea and removing ferrous iron (Holland, 1984). However, it has recently been suggested that the deep ocean remained anoxic until the late Proterozoic and that sulphide was responsible for removing iron from deep ocean water (Canfield, 1998). Simple models suggest that when atmospheric oxygen was an order of magnitude lower than now, if ocean nutrient concentrations and marine productivity were not greatly reduced, the demand for oxygen to respire sinking organic matter would have greatly exceeded the supply of oxygen (via diffusion from the surface and sinking of oxygen-rich surface waters). Hence all of the oxygen would have been removed from ocean water below the well-mixed surface layer.

Trends in the carbon isotope composition of sedimentary organic matter and carbonate through the Proterozoic suggest that the organic carbon reservoir grew in size, relative to the carbonate reservoir (DesMarais *et al.*, 1992). A second significant increase of oxygen has been postulated in the late Proterozoic. Carbon isotope studies suggest high organic carbon burial rates 1.1–0.8 Gyr ago (DesMarais *et al.*, 1992) and 0.8–0.7 Gyr ago (Knoll *et al.*, 1986). The evolution of non-photosynthetic sulphide-oxidizing bacteria (*Beggiatoa*) and an increase in sulphur-isotope fractionation suggest that oxygen increased from below to above 5–18% PAL (0.01–0.04 atm) during the interval 1.05–0.64 Gyr ago (Canfield and Teske, 1996). This appears to conflict with the estimate that pO_2 had already exceeded 0.03 atm, 2.0 Gyr ago (Holland and Beukes, 1990). Hence the suggestion above that oxygen peaked ~ 2.0 Gyr ago before relaxing to a lower concentration for most of the Proterozoic. A further increase in oxygen has been suggested in a relatively brief interval just before the Cambrian boundary 0.6–0.57 Gyr ago (Logan *et al.*, 1995).

The first large multi-cellular creatures, the Ediacara, appeared near the end of the Proterozoic, ~ 0.6 Gyr ago. They have conventionally been interpreted as metazoans adapted to low dissolved oxygen concentrations. If it was a metazoan, *Dickinsonia* would only have required pO_2 ~ 0.002 atm (Runnegar, 1991). This would be consistent with parts of the ocean remaining anoxic right until the end of the Proterozoic (Canfield, 1998). However, the evolutionary scenario is counterintuitive – the first large metazoans are unlikely to evolve in the most oxygen-depleted habitats. The alternative hypothesis is that the Ediacara were phototrophs and therefore generated their own O_2 supply (McMenamin, 1998). Evidence from paleosols suggests that atmospheric oxygen had been >0.002 atm since 2.0 Gyr ago (Rye and Holland, 1998).

Concentrations of oxygen in the first 200 million years of the Phanerozoic eon (0.57–0.37 Gyr ago) are poorly constrained. The fauna appearing in the Cambrian required dissolved $[O_2]$ >45 µmol kg^{-1} which in turn demands atmospheric pO_2 >0.02 atm, but it could have been much higher (Holland, 1984). Geochemical models driven by the abundance of reduced matter in different rocks or by carbon and sulphur isotope records predict pO_2 ~ 0.1–0.2 atm (Berner, 1989; Berner *et al.*, 2000), but these should be treated with caution because the data driving them are scarce in this interval and they are probably missing key feedback mechanisms. The colonization of the land surface by plants, which began ~ 420 Myr ago (million years ago) caused large amounts of organic carbon to be buried on the continents for the first time (Berner and Canfield, 1989) and produced lignin, which is difficult to biodegrade and hence tended to be buried in continental and marine sediments. Plants also amplified the rate of rock weathering, enhancing the supply of phosphorus to the land surface and oceans, thus tending to increase productivity and organic carbon burial (Lenton and Watson, 2000). These mechanisms should have increased atmospheric O_2.

By ~ 370 Myr ago the first fossilized charcoal appeared (Rowe and Jones, 2000) suggesting that oxygen had risen sufficiently high to sustain fires. Combustion experiments indicate that fire cannot be sustained in dry paper at 17% O_2 (Watson, 1978). Burning wood is extinguished when O_2 is reduced to 13.2–19%, depending on the orientation of the fuel (Rasbash and Langford, 1968). The charcoal record has been continuous since 350 Myr ago, indicating that O_2 has been at least $\sim 15\%$ of the atmosphere since then (early Carboniferous). Combustion experiments also indicate that small increases in O_2 above 21% cause large decreases in the 'ignition energy' required to start and sustain fire (Watson, 1978). This translates into a rapid increase in fire probability with rising oxygen. However, forests have been widespread throughout the past 350 million years, indicating that fires have never been so frequent as to prevent forest regeneration. This suggests that O_2 has never risen above some critical upper limit, which has been estimated as $\sim 25\%$, using terrestrial ecosystem models (Watson, 1978; Lenton and Watson, 2000).

Data on the abundance and organic carbon and pyrite sulphur contents of different sediment types have been used to force a geochemical model which predicts that atmospheric oxygen rose to $\sim 35\%$ in the Carboniferous (~ 300 Myr ago), dropped to $\sim 15\%$ in the Triassic (~ 200 Myr ago) and rose again to $\sim 27\%$ in the Cretaceous–early Tertiary (~ 50 Myr ago) (Berner and Canfield, 1989). Early models driven by carbon and sulphur isotope records tended to predict even larger, unrealistic oxygen variations, including negative values (Berner, 1987; Berner, 1989; Lasaga, 1989). Recognition that carbon and sulphur isotope fractionation are sensitive to oxygen concentration can return oxygen variations between $\sim 35\%$ and $\sim 15\%$ (Berner *et al.*, 2000). These geochemical models lack sufficient negative feedback to further reduce the oxygen variations. Their high O_2 predictions need revision downwards because they are inconsistent with the continuous existence of forests. Stabilizing mechanisms (Lenton and Watson, 2000) that could have regulated oxygen close to 21% of the atmosphere throughout the past 350 Myr are discussed further in section 3.6.

3.4 Oxygen toxicity

Despite oxygen being one of the most abundant biological products at the Earth's surface, its metabolic products are extremely toxic to life. The following explanation of oxygen toxicity draws heavily on the review by Fridovich (1977). The basis of oxygen's toxicity to biology lies in its physics and chemistry. Unusually for a gas, dioxygen is paramagnetic, having two unpaired electrons with parallel spins. This presents a barrier to chemical reaction with any organic reductant that has a pair of electrons to give up to O_2. The incoming electron pair cannot be accommodated in the partially filled orbitals on O_2. Hence the organic donor must undergo relatively slow spin inversion before it can react. This spin restriction is eliminated in electronically excited singlet oxygen ($^1O_2$), making it vastly more reactive than ground state oxygen. If atmospheric oxygen were as reactive as singlet oxygen life would be impossible and any accumulation of organic matter would be unlikely.

The spin restriction can be circumvented in two ways. Combining O_2 with a paramagnetic transition metal is the course followed by many oxidases containing, for example, Cu^{2+} and Fe^{2+}. Alternatively, the univalent pathway of adding electrons one at a time is quite common. This explains oxygen's toxicity, because the intermediates produced are much more reactive than O_2 itself. In the reduction of O_2 to $2H_2O$, the superoxide anion radical (O_2^-), hydrogen peroxide (H_2O_2) and the hydroxyl radical ($OH\cdot$) are all produced. Superoxide always gives rise to hydrogen peroxide in the pH range of biochemistry and together they interact to produce both $OH\cdot$ and singlet $^1O_2$ within cells. The hydroxyl radical ($OH\cdot$) reacts rapidly with virtually all organic compounds, giving reason enough for oxygen toxicity. In addition, singlet $^1O_2$ is a powerful oxidant and more discriminating, reacting rapidly at carbon/carbon double bonds, attacking polyunsaturated fatty acids and, by a free radical chain reaction, causing disproportionately great damage to cell membranes.

Given that $OH\cdot$ and singlet $^1O_2$ are so dangerous to living things, organisms should avoid producing them, and where their production is unavoidable the consequences must be minimized. Production can be avoided by multivalent oxygen reduction (avoiding the univalent pathway). Enzymes to achieve this use paramagnetic transition metals, or special organic substances such as flavins, at their reactive sites. Most of the oxygen consumed by aerobes is reduced by cytochrome c oxidase, which contains iron (heme) and copper at its active site and reduces O_2 to water. Several copper-containing oxidases achieve the same reduction. Many flavin-containing oxidases reduce O_2 to H_2O_2 without producing intermediates.

Unfortunately some production of harmful intermediates appears to be unavoidable in respiring cells, and active defences are required against this. The accumulation of H_2O_2 is prevented by two related classes of iron-containing enzymes, the catalases and the peroxidases, which catalyse its reduction to $2H_2O$. The accumulation of O_2^- is prevented by enzymes called superoxide dismutases that catalyse an oxidation-reduction reaction in which O_2^- is both electron donor and electron acceptor and H_2O_2 is formed. Oxygen also enhances the lethality of ionizing radiation, and this deleterious effect is inhibited by superoxide dismutases and catalases.

Superoxide dismutases, catalases and peroxidases provide the 'frontline defences' against oxygen toxicity, but even in their presence a low level of oxidative attack occurs. Compounds called antioxidants act in the 'rear guard' to minimize the damage it causes. The free radical chain reaction triggered by singlet $^1O_2$, which oxidizes polyunsaturated lipids, can be broken by antioxidants that react with the chain-propagating radicals. Vitamin E (α tocopherol) is one such compound, deficiency of which causes muscular dystrophy and reproductive failure in humans.

There are mechanisms to repair damage done by oxygen radicals and their reactive progeny, but some of the damage is irreversible, and can be viewed as chronic, low-level, cumulative oxygen toxicity. This is the chemical cause of ageing – the gradual wearing down of organisms. Hence it has been remarked of oxygen that 'as little as one breath is known to produce a life-long addiction to the gas ... which invariably ends in death' (Campbell, quoted by Gilbert, 1981).

3.5 Influences of oxygen on evolution

The capacity to detoxify reactive oxygen species must have evolved prior to oxygenic photosynthesis and aerobic metabolism, or the first aerobic organisms would have poisoned themselves. This poses something of a puzzle as to what drove evolution of protection from O_2 toxicity before there were significant amounts of O_2 in the environment (McKay and Hartman, 1991). Carotenoids (corrin derivatives), which protect against H_2O_2, are found in ancient anaerobic bacteria and do not require O_2 for their production, yet protect against its toxic effects (Gilbert, 1981). Cytochrome oxidase, which catalyses the reduction of oxygen to water, also arose prior to oxygenic photosynthesis (Castresana *et al.*, 1994; Castresana and Saraste, 1995). Such occurrences suggest that there was another source of reactive oxygen species on the early Earth, the nature of which is uncertain at present. Photochemical production of oxidants such as H_2O_2 in surface waters (McKay and Hartman, 1991) is now thought to be unlikely (according to modelling by J.F. Kasting, personal communication).

After the origin of oxygenic photosynthesis, localized accumulation of O_2 would have applied a common selection pressure to a varied biota, driving the evolution of a variety of oxygen defences. After the corrin derivatives, the more oxidized porphyrins including catalases and peroxidases appeared (Gilbert, 1981). These occur in most bacteria except methanogenic bacteria and anaerobic clostridia.

Superoxide dismutases (SODs) must have evolved at an early stage and are present in all organisms except some oxygen-sensitive obligate anaerobes (Fridovich, 1981). There are three main types of SODs, those containing copper and zinc, those containing manganese, and those containing iron. The manganese and iron SODs are found in prokaryotes and the mitochondria of eukaryotic cells, and may have appeared before oxygenic photosynthesis, because they are present in some ancient non-oxygenic photosynthesizers. Obligate anaerobic photosynthetic bacteria (green sulphur bacteria, purple sulphur bacteria) contain the iron SOD. Facultative aerobic photosynthetic bacteria (purple non-sulphur bacteria) contain the manganese SOD. This anaerobic/facultative aerobic distinction suggests the iron SOD may have evolved before manganese SOD. The cuprozinc SODs are characteristic of the cytoplasm of eukaryotic cells. The eukarya diverged from the bacteria and archaea early in the history of life (Knoll, 1992), hence the cuprozinc SOD is probably of similar antiquity to the other SODs (Runnegar, 1991). Eukaryotes later acquired mitochondria by symbiosis (Knoll, 1992), as indicated by the similarities between bacterial and mitochondrial SODs coupled with the great differences between mitochondrial and cytosolic SODs from the same species.

Figure 3.3 summarizes current evidence for the appearance of some important oxygen-producing and oxygen-consuming organisms. The potential for aerobic respiration arose before oxygenic photosynthesis, with the evolution of the key enzyme, cytochrome oxidase (Castresana *et al.*, 1994; Castresana and Saraste, 1995). However, there was essentially no O_2 in the environment to support aerobic respiration prior to the first oxygenic photosynthesizers. Their origin was discussed in section 3.3, and they probably performed aerobic respiration as the most rewarding means of utilizing the energy they had made available. Once O_2 became locally available, free-living aerobic respiring bacteria, which included the ancestors of mitochondria, could have thrived. Local accumulation of O_2 would also have triggered the evolution of further metabolic pathways utilizing its power as an oxidant. Methane oxidation is one such pathway that was apparently widespread by $\sim 2.8\,$Gyr ago (Hayes, 1994), occurring also in ponds on the land surface (Rye and Holland, 2000). Oxygenic photosynthesis provided the basis of microbial mat communities (Buick, 1992) that included a suite of organisms utilizing the oxygen and organic carbon produced.

The origin of eukaryotic cells with mitochondria (which all perform aerobic respiration) was at least facilitated by rising oxygen. Their O_2 requirement of $\sim 0.002\,$atm is harder

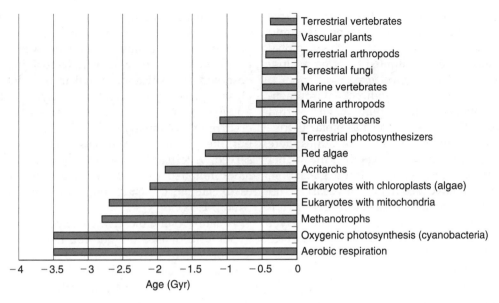

Figure 3.3 Major events in the evolution of oxygen-producing and oxygen-consuming organisms. Simplified from Raven (1997) with the addition of aerobic respiration (Castresana and Saraste, 1995), methanotrophs (Hayes, 1994; Rye and Holland, 2000), eukaryotes with mitochondria (Knoll, 1992) and small metazoans (Knoll, 1992). Acritarchs are thought to be cysts of green algae of the class Prasinophyceae. Dates of appearance are approximate and subject to revision. For example, the earliest appearance of eukaryotic algae was pushed back significantly with the discovery of 2.1-Gyr-old *Grypania* (Han and Runnegar, 1992). Recent genomic sequence analyses (Hedges *et al.*, 2001) suggest later origins for cyanobacteria, 2.6 Gyr ago, and eukaryotes with mitochondria, 1.8 Gyr ago.

to satisfy than that of free-living aerobic prokaryotes, because the mitochondria are encased within a larger cell that limits O_2 supply by diffusion (Runnegar, 1991). Sufficiently O_2-rich local environments likely existed in the late Archean, and the acquisition of mitochondria has been tentatively dated at 2.8–2.4 Gyr ago (Knoll, 1992). However, recent genomic sequence analyses put it more recently at 1.8 Gyr ago (Hedges *et al.*, 2001). Photosynthesizing eukaryotes (those with chloroplasts) are less likely to be directly O_2 limited because they have their own O_2 supply. However, when O_2 was at low Archean concentrations, fixed nitrogen would have been very scarce in the surface ocean (Knoll, 1992). Heterotrophs (including eukaryotes with mitochondria) could obtain nitrogen easily enough but autotrophs would have had to fix their own. Most prokaryote autotrophs can fix nitrogen but eukaryote chloroplasts do not. Rising O_2 may have lifted a barrier to their evolution by triggering widespread nitrification and thus

increased nitrate production. The earliest eukaryote fossils appear to be algae (Han and Runnegar, 1992) and are contemporaneous with the first major rise of atmospheric O_2 around 2.1 Gyr ago.

The rise of atmospheric O_2 2.2–2.0 Gyr ago must have disrupted obligate anaerobes at the Earth's surface, banishing them to anoxic environments, but it was probably not as catastrophic for anaerobes as previously suggested. Obligate anaerobic photosynthetic bacteria (green bacteria, purple sulphur bacteria) that could have been widespread under a reducing atmosphere and have formed the basis of many microbial mat communities would have had to migrate to deeper levels in the mats. However, if the deep ocean remained anoxic until the late Proterozoic (Canfield, 1998) obligate anaerobic heterotrophs could have continued to thrive there. Even anaerobic photosynthesizers may have been able to persist in the ocean, if the anoxic horizon was in the photic zone near the surface. The main beneficiaries of the rise in oxygen were the aerobes. The amount of free energy available to the biota was increased and aerobic heterotrophs could now achieve global significance, no longer limited to living in close proximity to photosynthesizers.

The rise in O_2 also had indirect environmental consequences that would have exerted strong selective pressure on the biota. It triggered loss of methane from the atmosphere and its greenhouse effect, cooling the planet and perhaps causing low-latitude glaciation, which could have decimated any early land biota. The methane-rich late Archean atmosphere probably had an organic haze layer that reduced the UV flux to the surface (Lovelock, 1995; Pavlov *et al.*, 2000). As O_2 rose this atmospheric shield would have disappeared, bathing the surface in extra UV. However, both prokaryotes (Margulis *et al.*, 1976) and early eukaryotes (Rothschild, 1999) had their own UV protection mechanisms. The organic haze was soon replaced by an effective ozone shield, once O_2 exceeded $\sim 0.1\%$ PAL (~ 0.0002 atm) (Margulis *et al.*, 1976). Early ideas that a lack of UV protection delayed the origin of life on land or of eukaryotes now seem unlikely (Rothschild, 1999).

During the mid-Proterozoic (2.0–1.0 Gyr ago) both oxygen concentrations and the biota appear to have been relatively stable. The next major evolutionary development was the radiation of phenotypic diversity in eukaryotes (Knoll, 1992). Fossils suggest this occurred 1.1–1.0 Gyr ago. Small, primitive metazoans (multi-cellular animals) are thought to have first become widespread around 1.0 Gyr ago. Trace fossils and possible fossilized metazoan faecal pellets are widespread from this time onwards. Early metazoans contributed to the demise of stromatolites through grazing and burrowing activity, augmented by declining levels of atmospheric CO_2 reducing carbon fixation rates (Rothschild and Mancinelli, 1990). High continental weathering rates suggest terrestrial photosynthesizers were present ~ 1.2 Gyr ago (Horodyski and Knauth, 1994). Some multi-cellular algae certainly existed in the oceans. A multi-cellular bangiophyte red alga aged 1.26–0.95 Gyr ago is the oldest eukaryote that can be confidently assigned to an extant phylum (Butterfield *et al.*, 1990). Multicellular eukaryotes typically have a higher oxygen demand than single cells. However, there is no indication of a rise in O_2 driving the observed diversification of eukaryotes. Indeed O_2 was probably sufficient by 2.0 Gyr ago. It is possible that eukaryote diversification occurred substantially earlier than 1.0 Gyr ago but the resulting organisms were not preserved. Alternatively, some other factor may have inhibited diversification, perhaps a late establishment of sexual population structures (Knoll, 1992).

A late-Proterozoic rise in oxygen is often hypothesized as a trigger for the Cambrian 'explosion', the diversification of large metazoans 0.54 Gyr ago (Knoll, 1996; Thomas, 1997; Knoll and Carroll, 1999). In order to attain a significant size, metazoans evolved a circulatory system with an oxygen carrier, haemoglobin. However, atmospheric oxygen probably rose above the concentration necessary for animal radiation (~ 0.02 atm) long before the first large metazoans evolved (Holland and Beukes, 1990; Canfield and Teske, 1996). Thus from current geochemical evidence it seems that pO_2 >0.02 atm was a necessary but not a sufficient condition for the evolution of large, metabolically active animals. Ecological drivers were probably at least as important (Erwin, 1993; Knoll and Carroll, 1999).

Oxygen is relatively insoluble in water and its solubility decreases with increasing salinity and temperature. Hence dissolved oxygen concentration limits the maximum potential size of some aquatic metazoa (Chapelle and Peck, 1999). The propensity of metazoa to suffer extinction events has been linked to their oxygen demand – the higher the oxygen demand, the more likely they are to be victims of any oxygen variability (McAlester, 1970). During the Phanerozoic, animals evolved to breathe air and thus benefited from a greater availability of oxygen that allowed them to become more metabolically active and made them less likely to suffer oxygen starvation than their aquatic counterparts. Terrestrial arthropods appeared ~ 420 Myr ago, followed by terrestrial vertebrates ~ 370 Myr ago. The transition from water- to air-breathing increased the potential for damage by reactive oxygen species, and vertebrates acquired unique oxygen defences, for example the antioxidant glutathione peroxidase (Gilbert, 1981).

A range of evolutionary changes among plants and animals has been tentatively linked to predicted O_2 variations over the last ~350 Myr (Robinson, 1989; Graham *et al.*, 1995; Dudley, 1998; Dudley, 2000). However, the variations in O_2 (Berner and Canfield, 1989) are probably significantly overestimated (Lenton and Watson, 2000). The postulated late-Carboniferous to early-Permian pulse of O_2 to higher than present levels broadly correlates with the diversification of insects and tetrapods, the origin of flight, insect gigantism, and size increases among aquatic invertebrates (Graham *et al.*, 1995). A postulated drop to lower than present O_2 levels at the end of the Permian broadly correlates with extinction of many tetrapod and insect lineages. It has long been noted that O_2 supply by diffusion limits body size in insects (Haldane, 1928). Increased O_2 partial pressure enhances diffusion dependent respiration, and it offers a greater energy supply for increases in metabolic rate, turnover and resource availability (Graham *et al.*, 1995; Dudley, 1998). Hence increased O_2 partial pressure has been put forward as a hypothesis to explain invertebrate gigantism in the Carboniferous, particularly giant flying insects, for example dragonflies (Protodonata) (Graham *et al.*, 1995; Dudley, 1998). A higher partial pressure of O_2 without a concomitant increase in the O_2 mixing ratio could have occurred if both atmospheric N_2 and O_2 increased during the Carboniferous. However, the burial of organic matter has a much weaker effect on N_2 than O_2, because the C/N burial ratio is >10 and the N_2 reservoir is larger than the O_2 reservoir. An alternative hypothesis (to avoid inferring highly flammable mixing ratios of O_2) is that the giant Carboniferous insects had special adaptations to concentrate oxygen.

The multiple historical origins of vertebrate flight in the late Jurassic and Cretaceous (150–100 Myr ago) have been linked to a postulated second rise in O_2 (Dudley 2000), although O_2 is not predicted to have peaked until the early Tertiary (50 Myr ago) (Berner and

Canfield, 1989). However, there has probably been sufficient O_2 for flight to occur throughout the last 350 Myr. O_2 appears not to have fallen below 15% in this interval, and 15 kPa corresponds to the oxygen partial pressure at 2.5 km altitude at present (Graham *et al.*, 1995). Bird and insect flight are observed well above this altitude (personal observations while mountaineering). As with the appearance of metazoans, sufficient O_2 opened a window of opportunity for the evolution of flight. However, evolution itself is driven by localized natural selection. For example, the evolution of flight may have been driven by its benefits as an escape from predation.

O_2-fuelled fires have shaped terrestrial ecosystems since the Carboniferous (Robinson, 1989; Scott and Jones, 1994), including triggering the evolution of fire tolerance and fire exploitation 'strategies'. Limited data on the charcoal content of coals (Robinson, 1989) and deep-sea cores (Herring, 1985) show some interesting variation in fire regimes over this time. If the mixing ratio of atmospheric oxygen has ever risen significantly above the present level this would have exerted, via increased fire frequency, stronger selective pressures on terrestrial plant ecology (Robinson, 1989). High-O_2 floras are expected to have thick bark as protection for the cambium, clonal reproduction and/or colonization through copious seeding, fire-proof seeds, investment in below-ground reproductive and storage tissues, restricted upper canopies, and dispensable leaves on low and mid-height plants (Robinson, 1989). Some of these features are observed in the fossils of Carboniferous swamp floras, and this has been taken as support for higher than present O_2 levels (Robinson, 1989). However, the dominant arborescent lycopods ('fern trees') had low reproduction rates and a morphology that looks very fire sensitive, while the climate was relatively dry, encouraging fires.

Oxygen directly affects C_3 plant growth by suppressing carbon assimilation, because O_2 and CO_2 compete for the oxygenase or carboxylase activity of the enzyme rubisco (Ribulose bisphosphate carboxylase/oxygenase). The effect is determined by the CO_2/O_2 ratio and temperature. At a given O_2 and temperature, a low CO_2 compensation points exist at which net CO_2 exchange is zero. For a given CO_2 and temperature, a high O_2 compensation point also exists at which net O_2 exchange is zero (Tolbert, 1994). Atmospheric CO_2 probably dropped by an order of magnitude after plants colonized the land surface ($\sim 420\,Myr$ ago) (Berner, 1998), making present levels of O_2 somewhat limiting to C_3 plants. This was a necessary condition for the much later evolution of C_4 plants ($\sim 13\,Myr$ ago) that use a CO_2 concentrating mechanism, but their evolution was more likely driven by declining CO_2 than rising O_2. C_3 plant Rubisco has a high selectivity for CO_2 relative to O_2, which probably evolved in response to declining CO_2/O_2 ratios (Raven, 1997). For the history of life prior to the Devonian ($\sim 400\,Myr$ ago) low O_2 and high CO_2 concentrations would not have significantly limited Rubisco activity, particularly as all extant cyanobacteria and most algae possess a CO_2 concentrating mechanism.

3.6 Oxygen regulation and the Gaia theory

Life has produced atmospheric oxygen, which is in turn one of the key environmental variables affecting life. This implies that there are feedback mechanisms between the activities of life and the oxygen content of the atmosphere. The Gaia theory posits that the Earth is a self-regulating system in which such feedback mechanisms control major environmental variables (Lovelock, 1995; Lenton, 1998). The stability of atmospheric oxygen over the past 350 Myr has often been taken as evidence for such regulation. The stepwise rise of oxygen since the origin of oxygenic photosynthesis, with periods of stability interspersed with intervals of rapid change is also consistent with oxygen being under feedback control. If so, a regulatory system has evolved, with a series of progressively higher 'set points', as oxygen was driven upwards by biotic activity.

Prior to 2.2 Gyr ago, atmospheric pO_2 was maintained below 0.0008 atm by oxygen

removal always exceeding the oxygen source. This was probably due to a combination of geochemical and biological control. The relative importance of biotic and abiotic factors has been the subject of some debate. Towe (1990) argued that a large flux of aerobic respiration was necessary to prevent the rise of oxygen, and hence aerobic respiration must have evolved early in the Archean. Kasting (1991) countered this, estimating that geochemical sources of reduced matter dominated the sink. Lovelock (1989) included the geochemical oxygen sink together with aerobic respiration and methane oxidation in a model that predicts oxygen regulation at very low concentrations during the Archean and a transition to pO_2 >0.001 atm around 2.2 Gyr ago, consistent with existing constraints (Figure 3.1).

The cause of the great oxidation event 2.2–2.0 Gyr ago is unresolved. Oxygen production must have exceeded oxygen consumption for O_2 to rise, and both a gradual decline in the oxygen sink and a more rapid increase in the oxygen source appear to have contributed to tipping the redox scales. As the Archean progressed, the reservoirs of reduced iron, sulphur, etc. in the ocean would have been gradually oxidized. Volcanic and metamorphic activity also declined with time, weakening the supply of reduced matter. Hence a decline in the oxygen sink without any change in the oxygen source could have started the oxidation event, as has been simulated (Lovelock, 1989). The oxidation event is then amplified by increased organic carbon burial.

Increased organic carbon burial could equally have triggered the oxidation event, given that the two are roughly coincident in the geological record (Karhu and Holland, 1996). The evolution of the first algae (photosynthesizing eukaryotes) may have predated the oxidation event, because a large alga (*Grypania*) was already present during it (Han and Runnegar, 1992). The first algae would have occupied new niches, and were better adapted to oxygen than cyanobacteria (Gilbert, 1981). Hence their evolution could have driven increased global primary productivity and organic carbon burial. Tectonic activity may also have generated particularly suitable geologic conditions for carbon burial at this time (DesMarais *et al.*, 1992).

An alternative hypothesis is that methanogens caused the great oxidation event (Catling, *et al.*, 2001). By creating a methane-rich atmosphere in the late Archean they greatly increased hydrogen loss to space resulting in net oxidation of the atmosphere/ocean-sediment system. Burial of organic carbon and oxidation of reduced matter are assumed to have been in balance until the oxidation event began. As oxygen rose, methanogen communities would have been disrupted because they are obligate anaerobes. Their contribution to the recycling of organic carbon may have been reduced and any delay before aerobic respiring organisms filled the carbon-recycling niche would have resulted in a greater fraction of primary production being buried.

Between 2.0 and ~ 0.8 Gyr ago, pO_2 probably remained between ~ 0.002 atm and ~ 0.02 atm. If the deep ocean remained anoxic during this interval then simple models suggest $pO_2 < 0.006$ atm (Kasting, 1991). The question arises as to what mechanisms can stabilize oxygen at these low concentrations? The need for stabilizing mechanisms was probably even greater then than it is now because the residence time of oxygen would have been significantly shorter. It seems likely that oxidative weathering was incomplete, hence there would have been negative feedback on the sink of oxygen – increasing atmospheric oxygen concentrations would generate increased oxygen removal. This mechanism is potentially fairly sensitive, because at low pO_2 the oxygen sink increases rapidly with small increases in pO_2 until the point is reached where the majority of incoming reduced matter ($\sim 10^{13}$ mol yr^{-1}) is oxidized (Holland, 1984). If the deep ocean were anoxic, much oxygen would also have been consumed within the water column, in the oxidation of up-welling reduced sulphur (Logan *et al.*, 1995; Canfield, 1998). However, this sink would have to have depended on oxygen production and/or pO_2 for it to stabilize oxygen. The 'set point', about which oxygen was regulated, may have been determined by the limiting effects of oxygen toxicity on growth (Lovelock, 1995), for

example of some cyanobacteria (Gilbert, 1981) and early eukaryotes (Knoll, 1992). Higher oxygen concentrations may not have been maintainable until improved oxygen tolerance had evolved.

Postulated rises in oxygen in the late Proterozoic ~ 0.8–0.57 Gyr ago have been linked to tectonics generating favourable conditions for organic carbon burial (DesMarais *et al.*, 1992) and to the evolution of metazoans with muscular, unidirectional guts, which packaged organic matter into rapidly sinking faecal pellets (Logan *et al.*, 1995). Prior to this evolutionary innovation, organic matter sank slowly through the water column, as finer material, and would have been more completely oxidized. With the advent of faecal pellets more organic carbon would have reached the sediments, more oxygen would have been released from surface waters and the deep ocean would have become more fully oxygenated. 'Oxygen happens' as K.J. Zahnle put it (personal communication). The resulting atmospheric oxygen concentrations in the early Phanerozoic (0.57–0.37 Gyr ago) are poorly constrained, but we can be confident that oxygen had approached the present level by ~ 370 Myr ago.

The stability of atmospheric oxygen over the past ~ 350 Myr is remarkable because the oxygen residence time is only ~ 3.25 Myr. Therefore, the whole oxygen reservoir has been replaced over 100 times, while its size has hardly varied. Processes such as the uplift of continents exposing more reduced matter should have tended to force changes in size of the oxygen reservoir. The only reasonable explanation is that some negative feedback mechanisms have been stabilizing oxygen.

The oxidative weathering sink of oxygen is saturated at present – all the reduced matter that is exposed is oxidized. Hence regulation of atmospheric oxygen is thought to involve negative feedback on the source of oxygen. The first suggestion was that the burial of organic carbon is enhanced under anoxic conditions, because anaerobic consumers are less efficient than their aerobic counterparts. If so, a decline in oxygen would be counteracted by an increase in the organic carbon burial source of oxygen. However, the observed effect is too weak to stabilize atmospheric oxygen (Betts and Holland, 1991).

The main control of the organic carbon burial flux in the ocean is the amount of marine productivity, which in turn depends on nutrient supply. A number of oxygen regulation mechanisms have been proposed involving phosphate as the control of productivity over long timescales. For example, it has been noted that the burial of phosphorus in organic matter (Van Cappellen and Ingall, 1996) and bound to iron minerals (Colman *et al.*, 1997) is less efficient under anoxic conditions. Hence, declining oxygen should cause more phosphorus to be recycled to the water column, fuelling more productivity and increased organic carbon burial. The problem with such mechanisms is that they are ineffective against rising oxygen, which readily tends to remove anoxia from the ocean, thus switching off the feedback (Lenton and Watson, 2000).

The main victims of rising oxygen would be plants on the land surface, especially forests, and it has been suggested that they play a role in oxygen regulation (Lovelock, 1995). A recent proposal (Lenton and Watson, 2000) involves the biological amplification of weathering, which is ultimately the only source of phosphorus to the land and ocean. Vascular plants amplify the rate of rock weathering by about an order of magnitude relative to primitive land biota (e.g. lichen and moss cover) and the effect is greatest for trees with their deep rooting systems (Bormann *et al.*, 1998). Increasing atmospheric oxygen tends to suppress vegetation by inhibiting photosynthetic carbon fixation. At the current CO_2 concentration, $\sim 27\%$ O_2 causes growth to cease in a crop plant (tobacco) (Tolbert *et al.*, 1995), while birch is less sensitive (Beerling *et al.*, 1998). More importantly, above about 21% oxygen, increasing fire frequency will tend to trigger ecological shifts from forest to faster regenerating ecosystems such as grassland. By these mechanisms, rising oxygen should suppress rock weathering and hence suppress the supply of phosphorus to the land and ocean, which in turn suppresses productivity and organic carbon burial. This mechanism is extremely effective at regulating against rising oxygen because of the high sensitivity of fire frequency to rising oxygen (Lenton and Watson, 2000). Furthermore, declining oxygen is counteracted by increased plant growth, more efficient rock weathering and increased phosphorus supply fuelling organic carbon burial.

If land plants are instrumental in regulating the oxygen content of the atmosphere at present, this may help constrain oxygen concentrations before plants began to colonize the land surface. Prior to land plants, the liberation of phosphorus from rock weathering would have been greatly reduced, tending to suppress global productivity and organic carbon burial. Furthermore, there was no lignin production and minimal burial of organic carbon on the continents. A weaker oxygen source implies a lower oxygen concentration, perhaps an order of magnitude lower ($O_2 \sim 0.02$ atm). This is consistent with the requirements of early metazoans (Figure 3.1). However, with more anoxia in the ocean, organic carbon burial would be more efficient and phosphorus recycling would be enhanced, tending to counteract reduced phosphorus input and maintain a significant marine organic carbon burial flux. Hence $O_2 \sim 0.05$–0.1 atm is suggested as a best guess for the interval ~ 0.6–0.4 Gyr ago. Further work should enable tighter constraints to be put on past concentrations of atmospheric oxygen.

3.7 Conclusion

Oxygen is one of the 'master variables' of the Earth system. Once oxygenic photosynthesis had arisen it was destined to transform the Earth's surface environment. The stepwise rise of atmospheric oxygen suppressed obligate anaerobes, and forced life to evolve numerous protections against oxygen toxicity. However, the great increase in the amount of free energy available to the biota more than outweighed these pitfalls. The progressive oxidation of the Earth's surface environment paved the way for single-celled then multi-cellular eukaryotes to dominate the biota. Metazoa were able to evolve and then become progressively larger and more metabolic active. Ultimately we owe our existence to myriad past photosynthesizers increasing the oxygen content of the atmosphere and a tight coupling between the evolution of the biota and its physical environment, which has prevented oxygen reaching dangerously flammable levels over the last $\sim$ 350 million years. The co-evolution of the biota and the oxygen content of the atmosphere over Earth history, the increase in free energy and biotic activity, and the emergent self-regulation of the coupled system support the Gaia theory (Lovelock, 1995; Lenton, 1998).

References

Beaumont, V. and Robert, F. (1999). Nitrogen isotope ratios of kerogens in Precambrian cherts: a record of the evolution of atmospheric chemistry? *Precambrian Research*, **96**: 63–82.

Beerling, D.J., Woodward, F.I., Lomas, M.R. *et al.* (1998). The influence of Carboniferous palaeoatmospheres on plant function: an experimental and modelling assessment. *Philosophical Transactions of the Royal Society of London, Biological Sciences*, **353**: 131–140.

Berner, R.A. (1987). Models for carbon and sulfur cycles and atmospheric oxygen: application to Paleozoic geologic history. *American Journal of Science*, **287**: 177–196.

Berner, R.A. (1989). Biogeochemical cycles of carbon and sulfur and their effect on atmospheric oxygen over Phanerozoic time. *Global and Planetary Change*, **75**: 97–122.

Berner, R.A. (1991). A model for atmospheric CO_2 over Phanerozoic time. *American Journal of Science*, **291**: 339–376.

Berner, R.A. (1998). The carbon cycle and CO_2 over Phanerozoic time: the role of land plants. *Philosophical Transactions of the Royal Society of London: Biological Sciences*, **353**: 75–82.

Berner, R.A. and Canfield D.E. (1989). A new model for atmospheric oxygen over Phanerozoic time. *American Journal of Science*, **289**: 333–361.

Berner, R.A., Petsch, S.T., Lake, J.A. *et al.* (2000). Isotope fractionation and atmospheric oxygen: implications for Phanerozoic O_2 evolution. *Science*, **287**: 1630–1633.

Betts, J.N. and Holland, H.D. (1991). The oxygen content of ocean bottom waters, the burial efficiency of organic carbon, and the regulation of atmospheric oxygen. *Palaeogeography, Palaeoclimatology, Palaeoecology* (Global and Planetary Change Section), **97**: 5–18.

Bormann, B.T., Wang, D., Bormann, F.H. *et al.* (1998). Rapid, plant-induced weathering in an aggrading experimental ecosystem. *Biogeochemistry*, **43**: 129–155.

Brocks, J.J., Logan, G.A., Buick, R. and Summons, R.E. (1999). Archean molecular fossils and the early rise of eukaryotes. *Science*, **285**: 1033–1036.

Buick, R. (1992). The antiquity of oxygenic photosynthesis: evidence from stromatolites in sulphate-deficient Archaean lakes. *Science*, **255**: 74–77.

Butterfield, N.J., Knoll, A.H. and Swett, K. (1990). A bangiophyte red alga from the Proterozoic of Arctic Canada. *Science*, **250**: 104–107.

Canfield, D.E. (1998). A new model for Proterozoic ocean chemistry. *Nature*, **396**: 450–453.

Canfield, D.E. and Teske, A. (1996). Late Proterozoic rise in atmospheric oxygen concentration inferred from phylogenetic and sulphur-isotope studies. *Nature*, **382**: 127–132.

Castresana, J. and Saraste, M. (1995). Evolution of energetic metabolism: the respiration-early hypothesis. *Trends in Biochemical Science*, **20**: 443–448.

Castresana, J., Luebben, M., Saraste, M. and Higgins, D.G. (1994). Evolution of cytochrome oxidase, an enzyme older than atmospheric oxygen. *European Molecular Biology Organization Journal*, **13**: 2516–2525.

Catling, D.C., McKay, C.P. and Zahnle, K.J. (2001). Biogenic methane, hydrogen escape, and the irreversible oxidation of early Earth. *Science*, **293**: 839–843.

Chapelle, G. and Peck, L.S. (1999). Polar gigantism dictated by oxygen availability. *Nature*, **399**: 114–115.

Colman, A.S., Mackenzie, F.T and Holland, H.D. (1997). Redox stabilisation of the atmosphere and oceans and marine productivity. *Science*, **275**: 406–407.

DesMarais, D.J. (1991). Microbial mats, stromatolites and the rise of atmospheric oxygen in the Precambrian atmosphere. *Global and Planetary Change*, **97**: 93–96.

DesMarais, D.J., Strauss, H., Summons, R.E. and Hayes, J.M. (1992). Carbon isotope evidence for the stepwise oxidation of the Proterozoic environment. *Nature*, **359**: 605–609.

Dudley, R. (1998). Atmospheric oxygen, giant Paleozoic insects and the evolution of aerial locomotor performance. *Journal of Experimental Biology*, **201**: 1043–1050.

Dudley, R. (2000). The evolutionary physiology of animal flight: paleobiological and present perspectives. *Annual Review of Physiology*, **62**: 135–155.

Duursma, E.K. and Boisson, M.P.R.M. (1994). Global oceanic and atmospheric oxygen stability considered in relation to the carbon cycle and to different time scales. *Oceanologica Acta*, **17**: 117–141.

Erwin, D.H. (1993). The origin of metazoan development: a palaeobiological perspective. *Biological Journal of the Linnean Society*, **50**: 255–274.

Fridovich, I. (1977). Oxygen is toxic! *BioScience*, **27**: 462–466.

Fridovich, I. (1981). Superoxide radical and superoxide dismutases. In: D.L. Gilbert, *Oxygen and Living Processes*. New York: Springer-Verlag, pp. 250–272.

Gilbert, D.L. (1981a). Perspective on the history of oxygen and life. In: D.L. Gilbert, *Oxygen and Living Processes*. New York: Springer-Verlag, pp. 1–43.

Gilbert, D.L. (1981b). Significance of oxygen on Earth. In: D.L. Gilbert, *Oxygen and Living Processes*. New York: Springer-Verlag, pp. 73–101.

Graham, J.B., Dudley, R., Aguilar, N.M. and Gans C. (1995). Implications of the late Palaeozoic oxygen pulse for physiology and evolution. *Nature*, **375**: 117–120.

Haldane, J.B.S. (1928). *Possible Worlds*. Harper & Row.

Han, T.-M. and Runnegar B. (1992). Megascopic eukaryotic algae from the 2.1-billion-year-old Negaunee iron-formation, Michigan. *Science*, **257**: 232–235.

Hayes, J.M. (1994). Global methanotrophy at the Archean-Proterozoic transition. In: S. Bengtson, *Early Life on Earth*. New York: Columbia University Press, pp. 220–236.

Hedges, S.B., Chen, H., Kumar, S. *et al.* (2001). A genomic timescale for the origin of eukaryotes. *BMC Evolutionary Biology*, **1**: 4.

Herring, J.R. (1985). Charcoal fluxes into sediments of the North Pacific Ocean: the Cenozoic record of burning. In: E.T. Sundquist and W.S. Broecker, *The Carbon Cycle and Atmospheric CO_2: Natural Variations Archean to Present*. Washington, DC: American Geophysical Union, pp. 419–442.

Holland, H.D. (1978). *The Chemistry of the Atmosphere and Oceans*: John Wiley.

Holland, H.D. (1984). *The Chemical Evolution of the Atmosphere and Oceans*. Princeton: Princeton University Press.

Holland, H.D. (1999). When did the Earth's atmosphere become oxic? A reply. *The Geochemical News*, **100**: 20–22.

Holland, H.D. and Beukes, N.J. (1990). A paleoweathering profile from Griqualand West, South Africa: evidence for a dramatic rise in atmospheric oxygen between 2.2 and 1.9 BYBP. *American Journal of Science*, **290-A**: 1–34.

Holland, H.D., Feakes, C.R. and Zbinden, E.A. (1989). The Flin Flon paleosol and the composition of the atmosphere 1.8 BYBP. *American Journal of Science*, **289**: 362–389.

Horodyski, R.J. and Knauth, L.P. (1994). Life on land in the Precambrian. *Science*, **263**: 494–498.

Karhu, J.A. and Holland, H.D. (1996). Carbon isotopes and the rise of atmospheric oxygen. *Geology*, **24**: 867–870.

Kasting, J.F. (1991). Box models for the evolution of atmospheric oxygen: an update. *Global and Planetary Change*, **97**: 125–131.

Kasting, J.F. (1993). Earth's early atmosphere. *Science*, **259**: 920–926.

Knoll, A.H. (1992). The early evolution of eukaryotes: a geological perspective. *Science*, **256**: 622–627.

Knoll, A.H. (1996). Breathing room for early animals. *Nature*, **382**: 111–112.

Knoll, A.H. and Carroll, S.B. (1999). Early animal evolution: emerging views from comparative biology and geology. *Science*, **284**: 2129–2137.

Knoll, A.H., Hayes, J.M., Kaufman, A.J. *et al.* (1986). Secular variation in carbon isotope ratios from Upper Proterozoic successions of Svalbard and East Greenland. *Nature*, **321**: 832–838.

Lasaga, A.C. (1989). A new approach to isotope modeling of the variation of atmospheric oxygen through the Phanerozoic. *American Journal of Science*, **289**: 411–435.

Lenton, T.M. (1998). Gaia and natural selection. *Nature*, **394**: 439–447.

Lenton, T.M. and Watson A.J. (2000). Redfield revisited: 2. What regulates the oxygen content of the atmosphere? *Global Biogeochemical Cycles*, **14**: 249–268.

Logan, G.B., Hayes, J.M., Hieshima, G.B. and Summons, R.E. (1995). Terminal Proterozoic reorganization of biogeochemical cycles. *Nature*, **376**: 53–56.

Lovelock, J.E. (1989). Geophysiology, the science of Gaia. *Reviews of Geophysics*, **27**: 215–222.

Lovelock, J.E. (1995). *The Ages of Gaia – A Biography of Our Living Earth*. Oxford: Oxford University Press.

Margulis, L., Walker, J.C.G. and Rambler, M. (1976). Reassessment of the roles of oxygen and ultraviolet light in Precambrian evolution. *Nature*, **264**: 620–624.

McAlester, L.A. (1970). Animal extinctions, oxygen consumption, and atmospheric history. *Journal of Paleontology*, **44**: 405–409.

McKay, C.P. and Hartman, H. (1991). Hydrogen peroxide and the evolution of oxygenic photosynthesis. *Origins of Life and Evolution of the Biosphere*, **21**: 157–163.

McMenamin, M.A.S. (1998). *The Garden of Ediacara: Discovering the First Complex Life*. New York: Columbia University Press.

Mojzsis, S.J., Arrhenius, G., McKeegan, K.D. *et al.* (1996). Evidence for life on Earth before 3,800 million years ago. *Nature*, **384**: 55–59.

Mudd, J.B. (1996). Biochemical basis for the toxicity of ozone. In: M. Yunus and M. Iqbal, *Plant Response to Air Pollution*. Chichester: John Wiley & Sons Ltd, pp. 267–283.

Pavlov, A.A., Kasting, J.F., Brown, L.L. *et al.* (2000). Greenhouse warming by CH_4 in the atmosphere of early Earth. *Journal of Geophysical Research*, **105**: 11981–11990.

Rasbash, D.J. and Langford, B. (1968). Burning of wood in atmospheres of reduced oxygen concentration. *Combustion and Flame*, **12**: 33–40.

Raven, J.A. (1997). The role of marine biota in the evolution of terrestrial biota: gases and genes. *Biogeochemistry*, **39**: 139–164.

Robinson, J.M. (1989). Phanerozoic O_2 variation, fire, and terrestrial ecology. *Palaeogeography, Palaeoclimatology, Palaeoecology* (Global and Planetary Change Section), **75**: 223–240.

Rosing, M.T. (1999). ^{13}C-depleted carbon microparticles in >3700-Ma sea-floor sedimentary rocks from West Greenland. *Science*, **283**: 674–676.

Rothschild, L.J. (1999). The influence of UV radiation on protistan evolution. *Journal of Eukaryotic Microbiology*, **46**: 548–555.

Rothschild, L.J. and Mancinelli R.L. (1990). Model of carbon fixation in microbial mats from 3,500 Myr ago to the present. *Nature*, **345**: 710–712.

Rowe, N.P. and Jones, T.P. (2000). Devonian charcoal. *Palaeogeography, Palaeoclimatology, Palaeoecology*, **164**: 331–338.

Runnegar, B. (1991). Precambrian oxygen levels estimated from the biochemistry and physiology of early eukaryotes. *Global and Planetary Change*, **97**: 97–111.

Rye, R. and Holland, H.D. (1998). Paleosols and the evolution of atmospheric oxygen: a critical review. *American Journal of Science*, **298**: 621–672.

Rye, R. and Holland, H.D. (2000). Life associated with a 2.76 Ga ephemeral pond?: evidence from Mount Roe #2 paleosol. *Geology*, **28**: 483–486.

Rye, R., Kuo, P.H. and Holland, H.D. (1995). Atmospheric carbon dioxide concentrations before 2.2 billion years ago. *Nature*, **378**: 603–605.

Schlesinger, W.H. (1997). *Biogeochemistry – An Analysis of Global Change*. London: Academic Press.

Schopf, J.W. (1993). Microfossils of the early Archean apex chert: new evidence of the antiquity of life. *Science*, **260**: 640–646.

Scott, A.C. and Jones, T.P. (1994). The nature and influence of fire in Carboniferous ecosystems. *Palaeogeography, Palaeoclimatology, Palaeoecology*, **106**: 91–112.

Summons, R.E., Jahnke, L.L., Hope, J.M. and Logan, G.A. (1999). 2-Methylhopanoids as biomarkers for cyanobacterial oxygenic photosynthesis. *Nature*, **400**: 554–557.

Thomas, A.L.R. (1997). The breath of life – did increased oxygen levels trigger the Cambrian Explosion? *Trends in Ecology and Evolution*, **12**: 44–45.

Tolbert, N.E. (1994). Role of photosynthesis and photorespiration in regulating atmospheric CO_2 and O_2. In: N.E. Tolbert and J. Preiss, *Regulation of Atmospheric CO_2 and O_2 by Photosynthetic Carbon Metabolism*. Oxford: Oxford University Press, pp. 8–33.

Tolbert, N.E., Benker, C. and Beck, E. (1995). The oxygen and carbon dioxide compensation points of C_3 plants: possible role in regulating atmospheric oxygen. *Proceedings of the National Academy of Sciences USA*, **92**: 11230–11233.

Towe, K.M. (1990). Aerobic respiration in the Archaean? *Nature*, **348**: 54–56.

Van Cappellen, P. and Ingall, E.D. (1996). Redox stabilisation of the atmosphere and oceans by phosphorus-limited marine productivity. *Science*, **271**: 493–496.

Watson, A.J. (1978). *Consequences for the Biosphere of Forest and Grassland Fires*. Department of Cybernetics, University of Reading.

Widdel, F., Schnell, S., Heising, S. *et al.* (1993). Ferrous iron oxidation by anoxygenic phototrophic bacteria. *Nature*, **362**: 834–836.

4

Chemistry of the early oceans: the environment of early life

John Raven and Keith Skene

ABSTRACT

See Plate 2
The earliest life on Earth was in the oceans, and it is to the early ocean that we must look to find the sources of energy and chemicals used by the earliest living organisms. The energy source for early organisms is widely held to be inorganic chemical reactions, involving interactions of the ocean with the Earth's crust. An alternative view is that the energy source was from organic chemical reactions in the ocean. Since the organic chemicals would have been produced by interaction of extraterrestrial inputs (meteorites, comets, solar radiation) with atmospheric components, this view focuses on ocean/atmosphere interactions as most significant in providing the energy for early life to the ocean.

Later the direct use of light energy by organisms at the ocean surface became the major energy transforming reaction providing energy for almost all living organisms.

The chemical elements which were used by the earliest organisms were determined by their availability to early organisms as well as by their physicochemical appropriateness for particular biological functions. Modification of ocean chemistry by the activities of living organisms, for example the accumulation of photosynthetically produced oxygen, changed the availability of several biologically essential elements; an example is the decreased availability of iron.

4.1 Introduction

Entities which have the properties of multiplication, variation and heredity can be said to be alive (Muller, 1966; Maynard Smith and Szathmáry, 1995). Such a definition of life does not specify the nature of the genetic material or of the phenotype, although it is inescapable that life always had a material basis and depended on energy dissipation. Our discussion of the chemistry of the ocean in relation to the origin and early evolution of life emphasizes how the early ocean could supply the energy and materials needed for extant life, and also of putative ancestral organisms and prebiotic components.

We begin by outlining the likely chemical composition of the Hadean Ocean, including changes to ocean chemistry due to interactions with the Earth's crust and with the atmosphere (as well as with extraterrestrial items arriving via the atmosphere). We then consider the ways in which the Hadean Ocean could have supplied the materials, and the energy, needed for the origin and early evolution of life. Finally, we briefly consider how early life could have modified the chemistry of the ocean, and how these changes could have fed back to the evolution of the early organisms.

4.2 The chemistry of the Hadean Ocean

The ocean at about the time of the evolution of life, i.e. some 3.9 billion years ago (Rosing, 1999), was probably somewhat less alkaline

Evolution on Planet Earth
ISBN 0-12-598655-6

than the present ocean, in the sense both of pH (pH 5.5) and of mol H^+-neutralizing capacity (Holland *et al.*, 1986; Williams and Fraústo da Silva, 1996; Russell and Hall, 1997). This could perhaps be correlated with the high CO_2 levels in the ocean and atmosphere (0.1–1.0 MPa; Williams and Fraústo da Silva, 1996; Russell and Hall, 1997). The early ocean in equilibrium with a near-neutral (in redox terms) atmosphere was reducing, with significant Fe^{2+}, Mn^{2+}, Ni^{2+} and S^{2-} concentrations; the low solubility of the transition metal sulfides would have limited the concentration of free sulfide in the (probable) presence of excess transition metal ions (Russell and Hall, 1997; cf. Cleaves and Miller, 1998). Russell and Hall (1997) also suggest that early seawater contained sufficient phosphate for early life. As to the bulk solutes in the early ocean, while it is widely held that the early ocean was somewhat less saline than the present ocean, there are also suggestions that the concentration of total salts was 1.5–2 times the present value (Knauth, 1998).

Inputs to the early ocean came from the crust via dissolution of exposed crust, and from hydrothermal vents, and from the atmosphere via atmospheric chemistry and extraterrestrial inputs. Inputs from exposed continental crust was not an option, since the earliest known example of such exposed continental crust comes from about 3.5 billion years ago.

The inputs from hydrothermal vents would have been warmer than the bulk early ocean (i.e. above 80°C), more alkaline (up to pH 9) and more reducing (a higher concentration of S^{2-}): Williams and Fraústo da Silva (1996); Russell and Hall (1997). This fluid would also contain NH_3 produced by reduction of N_2 using metallic iron or reduced iron oxides in localized low H_2O-regions in the lithosphere at high temperature (300–800°C) and pressure (100 MPa), providing perhaps 1–100 Gmol NH_3 per year (Table 4.1).

The early atmosphere could have produced $NO^{\cdot}$ by reactions of N_2 with H_2O and CO_2 energized by lightning or meteorite passage. Dissolution of these $NO^{\cdot}$-free radicals would produce NO_3^- and NO_2^-, which could be reduced to NH_3 in the ocean using Fe compounds (see Raven and Yin, 1998; Brandes *et*

Table 4.1 Fluxes of N in the early ocean which may be relevant for the origin and early evolution of life, and N fluxes in the early ocean. Data for N from the compilation in Raven and Yin (1998) and from Brandes *et al.* (1998).

Flux	Magnitude
Atmospheric N_2 conversion to $NO^{\cdot}$ by oxidization with CO_2 or H_2O energized by lightning or meteorite impacts; dissolution of $NO^{\cdot}$ and formation of NO_2^-; reduction of NO_2^- to NH_4^+ (in the early ocean)	≤ 50 Gmol NH_4^+ per year
Catalysis of NH_4^+ production by Ti (on early Earth and today)	≤ 4 Gmol NH_4^+ per year
N_2 reduction to NH_3 by metallic Fe or reduced iron oxides in localized low-H_2O, high temperature (300–800°C) and pressure (100 MPa), followed by efflux as NH_4^+ to the ocean in hydrothermal vents (on early Earth and today)	1–100 Gmol NH_4^+ per year
Biological N_2 fixation in extant ocean	~1 Tmol NH_4^+ per year
Global assimilation of combined N (mainly NH_4^+, NO_3^-) by primary producers in the extant ocean	~420 mol N per year
Global $NO_x^{\cdot}$ production by lightning in present atmosphere (over ocean *and land*) today	~1 Tmol $NO_x^{\cdot}$ per year

al., 1998) at rates similar to these suggested (above) for NH_3 input from hydrothermal vents (Table 4.1). Brandes et al. (1998) suggested that the oceanic NH_3 could produce an atmospheric NH_3 partial pressure adequate to permit a sufficient NH_3 greenhouse effect. This would have significantly offset the 'weak young Sun' and hence help maintain liquid water on the Earth's surface, provided that UV-induced breakdown of atmospheric NH_3 is restricted by high altitude UV-absorbing organic haze (Brandes et al., 1998; Chyba, 1998).

As for abiogenic production of organic-C, there is no obvious way in which organic-C could be produced in the bulk ocean. Production of organic-C would have involved interactions with the lithosphere and the atmosphere. Organic-C production in the early ocean in association with the lithosphere could have involved interaction of the early ocean with the iron sulphur minerals at the lithosphere/hydrosphere interface (Wächter-häuser, 1990), and/or the interaction of the alkaline, strongly reducing hydrothermal vent solution with the less alkaline, less reducing bulk seawater, possibly via an iron monosulfide membrane which forms spontaneously at the vent fluid/bulk seawater interface (Russell and Hall, 1997) (Table 4.2). The possibility of an atmospheric involvement seems less likely now, with the widespread assumption of a neutral atmosphere (predominantly $CO_2/N_2/H_2O$), than at the time of the Urey–Miller experiments (early 1950s) which were predicated on a reducing Hadean atmosphere (Broecker, 1985; Maynard Smith and Szathmáry, 1995). Thus, the production of significant quantities of protein amino acids and nucleic acid bases were produced from energization of a reducing atmosphere (H_2, CH_4, NH_3, H_2O) by an electric discharge mimicking the electrical discharge of a thunderstorm in the early atmosphere. However, when this experiment is repeated with a neutral ($CO_2/N_2/H_2O$) atmosphere, very little organic material, such as amino acids and nucleic acid bases, is produced. Possible additional sources of organic sources from above the ocean include the suggestion of Woese (1979) of chemical evolution in water droplets in clouds, and the suggestion of Oro (1961) of organic carbon delivery by comets and meteorites. The suggestion of water droplets as an important seat of organic compound production can, to the extent that the water droplets could have entrained mineral particles, be considered an extension of the hydrosphere/lithosphere interface (Maynard Smith and Szathmáry, 1995; cf. Cairns-Smith and Hartman, 1986; Wächterhäuser, 1990). Delivery of organic matter by impacting comets and meteorites (Oro, 1961) requires a relatively high total atmospheric pressure (1 MPa at the Earth's surface, mainly as CO_2) to limit destruction of organic matter en route to the surface (Maynard Smith and

Table 4.2   Fluxes of C in the early ocean which may be relevant for the origin and early evolution of life, and C fluxes in the present ocean. Data from Raven (1996), Falkowski and Raven (1997) and Jakosky and Shock (1998).

Flux	Magnitude
Chemolithotrophic inorganic-C reduction using reductant from hydrothermal vents on the early Earth (and today)	≤ 50 Gmol C per year
Photolithotrophic inorganic-C reduction in the extant ocean	~5 Pmol C per year
Chemolithotrophic inorganic-C reduction related to nitrification of NH_4^+ to NO_2^- and of NO_2^- to NO_3^- in the extant ocean (NH_4^+ comes mainly from viral lysis, grazing phytoplankton followed by excretion of NH_4^+; O_2 comes from photosynthesis)	~16 Tmol C per year

Szathmáry, 1995). With ten times the present atmospheric density organic carbon from comets might reach the Earth at a rate of as much as 0.1–1.0 Gmol C of organic matter per year (Chyba *et al.*, 1990) (cf. Table 4.2).

4.3 How could the early ocean supply material for the origin and early evolution of life?

Here we look at the chemistry of the early ocean (as discussed above) in the context of the need for elements (and their chemical forms) for life. The perceptive reader may have detected a bias in the direction of the elements known to be used in present-day life in that discussion of early ocean chemistry. At all events, the elements used in organisms must not only have the appropriate chemical properties, but also have been available at the time of the origin and early evolution of life (Williams and Fraústa da Silva, 1996).

The listing of essential elements given below is based on Marschner (1995) and Williams and Fraústa da Silva (1996); see also Raven and Smith (1981).

H: Essential for all organisms; obtained from H_2O

B: Essential for some organisms; obtained from $B(OH)_3$

C: Essential for all organisms; obtained from inorganic C ($CO_2/HCO_3^-/CO_3^{2-}$) and any organic C produced prebiotically

N: Essential for all organisms; obtained from NH_3/NH_4^+ produced prebiotically as described earlier. Biota subsequently evolved the capacity to use N_2 and NO_3^- (via their conversion to NH_3/NH_4^+)

O: Essential for all organisms; obtained from H_2O, inorganic C and (much later, and to a limited extent) O_2

Na: Essential for some organisms; obtained from Na^+ in seawater

Mg: Essential for all organisms; obtained from Mg^{2+} in seawater

Si: Essential for some organisms; obtained from $Si(OH_4)$ in seawater

P: Essential for all organisms; obtained from HPO_4^{2-} in seawater

S: Essential for all organisms; obtained from S^{2-} in seawater and, much later, from SO_4^{2-}, derived mainly via oxidation of S^{2-} by (*inter alia*) O_2

Cl: Essential for some organisms; obtained from Cl^- in seawater

K: Essential for all organisms; obtained from K^+ in seawater

Ca: Essential for all organisms; obtained from Ca^{2+} in seawater

Mn: Essential for all organisms; obtained from Mn^{2+} in seawater

Fe: Essential for all organisms; obtained from Fe^{2+} in seawater and, much later from Fe^{3+} in seawater following oxidation by (*inter alia*) O_2

Co: Essential for many organisms; obtained from Co^{2+} in seawater

Ni: Essential for most organisms; obtained from Ni^{2+} in seawater

Cu: Essential for all organisms; very scarce in early ocean (Cu_2S very insoluble), later available as Cu^{2+} in seawater

Zn: Essential for all organisms; obtained from Zn^{2+} in seawater

Se: Essential for most organisms; obtained from Se^{2-} in seawater, later available as SeO_3^{2-} and SeO_4^{2-} in seawater

Mo: Essential for all organisms; obtained from MoO_4^{2-} in seawater

I: Essential for some organisms; obtained as I^-, IO_3^- in seawater

The elements listed above are required over a very wide range of concentrations in biota, where 'required' is defined by the concentration in biomass which permits the organism to achieve its maximum metabolic and growth potential. In terms of number of atoms, H and O dominate through the occurrence of H_2O as more than half of the biomass of essentially all organisms. On the basis of H_2O-free biomass H, C and O dominate, with N, P, K, S, Mg and Ca comprising the other 'major' essential elements. The other essential elements are needed in much smaller amounts (10^{-3}–10^{-6} that of C).

4.4 How could the early ocean supply energy for the origin and early evolution of life?

Life requires a continual input of free energy for the growth and maintenance of organisms. The free energy transformed by organisms can, in principle, come from a very wide range of sources, for example mechanical energy (e.g. water flow over an attached organism) and osmotic potential energy (e.g. tidal changes in external salinity in an estuary). In the absence of significant evidence for such energy sources in extant organisms, we deal here with catalysis of the coupling of energy from light, or organic chemical, or inorganic chemical reactions, to essential processes in the origin of life, and the early evolution of life. In all cases energy is dissipated during the energy transformations; less energy is stored in the products than was present in the substrate if a net conversion of useful energy is to take place. In addition to coupling the exergonic reaction to the (pre)biologically significant endergonic reaction, catalysis is involved in speeding the exergonic reaction which is based on energetic disequilibrium.

Dealing first with photochemistry, solar radiation is not, of course, endogenous to the sea. There is a very weak radiation, some of which is at wavelengths which could be involved in profitable photochemistry, in the ocean as a result at least in part of sonoluminescence involving thermal excitation of Na^+ ions (Matula, 1999; Hilgenfeldt *et al.*, 1999). However, it is not certain that the radiation is quantitatively sufficient to support photosynthetic life (see Yurkov and Beatty, 1998). Furthermore, it is likely that at least some of this radiation comes today from the spontaneous oxidation of sulfide by molecular oxygen (Tapley *et al.*, 1999). This mechanism would not have been available to the earliest life forms with ready access to sulfide, especially at hydrothermal vents, but with negligible molecular oxygen available until biological production of this gas evolved. One of the inorganic sinks for this oxygen which delayed its accumulation in the biosphere was, of course, this photon-emitting oxidation of sulfide.

Solar radiation is clearly adequate to support photosynthesis amounting globally to some 4 Pmol inorganic-C converted to organic-C each year in the present ocean (Falkowski and Raven, 1997; Table 4.2). However, this productivity is confined to the upper 300 m of even the most transparent seawater as a result of the photon absorption properties of H_2O. Generally there are other absorbing and scattering materials in seawater which attenuate solar radiation and decrease the depth at which photosynthetic growth can occur. All natural waters, regardless of their O_2 content, attenuate UV radiation more than photosynthetically active radiation (Kirk, 1994a, b). While photochemical energization of the earliest living cells is neither likely nor widely advocated, it is possible that photochemistry in the ocean, and especially in the atmosphere, was important in generating particular chemical species related to the origin of life.

Turning to the energization of early life by organic chemical reactions (chemo-organotrophy), this requires a prebiotic generation of organic matter. This could result from atmospheric (energized by UV radiation, lightning or meteorite impacts), crustal (including hydrothermal vent/bulk seawater interactions, and reactions at the crust/seawater interface) or extraterrestrial (coming to Earth on comets) events. It also requires that some reaction of the organic substrate is exergonic, by an internal rearrangement of the organic substrate, including its dismutation, or by reaction with some other organic or inorganic molecule (see Broda, 1975). Current views are that this is not the most likely means of energizing the earliest cells (see Edwards, 1998).

The third possibility is that the earliest cells were chemolithotrophic, i.e. were energized by inorganic chemical reactions. This currently popular hypothesis is supported by molecular phylogenetic evidence suggesting that the root of the universal tree of life is in extant chemolithotrophic (and thermophilic) organisms (Pace, 1997). Suggestions have been made of chemically (Wächterhäuser, 1990) and geologically (Russell and Hall, 1997, 1999) explicit models of a chemolithotrophic energization of the origin of life and of the earliest organisms (Table 4.2).

The chemically explicit models of Wächterhäuser (1990) relate to surface chemistry in the Hadean Ocean which can drive reactions such as $FeS + H_2S + CO_2 \rightarrow FeS_2 + HCOOH$, and subsequently the reactions of the reverse tricarboxylic acid cycle. Russell and Hall (1997, 1999) base their mechanism of energy transduction on the hotter, more reducing, more alkaline solution emanating from hydrothermal vents reacting with the warm, less reducing, less alkaline bulk seawater. The reaction is suggested to occur across an iron monosulfide layer which forms spontaneously between the two solutions. Russell and Hall (1997, 1999) suggest that this 'membrane' could act as the place in which exergonic redox reactions (more reducing inner phase, less reducing outer phase) and H^+ fluxes (from the less alkaline outer phase to the more alkaline inner phase) could energize essential biosyntheses. How this occurred, and how the iron monosulfide 'membrane' became a lipoprotein biological membrane, needs further clarification (see Blobel, 1980; Koch and Schmidt, 1991; Edwards, 1998; Russell and Hall, 1999). The Russell and Hall (1997, 1999) scheme for a chemolithotrophic origin of life is shown in Figure 4.1 in the context of Mars 4.3 billion years ago (Early Noachian).

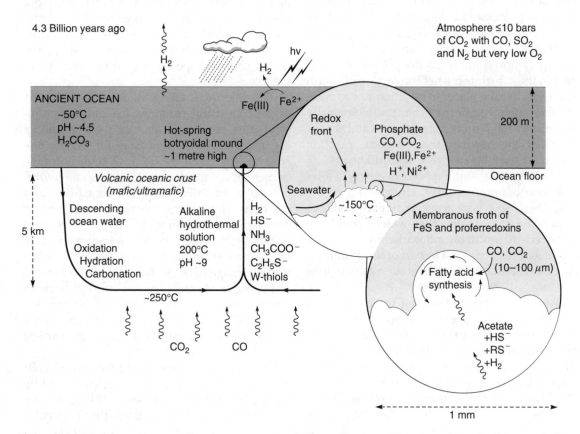

Figure 4.1 Model environment for the emergence of life on Earth or Mars (reproduced with permission from Russell, M.J. and Hall, Allen J. (1997/1999) On the inevitable emergence of life in Mars (Julian A. Muscot, ed.), *The Search for Life on Mars*. Proceedings of the 1st UK Conference, British Interplanetary Society, London) by chemolithotrophic mechanism, related specifically to the putative Early Noachian (Mars 4.3 billion years ago) ocean floor at a submarine alkaline hot spring. The mechanism proposes that the fatty acids which compose the organic membrane which replaced the iron monosulfide were generated in the iron monosulfide compartments comprising the sulfide mound.

This model of the energization of the early iron monosulfide 'membrane' and the early lipoprotein membranes in terms of an electrochemical potential gradient of H^+, involving chemiosmotic reactions (Koch and Schmidt, 1991; cf. Raven and Smith, 1981, 1982), is based on an early ocean which was acidic or near-neutral (Kasting, 1993; Russell and Hall, 1997, 1999). It is important to realize that there are alternatives to this notion of an acidic early ocean, e.g. the early soda ocean suggestions of Kempe and Degens (1985) and Kempe and Kazmierczak (1997). The suggestions as to the evolution of H^+ active transport across membranes and of chemiosmotic reactions of Raven and Smith (1981, 1982) were in fact based on a neutral or slightly alkaline early ocean, albeit not in the context of a chemolithotrophic origin of life and chemolithotrophic energization of the earliest cells. The lack of calcification of the earliest organisms accords better with an acidic rather than an alkaline early ocean.

An acidic early ocean would, in equilibrium with more than 0.1 MPa partial pressure of CO_2, have both a high concentration of CO_2 (in excess of $40 \, mol \, m^{-3} \, CO_2$). There would also be a high concentration of HCO_3^-, at least $4 \, mol \, m^{-3}$ with a pH of 5, i.e. at least twice the concentration in today's seawater with a pH of 8.0 in equilibrium with 33 Pa CO_2 in the atmosphere. This high CO_2 concentration in the early ocean (regardless of its pH, providing it is in equilibrium with at least 0.1 Ma CO_2 in the atmosphere) would be adequate to saturate the carboxylase enzyme(s) of any known autotrophic or anaplerotic inorganic carbon fixation (Raven, 1991, 1997a, b, c). This conclusion is independent of whether the enzymes use CO_2 or HCO_3^- as their immediate substrate (provided that the intracellular pH is maintained close to neutrality), or of whether there is a significant diffusion restriction of CO_2 entry (Raven, 1991, 1997a, b, c). At all events, CO_2 diffusion through water and the lipid component of lipoprotein bilayers would have been adequate to supply inorganic carbon to the carboxylases; no inorganic carbon concentrating mechanisms would have been needed.

Although much needs to be clarified, the chemolithotrophic energization of the earliest biota seems to be the best currently available hypothesis.

4.5 **How did early biota change ocean chemistry and how did these chemical changes influence the early evolution of life?**

We assume a chemolithotrophic origin of life, and chemolithotrophy as the earliest energetic coupling mechanism in biota. In quantitative terms it is likely that the rate of reductant input from hydrothermal vents would limit worldwide primary production each year to about 50 Gmole, at least inasmuch as input through vents is concerned (Jakosky and Shock, 1998). A parallel input of chemolithotrophic substrates could come from the sorts of reactions discussed by Wächterhäuser (1990). However, quantifying the productivity based on such energization is more complex, and will not be attempted here. To this 'new' productivity could be added 'recycled' productivity, which could result from the regeneration of inorganic reductant and inorganic carbon from 'respiratory' metabolism of the organic carbon and inorganic oxidant generated in chemolithotrophy. Such cycles of recycled chemolithotrophic productivity and 'respiration-based' productivity would run down, for example by sedimentation of reactants such as organic carbon (and inorganic oxidant?).

Thus there is likely to have been an incomplete recycling of the products of chemolithotrophy to regenerate reductant at a sufficient reducing redox potential to reduce inorganic carbon to sugars. However, additional productivity could result from an additional energy input which could generate a strong reductant from a weaker reductant. Such an energy input could come from photons in the process of photosynthesis, permitting growth by photolithotrophy. However, thus far we have been implicitly considering biota at great depth in the ocean, centred on hydrothermal vents. These biota would have not suffered from damage from the higher than present UV output from the young Sun, unscreened by the atmosphere due to the absence of O_2, and hence UV-absorbing O_3, in the atmosphere, due to the great depth of UV-absorbing water (Kirk, 1994a). However, despite the lower attenuation of photosynthetically active radiation than of damaging (UV-B) UV radiation by natural seawaters as has been mentioned earlier, no photosynthetically significant radiation from the Sun penetrates to 300 m even in the clearest waters.

As to why the biota should occur close enough to the ocean surface to permit photosynthesis, we can suggest that chemolithotrophs were exploiting inorganic reductants in near-surface sediments.

A plausible model for the evolution of photosynthesis from UV-screening protein-pigment systems is given by Mulkidjanian and Junge (1997). Here the UV was screened out by absorption of UV by membrane-associated proteins, with associated pigments (cyclic tetrapyrrols) able to accept excitation energy transferred from the protein, with energy dissipation by fluorescence, heat or photochemistry. The photochemical option for dissipation of UV energy could generate a strong reductant from a weaker reductant. Eventually, the energization of this process came to rely almost entirely on the absorption directly by the tetrapyrrol pigments of longer wavelength (>400 nm) radiation. In such organisms the UV-screening protein-pigment complexes evolved independently into two kinds of reaction centers of the RC1 and the RC2 types (Mulkidjanian and Junge, 1997). Both of these can bring about inorganic carbon reduction to sugars using reductant originally derived from weak reductants such as S^{2-}.

Ultimately even weak reductants such as Fe^{2+} and S^{2-} would be locally exhausted, as sedimentation of the organic product of primary production prevented the respiratory regeneration of Fe^{2+} from Fe^{3+} and S^{2-} from S^0 and SO_4^{2-}. Such local depletion of weak reductants means that continued photosynthetic generation of strong reductants which could reduce inorganic carbon to carbohydrates, required the substitution of an even

weaker reductant for the S^{2-} and Fe^{2+}. This even weaker reductant is H_2O which is, for all biotic purposes, infinitely available in seawater. The energy required to move an electron from H_2O to a reductant strong enough to reduce inorganic carbon to produce sugars is significantly greater than that needed for the use of a stronger reductant as electron donor. Accordingly all O_2-evolving organisms (i.e. those using H_2O as their electron donor) have both an RC1 *and* an RC2 reaction involved in transferring an electron from H_2O to CO_2, so that the energy from two photons (one used in the RC1, the other in the RC2) is used to transfer the 'electron'. In evolutionary terms what is needed here is a horizontal gene transfer so that RC1 and RC2 can be expressed in a single cell. Furthermore, the redox span covered by RC2 is shifted toward more oxidizing potentials, so that the oxidant generated by this RC2 can oxidize H_2O to produce O_2, but the reductant generated by this RC2 is certainly incapable of reducing inorganic carbon to sugars (Olson and Pierson, 1986; Trissl and Wilhelm, 1993). The remainder of the redox span to the redox level needed to reduce inorganic carbon to carbohydrates is covered by RC1. The ability to use H_2O as the ultimate electron donor for inorganic carbon fixation, with consequent O_2 production, increases the ability of photosynthetic organisms to oxidize the S^{2-} and Fe^{2+} in the ocean over and above the direct oxidation of S^{2-} and Fe^{2+} by organisms with RC1 or RC2. O_2 build-up in the atmosphere requires that the O_2 produced in photosynthesis is not entirely consumed in respiration of the other product of photosynthesis, i.e. organic carbon sedimentation and long-term storage of this organic carbon is needed. O_2 accumulation in the atmosphere also requires that the S^{2-} and Fe^{2+} sinks for O_2 have been oxidized; these S^{2-} and Fe^{2+} sinks are not just the pre-existing pools in the ocean, but also the continued input from hydrothermal vents. Finally, the presence of O_2 in the atmosphere permits oxidation of Fe^{2+} and S^{2-} exposed on the continental crust, first found some 3.5 billion years ago (Buick *et al.*, 1995; Raven, 1997a), again forming a sink for O_2 which must be satisfied before O_2 can further increase in the atmosphere.

The build-up of O_2 would diminish the availability of Fe and Mn to O_2-evolving photosynthetic organisms due to the insolubility of the Fe_2O_3 and MnO_2 which form under oxidizing conditions. However, Cu becomes much more available under these conditions, as the soluble Cu^{2+} instead of the very insoluble Cu_2S. The implications of this for the functioning, and subsequent evolution, of photosynthetic O_2-evolvers is discussed by Raven *et al.* (1999). Furthermore, organic matter can scavenge Mo from seawater, thus providing another feedback of the evolution of life on the chemistry of the oceans, in this case with implications for the capacity of organisms to acquire N for N_2 and NO_3^- via Mo-requiring nitrogenase and nitrate reductase enzymes (Holland *et al.*, 1986; Falkowski, 1997; Falkowski and Raven, 1997).

4.6 Conclusion

The early ocean provided all of the elemental requirements for the origin and early evolution of life, although some inputs from the atmosphere and crust were needed to provide the appropriate chemical species. The chemistry of life is determined by both the appropriate chemical properties of a given element and its availability. Some essential elements (e.g. P) may not have been abundant at any stage during the origin and early evolution of life.

Inorganic chemical reactions are the only energy sources for the origin of life which are endogenous to the early ocean, especially if the crust/ocean interface is acceptable as part of the ocean. These inorganic chemical energy sources (e.g. involving Fe and S) could allow a chemolithotrophic energization of the origin and early evolution of life. Additional energy inputs could come from the more alkaline and reducing solution from hydrothermal vents which is out of energetic equilibrium with the less alkaline, less reducing, seawater.

Early life modified the chemistry of the early ocean. In conjunction with the sedimentation of organic reduced products of chemolithotrophs (and possibly of the oxidized inorganic product), chemolithotrophy would decrease the availability of inorganic reductant for chemolithotrophy. This modification to the medium, together with use of some of the metabolic machinery involved in chemolithotrophy, provides both selection pressure and some part of the mechanism for the evolution of photolithotrophy.

References

Blobel, G. (1980). Intracellular protein topogenesis. *Proceedings of the National Academy of Sciences*, **77**: 1496–1500.

Brandes, J.A., Boctor, N.Z., Cody, G.D. *et al.* (1998). Abiotic nitrogen reduction on the early Earth. *Nature*, **395**: 365–367.

Broda, E. (1975). *The Evolution of the Bioenergetic Processes*. Oxford: Pergammon Press.

Broecker, W.S. (1985). *How to Build a Habitable Planet*. Palisades, New York: Eldigio Press.

Buick, R., Thornett, J.R., McNaughton, N.J. *et al.* (1995). Record of continental crust ~ 3.5 billion years ago in the Pilbana Craton of Australia. *Nature*, **375**: 574–577.

Cairns-Smith, A.G. and Hartman, M. (eds) (1986). *Clay Minerals and the Origin of Life*. Cambridge: Cambridge University Press.

Chyba, C. (1998). Buried origins. *Nature*, **395**: 239–330.

Chyba, C.F., Thomas, P.J., Brookshaw, L. and Sagan, C. (1990). Cometary delivery of organic molecules to the early Earth. *Science*, **249**: 366–373.

Cleaves, H.J. and Miller, S.L. (1998). Oceanic protection of prebiotic organic compounds from UV radiation. *Proceedings of the National Academy of Sciences*, **95**: 7260–7263.

Edwards, M.R. (1998). From a soup or a seed? Pyritic metabolic complexes in the origin of life. *Trends in Ecology and Evolution*, **13**: 178–181.

Falkowski, P.G. (1997). Evolution of the nitrogen cycle and its influence on the biological sequestration of CO_2 in the ocean. *Nature*, **387**: 272–275.

Falkowski, P.G. and Raven, J.A. (1997). *Aquatic Photosynthesis*. Malden, Massachusetts: Blackwell Science.

Hilgenfeldt, S., Grossmann, S. and Lohse, D. (1999). A simple explanation of light emission in sonoluminescence. *Nature*, **398**: 402–405.

Holland, H.D., Lazar, B., McCaffey, M. (1986). Evolution of the atmosphere and oceans. *Nature*, **320**: 27–33.

Jakosky, B.M. and Shock, E.L. (1998). The biological potential of Mars, the early Earth and Europa. *Journal of Geophysical Research*, **103**: 19359–19364.

Kasting, J.F. (1993). Earth's early atmosphere. *Science*, **259**: 920–926.

Kempe, S. and Degens, E.T. (1985). An early soda ocean? *Chemical Geology*, **53**: 95–108.

Kempe, S. and Kazmierczak J. (1997). A terrestrial model for an alkaline Martian atmosphere. *Planetary and Space Science*, **45**: 1493–1499.

Kirk, J.T.O. (1994a). Optics of UV-radiation in natural water. *Archiv für Hydrobiologie Beihefte*, **43**: 1–16.

Kirk, J.T.O. (1994b). *Light and Photosynthesis in Aquatic Systems*. Second Edition. Cambridge: Cambridge University Press.

Knauth, L.P. (1998). Salinity history of the Earth's early ocean. *Nature*, **395**: 554–555.

Koch, A.L. and Schmidt, T.M. (1991). The 1st cellular bioenergetic process – primitive generation of a proton-motive force. *Journal of Molecular Evolution*, **33**: 297–304.

Marschner, H. (1995). *Mineral Nutrition of Higher Plants*. Second Edition. London: Academic Press.

Matula, T.J. (1999). Inertial cavitation and single-bubble sonoluminescence. *Philosophical Transactions of the Royal Society of London A*, **357**: 225–249.

Maynard Smith, J. and Szathmáry, E. (1995). *The Major Transitions in Evolution*. Oxford: Oxford University Press.

Mulkidjanian, A.Y. and Junge, W. (1997). On the origin of photosynthesis as inferred from sequence analysis a primordial UV-protection as common ancestor of reaction center and antenna proteins. *Photosynthesis Research*, **51**: 27–42.

Muller, H.J. (1966). The gene material as the initiator and organizing basis of life. *American Naturalist*, **100**: 493–517.

Olson, J.M. and Pierson, B.K. (1986). Photosynthesis 3.5 thousand million years ago. *Photosynthesis Research*, **9**: 251–259.

Oro, J. (1961). Comets and the formation of biochemical compounds on the primitive Earth. *Nature*, **190**: 389.

Pace, N.R. (1997). A molecular view of microbial diversity and the biosphere. *Science*, **276**: 734–740.

Raven, J.A. (1991). Implications of inorganic C utilization: ecology, evolution and geochemistry. *Canadian Journal of Botany*, **69**: 908–924.

Raven, J.A. (1996). The role of autotrophs in global CO_2 cycling. In: M.E. Lidström and F.R. Tabita (eds) *Microbial Growth on C_1 Compounds*. Dordrecht: Kluwer Academic Publishers, pp. 351–358.

Raven, J.A. (1997a). The role of marine biota in the evolution of terrestrial biota: gases and genes. *Biogeochemistry*, **39**: 139–164.

Raven, J.A. (1997b). Inorganic carbon acquisition by marine autotrophs. *Advances in Botanical Research*, **27**: 85–209.

Raven, J.A. (1997c). Putting the C in phycology. *European Journal of Phycology*, **32**: 319–333.

Raven, J.A. and Smith, F.A. (1981). H^+ transport and the evolution of photosynthesis. *Biosystems*, **14**: 95–111.

Raven, J.A. and Smith, F.A. (1982). Solute transport at the plasmalemma and the early evolution of cells. *Biosystems*, **15**: 13–26.

Raven, J.A. and Yin, Z.-H. (1998). The past, present and future of nitrogenous compounds in the atmosphere and their interactions with plants. *New Phytologist*, **139**: 205–219.

Raven, J.A., Evans, M.C.W. and Korb, R.E. (1999). The role of trace metals in photosynthetic electron transport in O_2-evolving organisms. *Photosynthesis Research*, **60**: 111–149.

Rosing, M.T. (1999). ^{13}C-depleted carbon microparticles in >3700 Ma sea-floor sedimentary rock from West Greenland. *Science*, **283**: 674–676.

Russell, M.J. and Hall, A.J. (1997). The emergence of life from iron monosulphide bubbles at a submarine hydrothermal redox and pH front. *Journal of the Geological Society, London*, **154**: 377–402.

Russell, M.J. and Hall, A.J. (1999). On the inevitable emergence of life on Mars. *British Interplanetary Society Proceedings*, 26–36. ISSN 0007-09-4X.

Tapley, D.W., Buettner, G.R. and Shick, J.M. (1999). Free radicals and chemiluminescence as products of the spontaneous oxidation of sulfide in seawater, and their biological implications. *Biological Bulletin*, **196**: 52–56.

Trissl, H.-W. and Wilhelm, C. (1993). Why do thylakoid membranes from higher plants form grana stacks? *Trends in Biochemical Science*, **18**: 415–418.

Wächterhäuser, G. (1990). Evolution of the first metabolic cycles. *Proceedings of the National Academy of Sciences*, **87**: 200–204.

Williams, R.J.P. and Fraústo da Silva, J.J.R. (1996). *The Natural Selection of the Chemical Elements. The Environment and Life's Chemistry*. Oxford: Clarendon Press.

Woese, C.R. (1979). A proposal concerning the origin of life on Earth. *Journal of Molecular Evolution*, **13**: 95–101.

Yurkov, V. and Beatty, J.T. (1998). Isolation of aerobic anoxygenic photosynthetic bacteria from black smoker plume waters off the Juan de Fuca ridge in the Pacific Ocean. *Applied and Environmental Microbiology*, **64**: 337–349.

5

The role of carbon dioxide in plant evolution

William G. Chaloner

ABSTRACT

See Plates 6–9

Carbon dioxide is a very minor component of the atmosphere, currently at about 0.03% (*c*. 350 parts per million). It is, together with water vapour, the major greenhouse gas implicated in the enhancement of the 'greenhouse effect' in causing 'global warming'. In this context it is sometimes referred to as an atmospheric pollutant, although, in fact, from early in Earth history it has been a natural component of the atmosphere, and is the source of the carbon content of the organic matter of virtually all biomass on Earth.

Photosynthetic plants, algae and bacteria take up carbon dioxide using light energy and chlorophyll as a catalyst, converting it to carbohydrate, and subsequently to a wide range of organic substances and releasing oxygen. Some of these products of biosynthesis are broken down in plant and animal respiration, consuming oxygen and releasing carbon dioxide.

Early in Earth history, long before the appearance of living systems, carbon was cycled from the atmosphere into the geosphere as carbonate sediments. With the appearance of photosynthesis, this cycling of carbon was greatly accelerated, the more so when plant life colonized the land. The burial of organic carbon (as coal and in marine sediments) increased the oxygen content of the atmosphere, and resulted in a dramatic fall in the atmospheric carbon dioxide through Devonian and into Carboniferous time.

Terrestrial plant life responded to this change, in modifying the density of gas-exchange pores or stomata on the surface of their photosynthetic tissue. Changes in stomatal density observed in land plant fossils are compatible with the synchronous changes in global atmospheric carbon dioxide calculated from physical and geochemical evidence. Further, the evolution of large, planated (thin, laminate) leaves by late Devonian time appears to have been linked to the falling level of carbon dioxide, brought about largely by terrestrial photosynthesis. Land plant life can rightly be said to have evolved in response to changes in its environment induced by the plants themselves.

Changes in oxygen content of the atmosphere are closely linked to those in carbon dioxide; fossil charcoal gives evidence of the occurrence of wildfire throughout the geological column, since the time when plant life produced biomass fuel and there was oxygen enough for it to burn. This puts certain constraints on the levels of oxygen that may have existed in the past. The existence of wildfire has in turn influenced the evolution of 'fire ecology' in those plant communities particularly prone to burning.

5.1 Introduction

Although carbon dioxide forms only between 0.03 and 0.04% of the atmosphere, it is the ulti- mate source of all the carbon in the organic matter which makes up the living plants, ani- mals, fungi, protists and bacteria of the 'bio- sphere', including ourselves. Green plants and the cyanobacteria take the carbon dioxide

Evolution on Planet Earth
ISBN 0-12-598655-6

from the air in the process of photosynthesis, using solar energy with chlorophyll pigments as catalysts, producing carbohydrates and releasing oxygen:

$$CO_2 + H_2O \rightarrow CH_2O + O_2$$
(light energy, chlorophyll)

These carbohydrates may later be metabolized into lipids and proteins, and this wide range of plant products may then be consumed by animals, fungi, protists and bacteria which incorporate the carbon into their own biomass. Most organisms, including the plants themselves, eventually break down some of the carbon compounds of their biomass by oxidizing them in the process of respiration, using atmospheric oxygen and returning the carbon to the atmosphere as carbon dioxide:

$$CH_2O + O_2 \rightarrow CO_2 + H_2O$$

The two processes of photosynthesis and respiratory oxidation are the core features of the terrestrial carbon cycle. The two equations are, of course, the same, each one simply the reverse of the other. As will be seen below, they are not the whole story, although they are the major part of it.

This terrestrial carbon cycle has its counterpart in the marine realm. Marine organisms generally depend on photosynthesis by phytoplankton (microscopic algae), as well as the biomass produced by attached seaweeds growing in the littoral zone, coupled with some detrital terrestrial biomass washed off the land. This marine bioproductivity by phytoplankton and seaweeds is derived largely from dissolved carbon dioxide in the ocean water.

Just as in terrestrial biota, the animal life of the oceans from whales to single celled animals lives directly or indirectly on this photosynthetic production of biomass by marine plant life. The terrestrial and marine parts of the carbon cycle are linked via the shared access to atmospheric carbon dioxide, and such movement of biomass as occurs between land and sea.

It has been known since the last century that carbon dioxide is a major 'greenhouse gas', that is, its capacity to absorb and retain solar energy

re-radiated from the Earth's surface far exceeds that of the other major atmospheric components. Indeed, if it were not for the combined 'greenhouse effect' of carbon dioxide and water vapour in the atmosphere, the average temperature of the Earth would be some 20°C colder than at present.

Over the last twenty years or so the role of carbon dioxide in the Earth's climate system has received an extraordinary amount of scientific attention and general public interest. In this time two features of change in the atmospheric level of that gas have had an enormous influence on the thinking of scientists concerned with global change.

First came the evidence from observations of atmospheric composition on Mona Loa in Hawaii that the level of carbon dioxide in the atmosphere fluctuates between summer and winter, but that year after year it has been rising steadily since records were first obtained in the 1960s. Further, evidence from ice cores obtained by drilling into the Antarctic ice cap showed that air trapped in the snow fall showed how changes in atmospheric carbon dioxide over the last 200 000 years had paralleled the temperature changes over the same period. The temperature over the region from which the snowfall is derived can be determined from the oxygen isotopes present in the water forming the ice.

When the global temperature had been at it lowest, as at the peak of the last glacial phase, around 20 000 years ago, carbon dioxide had also been low. During the last warm phase, some 120 000 years ago, the carbon dioxide as now was at a high level. The ice core data also showed that the present rising carbon dioxide figures were merely the latest phase of a steep rise which had started around 150 years ago at the time of the industrial revolution, and is generally regarded as caused by the burning of fossil fuel. This added to the atmospheric carbon dioxide by returning the products of combustion of organic carbon long buried as coal or hydrocarbon.

Many climatologists believe that the ice core data indicated that if we continued to burn fossil fuel at present rates, the enhanced greenhouse effect would result in a global rise in temperature. The actual level and effect of such global warming is the subject of lively debate and much concern ever since this proposition began to receive wide acceptance.

A further feature of the carbon cycle is that weathering of certain minerals in rocks can also be a pathway for carbon dioxide to be removed from the atmosphere. By this means carbon (as carbon dioxide) in the atmosphere can be carried into the oceans and buried in deep water sediments in the form of calcium carbonate, so taking it out of the day-to-day carbon circulation. The details of this process are considered later in this chapter.

Also, the occurrence of wildfire (forest or grassland fires) represents a short-cut version of the biological process of respiration. The simple chemical formula for combustion is exactly the same as that for respiration, although the details of the two processes are very different. Both, however, involve the oxidation of organic matter, using atmospheric oxygen and releasing carbon dioxide and water vapour. (The history of wildfire through geological time and its relationship to changes in the oxygen content of the atmosphere will be considered in the final section of this chapter.)

In the course of Earth's history there have been major changes in the composition of the atmosphere and in particular in its oxygen and carbon dioxide content (Figure 5.1; Kasting, 1987). The rise of photosynthesis brought about the most radical change from a carbon dioxide-rich early atmosphere, low in oxygen, to our present oxygenated atmosphere with a very low carbon dioxide level.

Oxygenic photosynthesizers (plants, algae and cyanobacteria) in their physiological environment brought about this drastic change, but they in turn are seen to have responded to that change. Changes in the carbon dioxide level affected climate, through enhancing or diminishing the greenhouse effect. Of course the plant life was in turn dependent on the climate

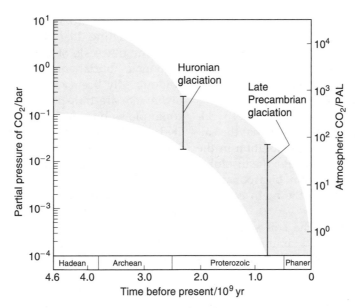

Figure 5.1 Variation in atmospheric carbon dioxide through the course of Earth history, derived from modelling the contribution of carbon dioxide, ozone and water vapour to the greenhouse effect, and reconciling this with geological evidence of global temperature. The bars for each of the two Precambrian glaciations shown represent the range of possible values, as does the shaded band for the range of carbon dioxide values. This implies that drawdown of atmospheric carbon dioxide predated the origin of photosynthesis in the Archean, and progressed increasingly rapidly into the Huronian glaciation. The drawdown was greatly accelerated in the Phanerozoic (Phaner) with the rise of terrestrial vegetation (Reproduced with permission from Francis, P. and Dise, N. (1997) *Atmosphere, Earth and Life*, Open University, based on the work of Kasting, 1992 and Schopf and Klein, 1992).

and thus adapted to it, migrated, or in some cases was driven eventually to extinction over time.

In this chapter we examine some of the interactions between the atmospheric carbon dioxide and the plant life which depends upon it. In particular, we look at the ways in which plants have evolved to accommodate the changes in the environment which they themselves have been instrumental in bringing about.

5.2 The origin of the carbon cycle

When we consider the nature of the carbon cycle in the early stages of Earth history, we need to return to those processes involving carbon dioxide that can take place *without* the intervention of plants or animals. At the present time any mineral matter in the soil containing calcium silicate can be attacked in the weathering process by the action of atmospheric carbon dioxide. Calcium silicate is a component of a wide group of feldspars, among the commonest of minerals in the Earth's crust, and so, of course, of the soils derived from them.

The result of this interaction between calcium silicate and carbon dioxide is that the calcium ions are leached out into solution in the soil water as calcium bicarbonate; the insoluble silica is left behind. Calcium bicarbonate in solution in groundwater is the main cause of 'water hardness'.

As the bicarbonate in the soil solution is eventually washed out and drained into the oceans, it can then be used by marine organisms – both phytoplankton and animal life – to form shells and other protective structures. On the death of the organism these protective structures may then fall to the sea bed, taking the calcium carbonate 'out of circulation' from the carbon cycle, at least for the time being.

The net result of this movement of bicarbonate ions is the transfer of atmospheric carbon dioxide into carbonate ions 'locked up' in marine sediments on the floor of the ocean. There is, of course, a parallel movement of calcium

from weathered igneous rock feldspars into calcium carbonate on the ocean floor:

$$CaSiO_3 + CO_2 \rightarrow CaCO_3 + SiO_2$$

This represents the pathway of carbon transport from CO_2 in the soil (ultimately from the atmosphere) into buried carbonate on the sea floor. The process, however, need not involve living organisms, even though it is described in those terms above. The precipitation of calcium carbonate on the sea floor can occur simply as a physical process. Typically, evaporation of warm shallow seawater raises the concentration of calcium and bicarbonate ions to the level at which calcium carbonate can be precipitated.

Like photosynthesis this pathway of silicate weathering is an important route of carbon dioxide drawdown from the air to sediment on the ocean floor (or continental surface, in the case of limestone forming in a freshwater lake). However, unlike photosynthesis, this drawdown can occur in the absence of living organisms. An important difference is that silicate weathering does not release oxygen.

In the timescale of geological processes this carbonate burial on the ocean floor has undoubtedly been very significant. Carbonate sediments are transported on the oceanic tectonic plates on which they have been deposited, and may eventually be subducted beneath the margins of continental plates, as is occurring today along the western margin of North America.

Such oceanic sediments are drawn deep into the upper part of the Earth's mantle, where the rising temperature breaks down the calcium carbonate and releases the carbon dioxide into the ensuing molten magma. Upward migrating 'plumes' of this lighter mantle material may then emerge at the surface as volcanic eruptions, once more releasing the carbon dioxide into the atmosphere.

This scenario means that the carbon dioxide emissions from a volcano, such as Mount St Helens in Washington, USA, which erupted in May 1980, are simply the return to the atmosphere of carbon dioxide originally drawn from the air, and deposited as carbonate or organic carbon on the ocean floor.

This digital artist's concept shows the International Space Station after assembly is completed in 2004. In total, 16 countries are cooperating to provide a state-of-the-art complex of laboratories in the weightless environment of Earth orbit. With this facility advances will be made on assessing the effect of gravity on evolution (photo credit: courtesy of NASA, http://spaceflight.nasa.gov/gallery/images/station).

Plate 1 *Synechococcus*, a widespread cyanobacterium on Earth today, is likely to be similar to some of the first cyanobacteria (Lynn Rothschild). **See Chapter 1**

Plate 2 Recent work suggests that the earliest organisms, or at least the last common ancestor for life, may have arisen in a hot environment. It is known that the surface of the early Earth suffered heavy bombardment prior to the origin of life, which would have heated the surface. It is therefore possible that at the time life arose, early earth looked similar to the boiling water one sees today in Shell Geyser in Yellowstone National Park, USA (Lynn Rothschild). **See Chapters 1 and 4**

Plate 3 Microbial mats composed of the cyanobacterium *Lyngbya aestuarii*. These nitrogen-fixing organisms can form extensive mats. Photos taken in Guerrero Negro, Baja California Sur, Mexico. Photo credit: Lynn Rothschild.
See Chapter 2

Plate 4 Forest fire in Australia. Combustion of organic matter requires at least ~15% oxygen in the atmosphere. Above ~25% oxygen, fires would be so frequent that they would prevent the regeneration of forests. (Photo credit: Pat Barling, volunteer bushfire-fighter.)
See Chapter 3

Plate 5 Lower Carboniferous charcoal beds in Donegal, Ireland. Charcoal indicates forest fires and hence oxygen >15% of the atmosphere about 350 million years ago. (Photo credit: Tim Jones.) **See Chapter 3**

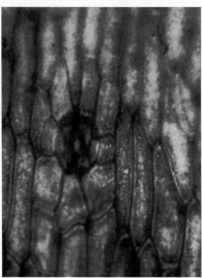

Plate 6 A stoma on the surface of the early Devonian plant *Rhynia*, preserved in three dimensions in volcanically derived silica. Stomatal density on early Devonian plants is generally low, in the high carbon dioxide atmosphere; only rarely will more than one stoma be encountered in a field of this size (compare Plate 7). (Photo credit: Bill Chaloner). **See Chapter 5**

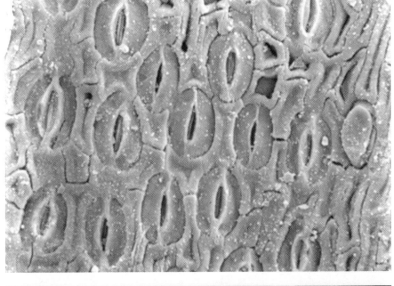

Plate 7 Stomata on the lower surface of a leaf of the late Carboniferous conifer *Swillingtonia*, preserved as charcoal. Stomatal density in this species is very high, at a time of low atmospheric carbon dioxide level, and contrasts strikingly with the sparse stomatal density of the early Devonian *Rhynia* (Plate 6). (Photo credit: Bill Chaloner). **See Chapter 5**

Plate 8 A leaf of *Gingko huttoni* from the Middle Jurassic of Yorkshire, preserved as a compression fossil in deltaic mudstone. The coin inserted between it and the matrix illustrates the coherence of the fossil, of which the cuticle has survived with little chemical change. (Photo credit: Bill Chaloner). **See Chapter 5**

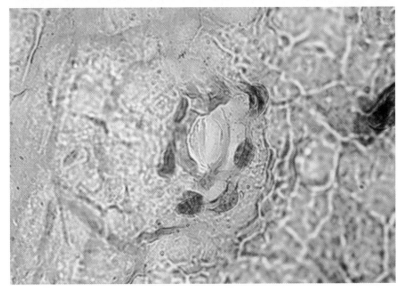

Plate 9 Cuticle of fossil *Gingko huttoni* from the Yorkshire Jurassic, showing a stoma. This was prepared from a specimen of the type shown in Plate 8. Stomatal counts on such material produced the data shown for the Middle Jurassic in Figure 5.5, item 4. (Photo credit: Bill Chaloner). **See Chapter 5**

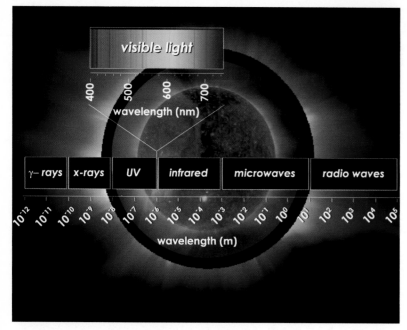

Plate 10 The solar spectrum superimposed on a false-color image of the Sun. The Sun's outer atmosphere is as it appears in ultraviolet light emitted by electrically charged oxygen flowing away from the Sun to form the solar wind (region outside black circle), and the disc of the Sun in light emitted by electrically charged iron at temperatures near two million degrees Celsius (region inside circle). This composite image taken by two instruments (UVCS, outer region and EIT, inner region) UVCS/EIT composite image. Source: Solar & Heliospheric Observatory (SOHO). SOHO is a project of international cooperation between ESA and NASA.
See Chapter 6

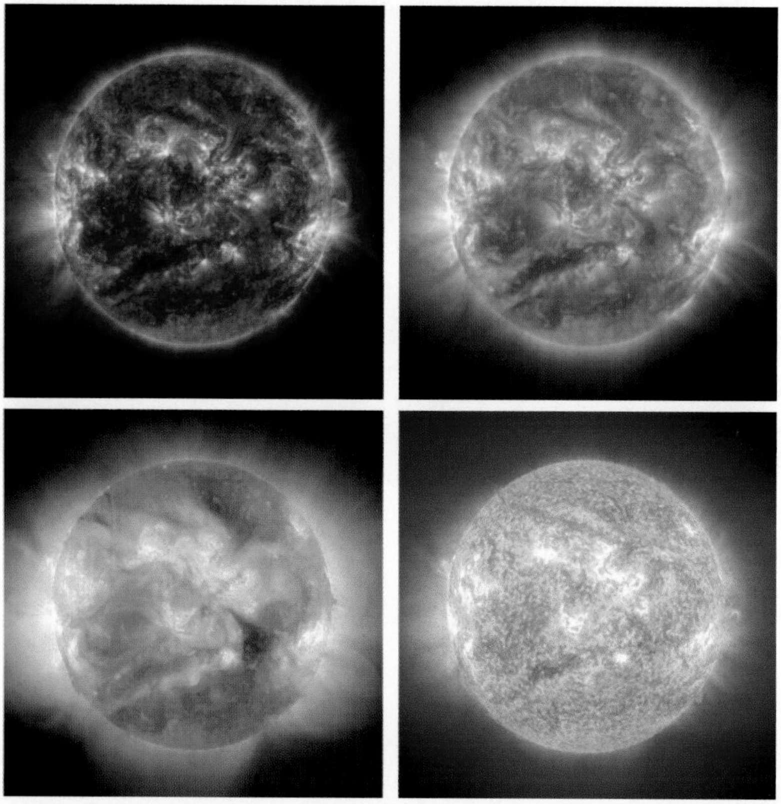

Plate 11 Four images taken on 18 January 2002 by the Extreme-Ultraviolet Imaging Telescope (EIT) on board the Solar and Heliospheric Observatory (SOHO) spacecraft, which is orbiting the Sun just ahead of the Earth. SOHO was launched in 1995 and will continually monitor the Sun for several years. These false-color pictures were taken in four emission bands. Upper left (blue) is the FE IX-X 171 Å emission showing the solar corona at a temperature of about 1.3 million K. Upper right image is the FE XII 195 Å emission showing the solar corona at a temperature of about 1 million K. Lower left image (yellow) is the FE IXV 284 Å emission. The lower right image (orange) was taken in the ultraviolet light using the Hell line 304 Å emission. This material is 16 000 to 80 000 K. Source: Solar & Heliospheric Observatory (SOHO). SOHO is a project of international cooperation between ESA and NASA.
See Chapter 6

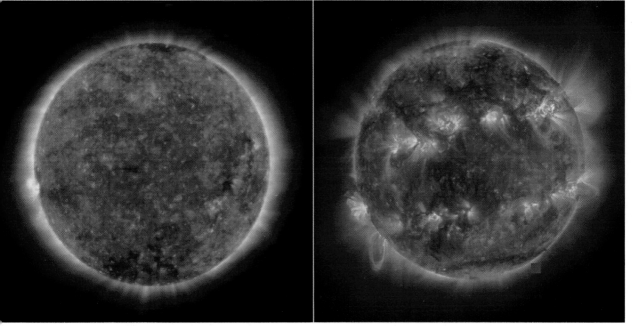

Plate 12 A comparison of two Extreme-Ultraviolet Imaging Telescope (EIT) images almost two years apart illustrates how the level of solar activity has increased significantly. The Sun attained sunspot maximum in the year 2000. These images are captured using Fe IX-X 171 Å emission showing the solar corona at a temperature of about 1.3 million K. Many more sunspots, solar flares, and coronal mass ejections occur during the solar maximum. The numerous active regions and the number/size of magnetic loops in the recent image show the increase. Source: Extreme-Ultraviolet Imaging Telescope (EIT) on board the Solar & Heliospheric Observatory (SOHO). SOHO is a project of international cooperation between ESA and NASA. **See Chapter 6**

Plate 13 The history of life on Earth. The ingredients of life (e.g. organics and water) were brought to the Earth via planetesimals and comets; life started as 'microbial world' about 4 billion years ago; the first eukaryotes appeared about 2 billion years ago; today advanced technologies enable us to explore our solar system and the universe (artist's view, NASA). **See Chapter 7**

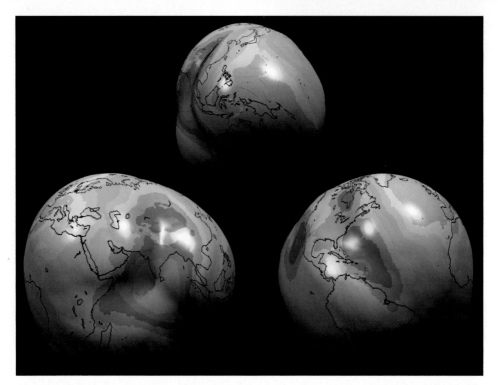

Plate 14   The GRACE (Gravity Recovery and Climate Experiment) spacecrafts were launched 17 March 2002. They will map the variations in gravity over the Earth's surface. Image credit: Center for Space Research at The University of Texas at Austin. For further information, see http://www.csr.utexas.edu/grace/ **See Chapter 9**

Plate 15   The moon is the only extra-terrestrial body on which humans have experimented with locomotion. Gravity is only one sixth of that on Earth, and walking is difficult because of the high Froude number. Lunar astronauts have adopted a variety of unusual gaits, often including hopping and loping. (NASA) **See Chapter 10**

This movement of carbon – from the air into marine sediments and by subduction and volcanicity back into the atmosphere – represents a very slow part of the present-day carbon cycle. Cycling of carbon in this way, however, must have preceded life on Earth. The precipitation of the calcium carbonate would have had to have occurred through inorganic means in sites where high evaporation caused saturation of the ocean water with calcium bicarbonate. However, if the early Precambrian atmosphere, as is now accepted (Kasting, 1987), had over 1000 times its present level of carbon dioxide, the scenario is not unreasonable (Figure 5.1).

There are further processes by which carbon dioxide dissolved in the ocean could have been incorporated into submarine magma. Ocean water can be drawn into the mid-oceanic vents (such as those along the mid-Atlantic ridge) and the carbon dioxide dissolved in the water incorporated into the basaltic magma to produce 'carbonated basalt'. Again by subduction over time this carbon would have been returned to the atmosphere through volcanicity.

Since these several versions of an 'abiotic carbon cycle' would have preceded life on Earth, it is worth noting that the carbon cycle was not initiated by the activity of the biosphere, although clearly the rise of photosynthetic organisms drastically altered the mechanisms of the cycle.

Somewhere around 3500 million years ago – early in the 4500 million year history of our Earth – organisms appeared in the sea which were capable of photosynthesis. (The origin of these organisms and their impact on the nature of Earth's atmosphere are dealt with in Chapters 1, 2 and 6.)

The earliest oxygenic photosynthetic organisms were single-celled cyanobacteria, some of which left a good fossil record of their presence and activities in the form of sizeable reefs of calcium carbonate. There may have been other forms of the same group of photosynthetic bacteria living in the sea as phytoplankton, but they left virtually no fossil record. Slowly these phototrophs began to change the atmosphere, removing carbon dioxide and replacing it with oxygen. There is significant physical evidence to suggest that the early Earth's atmosphere was lacking in oxygen and rich in carbon dioxide, and that these organisms slowly brought about the oxygenation of the atmosphere.

Aside from the purely chemical affects this had on the stability of certain minerals and weathering processes, it would have had an enormous effect on the microorganisms themselves. It is a reasonable supposition that other bacteria, living on the products of the cyanobacteria, would have existed at that time, functioning as anaerobes – organisms adapted to an oxygen-free environment. Such bacteria still exist, although they are restricted to those few and limited environments in which oxygen is very scarce or absent.

Those early anaerobes would have been greatly disadvantaged by the increasing availability of oxygen produced by their photosynthetic contemporaries. For them oxygen would have constituted a pollutant – a toxic component introduced by the activity of another organism and foreign to their normal environment. Had they been able, those early anaerobes might have protested vociferously at the pollution of their environment by those oxygen-producing cyanobacteria. Fortunately for evolution no steps were taken to curtail this oxygen production. Thus the Earth's atmosphere became oxygenated, and the anaerobes retreated to special environments where the availability of oxygen is very low.

For a long period in the latter half of Precambrian time – probably from 2500 until 600 Ma ago – the main site of photosynthetic oxygen production must have been in the ocean by the activity of phytoplankton living in the uppermost illuminated levels. This bioproductivity, in turn, must have fuelled the explosion of invertebrate animal life which took place in the Cambrian (Gould, 1989). Photosynthetic microorganisms must also have become adapted to living on land in wet surfaces, such as wetland areas or sites in the proximity of the splash from waterfalls, although there are no fossil records of such ephemeral terrestrial occupants.

The photosynthetic uptake of carbon and the eventual burial of some small fraction of it in sediments on the ocean floor would nevertheless have continued to take carbon 'out of circulation', at least in the geological short term. This drawdown of carbon dioxide would have continued to reduce the reservoir of that gas in the atmosphere.

Berner (1998) has modelled the changes in the carbon dioxide in the atmosphere over the Phanerozoic – the time since sizeable invertebrate animals appear in the fossil record. This model uses a variety of sources as evidence, including the actual observable level for each time interval of 'carbon burial' in sedimentary rocks as coal or other organic carbon, as well as in the form of carbonate. He also uses the evidence from the fractionation of stable carbon isotopes preserved in ancient soils and in ocean floor sediments, both of which carry somewhat obscure responses to the atmospheric carbon dioxide level.

On the basis of this and other evidence, Berner has produced the carbon dioxide curve shown in Figure 5.2. In spite of broad error bands above and below the curve, important trends are evident. Some 600 million years ago, the atmospheric carbon dioxide was somewhere between 10 and 20 times its present level. Then, starting about 450 million years ago, the combined effect of both marine and terrestrial photosynthesis decreased the carbon dioxide level. This trend continued until by 300 million years ago, atmospheric carbon dioxide reaching something close to its present-day level.

There were two major contributors to this removal of carbon from the atmosphere. The burial of dead organic matter consisting of both plants (phytoplankton) and animals in deep water sediments (much as occurs today) was coupled with the continuing accumulation of calcium and magnesium carbonate (as limestone and dolostone) from the shells and 'skeletal material' of molluscs, corals and other invertebrates in shallow water sediments. But

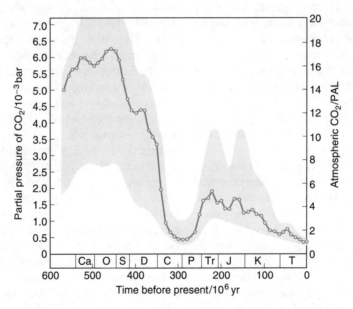

Figure 5.2 Variation in the atmospheric carbon dioxide level through the Phanerozoic, derived from Berner's GEOCARB model. The left-hand axis scale is in partial pressure units, that at the right hand is expressed as a multiple of the present atmospheric level (PAL). The shaded area represents the range of approximate error based on sensitivity analyses. The geological periods are abbreviated on the timescale (Ca, Cambrian to T, Tertiary). The best estimate figure shows the Silurian value at a peak of some 18 times the present level. The photosynthetic fixation by terrestrial plants, and silicate weathering accelerated by increasingly elaborate rooting systems, produced the steep fall through the Devonian and Carboniferous, reaching a level comparable to that of the present at the Carboniferous/Permian transition. Reproduced with permission from Francis, P. and Dise, N. (1997) *Atmosphere, Earth and Life*, Open University. See also Berner (1998) and references there cited.

the final massive fall in carbon dioxide was the result of the invasion of the land by the photosynthetic plants, which is the subject of the next section.

The 'greening of the land'

During the latter part of the Lower Paleozoic there is increasing fossil evidence that plants were beginning to show adaptations to life in a terrestrial habitat. Plants living on the land need a number of modifications from the structure of any seaweed known to survive life on a land surface. First, they need to have a structural rigidity to enable them to stand upright, and perhaps more important, grow up above their neighbours in the competition for light for photosynthesis. Seaweeds have no need for such 'mechanical tissue' to give them rigidity, as the dense seawater in which they grow gives them adequate flotation to keep them upright. Small land plants can achieve relative rigidity from the turgor of their tissues; the stalk of a dandelion flower can stand up in just this way, but on wilting (from loss of water) it will flop over on its side. Larger plants need rigid cell walls to stand upright, regardless of the availability of water as in woody trees and shrubs. All land plants also need a plumbing system – water-conducting tissue, broadly referred to as xylem – to transport water from the soil (its main source of water on land) to other parts of the plant. Early plants on land seem to have combined the use of the rigid cells of their plumbing system, the xylem with the turgor of their other tissues in order to stand upright. Later, as land plants became much larger, the water conducting cells took over the main mechanical role, as seen today.

Plants living on land have other important adaptations. One of these is a cuticular covering – a lipid-based layer over their outer surface or cuticle – which reduces the loss of moisture by evaporation from the outermost cells. This retention of water within the plant is evidently a problem only for plants exposed permanently to the air. Broadly speaking, plants living in water (seaweeds and fresh-water aquatic plants) neither need nor have such a cuticle. But since virtually all land plants need to take in carbon dioxide (and incidentally allow the outward diffusion of oxygen) in the course of photosynthesis, these plants have microscopic pores in their epidermis and the cuticle covering it to allow gas diffusion in and out. Since water vapour can also diffuse through these pores, so bypassing the cuticular waterproof covering, land plants have the means of closing the pores when water loss by transpiration threatens them with fatal dehydration. Each pore in the epidermis can be closed by two guard cells encircling the pore; the combination of pore and guard cells is called a stoma (meaning, appropriately, a 'mouth' with 'lips' which can close). (See the stomata typical of various early land plants shown in the plate section, Plates 6 and 7). All land-adapted plants also produce minute spores which are protected by a remarkably tough outer layer, the exine, which plays a role in their reproduction. Most seaweeds and microscopic planktonic plants (algae) reproduce in water by means of motile cells that can swim. For some of these algae such swimming cells are gametes involved in sexual reproduction, while in others such swimming cells can give rise to a new adult directly. But on land such dispersal by swimming cells is impossible, and primitive living plants like ferns and horsetails produce wind-borne spores which serve to colonize new habitats. These plants and their later derivatives do retain a motile stage in their life cycle, but this stage only operates within soil moisture as a means of sexual reproduction. The formation of resistant air-borne spores in the past was clearly an important step in land plant adaptation.

There is no fossil evidence of plants with xylem, a cuticle with stomata and spore-bearing structures until the end of the Silurian. The earliest fossil plants showing all these features are minute, only a few centimetres in height, typified by the genus *Cooksonia*. This genus occurs in many parts of the world, in rocks close to the Silurian/Devonian boundary. Fossils show fragments of cuticle (but lacking stomata), fragments of cells similar to the water-conducting xylem of later plants, and many types of resistant spores, from the Ordovician onwards. These microscopic fossils give us a strong hint that plants were evolving adaptations to life on land. But *Cooksonia* and its contemporary *Uskiella* together constitute the earliest (basal Devonian) record of land plants in which all four of these key attributes of vascular plants (cuticle, stomata, xylem and spores) are present (Kenrick and Crane, 1997a, b). However, it is not until a little later in the Lower Devonian that there is evidence of vascular plants (e.g. *Rhynia*) with all four features in a single plant. (For a fuller discussion of the early radiation of land plants see Bateman *et al.* (1998), Kenrick and Crane (1997a, b) and Niklas (1997).)

From the earliest Devonian onwards (about 400 million years ago) plant life on land diversified, and by the end of the Devonian large trees capable of forming forests were distributed globally. The photosynthetic drawdown of carbon dioxide into carbohydrates increased enormously; by Carboniferous time many plants were adapted expressly to cope with the somewhat acid and anaerobic substrate represented by the extensive wetlands of that period. On falling to the swamp floor, the biomass containing the carbon they had fixed did not decompose to return the carbon to the atmosphere, since the acidic swamp floor suppressed microbial activity. That biomass passed into peat and thence coal, much of its carbon not to be returned to the atmosphere until mined and burnt in the last century or so. But this terrestrial carbon burial added to the rate of fall of the atmospheric carbon dioxide previously driven only by burial of marine detrital biomass and the carbonate burial of marine shells and skeletons. The atmospheric carbon dioxide fell to close to its present low level by the end of the Carboniferous (Berner, 1998 – see Figure 5.2).

One of the interesting features of Berner's carbon dioxide curve is that the two low levels correspond to the two major global ice ages of the Phanerozoic – in the Permo-Carboniferous and the Quaternary periods. Berner has argued, with good reason, that the diminished greenhouse effect of the lower carbon dioxide is likely to have been a significant factor in the onset of those two global glacial phases. This is not to claim that weakening of the greenhouse effect from falling carbon dioxide levels, largely controlled by photosynthetic carbon fixation, was the sole cause of these ice ages. Other global changes relating to the tilt of the Earth's axis and eccentricity of its orbit also seem to be involved in the onset of major glacial phases, at least for the Pleistocene glaciations or Milankovitch cycles (see, e.g., Bradley, 1985). It appears, however, that plant life on the land, by its impact on the greenhouse effect, played a significant role in modifying the global climate, which was in turn a major element of the environment to which land plant life had to adapt. The steep climatic gradient from poles to equator, which seems to have accompanied the Carbo-Permian glaciation, gave rise to the first clearly defined regional floras (the equivalent of present-day biomes) that we can recognize in the fossil record (Chaloner and Meyen, 1973).

The existence of a Devonian land flora with its photosynthetic capacity, producing biomass on the land to an unprecedented extent, must have given an enormous boost to the organisms responsible for decomposition in the terrestrial setting. In the marine realm fungi, protozoa and bacteria must have played a major role from the earliest times in the biodegradation of dead organisms, just as they do now. But new species must have evolved to function in the same way in the very different setting of terrestrial soils. The products of land plants, comminuted by detritivore invertebrates in the soil, would then have been degraded back to carbon dioxide by the activity of the soil microflora. Equally, microorganisms must have assailed the living plants as pathogens; we can see fungi within the tissues of the earliest land plants (Taylor and Taylor, 1993). But it is difficult to distinguish in fossils between those that were pathogens and fungi which had entered the plant tissue after death. The activity of these microorganisms returning the fixed carbon to the atmosphere played an increasing role in carbon cycling on land as the terrestrial flora diversified.

A further by-product of the increasing complexity of land plant life through the course of the Devonian was the accelerated weathering of silicate minerals. As land plants became larger and more differentiated structurally, they developed root systems capable of penetrating deeper into the soil and subsoil. These more extensive root systems in turn carried organic matter deeper into the soil-to-bedrock continuum. This would also have contributed to the break-up of the underlying rock as the growing roots penetrated and expanded in the rock crevices. The acid humic material produced by the biodegradation of dead or dying roots and the products of their decay all brought carbon dioxide at a low pH into close juxtaposition with silicate minerals. Then, as now, this must have increased the rate of weathering with the ensuing drawdown of carbon as bicarbonate into the soil moisture and thence into the runoff. Figure 5.3 illustrates the fossil evidence we have from the work of Algeo and Scheckler (1998) on this expansion of the subterranean role of land plants through the course of Devonian time.

Moulton and Berner (1998) have measured the impact of modern vegetation cover on this silicate weathering process by comparing the release of calcium and magnesium ions from a vegetated and unvegetated area of the same igneous rock outcrop. Results from these and other experiments indicate that the rate of weathering release of calcium and magnesium to streams and to vegetation is two to five times the rate of release on 'barren' rock surface uncolonized by vascular plants. This is one of the most direct and telling indications of the impact that silicate weathering must have had on the drawdown of carbon dioxide ensuing from the land plant colonization of the Earth between 400 and 350 Ma.

In summary, the expanding land flora of the Devonian and Carboniferous led to a primary terrestrial production rate, driven by land plant photosynthesis, previously unmatched in Earth history. This was accompanied by a greatly accelerated silicate weathering drawdown of carbon dioxide. The development of extensive areas of coal swamps in Carboniferous time meant that much of this biomass production by land vegetation came to be buried as peat, rather than recycled back into the atmosphere.

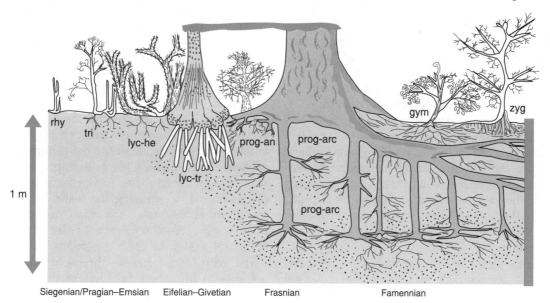

Figure 5.3 Evolution of the roots of terrestrial vascular plants; these are shown as a montage of different plants selected from examples from early (Siegenian/Pragian) to late Devonian (Famennian). The increasing depth of rooting through the course of the Devonian accelerated the access of organic matter and carbon dioxide to silicate minerals in the bedrock material. This in turn increased the drawdown of atmospheric carbon dioxide by silicate weathering. Reproduced with permission from Algeo and Scheckler (1998) *Phil. Trans. Roy. Soc. Lond.* B353, p. 116); for a key to the abbreviated plant names, see that source.

A further pathway of the carbon cycle, harder to quantify, must have been the increasing transport out to sea of terrestrial plant detritus, part of which would also ultimately have been buried, contributing to the long-term store of kerogen in marine sediments. All of these processes contrived to produce the very steep fall in atmospheric carbon dioxide shown in Figure 5.2, between the Silurian and end of the Carboniferous.

5.4 Atmospheric changes and evolutionary responses

So far we have merely noted some of the processes involved in the working of the carbon cycle through the course of Earth history, and how components of the cycle have changed as land plants have evolved. We consider some of the ways in which those changes have affected plants, and in particular, how plants have evolved to adapt to the changes.

We have already seen that stomata function as the route by which carbon dioxide is taken up by land plants. Huge numbers of stomata are present on the underside of an average leaf; their density is recorded in terms of the number of stomata per square millimetre running to several hundred. As you would expect, the number varies considerably between species, and even between those growing as members of the same community in similar environments. As the evidence of rising atmospheric carbon dioxide came to the fore with concerns over its possible association with enhanced greenhouse warming, Woodward (1987) embarked on an investigation to test whether the rising carbon dioxide had any effect on stomatal density.

He selected several native British trees, and using pressed (dried) specimens collected between 50 and 200 years ago, he measured their stomatal density. He then measured the stomatal density of the same species of living plants selected as being as close as possible to the site of the original pressed specimens. Because the stomatal density is significantly different between species, he normalized his results by treating the present-day density as 100% for each species. Dividing the stomatal density of the old material by the present-day value, he obtained a percentage figure which is comparable between the several species.

While the data was somewhat scattered, he found that going back in time the stomatal density rose to something in the order of 140% of the present value for each of the species studied. It appeared that this was the response of the trees involved to the rising carbon dioxide levels over this period (as documented in the ice core record, Figure 5.4). To confirm this finding, Woodward grew a selection of the same species under reduced carbon dioxide atmospheres for some weeks, observing that the leaves thus produced showed the same trend of raised stomatal densities.

It appears, then, that leaves have responded to the rising carbon dioxide of the last 100 years by reducing the number of stomata per unit area. In simplistic terms, the plants are taking advantage of the greater availability of carbon dioxide to reduce the number of stomata, thus enabling them to maintain the same level of carbon uptake, but with a greater economy in water through the reduced transpirational loss via the stomata.

Plant physiologists interested in the water economy of plants look at the 'water use efficiency' of the photosynthetic process. This is expressed as the number of carbon dioxide molecules taken in divided by the number of water molecules lost in the process. The rising ambient carbon dioxide of the last century has, in these terms, enabled plants to improve their water use efficiency. They fix the same amount of carbon, while reducing the water loss that inevitably accompanies the open stomata. In other words, they can tolerate a reduced availability of water in their environment, compared with their tolerance threshold under a lower level of carbon dioxide. For these trees at least, there appears to be some compensation in an elevated level of carbon dioxide in improving their tolerance of water stress.

Woodward's experiment suggested the possibility that counting the stomata on the leaves of ancient fossil plants might tell us something about changes in carbon dioxide in the geo-

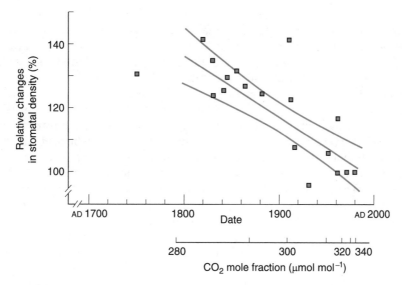

Figure 5.4 Changes in stomatal densities of herbarium specimens of eight British tree species over 200 years. The stomatal data are brought onto a common scale by expressing the stomatal density of all eight species as a percentage of their respective present-day values. Thus, over 200 years there has been a mean fall in these terms from 140% of the modern mean density at around 1800, to the 100% value of the present day. The contemporaneous values of atmospheric carbon dioxide derived from ice core studies are shown below. Reproduced with permission from Woodwards, F. I. (1987) Stromatal numbers are sensitive. *Nature* 327.

logical past. Van der Burgh *et al.* (1993) studied a sequence of leaves of *Quercus petraea* through the Miocene/Pliocene sequence in the Lower Rhine Embayment. Since the fossils are manifestly close to an extant species of *Quercus*, those authors were able to calibrate the palaeoatmospheric carbon dioxide level represented by stomatal indices in the fossils from those seen in leaves of this species collected between 1850 and the present, over which time the carbon dioxide has risen from 260 to 380 ppm.

They were able to show that their stomatal indices correlated with fluctuations in the carbon dioxide level which were consistent with temperature fluctuations derived from pollen data. In the following year, Beerling and Chaloner (1994) demonstrated that stomatal density measurements of Quaternary fossil *Salix herbacea* leaves pointed to a carbon dioxide excursion coinciding with the Late Glacial climatic oscillation, long recognized in many European sites. Wagner *et al.* (1999) used the stomatal density of a series of fossil *Salix* leaves to demonstrate a carbon dioxide fluc-

tuation coinciding with the Preboreal Oscillation in climate.

These late Tertiary and Quaternary observations demonstrated the potential of stomatal density in plant fossils for giving us a new source of data on atmospheric carbon dioxide over a much longer time span – indeed for the entire period since plant life colonized the land. At the very least, we should be able to see whether the changes in carbon dioxide levels shown in Berner's curve (Figure 5.2), derived from purely physical sources of evidence, agreed with the changes in stomatal density seen in fossil leaves.

Accordingly, Dr McElwain and I explored a series of fossil plants ranging in age from the Devonian to the Eocene (McElwain and Chaloner, 1996; Chaloner and McElwain, 1997; McElwain, 1998) measuring their stomatal densities. Unlike the material that Woodward had worked on, our plants were long extinct, so that we could not compare them with the same species growing under present-day carbon dioxide levels. Instead, we had to compare them with what we called the 'nearest living equivalent species', that is, plants which in our judgement were as closely related to the fossil as possible, and grew in comparable habitats.

Of course this means that whatever measure we derive from such observations will only give a broad guide to the trends of carbon dioxide change. Furthermore, only a limited number of plant fossils are preserved in such a way that a measure of their stomatal density can be made. But at least the fossils give a biological version of the story of carbon dioxide change with which to compare the curve derived solely from physical data.

There are in fact three main types of preservation in which such measurements are possible, and some of these are illustrated in the plate section, Plates 6 to 9. The most abundant type which yield stomatal data are so-called 'compression' fossils, where the leaf tissue has been buried in anaerobic mud so that biological decay was greatly retarded (Chaloner, 1999). In that setting the bulk of the leaf tissue has often become altered to a coal-like residue, sandwiched between the two cuticles, which had covered the leaf surface (Plate 8).

That coal-like material can be removed by suitable oxidation, leaving the two cuticles intact. The fossil cuticle normally shows the outlines of the epidermal cells and stomata as ridges on the inner face of the cuticle, and thus the stomatal guard cells can be clearly seen (Plate 9). Much rarer is the preservation of the plant tissue in an uncompressed state, infiltrated with mineral matter before any significant biodegradation has taken place.

This very unusual preservation occurred in the Lower Devonian silicified peat, which outcrops in Rhynie in Aberdeenshire, Scotland. There, a number of plants growing in a wetland habitat were inundated by boiling water from which silica was later precipitated, preserving the plant tissue in amazing detail. The stoma and adjoining epidermal cells of the plant (*Rhynia*) in Plate 6 from the Rhynie site were preserved in this way.

Finally, plant material may be preserved as charcoal, following forest fire ('wildfire'). The stomata on the fossil leaf shown in Plate 7 were preserved in this way; the leaves of this conifer were probably lying in the litter of fallen plant debris when a forest fire swept through the vegetation in Carboniferous time. Such fire then, as now, destroyed much of the biomass by combustion; but material lying in the litter, with restricted access to the air, may become charred. This means that its composition was modified by being raised to several hundred degrees centigrade with the exclusion of air preventing actual burning – a process called pyrolysis.

Such a residue of charred plant litter is readily recovered on the site of modern forest fires. The charred tissue is remarkably robust both physically and chemically, because unlike the original tissue it does not decompose readily. Its cellular detail is preserved, and small fossil organs may be readily transported by water and redeposited far from the site where they became charred.

We examined a limited number of Devonian, Carboniferous, Permian, Jurassic and Eocene plants to obtain stomatal densities, and then plotted the stomatal density expressed as a ratio of the nearest living equivalent (NLE) plant (see McElwain and Chaloner, 1996, Chaloner and McElwain, 1997 and McElwain, 1998). We divided the fossil stomatal density into that of the NLE to give a parameter that we call the stomatal ratio. If the fossil density is less than that of the NLE, then that ratio will be higher than 1, and corresponds to an elevated carbon dioxide level at the time of growth of the plant (Figure 5.5), compared with the present day. The methodology and interpretation of stomatal density measurement is discussed in Beerling (1999) and in Poole and Kürchner (1999).

For the early Devonian plants (*Rhynia*, *Sawdonia*) that we studied, we obtained stomatal ratios of between 5 and 6, while the Carboniferous and Permian conifers from which we were able to obtain data had values between 1 and 0.5. Taking the low stomatal ratios as indicating low carbon dioxide levels (following the trend of Woodward's observations) this corresponds with low carbon dioxide in the Carboniferous and Permian, and high levels in the early Devonian, which is conformable with Berner's curve (Figures 5.2 and 5.5).

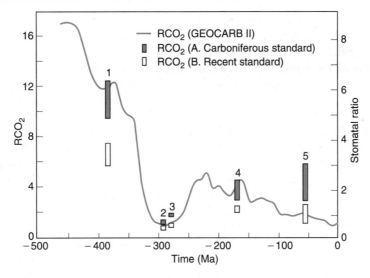

Figure 5.5 Estimates of palaeoatmospheric carbon dioxide levels derived from stomatal ratios (solid black bars) from five groups of fossil plants, superimposed on Berner's estimates (see Figure 5.2). The left-hand scale (RCO_2) is the ratio of the estimated level at any time to that of the present day; the right-hand scale (stomatal ratio) is the stomatal density of a given fossil to that of its 'nearest living equivalent'. For details of this treatment, and the identity of the fossil used, see McElwain (1998). The 'Carboniferous standard' establishing the equivalence of the stomatal ratio to the carbon dioxide scale is based on accepting Berner's value of the Carboniferous carbon dioxide level as being approximately as now, and so placing the Carboniferous fossil stomatal value (point 2) on Berner's curve. The 'Recent standard' is based on taking the stomatal ratio of 1 as equal to 1 RCO_2 unit. Reproduced with permission from McElwain (1998) *Phil. Trans. Roy. Soc. Lond.* B353, p. 93.

Other aspects of early vascular plant stomata are discussed in Edwards *et al.* (1998), who emphasize the importance of water stress in influencing stomatal density, and list the stomatal parameters for a wide range of early Devonian vascular plants. They record stomatal data for the Devonian *Drepanophycus spinaeformis*, a probable lycopod with microphyllous leaves. This species is noteworthy for its longevity from Lower to Upper Devonian, and they observe from the very limited data available that for this species 'there is a trend of increasing (stomatal) frequencies from Lower to Upper Devonian'. This is again consistent with Berner's postulated falling carbon dioxide over this interval.

The observations of stomatal density of extinct plants do not, of course, give us the means of reading an exact value for the palaeo-ambient carbon dioxide level directly. Unfortunately we have no direct basis for equating a stomatal ratio value with a carbon

dioxide value. However, we have attempted to derive that equivalence from two sources.

The first is based on placing the Carboniferous and Permian values for the stomatal ratio (Carboniferous *Swillingtonia*, Plate 7, of 0.58 and the Permian *Lebachia*, ratio value 0.79), as close as possible to the Berner carbon dioxide value at that time. This then set the two scales of one stomatal ratio unit being equivalent to two RCO_2 units. (RCO_2 of Berner (1998) is the ratio of the palaeo-carbon dioxide value to PAL, present atmospheric level, of 350 ppm.) Assuming that value of the equivalence places the error bar for the Devonian data onto the Berner curve (Figure 5.5, 'Carboniferous standard' values).

An alternative equivalence between stomatal ratios and carbon dioxide levels can be arrived at by assuming that a stomatal ratio of 1.0 (fossil and nearest living equivalent have equal stomatal density) should correspond with an RCO_2 value of one (i.e.

Carboniferous fossil and extant carbon dioxide levels, of approximately 350 ppm, were equal). This gives the significantly different equivalence of one stomatal ratio unit equal to one RCO_2 unit (see the 'Recent standard' data bars shown in Figure 5.5).

The five groups of stomatal data shown in Figure 5.5 clearly show the same trend as Berner's carbon dioxide curve. Their fit to that curve is understandably best when standardized by his Carboniferous carbon dioxide value, but even the lower values from the 'Recent standard' are of course showing the same trend of change. We cannot claim to have a precise 'carbon dioxide barometer' by treating the fossil stomatal ratio as a proxy measure, because all the fossil plant species that we are dealing with are now extinct. Further, the remoteness of these fossils from any living relative or 'nearest living equivalent' increases as we go back in time. Yet in spite of this proviso we can at least claim that these stomatal results are consistent with the trend of Berner's curve.

In terms of an evolutionary adaptation to their environment, the stomatal results do suggest that the response of land plants to changing carbon dioxide levels from the Devonian to the present day follows the trend seen over the short-term changes reported by Woodward. Plants have evolved in response to the atmospheric changes for which they themselves were largely responsible.

5.5 The rise of C4 plants and 'carbon dioxide starvation'

There are two distinct processes of photosynthetic carbon fixation represented among living plants. In the most widespread (and in evolutionary terms, the most ancient) form, the first recognizable product of the uptake of carbon dioxide is a C3 carbon compound, and on this basis it is referred to as the so-called C3 photosynthesis mechanism. In contrast, C4 plants employ a process in which the carbon dioxide is first taken up by a three-carbon receptor, so that a four-carbon molecule

results. This is then transported to cells in a special sheath around the vascular strand in the leaf, where the carbon is released again to be fixed by the more ancient C3 process.

Plants with this C4 strategy are seemingly better adapted to warm environments with high light intensity, this form of photosynthesis being particularly prevalent among tropical grasses. Certainly, the C4 mechanism is more common among plants in low latitude settings than at high latitudes. (For a brief but good account of these two strategies and references to relevant sources see Moore, 1994 and Cerling *et al.*, 1998.)

An important feature of the C4 pathway is that it is able to sustain photosynthesis at lower CO_2 concentrations than the C3 process. This, of course, raises the important question of how far changes in atmospheric carbon dioxide have changed the competitive ability of C3 versus C4 plants in the same habitats for as long as both have existed through the course of geological time.

The present 'natural' level of atmospheric carbon dioxide in the current interglacial (and from the ice core results, the earlier interglacials) represents a state of 'carbon dioxide starvation' for C3 plants in the warmer parts of the world. It is clear, at least, that these conditions favour C4 grasses at the expense of C3 grasses in low latitudes (Cerling *et al.*, 1998).

Because of differences in the fractionation of carbon isotopes in the two carbon fixation pathways, the stable isotopic composition of the carbon of C3 biomass is easily distinguished from that of C4 biomass. A particularly fruitful way of assessing changes in the contribution of the two processes to global biomass has been the study of enamel in fossil mammalian teeth, which appear to protect the isotopic signal from diagenetic change in the course of fossilization.

Although the exact time of origin of C4 photosynthesis cannot be securely established, the isotopic signal suggests that by the end of the Miocene there was a marked expansion of global C4 biomass, perhaps triggered by the atmospheric carbon dioxide falling below about 500 ppm (see Figure 5.2 and Cerling *et al.*, 1998). This appears to be linked with the increased aridity of the climate in tropical areas, and so the spread of grasslands, which in turn correlates with major changes in the nature of the contemporaneous faunas.

It is not feasible in this context to explore further these interactions of atmosphere, climate, flora and fauna which are discussed fully in Cerling *et al.* (1998). However, in terms of plant/environment interactions alone it appears that the falling carbon dioxide level through the course of Tertiary time has favoured the selection of C4 versus C3 plants, at least in lower latitudes. The ultimate cause of that carbon dioxide fall is certainly not due solely to terrestrial plant carbon fixation, since accelerated silicate weathering associated with mountain uplift is an important contributory factor (see, e.g., Francis and Dise, 1997). But the interplay of C3 versus C4 photosynthesis appears to be a clear-cut instance of a global atmospheric change in which plants are causally involved, having played a role in selecting one group of plants at the expense of another.

5.6 The evolution of the laminate leaf

A further aspect of the impact of changing carbon dioxide levels on land plants relates to the shape of the photosynthetic structures that they have evolved. Apart from some plants of very arid environments, the great majority of living land plants have relatively thin, flat 'laminate' leaves as their main sites of photosynthesis.

These leaves are normally set more or less horizontally to maximize their potential for receiving incident light. Their thin texture keeps the diffusion pathway for the carbon dioxide from the stomata to the site of photosynthesis in the upper part of the leaf tissue to a minimum. Such thin, laminate leaves in a vast diversity of sizes and shapes characterize most land plants from the Carboniferous to the present day.

However, all the early Devonian plants had 'naked' photosynthetic axes – unclothed, that is, by leaves! The early Devonian plants such as *Cooksonia*, *Rhynia*, *Aglaophyton* (see Plate 6 and Figure 5.6) and most of their contempor-

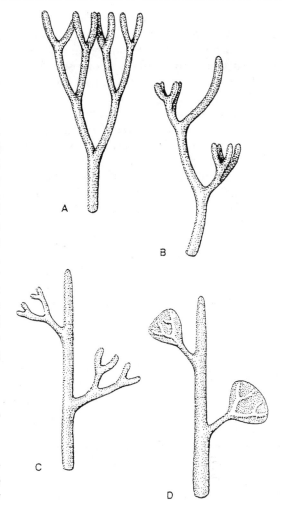

Figure 5.6 The origin of the megaphyll leaf, as visualized by Zimmermann (1959). This interprets the origin of most modern planated (thin, laminate) leaves as being derived from the modification of a dichotomously branched cylindrical photosynthetic axial system represented by the early Devonian plant *Rhynia* (A). Evolutionary steps involved overtopping of lateral branches (B), their becoming flattened (planated, C) and eventually 'webbed' (D). This produced the typical 'megaphyll leaves' of most modern plants, each with its own branching vein system. Reproduced with permission from Gifford, E.M. and Foster, A.S. (1988). *Morphology and Evolution of Vascular Plants*, W.H. Freeman.

aries had simple cylindrical upright stems – technically 'axes', since there is no stem/leaf differentiation. Visually they must have resembled a bunch of photosynthetic pencils.

Leaves with a branching vein system (as seen in modern plants) do not appear until the end of the Middle Devonian, and do not become a significant component of fossil floras until the Late Devonian. This is roughly some 25 million years after the appearance of the earliest land plants with their photosynthetic axes (see treatment of *Archaeopteris* and allied forms in Gensel and Andrews, 1984; Meyer-Berthaud *et al.*, 1999).

The conventional view of this evolution of the 'megaphyll leaf' is that progressive modification of a repeatedly forking photosynthetic axis produced a lateral system, which was overtopped by the growth of what was destined to become 'the stem'. This lateral system of forking photosynthetic cylindrical structures then became flattened, and the gaps between the branches 'webbed in' (like the feet of a duck) to produce a typical thin, flattened planar structure (Figure 5.6). This evolutionary sequence, first advocated by Zimmermann in his 'telome theory' (Zimmermann, 1959) is well explained in Gifford and Foster (1989).

Just why it took 25 million years for this rather simple adaptation to take place is rarely discussed. However, the steep fall of carbon dioxide in the course of the Devonian may well be a significant part of the story. While the early Devonian carbon dioxide was of the order of ten times its present level (Figure 5.2) a simple green cylinder, with rather sparse stomata on its outer surface, may have been an effective photosynthetic organ. Given a high carbon dioxide concentration, the low surface-to-volume ratio and the low stomatal density of *Rhynia* and related plants (Figs 5.6A, 5.7) evidently represented optimal adaptation to that environment.

For those early vascular plants with a slender vascular supply, water economy (achieved with low stomatal density) may have been more important than high bioproductivity. The falling carbon dioxide in the latter part of the Devonian may then have been the selective pressure which drove them to make photosynthetic structures able to function efficiently in that environment. For it is not until the very end of the Devonian that we find plants like *Archaeopteris* making sizeable (1 cm^2) leaves

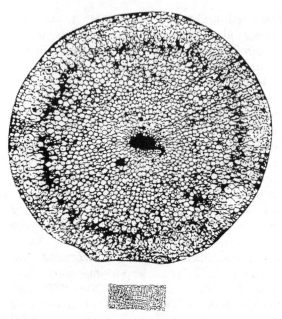

Figure 5.7 A comparison of the photosynthetic tissue of a transverse section of the upright axis of an early Devonian vascular plant, *Rhynia*, with a modern thin laminate leaf of *Syringa*, at the same scale. Intercelluar spaces as a pathway for carbon dioxide diffusion extend into the *Rhynia* axis about four times the depth of the diffusion pathway between the stomata on the lower surface of the *Syringa* leaf, and the main photosynthetic tissue in the upper part of the leaf. In the high carbon dioxide levels of the Early Devonian, cylindrical photosynthetic axes such as *Rhynia* effected adequate carbon fixation. The selective pressure to produce the more photosynthetically efficient thin megaphyll leaf, with its greater surface-to-volume ratio, did not come into play until the late Devonian. By that time the atmospheric carbon dioxide, driven largely by terrestrial photosynthesis, fell to much lower levels and such leaves became widely adopted.

with a branching venation system, as a dominant component of the flora.

An important factor in the evolution of the planated megaphyll leaf, which appears to have been critical, is the role of transpiration in cooling such leaves, when exposed to high levels of insolation. Modelling of the photosynthetic efficiency and leaf temperature of Devonian plants throws a new light on the factors which must have influenced changes in both stomatal density and leaf form, as the atmospheric carbon dioxide level fell by about

90% of its early Devonian value (Beerling *et al.*, 2001a). The low stomatal density of early Devonian upright 'stem plants' (*Rhynia* among others) allowed adequate carbon dioxide uptake in the high CO_2 atmosphere, but even when all the stomata were fully open the low stomatal density would have given quite a low transpiration rate. Beerling *et al.* show that if planated leaves with the same low stomatal density had developed in the early Devonian they would have reached lethal temperature at low palaeo-latitude and near-lethal temperature at high palaeo-latitudes. This would have been the result of their intercepting solar energy more effectively than the ancestral upright cylindrical form. Megaphyll planated leaves only became an effective improvement on the photosynthetic axes when the CO_2 level had reached the point that high stomatal densities became essential for adequate carbon uptake. The planated leaf then had the means of effecting a high level of transpirational cooling, as an incidental side-effect of the high stomatal density which had been driven by the falling CO_2. This was eventually to make the megaphyll leaf the optimal form of photosynthetic organ in all but the driest habitats. Some have questioned the relevance of transpirational cooling to the story of leaf evolution, suggesting that plants avoid water loss at all costs, and close their stomata under conditions of high insolation (Tanner, 2001). Counter to this line of thought, it can be shown that stomatal densities are generally highest in tropical plants and lowest in those of high latitudes (Beerling *et al.*, 2001b) suggesting a causal relationship between high stomatal density and exposure to high insolation.

If the interpretation of the origin of the laminate megaphyll leaf suggested by the modelling of Beerling *et al.* (2001a) is correct, it means that in this instance too, land plants were being driven to make an evolutionary adaptation to changes in their environment which they themselves had produced. In the evolution of the leaf as in the increasing density of stomata, plants were adapting to the lowered level of carbon dioxide which they had brought about by their own ecological success in scouring the atmosphere of their only source of carbon.

5.7 Wildfire and oxygen

We noted at the start of this chapter that the oxygenation of the Earth's atmosphere is attributed to the photosynthetic activity initially of bacteria and subsequently of plants. As we know from the present-day carbon cycle, photosynthesis by living plants, followed by their eventual consumption by herbivores or their death and decay, involves fixed carbon returning to the atmosphere as carbon dioxide. Photosynthesis simply cycles the carbon through the biosphere and back to the atmosphere.

However, when any of that 'fixed' carbon (organic material) becomes buried in ocean sediments, for example as coal, and thus is 'taken out of circulation', the oxygen that was generated in the photosynthesis remains in the atmosphere. Therefore, we see the process of 'carbon burial' as being an important step in the oxygenation of the atmosphere (Berner and Canfield, 1989). Similarly, carbon burial has had an important role in the ups and downs of atmospheric carbon dioxide through geological time. Estimating its extent is an essential element in the kind of carbon cycle modelling that contributes to Berner's carbon dioxide curve (Figure 5.2).

Once plant life was well established on the land, the potential ignition of dead and living plant material by lightning strike to produce wildfire (forest fire, grass fire) became a significant feature of terrestrial ecology. We have a record of charcoal (more or less equivalent to the fusain of coal petrologists) from the Devonian to the present day (Cope and Chaloner, 1985; Robinson *et al.*, 1997).

In terms of atmospheric history alone this record of wildfire is important. From experimental work on the conditions under which wildfire can be ignited by lightening, it appears that there is both an upper and lower limit to the oxygen concentration under which fire can be started. This 'fire window' lies between 13 and 30% oxygen, as compared with the present level of 21% (Jones and Chaloner, 1991).

In other words, the evidence of fossil charcoal suggests that the oxygen content of the atmosphere must have lain within these limits since Devonian time. Berner and Canfield (1989) have modelled changing oxygen levels through the Phanerozoic and these broadly show a converse pattern to the carbon dioxide levels shown in Figure 5.2.

Insofar as burial of carbon (and its release though weathering of buried coal and volcanic return of subducted carbon-bearing sediments) controls oxygen levels, this is as would be expected. It may be that fire and the vegetation that fuels it (and of course the carbon dioxide/oxygen exchange that it effects) may be one of the feedback mechanisms which have regulated the Earth's (relatively steady) oxygen level through the Phanerozoic (Lovelock, 1995; Francis and Dise, 1997).

It is interesting to speculate on when in Earth history the oxygen first reached an adequate level for fire to occur, and the very first ignition of plant biomass took place. This 'first wildfire' must no doubt have long preceded the first record of fossil charcoal, and indeed may well have occurred before plants colonized the land. It might, for example, have been a fire ignited in a pile of dead seaweed, piled up on a lee shore in Precambrian time!

As remarked at the start of this chapter, the process of combustion has much the same role as respiration in the process of carbon cycling – it converts fixed carbon in the form of plant biomass back into carbon dioxide. The reaction is very rapid in the case of fire, and the energy produced is dissipated entirely as heat.

In this it obviously differs drastically from the process of respiratory oxidation of biomass in the body of a herbivore, where much of the energy released may be used in muscular contraction (movement) or in other metabolic processes. Some also, of course, may be released as heat to maintain body temperature of the herbivore.

The ecological effects of fire are obviously of immense importance. Many plants have adaptations to a 'fire ecology', such as releasing their seeds only after exposure to the high temperature generated by forest fire. This enables them to withhold their seeds until the open habitat created by fire becomes available to them.

These and the many other aspects of fire ecology are discussed in Gill *et al.* (1981), and further consideration of them is beyond the scope of this chapter. But what is relevant here is the fact that fire itself only became possible as plants produced flammable biomass and an oxygen level which could sustain combustion. In this sense, we see once more that plant life, by modifying its environment, created a hazard (fire) which is of its own making. Again, we see how plants (in their adaptation to fire-prone communities) have evolved to respond to changes that they themselves have produced.

References

Algeo, T.J. and Scheckler, S.E. (1998). Terrestrial-marine teleconnections in the Devonian: links between the evolution of land plants, weathering processes and marine anoxic events. *Philosophical Transactions of the Royal Society, London*, B 353: 113–130

Bateman, R.M., Crane, P.R., DiMichele, W.A. *et al.* (1998). Early evolution of land plants: phylogeny, physiology and ecology of the primary terrestrial radiation. *Annual Review of Ecology and Systematics*, 29: 263–292.

Beerling, D.J. (1999). Stomatal density and index: theory and application. In: T.P. Jones and N.P. Rowe (eds) *Fossil Plants and Spores: Modern Techniques*. Geological Society, London, pp. 251–265.

Beerling, D.J. and Chaloner, W.G. (1994). Atmospheric CO_2 changes since the last glacial maximum: evidence from the stomatal density of fossil leaves. *Review of Palaeobotany and Palynology*, 81: 11–17.

Beerling, D.J., Osborne, C.P. and Chaloner, W.G. (2001a). Evolution of leaf-form in land plants linked to atmospheric CO_2 decline in the Late Palaeozoic era. *Nature*, 410: 352–254.

Beerling, D.J., Osborne, C.P. and Chaloner, W.G. (2001b). Do drought-hardened plants suffer from fever ... a response. *Trends in Plant Science*, 6: 507–508).

Berner, R.A. (1998). The carbon cycle and CO_2 over Phanerozoic time: the role of land plants. *Philosophical Transactions of the Royal Society, London*, B 353: 75–82.

Berner, R.A. and Canfield, D.E. (1989). A new model for atmospheric oxygen over Phanerozoic time. *American Journal of Science*, 289: 333–361.

Bradley, R.S. (1985). *Quaternary Palaeoclimatology*. London: Allen & Unwin, 472pp.

Cerling, T.E., Ehleringer, J.R. and Harris, J.M. (1998). Carbon dioxide starvation, the development of C4 ecosystems, and mammalian evolution. *Philosophical Transactions of the Royal Society*, **B** 353: 159–171.

Chaloner, W.G. (1999). Plant and spore compression in sediments. In: T.P. Jones and N.P. Rowe (eds) *Fossil Plants and Spores: Modern Techniques*. Geological Society, London, pp. 36–40.

Chaloner, W.G. and McElwain, J.C. (1997). The fossil plant record and global climatic change. *Review of Palaeobotany and Palynology*, **95**: 73–82.

Chaloner, W.G. and Meyen, S.V. (1973). Upper Palaeozoic floras of northern continents. In: A. Hallam (ed.) *Atlas of Palaeobiogeography*. Amsterdam: Elsevier, pp. 169–186.

Cope, M.J. and Chaloner, W.G. (1985). Wildfire: an interaction of biological and physical processes. In: B.H. Tiffney (ed.) *Geological Factors and the Evolution of Plants*. Yale University Press, pp. 257–277.

Edwards, D., Kerp, H. and Hass, H. (1998). Stomata in early land plants: an anatomical and ecophysiological approach. *Journal of Experimental Botany*, **49**: 255–278.

Francis, P. and Dise, N. (1997). *Atmosphere, Earth and Life*. Milton Keynes, UK: The Open University.

Gensel, P.G. and Andrews, H.N. (1984). *Plant Life in the Devonian*. New York: Praeger.

Gifford, E.M. and Foster, A.S. (1989). *Morphology and Evolution of Vascular Plants*. New York: Freeman.

Gill, A.M., Groves, R.H. and Noble, I.R. (eds) (1981). *Fire and the Australian Biota*. Canberra: Australian Academy of Science.

Gould, S.J. (1989). *Wonderful Life*. New York: W.W. Norton.

Jones, T.P. and Chaloner, W.G. (1991). Fossil charcoal, its recognition and palaeoatmospheric significance. *Palaeogeography, Palaeoclimatology, Palaeoecology*, **97**: 30–50.

Kasting, J.F. (1987). Theoretical constraints on oxygen and carbon dioxide concentrations in the precambrian atmosphere. *Precambrian Research*, **34**: 205–229.

Kasting, J.F. (1992). Proterozoic climates: the effect of changing carbon dioxide concentrations. In: J.W. Schopf and C. Klein (eds) *The Proterozoic Biosphere, a Multidisciplinary Study*. Cambridge University Press.

Kenrick, P. and Crane, P.R. (1997a). The origin and early evolution of plants on land. *Nature*, **389**: 33–39.

Kenrick, P. and Crane, P.R. (1997b). *The Origin and Early Diversification of Land Plants*. Washington: Smithsonian Institution Press.

Lovelock, J. (1995). *Gaia – A New Look at Life on Earth*. Oxford: Oxford University Press.

McElwain, J.C. (1998). Do fossil plants signal palaeoatmospheric CO_2 concentration in the geological past? *Philosophical Transactions of the Royal Society*, **B 353**: 83–96.

McElwain, J.C. and Chaloner, W.G. (1996). The fossil cuticle as a skeletal record of environmental change. *Palaios*, **11**: 376–388.

Meyer-Berthaud, B., Scheckler, S.E. and Wendt, J. (1999). *Archaeopteris* is the earliest known modern tree. *Nature*, **398**: 700–701.

Moore, P.D. (1994). High hopes for C4 plants. *Nature*, **367**: 322–323.

Moulton, K.L. and Berner R.A. (1998). Quantification of the effect of plants on weathering: studies in Iceland. *Geology*, **26**: 895–898.

Niklas, K.J. (1997). *The Evolutionary Biology of Plants*. University of Chicago Press.

Poole, I. and Kürschner, W.M. (1999). Stomatal density and index: the practice. In: T.P. Jones and N.P. Rowe (eds) *Fossil Plants and Spores: Modern Techniques*. Geological Society, London, pp. 257–260.

Robinson, J.M., Chaloner, W.G. and Jones, T.P. (1997). Pre-Quaternary records of wildfire. In J.S. Clark *et al.* (eds) *Sediment Records of Biomass Burning and Global Change*. Berlin: Springer, *NATO ASI Series I Global Environmental Change*, **51**: 253–270.

Schopf, J.W. and Klein, C. (1992). *The Proterozoic Biosphere, A Multidisciplinary Study*. Cambridge University Press.

Tanner, W. (2001). Do drought-hardened plants suffer from fever? *Trends in Plant Science*, **6**: 507.

Taylor, T.N. and Taylor, E.L. (1993). *The Biology and Evolution of Fossil Plants*. New Jersey: Prentice Hall.

Van Der Burgh, J., Visscher, H., Dilcher, D.L. and Kurschner, W.M. (1993). Palaeoatmospheric signatures in Neogene fossil leaves. *Science*, **260**: 1788–1790.

Wagner, F., Bohncke, S.J.P., Dilcher, D.L. *et al.* (1999). Century-scale shifts in early Holocene atmospheric CO_2 concentrations. *Science*, **284**: 1971–1973.

Woodward, F.I. (1987). Stomatal numbers are sensitive to increases in CO_2 from pre-industrial levels. *Nature*, **327**: 617–618.

Zimmermann, W. (1959). *Die Phylogenie der Pflanzen*. Stuttgart: Gustav Fischer Verlag, 2nd edn.

PART 2
Radiation

6

The sun: the impetus of life

Lynn J. Rothschild

ABSTRACT

See Plates 10-12
The Sun is vital for life on Earth, its physiology, ecology and evolution. Its energy influences climate, and is used as the ultimate source of energy for nearly all life. The direct influence of the Sun on individual organisms and metabolic processes depends on the spectral nature of the radiation used, its total flux and the timing of exposure. All of these parameters have changed over geological time, and radiation received by any given organism also varies daily, seasonally and annually. In this chapter, the evolutionary impact of solar radiation is examined from the perspective of different spectral ranges, including UV radiation, visible radiation, and infrared radiation.

The Sun is instrumental in prebiotic chemistry, the physical, chemical and then biogeochemistry of the Earth. It is been both a vital resource during the course of evolution, as well as a hazardous substance. Some day we must move beyond the immediate influence of our star or be consumed by it, but we will always bear the unmistakable imprint of the Sun.

6.1 Introduction

The Sun is the most critical extra-terrestrial body for terrestrial evolution for without it there would be no solar system, no planets, and no life. The Sun has played a vital role as a source of energy for prebiotic and biosynthetic processes, and in the maintenance of habitable temperatures. Through gravitational interactions of the Earth with the Sun and the Moon, the rotational period of the Earth has slowed, and tides are created twice daily. The tidal cycle is important to coastal species, and in fact may have been the cradle of life on early Earth.

The best-known role of the Sun in terrestrial biology is as a source of radiation. Yet, solar radiation is a double-edged sword. Earth's biota relies almost exclusively on solar radiation to keep temperatures clement and to power photosynthesis. 'Visible' radiation is critical to vision, photosynthesis, and as a metabolic signal. The Earth is also exposed to solar radiation in longer and shorter wavelengths, the infrared and ultraviolet (UV) regions respectively. Such radiation can be beneficial to life (e.g. for infrared: heat, bacterial photosynthesis, plant morphogenesis, and for UV: vitamin D synthesis, photorepair, circadian rhythms).

Yet, UV radiation is more often damaging to organisms because both proteins and nucleic acids have a maximum absorption in wavelengths in the UV region. The solar radiation flux that organisms are exposed to changes diurnally, seasonally, during its life cycle, and most likely during geological time. For these reasons, solar radiation has provided an evolutionary opportunity as well as a challenge to life on Earth. In this chapter I explore the pervasive role of the Sun on biotic evolution, with an emphasis on the role of solar radiation.

Evolution on Planet Earth
ISBN 0-12-598655-6

6.2 The Sun: vital facts

The Sun is a 4.6×10^9-year-old main sequence star, worshipped by the Greeks as the young god Helios, and by the Romans as the god Sol. The Sun was born by the accretion of the protostellar core, covered by a cloud of dust (Canuto *et al.*, 1983). The protostellar Sun would have emitted primarily infrared radiation. About 10^5 years after formation, the Sun entered a stage of protostellar evolution characterized by T-Tauri stars. By this time the Sun was visible, its photosphere and chromosphere were present, but there were relatively little dust and gas. About 12 million years (Myr) after formation, it entered the radiative phase.

Currently the Sun is a main sequence star – a phase it entered 50–100 Myr after formation. The structure of the present-day Sun is illustrated in Figure 6.1. At the center is the core – the place where nuclear reactions consume hydrogen to form helium which results in the release of energy that eventually leaves the Sun as light.

The temperature at the very center of the Sun is about 15 000 000°C and the density is about 150 g cm^{-3} (about ten times the density of gold). The temperature and density decrease further from the center of the core to its outer edge, 25% of the distance to the surface. From the edge of the core to 70% of the way to the surface is the radiative zone, a zone containing energy in the form of radiation (photons). From the inner to outer parts of the radiative zone the temperature drops from 7 000 000 to 2 000 000°C, while the density drops from 20 g cm^{-3} (about the density of gold) down to 0.2 g cm^{-3} (less than the density of water).

The interface layer is a thin layer between the radiative and convection zones. It is now thought that the Sun's magnetic field is generated by a magnetic dynamo in this layer. The convection zone is the outer-most layer, which extends from a depth of about 200 000 km up to the visible surface. At the base of the convection zone the temperature is about 2 000 000°C, a temperature low enough for the heavier ions (e.g. carbon, nitrogen, oxygen, calcium, and iron) to retain some of their electrons. This makes the material more opaque so that it is harder for radiation to get through, and traps heat that ultimately makes the fluid unstable so that it starts to 'boil' or convect. These convective motions carry heat quite rapidly to the surface. The fluid expands and cools as it rises, so

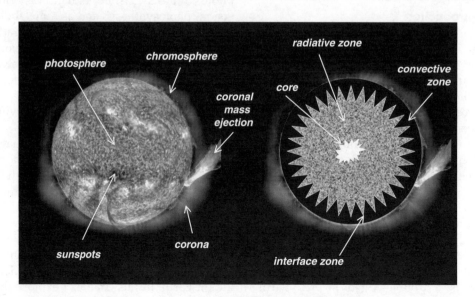

Figure 6.1 The structure of the Sun. The view on the left highlights visible portions of the Sun, while the panel on the right represents a cut-away view of the Sun to illustrate the internal regions.

that at the visible surface the temperature has dropped to 5700°C and the density is only $2 \times 10^{-7}\,\mathrm{g\,cm^{-3}}$ (1/10 000th the density of air at sea level).

The photosphere is the visible portion of the Sun and represents a layer about 100 km in depth. The chromosphere is the irregular layer above the photosphere. In it the temperature rises from about 6000°C at its base to 20 000°C in the outer regions. The transition region of the Sun's atmosphere separates the chromosphere from the much hotter corona which is the Sun's outer atmosphere. Coronal features include helmet streamers (large cap-like structures with long, pointed peaks overlying sunspots and active areas, and formed between areas of the solar wind emission), polar plumes (also formed by the action of the solar wind, but in association with 'open' magnetic field lines at the Sun's poles), coronal loops (found around sunspots and associated with closed magnetic field lines that connect magnetic regions on the solar surface), and coronal holes (regions where the corona is dark and thought to be the source of solar wind). Coronal mass ejections are huge (up to $10^{13}\,\mathrm{kg}$) masses of gas ejected from the Sun over the course of several hours. 'Halo events' are coronal mass ejections aimed at the Earth. (For further information see http://science.msfc.nasa.gov/ssl/pad/solar/.)

The Sun has used up about half of its hydrogen, and will thus burn 'peacefully' for another 5 billion years (Table 6.1). Its mass is $1.989 \times 10^{30}\,\mathrm{kg}$ (more than 99.8% of the mass of our solar system). The temperature of the Sun varies depending on location from approximately 15 600 000 K (core) to 5800 K (surface) to 3800 K (sunspots). Its current composition is 92.1% H and 7.8% He by number of atoms; the remaining 0.1% consists of elements with mass greater than helium. The core of the Sun produces 3.86×10^{33} ergs (3.86×10^{26} joules) of energy per second as gamma rays.

As the energy travels outwards it is absorbed and re-emitted at increasingly lower temperatures, so that by the time it reaches the surface, hundreds of thousands of years later, it is primarily visible light and infrared radiation (heat). The Sun's major exports are radiation in the ultraviolet, visible and infrared regions of the electromagnetic spectrum (Figures 6.2 and 6.3). At the distance of the Earth (1 AU), total solar irradiance is $1353\,\mathrm{W\,m^{-2}}$ or $19.4\,\mathrm{k\,cal\,min^{-1}\,m^{-2}}$ the solar constant (Thekaekara, 1970). Referring to Figure 6.3, the solar spectrum is the spectrum at 1 AU. The standard solar constant at zero air mass (outside of the Earth's atmosphere) is from ASTM Designation: E 490 (September 1999 draft revision), which explains how the data were obtained. The Upper Atmosphere Research Satellite (UARS)/ATLAS-2 spectrum is used between 0.1195 and >0.3795 μm. The values are averages of two different instruments, the Solar Ultraviolet Spectral Irradiance Monitor (SUSIM) and the Solar Stellar Irradiance Comparison Experiment (SOLSTICE). These data were obtained in April 1993 during a period of moderate solar activity. Thuillier's

Table 6.1	**The Sun: vital facts.**
Age	4.5×10^9 years old
Mass	1.989×10^{30} kg
Diameter	1 390 000 km
Temperature	5800 K (surface)
Composition (mole per cent)	92.1% H and 7.8% He, 0.1% metals
Energy output	3.86×10^{26} joules of energy per second
Products exported to Earth	Charged particles (solar wind, solar flare), radiation, X-rays

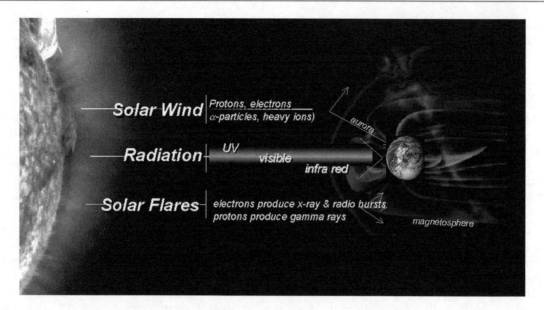

Figure 6.2 Exports of the Sun. The major exports of the Sun are radiation, solar wind and solar flares.

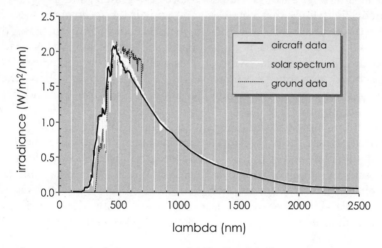

Figure 6.3 The solar spectrum is the spectrum at 1 AU – for details see text.

spectrum is used above 0.3797 µm. In the 0.41 to 0.825 µm range, the values are from the McMath Solar Telescope at Kitt Peak, Arizona. In the 0.825 to 4.0 µm range, the values are from the high-resolution solar atlas. The composite spectral irradiance data were then scaled by a factor of 0.99745 to force the integrated total irradiance equal to the solar constant. Aircraft data match closely to the standard solar spectrum. These data were obtained in August 1967 from NASA air-

craft measurements at 38 000 feet, atmospheric pressure = 208 mb (NASA TR R-351, 1970). The surface data were obtained from the NASA Ames Research Center Plant Laboratory and were measured at NASA Ames Research Center on 1 July 1996 at solar noon by the Optronic Laboratories, Inc. OL-752 (High Accuracy UV-Visible) Spectroradiometer. Note that this surface data is higher than the extraterrestrial in some portions of the spectrum. This is due to the fact that the measurements were made closer to the summer solstice.

In addition to radiation, the Sun exports solar wind, a low density stream of charged particles (90–95% protons, 5–10% α-particles and a relatively small number of heavier ions) traveling through the solar system at about $450 \, \mathrm{km \, sec^{-1}}$. During relatively quiet periods the solar wind contains 1 to 10 protons per $\mathrm{cm^3}$ moving outward from the Sun at 350 to $700 \, \mathrm{km \, sec^{-1}}$ resulting in an ion flux of 10^8 to 10^9 ions per $\mathrm{cm^2 \, sec^{-1}}$, each ion having an energy equal to at least 15 electron volts. The NASA/ESA *Ulysses* mission (launched from the Space Shuttle *Discovery* on 6 October 1990: http://science.msfc.nasa.gov/ssl/pad/solar/ulysses.htm) has provided us with our current view of the solar wind.

Occasionally there are periods when the solar wind decreases to a fraction of its normal speed and density. The most recent, drastic, long-term decrease was observed on 10–12 May 1999. During that time, NASA's ACE (Advanced Composition Explorer) spacecraft (launched 25 August 1997 – http://www.srl.caltech.edu/ACE/ace_mission.html) and Wind (http://xsl.msfc.nasa.gov/NASA.Projects/Space.Science/Sun-Earth.Connection/Wind.Mission/.index.html) (the site is case sensitive). Spacecraft each observed a drop in the density of the solar wind of >98%.

ACE also noted a 99.9% drop in the density of helium in the solar wind but not heavier ions as they seemingly could not escape the Sun's gravity. As a result of the decrease in the solar wind, the Earth's magnetic field swelled to five to six times its normal size, and produced an unusually intense auroral display at the North Pole.

During periods of solar flares, solar wind increases substantially. Auroral displays (e.g. the northern lights) result from part of the solar wind being captured by the Earth's magnetic field and conducted downward toward the magnetic poles where the charged particles collide with oxygen and nitrogen atoms. Electrons are stripped from the oxygen and nitrogen, resulting in ions that emit visible radiation at characteristic red or greenish-blue wavelengths.

If the solar wind reached the surface of the Earth, it would seriously damage vital biological molecules. Fortunately, the Earth's magnetosphere deflects most of the low energy particles around the planet, including those in the solar wind, while high energy particles are absorbed by the atmosphere with mass shielding of $1000 \, \mathrm{g \, cm^{-2}}$.

But it is possible that during the periods of magnetic reversal which occur every 200 000 to 300 000 years, the low energy particles could penetrate the magnetosphere and reach the Earth. It seems likely that this would have a major influence on evolution, perhaps in the form of a mass extinction if the charged particles were damaging enough, or an evolutionary radiation if the solar wind increased mutation rates. To date, a correlation between extinction events and magnetic reversals has yet to be proven (Plotnick, 1980; Lutz, 1985, Raup, 1985).

Solar flares are a localized but intense eruption. This sudden, intense brightening of the Sun's chromosphere develops in a few minutes and may last a few hours. Initially most of their energy is in the form of electrons and protons. The electrons produce X-ray bursts and radio bursts, while the protons produce gamma rays. In about two days the charged particles reach the Earth and may cause electrical storms or auroras. Because these events are very short-lived, they are unlikely to have an effect on evolution on Earth, though such events pose a concern for astronauts and indeed evolution of life forms in the space environment (see, e.g., Chapter 7).

There are many unanswered questions concerning possible links between the Sun and terrestrial evolution, one such being: How does the variability of the Sun affect the Earth's climate? Sunspot activity is on a 9.5–11-year cycle, with a solar maximum (maximum of sunspot activity) occurring in 2000. Along with an increase in sunspots, there is also an increase in coronal mass ejections and high energy solar flares, as well as fluctuations in the Sun's magnetic field.

During these periods of intense solar activity, geomagnetic storms occur, which produce spectacular aurora borealis and aurora australis (northern and southern lights, respectively). Geomagnetic storms can also disrupt radio transmissions and affect power grids. While this may be of interest to *Homo sapiens 'technologica'*, it is unlikely to have an influence on evolution.

There was, however, a prolonged period of low sunspot activity from 1645 to 1715 called the Maunder Minimum (after E.W. Maunder who investigated the observational records of sunspot activity in the late seventeenth century) that coincided with the Little Ice Age. Could the Maunder Minimum have caused the Little Ice Age or was it just a coincidence? During that time, the Sun was fainter by $0.4 \pm 0.2\%$, and the mean annual temperature in the northern temperate zone was 1–2°C lower than normal. A reduction in irradiance of just 0.4%, within the possible range for that time, would have been enough to produce the Little Ice Age (Baliunas and Jastrow, 1993).

The Sun emits X-rays. The longer wavelength ('soft') X-rays produce the layers of the ionosphere that make short-wave radio communication possible, whereas the shorter wavelength ('hard') X-ray pulses from solar flares ionize the lowest ionospheric layer, resulting in radio fadeouts. Radio communication *per se* has not had an effect on evolution (although radio communication has undoubtedly influenced civilization).

Is the Sun really necessary for life? The Sun is essential because without it there would be no solar system with discrete planets. The Earth/Sun combination of atmosphere and radiation allows the Earth to be clement. Parenthetically, other stars are just as essential because all elements on Earth today, except for hydrogen, helium, and traces of lithium and beryllium, were produced either by the nuclear furnaces at the center of stars or in supernovae explosions in a process called nucleosynthesis. Thus, we and our world are primarily 'stardust'.

Suppose we rephrase the question: Would life be possible without solar radiation? Perhaps in the beginning it might have been. There are organisms in hydrothermal vents and the subsurface that seem to be completely divorced from processes on the Earth's surface

(e.g. Gold, 1999). The tectonic activity that creates hydrothermal vents comes from the internal heat of the Earth (geothermal heat). This heat is derived from (i) the radioactive decay of potassium, uranium and thorium in the Earth's interior, and (ii) residual heat from the Earth's accretion and bombardment (see Nisbet and Sleep, this volume, Chapter 1).

All other ecosystems (and possibly these as well) rely ultimately on solar energy. The Sun keeps the Earth from freezing over. Geothermal heat flux at the surface is $0.1\,W\,m^{-2}$ compared to a net absorbed solar flux of about $240\,W\,m^{-2}$ (Kump *et al.*, 1999). Until very recently, it was thought that the radiation from stars were the only way to keep a planet from freezing over in the long term, and thus proximity to a star was essential to define a planet as 'habitable'. Since the discovery that a moon of Jupiter, Europa, is likely to have a liquid water ocean, we now realize that tidal friction (in the case of Europa, friction between Europa and the other moons of Jupiter) can create enough energy to maintain liquid water and the possibility of life forms.

6.3   Radiation

The solar spectrum is made up of radiation extending from the extremely short wavelength gamma rays and X-rays to the much longer microwaves and radiowaves (Figure 6.4). However, not all these wavelengths reach the Earth's surface. The solar radiation that reaches the Earth's surface today is approximately 7% ultraviolet, 48% visible and 45% infrared. Thus, these are the regions of potential interest to biological evolution.

Solar radiation flux has changed through time, as has the Earth's atmospheric composition which attenuates radiation; thus, surface fluxes of solar radiation have changed. As the Sun has progressively burned its hydrogen to helium, solar luminosity has increased from about 71% of its present level since the formation of the Earth 4.6 Ga (10^9 years ago) reaching essentially modern levels by 600 Ma (10^6 years ago) (Gough, 1981; Kasting, 1987).

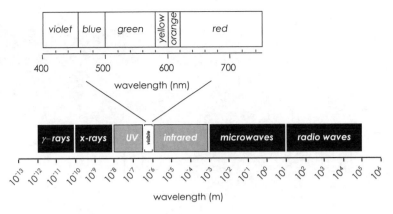

Figure 6.4 Components of the solar spectrum.

Superimposed on this increasing luminosity is a change in the spectral composition of solar radiation. Stars produce relatively more UV radiation when young. In the case of our young Sun during its pre-main sequence phase the UV contained up to 10^4 times more radiation below 210 nm than at present (Canuto *et al.*, 1982, 1983). Even when the Sun entered the main sequence about 4.4 Gyr ago, the far-UV (below 200 nm) luminosity was still 100 times its current value (Canuto *et al.*, 1983).

The obliquity of the Earth – the orientation of the spin axis relative to the orbital plane – has changed over time, which has changed the amount of solar radiation reaching a given spot on the Earth. Consequently, there have been changes in the zonation of insolation and the extremes of the seasons.

The obliquity of the Earth is influenced by the Moon, Sun and other planets in the solar system, and is relatively stable around a mean value of $23.3° \pm 1.3°$ (Laskar *et al.*, 1993). Even small changes in the Earth's obliquity cause major climatic shifts, such as those associated with the Milankovitch cycles.

The Earth's atmosphere is thought to have always contained enough carbon dioxide to attenuate radiation below 200 nm (peak absorption of $CO_2 = 190$ nm). Levels of atmospheric molecular oxygen (O_2) have risen from essentially zero when life arose to modern levels of almost 21% by the time of the Cambrian period 560 million years ago, mostly as a by-product of oxygenic photosynthesis.

Oxygen leads to the formation of ozone (O_3) which strongly attenuates wavelengths below 315 nm. This suggests that surficial UV fluxes were higher prior to the formation of an ozone 'shield', although Kasting *et al.* (1989) have suggested that an elemental sulfur vapor in the early anaerobic atmosphere could have provided some protection from UV radiation. Thus, over geologic time, the solar radiation in the range of 200 to 315 nm that has reached the Earth has probably declined dramatically.

In addition to changes over geologic time, the solar radiation flux varies diurnally, seasonally and with location, as well as randomly. At sunrise and sunset when the Sun's elevation is <10°, the visible part of the solar spectrum is enriched in blue (380 to 500 nm) and far-red (700 to 795 nm) light (see Salisbury and Ross, 1992: 566). Blue is enriched because much of the irradiation comes from the blue sky – blue due to atmospheric scattering which removes the shorter wavelengths. The far-red light enrichment results from the long path the light must travel through the atmosphere.

Solar radiation varies seasonally because of the rotation of the Earth (Figure 6.5). Surface flux varies with altitude and latitude (Figure 6.5), depth in the water column, water chemistry, and physics. At higher altitudes and lower latitudes the peak flux experienced is higher. However, at high altitudes the total daily flux may be even higher than that found at low latitudes during the summer when radiation is received for up to 24 h. Solar flux also has

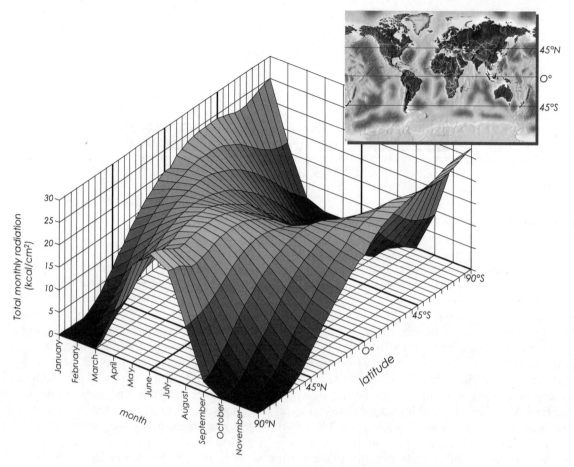

Figure 6.5   Latitudinal and seasonal variations in solar insolation. Data from the *CRC Handbook of Chemistry and Physics* (72nd edition. D.R. Lide, Editor-in-Chief, CRC Press, Boca Raton, 1991–1992).

an unpredictable element because it is affected by interfering factors such as clouds and atmospheric pollutants.

Just as the spectral composition of the radiation reaching the Earth has changed over geological time, so has day length. Rotational friction between the Earth and Moon, and secondarily between the Earth and Sun, has slowed the rotational period of the Earth (Darwin, 1880) from about 5 h when the Earth formed 4.5 Ga (billion years ago) to 15 h by 3.5 Ga, and 20 h by the end of the Precambrian at 0.6 Ga (Walker *et al.*, 1983). Without the Moon day length would now be substantially shorter – perhaps half of what it is today. Shorter day lengths would have profound implications for life, ranging from changes in the number of

continuous hours available for activities that required solar radiation (hunting, photosynthesis) to a decrease in the number of continuous hours of exposure to harmful UV radiation. This shorter day may have allowed the colonization of more areas with high fluxes of UV radiation because the longer dark period would have allowed repair of UV damage (e.g. to DNA) to take place in the absence of further damage.

6.4   The solar spectrum in evolution

The effect of solar radiation reaching the Earth depends on its spectral quality, with respect to its role in biological and pre-biological evolution (summarized in Table 6.2). This includes infrared, visible and ultraviolet radiation, all of which are types of non-ionizing radiation in the spectral range that reaches the Earth.

One of the mysteries of prebiotic evolution is why terrestrial organisms use laevorotatory (left-hand, L) amino acids and dextrorotatory (right-hand, D) sugars. Perhaps they represent a random 'choice' early in prebiotic chemistry that became hardwired into the biochemical machinery of life. An alternative explanation, originally suggested in 1874 by Le Bel, is that

circularly polarized light can produce an enantiomeric excess (Bonner *et al.*, 1999). The current suggestion is that dust grains within the molecular clouds where new solar systems are forming can generate circularly polarized light (Bailey *et al.*, 1998; Clark, 1999; Podlech, 1999). The ultraviolet radiation circularly polarized in this fashion could preferentially destroy one enantiomer over the other, thus influencing the molecular composition of life forms within the solar system.

This implies that any carbon-based life form within our solar system would be more likely to use L amino acids and D sugars, but that the carbon compounds within other solar systems might use a different suite of enantiomers depending on the enantiomeric products produced. In this way, the pre-solar nebula that ultimately formed the Sun would have had a profound affect on prebiotic chemistry even prior to the formation of the Sun or the Earth.

6.4.1 *Infrared radiation*

In 1800 William Herschel discovered that sunlight dispersed into its colors through a prism includes infrared radiation. Infrared radiation is the portion of the solar spectrum that elevates temperature but whose wavelengths are longer than visible light (750 nm), and therefore are not clearly visible to humans. Infrared radiation (in the wavelengths 2.5 to 25 μm) is strongly absorbed by water and therefore by the atmosphere. Perhaps for this reason it is not as important in evolution as the near infrared radiation (750 nm to 2.5 μm) or other parts of the terrestrial spectrum.

Arguably the most important influence of infrared radiation in evolution is its thermal content. On a planetary scale, infrared radiation from the Earth is reflected back to Earth by the 'greenhouse' gases in the atmosphere, thus heating the planet. On the Earth, the 'greenhouse' effect allows the presence of liquid water, a presumed prerequisite for habitability.

Heating of a surface leads to convection. As a result, there is convective movement within the atmosphere and bodies of water which affect life. For example, convection in bodies

Table 6.2 Summary of the effects of solar radiation on evolution.

Region	Impact on evolution
Ultraviolet (200 to 400 nm during early evolution; 290 to 400 nm today)	Prebiotic synthesis of organics
	Vision (insects)
	Vitamin D synthesis
	Photorepair of DNA damage
	Damage to biomolecules (e.g. DNA, protein), thus effects from inhibition of metabolic processes; can lead to death
	Driver for molecular evolution
	Driver for origin of sex
	Selective agent which may have influenced evolutionary steps from transition to land to human skin color
Visible (400-750 nm)	Vision, phototaxis
	Photosynthesis
	Plant cues (e.g. phytochromes and cryptochromes)
	Circadian rhythms in animals (cryptochromes)
Near Infrared (750 nm to 2.5 μm)	Heat
	Sensing of prey
	Light harvesting by bacteriochlorophylls

of water affects the distribution of organisms and nutrients.

In such a discussion of infrared radiation and evolution it should be noted that individual organisms also radiate infrared radiation. The snakes *Python molurus* and *Python regius* exploit this by using their pit organ to sense the infrared (8–12 μm) reflected from prey (Grace *et al.*, 1999).

Near infrared (700 to 900 nm) may also be used for cryptic coloration, such as by several glass frogs (Schwalm *et al.*, 1977). These neotropical frogs sit on leaves which are green in the visible range and reflect near infrared. Thus, to a predator that uses visible or infrared vision, such as birds and snakes, the frogs would be indistinguishable from the leaves.

It is possible that infrared reflectance in glass frogs may also play a role in thermoregulation as near infrared would normally result in heat gain if adsorbed by the skin, and re-

flectance of these photons would thus prevent excessive heat gain (Schwalm *et al.*, 1977).

Chlorophylls are used by plants, algae, and cyanobacteria to harvest visible light for photosynthesis. Bacterial chlorophylls absorb not only in the visible region, but also in the near infrared (Figure 6.6). Infrared radiation contains less energy than visible radiation, so it is difficult to understand why bacteriochlorophyll should rely so heavily on the IR region. Nisbet and colleagues (Nisbet *et al.*, 1995) have suggested that bacteriochlorophylls arose as heat (that is, near infrared) sensing devices for early organisms that lived near hydrothermal vents.

Today organisms with bacterial chlorophyll (all photosynthetic bacteria except the cyanobacteria) often live under oxygenic photosynthesizers and so have more limited access to radiation between 400 and 700 nm. It is plausible that bacterial photosynthesis arose

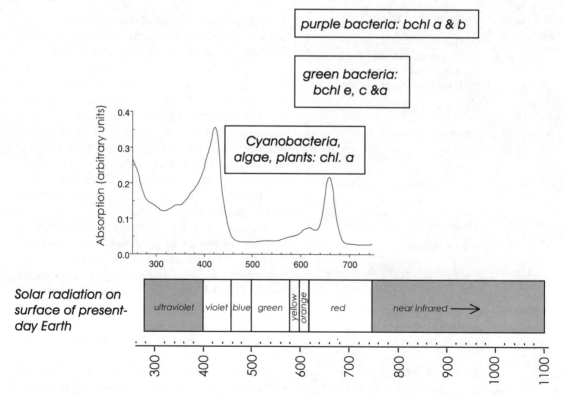

Figure 6.6 Spectra of chlorophylls and bacteriochlorophylls. Note that while the peak absorption for chlorophyll *a* is in the visible region, bacteriochlorophylls also absorb radiation in the near infrared.

in organisms that were in some way shielded from full solar radiation and thus were forced to use these lower energy wavelengths.

Plants absorb very little near infrared so they appear very bright when photographed with infrared film. In contrast, plants absorb virtually all of the thermal radiation. At normal temperatures, everything emits far infrared. Thus, about 50% of what a plant receives may be far infrared.

6.4.2 *Visible radiation*

Almost half of the solar radiation reaching the Earth is visible. Visible radiation spans the range between UV radiation, which is so energetic that it is often damaging or inhibiting to biological systems, and infrared radiation, which is not very energetic. This window of radiation has good penetration into water. For these reasons, many organisms have used radiation in this range for photosynthesis, for vision, and for environmental cues.

Visible light coincides broadly with photosynthetically active radiation (PAR). For eukaryotes and cyanobacteria, photosynthetically active radiation is defined as radiation from 400 to 700 nm. The absorption peaks for chlorophyll *a* in 100% methanol are 433 and 665 nm, and the ratio of the blue to the red peak is 1.21 (Figure 6.6).

With other chlorophylls such as *b*, *c* and *d*, and especially with accessory pigments such as carotenoids and phycobiliproteins, photosynthetic organisms, especially aquatic ones, have extended the range of usable wavelengths. Several reasons have been hypothesized for this (Rothschild, 1999). One is the good penetration of this range into the water column, and because the early photosynthesizers were aquatic, this was the range most available. At that time surface fluxes of UV radiation may have been substantially higher than they are today, so it may have been even more critical for photosynthetic organisms to live lower in the water column or under materials that attenuated UV radiation.

Additionally, UV radiation is too damaging to use for such a basic metabolic process. In fact, UV radiation is known to inhibit photo-

synthesis through several sites in the photosynthetic machinery (see Cockell and Rothschild, 1999). Infrared radiation is not very energetic, although some bacteria can utilize near infrared radiation in photosynthesis (discussed above).

Considering the fundamental role of photosynthesis for plants, PAR is an ideal eco-physiological signal for many plant processes (Salisbury and Ross, 1992). Response to the relative length of the day and night is termed photoperiodism. Photoperiodism influences virtually every aspect of plant growth and development, including seed germination, bud break in woody plants, stem elongation, flowering, vegatative and possibly sexual reproduction. Photoperiodism also influences animal behavior such as insect life cycles, breeding time, fur color change and migration in many birds.

The two major classes of light receptors in plants are the phytochromes which absorb in the red to far red region, and cryptochromes, which absorb in the UV to blue region. It is striking that the outer regions of the visible spectrum are used to signal metabolic processes. I would suggest that this is because on the Earth's surface these portions of the solar spectrum are enriched at sunrise and sunset (see above), thus providing a vital regulatory switch that is triggered by the onset and the conclusion of the light period.

Phytochromes are found in angiosperms, gymnosperms, liverworts, mosses, ferns, and some green, red and brown algae. Angiosperm phytochromes have two interconvertible forms, a red-absorbing blue form (P_r) with a peak absorption at 666 nm and a far red-absorbing olive colored form (P_{fr}) with a peak absorption of 730 nm. The peak absorption is slightly different in other taxa (Salisbury and Ross, 1992). In angiosperms the phytochromes are the receptors used in photoperiodism as well as other plant responses such as germination and flowering.

In plants, several photoreceptors are found for the UVA and blue regions of the spectrum. The cryptochromes absorb primarily in the blue wavelengths (400–500 nm). They are structurally similar to the DNA repair enzyme photolyase but do not repair DNA (Ahmad *et al.*,

1997). Instead, cryptochromes are involved in mediating growth, flowering time and phototropism in plants (see Thresher *et al.*, 1998).

Cryptochromes are found in the fruit fly *Drosophila* and mammalian cryptochromes recently have been found in mice and human eyes. Both plant (Somers *et al.*, 1998) and animal (Thresher *et al.*, 1998) cryptochromes are photoreceptors used to mediate circadian rhythms, suggesting that cryptochrome-mediated circadian rhythms are an ancient and evolutionarily conserved process.

Another photoreceptor that absorbs in the UVA to blue region of the spectrum is the autophosphorylating flavoprotein product of the *NPH1* gene that may be an important mediator of phototropism in plants (Christie *et al.*, 1998).

For many organisms light is important for vision. Humans can see light from 400 nm to far red wavelengths up to 760 nm. But 'visible' light is an anthropomorphic term because there are some animals that can see in the UV (e.g. bees) and infrared (e.g. snakes) regions of the spectrum (Table 6.3).

A variety of primarily photosynthetic prokaryotic and eukaryotic microbes display phototactic responses (see Castenholz, 1982). In prokaryotic taxa (e.g. cyanobacteria, flagellated purple bacteria, *Halobacterium*, non-pigmented *Beggiatoa*) the site of photoreception varies. Many algae, including members of the well-known genera *Euglena* and *Chlamydomonas*, contain an eyespot (stigma). In *Euglena* the eyespot contains carotenoids with absorption spectra in the range of 360–520 and 660–675 nm (Bold and Wynne, 1985).

These organisms are positively phototactic to low intensity radiation and negatively phototactic to high intensity light and darkness. It is thought that the eyespot is a light receptor. Alternatively, it could be used to shade the flagellar swelling which is then the actual site of photoreception (Bold and Wynne, 1985: 289–290). Among microbes light detection is clearly polyphyletic. Many invertebrates and vertebrates have eyes, and in invertebrates light-sensitive organisms have evolved 40–60 times (Ridley, 1996: 343).

Table 6.3 Animal vision. Data cited in text and from http://ls.la.asu.edu/askabiologist/research/seecolor/atable.html.

Animal	Spectral perception
Spiders (jumping spiders)	UVR and green
Most insects	UVR, blue, yellow (300–650 nm)
Crustaceans (crayfish)	Blue and red
Cephalopods (octopi and squids)	Blue only
Fish	Most see just two colors
Amphibians (frogs)	Most see some color
Reptiles (snakes)	Some color and infrared
Birds	Five to seven colors
Mammals (cats)	Two colors but weakly
Mammals (dogs)	Two colors but weakly
Mammals (squirrel)	Blues and yellows
Mammals (primates – apes and chimps)	Blue through red (400–760 nm)
Mammals (African monkeys)	Blue through red (400–760 nm)
Mammals (South American monkeys)	Cannot see red well
Mammals (humans)	Blue through red (400–760 nm)

Both visible and ultraviolet radiation are sensed by most insects, but they have little or no ability to detect red (Proctor *et al.*, 1996 and references therein). Thus, while humans can detect radiation from 400 to 760 nm, radiation from 300 to 650 nm is the 'visible' range for most insects.

Flowering plants have evolved to exploit this limitation by producing floral guide marks, many of which are in the UV region. For example, both marsh marigold (*Caltha palustris*) and buttercups (*Ranunculus*) show a similar pattern of guide marks that are only visible in the UV part of the spectrum. These marks accent the appearance and structure of the flower, and are thus used as 'close-in signals', especially in flowers with large petals or massed inflorescences, thus allowing the insect pollinator to orient more rapidly which results in a lower expenditure of energy.

Within this range, insects are able to discriminate colors (Proctor *et al.*, 1996 and references therein). For example, some butterflies are even able to distinguish between subtle differences in yellow, such as between 540 and 590 nm. Honeybees, bumblebees and other red-blind insects can distinguish between blue–green and ultraviolet. They can also recognize a color made by mixing opposites ends of their visual spectrum, yellow and ultraviolet, a color known as 'bee purple' in analogy with the color we perceive by mixing opposite ends of our visible spectrum, red and blue.

Bees are especially sensitive to ultraviolet radiation, a fact exemplified by the observation that bees can distinguish between pure yellow and a mixture of 98% yellow plus 2% ultraviolet. Thus, two flowers that look yellow to us may appear different in color to bees if they also reflect even a small bit of ultraviolet radiation. Spectral discrimination may be extremely good, with bees able to discriminate between wavelengths with as little as a 4.5 nm separation.

Clearly the ability to sense visible light has played an important role in evolution. The evolutionary pressure to evolve photoreception in the photosynthetic taxa is to determine the optimal location and time for photosynthesis, whereas in animals it is to locate food or detect danger such as predators.

6.4.3 *Ultraviolet radiation*

Ultraviolet radiation (UVR) is potentially the most interesting region of the solar spectrum with regards to evolution, and is both a selective agent and a mutagen. UVR is traditionally divided by physicists into four spectral regions:

near – 400–300 nm, middle – 300–200 nm, far – 200–100 nm, and extreme – below 100 nm. Biologists also divide UVR into four regions based on its interactions with biological materials: UVA – 400–315 nm, UVB – 315–280 nm, UVC – 280–100 nm, and vacuum – 200 nm. (The definitions of UVA, UVB and UVC are set by the Commission Internationale d'Eclairage (CIE), and deviate from the traditional definitions where, for example, UVC is defined as 280–200 nm.) Note that while the UVR that reaches the surface of the Earth today (290–400 nm) is non-ionizing, at shorter wavelengths such <200 nm, UVR is ionizing.

Solar radiation fluxes on the Earth's surface have changed over geologic time, partly as a result of UV-catalyzed photochemical reactions. Depletion of stratospheric ozone has resulted in elevated UVB fluxes. The initial concern over rising UVB was medical because UVR is the primary cause of skin cancer, possibly because it is an immunosuppresant (Ullrich, 1996).

UVR, however, has positive effects on the human body because it photolyzes the conversion of provitamin D_3 to previtamin D_3 in the skin, a critical step in the production of vitamin D. Vitamin D is vital for the calcification of cartilage and bone, and its deficiency leads to rickets in children and osteomalacia in adults. Additionally, UVR is used as a therapeutic agent for psoriasis.

More recently, there has been concern that elevated levels of UVB radiation damages plant and planktonic communities (e.g. Caldwell and Flint, 1994; Häder 1993; Vincent and Roy, 1993). UVA and blue light trigger photorepair of DNA damage.

The perception of UVR by insects has played a critical role in the co-evolution of plant/insect pollination systems. The role of UV radiation in early eukaryotic (protistan) evolution has recently been reviewed (Rothschild, 1999).

Ultraviolet radiation (UVR) was important in habitat creation and prebiotic chemistry long before life arose. It initiated the photochemical processes that led to the formation of O_2 and O_3 in the prebiological paleoatmosphere. The formation of a stratospheric ozone layer has been critical in evolution because ozone very strongly absorbs radiation from 220 to 330 nm, thereby providing protection from an extremely hazardous portion of the solar spectrum.

Today UVR at <242 nm initiates the formation of O_3 from O_2 in the stratosphere. Circularly polarized UVR may have produced a disequilibrium in biomolecular chirality, thus creating the evolutionary pressure towards the L-amino acids and D-sugars found in terrestrial organisms.

UVR was an energy source in the prebiological synthesis of organic compounds (Chyba and Sagan, 1992). For example, UV irradiation of ice analogs of interstellar and cometary ices destroys several species (particularly methanol) but creates others such as CO, CO_2, HCO, formaldehyde, methane, ethanol, formamide, acetamide, nitriles and more complex organic compounds after warming (Bernstein *et al.*, 1995).

Similarly, UV irradiation of polycyclic aromatic hydrocarbons in ices produces a rich array of organic materials including alcohols, quinones and ethers (Bernstein *et al.*, 1999). Organics produced through such photochemical reactions may well have reached the early Earth on meteorites or comets.

The origin of photosynthesis itself may be linked to UV photochemistry. Photosynthesis may have evolved from a mechanism to protect cells from UV radiation. Larkum (1991) proposed that the original function of photosynthetic reaction centers (intrinsic membrane polypeptides associated with pigments) may have been UVR protection with the reaction centers subsequently evolving chlorophyll-carrying antenna proteins. Mulkidjanian and Junge (1997) propose that the original reaction center was formed by the dimerization of pigment-carrying antenna proteins which themselves were ancient UV protectors.

Second, UV photochemistry could have been critical in the biosynthetic pathway of chlorophyll. An early light-driven redox pump could have been based on porphyrins or porphyrinogens (Larkum, 1991). The photoreaction itself could have been involved in the oxidation of uroporphyrinogen to uropor-

phyrin driven by UVR at <230 nm (Mercer-Smith *et al.*, 1985). The next step in chlorophyll biosynthesis, the formation of coproporphyrin from uroporphyrin, could also have been a process driven by light and near UVR. Thus, Larkum (1991) argued that UVR was a driving force in the evolution of chlorophylls.

While the high energy of UVR played a synthetic role in prebiotic chemistry, it is particularly harmful to living organisms because the peak absorption of three classes of vital biomolecules in the UV region of the spectrum. The peak absorption of DNA and RNA is approximately 260 nm, and that for protein 280 nm. As a result, there are many biological processes that are inhibited by UVR and only a few that are enhanced by it.

It is likely that ever since early evolution, prior to the accumulation of atmospheric oxygen, organisms were exposed to UVR. This assumption is reinforced by the fact that obligately and facultatively anaerobic prokaryotes show intrinsic resistance to UVR and are able to repair UV-induced DNA damage by photoreactivation (Rambler and Margulis, 1980; Pierson *et al.*, 1993). When these biological effects of UVR are reviewed against the backdrop of environmental UVR levels, we can begin to determine its role in biological evolution.

Evolution requires heritable changes in the gene pool of the population. This process is initiated through genetic changes at the level of the individual, and perpetuated through the population by mechanisms such as natural selection. UVR has a role in the generation of DNA damage whose repair can lead to a heritable change in the genome, that is, a mutation.

The major lesions produced directly by UV radiation are photoproducts such as cyclobutane pyrimidine dimers (especially thymine dimers), (6-4) dipyrimidine dimers and their Dewar isomers, and to a lesser extent dipurine adducts (Sancar and Tang, 1993). UVR also may cause double-stranded breaks, possibly during excision repair when a break occurs in the single-stranded region of the DNA (Smith, 1998).

UVR (280–400 nm) is involved in the photochemical formation of hydrogen peroxide, itself a DNA damaging agent (Cooper and Lean,

1992). Hydrogen peroxide produces reactive oxygen species, which cause double- and single-strand breaks, base and sugar modifications such as the production of 8-hydroxyguanine, and DNA-protein crosslinks. Such lesions may cause base substitutions, and much less frequently, insertions and deletions (Newcomb and Loeb, 1998).

In most taxa (but not including humans), cyclobutane thymine dimers may be repaired through monomerization of the dimer by the enzyme photolyase, a reaction that itself requires energy from the UVA or short-wave visible range up to 480 nm. Excision repair is a general repair mechanism for DNA damage (excluding double-stranded breaks when both strands are damaged) where the damaged portion of the DNA is removed and the complementary strand used as a template for resynthesis (Figure 6.7).

Mutation results if the resynthesis is inaccurate (DNA synthesis has an intrinsic error rate of ~1 per 10^9 to 10^{10} base pairs copied), or if other DNA damage occurs while single-stranded (e.g. deamination or production of a double-strand break). For phages and episomes the majority (75%) of mutations are base substitutions, mostly $GC \rightarrow AT$, followed by frameshifts (15%), tandem double-base substitutions (4%) and deletions (7%), though the last figure may be an overestimate (Livneh *et al.*, 1993). The mutational effect of UVR on chromosomal genes is similar where 43% of the mutations were transitions, 38% transversions, and 7% frameshifts (Livneh *et al.*, 1993).

How likely is UV-induced DNA damage to affect evolution? We know repair of this damage is exceedingly important to organisms, even under present UVR regimes. Loss of a single UV-induced DNA damage repair pathway, such as photoreactivation, may not be critical for survival in microbes when natural sunlight levels are as high as 18 μW cm^{-2} midday, such as occurs in the San Francisco Bay

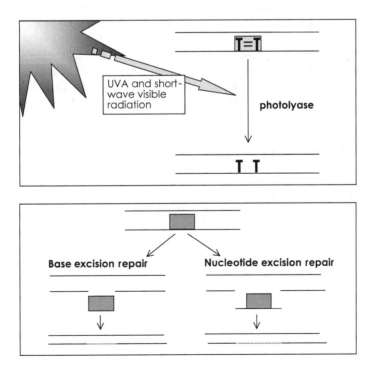

Figure 6.7 DNA repair. Top panel, photoreactivation, a DNA repair method that requires light. Bottom panel, excision repair, a generalized method to repair DNA damage on a single strand of DNA. During excision repair, the damaged portion of the DNA and some of the flanking regions are excised and new DNA synthesized using the undamaged strand as a template.

area in April (Garget and Rothschild, unpublished results), but deficiencies in multiple pathways in *Escherichia coli* and yeast are critical (Harm, 1969; Resnick, 1970).

In humans, defects in excision repair result in the disorders xeroderma pigmentosum, Cockayne's syndrome and trichothiodystrophy (TTD). Exposure to sunlight results in skin and eye cancers which can be fatal. Further, data from our laboratory (e.g. Rothschild and Cockell, 1999) has shown a substantial increase in DNA synthesis in response to UVR, a result which we have interpreted to suggest greatly enhanced levels of excision repair in response to UV-induced DNA damage.

UV-induced DNA damage should thus be an important driver of molecular evolution (Rothschild, 1999). First, unicellular organisms should be most susceptible since they do not sequester their germline into specialized gametes, so mutations are most likely to be passed on to subsequent generations (ciliated protists are an obvious exception since they sequester their germline in the micronucleus within the confines of a single cell).

Second, primarily asexual taxa should be more affected because sexual processes are likely to mask UV-induced mutational changes. Note that while there are few strictly asexual taxa, many unicellular taxa are effectively asexual because sexual recombination is such a rare event. Studies of obligately asexually reproducing dandelions supports the view that mutation influences the level of genotypic variation in asexual organisms (King and Schaal, 1990). Third, UV is most likely to affect photosynthetic organisms since by definition they must be exposed to solar radiation.

Most intriguing is the notion that sexual processes arose as a response to UV-induced DNA damage (Rothschild, 1999). Sexual processes result in genetic recombination creating genetic novelty. But sexual processes can also result in the removal of deleterious mutations from a population, whereas asexual populations of finite size exhibit a 'ratchet' mechanism, i.e. all members of the population are loaded with mutations and unloaded variants cannot be restored (Muller, 1964).

Experimental evidence supports the idea that DNA damage can induce sex in prokaryotes (bacterial transformation, induction of lytic cycles, recombination among λ bacteriophages) and eukaryotes (yeast). There is a mechanistic and biochemical link between recombinational repair, a mechanism used to repair double-stranded breaks, and meiosis. Thus, while the current selection pressure to maintain sex may not be related to DNA damage repair, it may have arisen as a response to damage from solar radiation.

The great protistan radiation, which witnessed the origin of many diverse phyla about 1 billion years ago, was thought to have been caused by a substantial change in the environment. This was also a period when oxygen levels were beginning to rise while UV flux was still high relative to today. Thus, I have suggested (Rothschild, 1999) that the protistan radiation could have been a response to elevated levels of oxidative DNA damage.

From the preceding discussion it is clear that UVR has played a role in evolution as a mutagen, which may have had all sorts of consequences ranging from the origin of repair mechanisms to sexual processes. But, because of the damaging nature of UVR to both the genome and other biomolecules, UVR has also acted as a selective agent in evolution.

The evolution of the genome itself could have been shaped by the need for organisms to minimize UVR-induced DNA damage. Because thymine dimers are a major site of UV-induced DNA damage, selection should minimize their occurrence in organisms exposed to UVR. An early attempt to verify this hypothesis (Lesk, 1973) did not support it, but with improved methodologies and available sequence information it should be revisited.

Because DNA itself absorbs UVR, it could have provided some role in UVR protection, thereby driving genomic evolution. Sagan (1973) suggested that early photosynthetic organisms protected their genomes by surrounding themselves with purines and pyrimidines, an innovation that eventually led to the origin of eukaryotes. While this may not have happened exactly as Sagan hypothesized, perhaps nucleic acids were used as a UV-protectant in some manner.

For example, the idea of increased ploidy – multiple copies of the genome as a way of coping with UV-induced DNA damage – is appealing. Haploid bacteria and yeast cells are more susceptible to radiation damage than dipoid cells, and human tumors with stem cell lines of higher ploidy are often highly resistant to radiotherapy (Martel, 1992). This suggests that if there are additional copies of the genome, the cell can tolerate more damage. Additional copies of a chromosome can be used for interchromosomal recombinational repair. Interestingly, both mechanisms have been suggested and provisionally rejected (Battista, 1998) for explaining the unusually resistant nature of the bacterium *Deinococcus radiodurans* to most forms of DNA damage.

The transition of life to land occurred well before the formation of the present ozone 'shield', so UVR undoubtedly affected this transition. There is no direct record of cyanobacteria and algae on land in the Early Paleozoic, although they were probably there but not fossilized. The first direct evidence for terrestrial organisms comes from microfossils found in 1.2 billion-year and 800 million-year paleokarst (Horodyski and Knauth, 1994).

The danger of high UV flux may have limited organisms to the water before this time. However, there are microbial communities that live in the intertidal or shallow water benthic communities, and probably have well into the Precambrian (Walter *et al.*, 1992). Such communities even today protect themselves from UVR by a variety of means including physical and chemical protection (e.g. living under a UV-attenuating substance or under a dead layer of cells), high levels of UV-absorbing pigments or apparently high rates of DNA damage repair (Margulis *et al.*, 1976; Pierson, 1994; Rothschild, 1999; Rothschild and Cockell, 1999).

UVR determines ecosystem dynamics even today, and must have been a factor in evolution. The effect of UVB radiation on terrestrial ecosystems is well known, and includes decreased primary productivity, altered plant species composition, and altered secondary chemistry which in turn affects herbivory, litter decomposition and biogeochemical cycles (Caldwell and Flint, 1994), with similar effects seen in the aquatic environment (Häder, 1993). For example, grazing by the marine heterotrophic nanoflagellated protist *Paraphysomonas badaiensis* on the cyanobacterium *Synechococcus* is affected by changes in grazing due to loss of viability of the predator (Oches and Eddy, 1998).

UVR also affects food web structure and function by differential susceptibility to UVR damage. Bothwell *et al.* (1994) showed an increase in a herbivore (Chrinomid fly larvae) in the absence of UVB and a decrease in its food source, an algal (diatom) community. This demonstrated the importance of ecological studies in determining the ultimate evolutionary implications of UVR sensitivity.

It is quite possible that UVR provided the selective pressure to evolve one of the most visible aspects of human evolution – skin color (Jablonski, 1992; Jablonski and Chaplin, 2000 and references therein). It has been known for several decades that skin color, as determined by reflectance spectrometry, closely matches latitude, as does solar UVR. Human skin is protected from UVR injury by two major defenses, the pigment melanin and the thickness of the most superficial layer of the skin (the stratum corneum).

Darwin (1871) summarized potential selective reasons for differences in human skin color, including protection from sunburn, but hesitated to endorse any of the prevailing hypotheses. But protection from sunburn, and as we now know, skin cancer, would in most cases not be a serious selective force until well into or past reproductive age.

Other current hypotheses have similar weaknesses (reviewed in Jablonski and Chaplin, 2000). For this reason, Jablonski (1992) proposed a novel hypothesis: UVR, especially in the longer UVA region, destroys folate, a B group vitamin vital for neural tube formation in early embryos. Thus, she suggests that UVR is a selective agent for the evolution of darker skin as mediated through UVR-induced destruction of folate in mothers leading to the production of neural tube defects in the embryo.

More recently, Jablonski and Chaplin (2000) have developed a more comprehensive scenario. Using skin reflectance data, clinical and epidemiological studies, and data on UVR levels extracted from the NASA TOMS data set, they suggest that skin pigmentation represents a fine balance between the competing selective pressures of allowing sufficient UVR penetration to allow synthesis of previtamin D_3 (a precurser to vitamin D_3 which is essential in calcium absorption) but prevent photodamage resulting in folate degradation, sunburn and skin cancer.

6.5 Mars?

Mars is 1.5 times as far from the Sun as the Earth, and because the intensity of light is a function of the inverse of the square of the distance, the solar radiation flux at Mars should be 44% of that of the Earth. The mean solar constant for Earth is $1366.1 \, W \, m^{-2}$, which means $588.6 \, W \, m^{-2}$ for Mars.

The amount of solar radiation that actually reaches the surface will depend on a variety of factors including atmospheric composition, cloud cover, dust storms, season and latitude (Sagan and Pollack, 1974). Present-day Mars has a thinner atmosphere than that of the Earth, resulting in more direct penetration of radiation to the surface. The atmosphere of Mars has a total atmospheric pressure of 6 mb, 95% of which is CO_2. Thus, radiation below 200 nm does not reach the surface.

In contrast, the atmosphere is only 0.13% oxygen, so the attenuation in the UVB region is negligible. Both UVC and UVB radiation will be greater on the surface of Mars than on present-day Earth, but UVA will be slightly less. UVA is not significantly attenuated by the Earth's atmosphere. Thus, the Sun–Mars distance more than compensates for the even less efficient attenuation of the UVA by the Martian atmosphere (see Rothschild and Cockell, 1999).

To calculate the UV flux on early Mars, the assumption of a 25% less luminous Sun is

made (Kasting, 1987, 1993). It is possible that the early Martian atmosphere was significantly thicker than today, which would have allowed the presence of liquid water. If life arose on Mars, it too would have to cope with high UV fluxes, in fact, much higher than on present-day Earth, and would have required screening and avoidance strategies (Wynn-Williams and Edwards, 2000).

In summary, in a very real sense we are the product of stars. In particular, our Sun has been instrumental in prebiotic chemistry, the physical, chemical and then biogeochemistry of the Earth. From the salad we eat and the oil we burn to the air we breathe, we are utterly dependent on solar radiation. It has been both a vital resource during the course of evolution, as well as a hazardous substance. Some day we must move beyond the immediate influence of our star or be consumed by it, but we will always bear the unmistakable imprint of the Sun.

Acknowledgements

I am indebted to David D. Wynn-Williams, Gerda Horneck, and Jacqueline Garget for extraordinarily thorough reviews of the manuscript. Jay Skiles, Hector d'Antoni and Tom Woods provided valuable spectral data for figures. Dana Rogoff and Beth Grode contributed pigment scans. I am grateful to all.

References

Ahmad, M., Jarillo, J.A., Klimczak, L.J. *et al.* (1997). An enzyme similar to animal type II photolyases mediates photoreactivation in *Arabidopsis. Plant Cell*, **2**: 199–207.

Bailey, J., Chrysostomou, A., Hough, J.H. *et al.* (1998). Circular polarization in star-formation regions: implications for biomolecular homochirality. *Science*, **281**: 672–674.

Baliunas, S. and Jastrow, R. (1993). Evidence on the climate impact of solar variations. *Energy*, **18**: 1285–1295.

Battista, J.R. (1998). DNA repair in *Deinococcus radiodurans*. In: J.A. Nickoloff and M.F. Hoekstra (eds) *DNA Damage and Repair, Volume 1: DNA Repair in Prokaryotes and Lower Eukaryotes*, New Jersey: Humana Press, pp. 287–303.

Bernstein, M.P., Sandford, S.A., Allamandola, L.J. and Chang, S. (1995). Organic compounds produced by photolysis of realistic interstellar and cometary ice analogs containing methanol. *The Astrophys. J.*, **454**: 327–344.

Bernstein, M.P., Sandford, S.A., Allamandola, L.J. *et al.* (1999). UV irradiation of polycyclic aromatic hydrocarbons in ices: production of alcohols, quinones, and ethers. *Science*, **283**: 1135–1138.

Bold, H.C. and Wynne, M.J. (1985). *Introduction to the Algae*, second edition. Englewood Cliffs, New Jersey: Prentice-Hall, Inc.

Bonner, W.A., Rubenstein, E. and Brown, G.S. (1999). Extraterrestrial handedness: a reply. *Origins Life Evol. Biosphere*, **29**: 329–332.

Bothwell, M.L, Sherbot, D.M.J. and Pollock, C.M. (1994). Ecosystem response to solar ultraviolet-B radiation: influence of trophic-level interactions. *Science*, **265**: 97–100.

Caldwell, M.M. and Flint, S.D. (1994). Stratospheric ozone reduction, solar UV-B radiation and terrestrial ecosystems. *Climatic Change*, **28**: 375–394.

Canuto, V.M., Levine, J.S., Augustsson, T.R. and Imhof, C.L. (1982). UV radiation from the young sun and oxygen and ozone levels in the prebiological paleoatmosphere. *Nature*, **296**: 816–820.

Canuto, V.M., Levine, J.S. and Augustsson,T.R. *et al.* (1983). The young Sun and the atmosphere and photochemistry of the early Earth. *Nature*, **305**: 281–286.

Castenholz, R.W. (1982). Motility and taxes. In: N.G. Carr and B.A. Whitton (eds) *The Biology of Cyanobacteria*. Berkeley: University of California Press, pp. 413–439.

Christie, J.M., Reymond, P., Powell, G.K. *et al.* (1998). *Arabidopsis* NPH1: a flavoprotein with the properties of a photoreceptor for phototropism. *Science*, **282**: 1698–1701.

Chyba, C. and Sagan, C. (1992). Endogenous production, exogenous delivery and impact-shock synthesis of organic molecules: an inventory for the origins of life. *Nature*, **355**: 125–131.

Clark, S. (1999). Polarized starlight and the handedness of life. *American Sci.*, **87**: 336–343.

Cockell, C.S. and Rothschild, L.J. (1999). The effects of ultraviolet radiation A and B on diurnal variation in photosynthesis in three taxonomically and ecologically diverse microbial mats. *Photochem. Photobiol.*, **69**: 203–210.

Cooper, W. and Lean, D. (1992). Hydrogen peroxide dynamics in marine and fresh water systems. In: W.A. Nierenber (ed.) *San Diego: Encyclopedia of Earth System Science*, Vol. 2, Academic, pp. 527–535.

Darwin, C.R. (1871). *The Descent of Man and Selection in Relation to Sex*. John Murray, London.

Darwin, G.H. (1880). On the secular change in the elements of the orbit of a satellite revolving about a tidally distorted planet. *Phil. Trans. Roy. Soc. London*, **171**: 713–891.

Gold, T. (1999). *The Deep Hot Biosphere*. New York: Springer-Verlag, p. 235.

Gough, D.O. (1981). Solar interior structure and luminosity variations. *Solar Phys.*, **74**: 21–34.

Grace, M.S., Church, D.R., Kelly, C.T. *et al.* (1999). The Python pit organ: imaging and immunocytochemical analysis of an extremely sensitive natural infrared detector. *Biosens. Bioelectron.*, **14**: 53–59.

Häder, D.P. (1993). Risks of enhanced solar ultraviolet radiation for aquatic ecosystems. *Progr. Phycol. Res.*, **9**: 1–45.

Harm, W. (1969). Biological determination of the germicidal activity of sunlight. *Radiation Res.*, **40**: 63–69.

Horodyski, R.J., Knauth, L.P. (1994). Life on land in the Precambrian. *Science*, **263**: 494–498.

Jablonski, N.G. (1992). Sun, skin colour and spina bifida: an exploration of the relationship between ultraviolet light and neural tube defects. *Proc. Australas. Soc. Hum. Biol.*, **5**: 455–462.

Jablonski, N.G. and Chaplin, G. (2000). The evolution of human skin coloration. *J. Human Evo.* **39**: 57–106.

Kasting, J.F. (1987). Theoretical constraints on oxygen and carbon dioxide concentrations in the Precambrian atmosphere. *Precambr. Res.* **34**: 205–229.

Kasting, J.F. (1993). Earth's early atmosphere. *Science*, **259**: 920–926.

Kasting, J.F., Zahnle, K.J., Pinto, J.P. and Young, A.T. (1989). Sulfur, ultraviolet radiation, and the early evolution of life. *Origins Life Evol. Biophere*, **19**: 95–108.

King, L.M. and Schaal, B.A. (1990). Genotypic variation within asexual lineages of *Taraxacum officinale*. *Proc. Natl. Acad. Sci. USA*, **87**: 998–1002.

Kump, L.R., Kasting, J.F. and Crane, R.G. (1999). *The Earth System*. Prentice-Hall. 368pp.

Larkum, A.W.D. (1991). The evolution of chlorophylls. In: H. Scheer (ed.) *Chlorophylls*. Boston, Massachusetts: CRC Press, pp 367–383.

Laskar, J., Joutel, F. and Robutel, P. (1993).

Stabilization of the Earth's obliquity by the Moon. *Nature*, **361**: 615–617.

Lesk, A.M. (1973). On hypothesized selective pressure by u.v. on DNA base composition. *J. Theor. Biol.*, **40**: 201–202.

Livneh, Z., Cohen, O., Skaliter, R. and Elizur, T. (1993). Replication of damaged DNA and the molecular mechanism of ultraviolet light mutagenesis. *Crit. Rev. Biochem. Mol. Biol.*, **28**: 465–513.

Lutz, T.M. (1985). The magnetic reversal record is not periodic. *Nature*, **317**: 404–407.

Margulis, L., Walker, J.C.G. and Rambler, M. (1976). Reassessment of roles of oxygen and ultraviolet light in Precambrian evolution. *Nature*, **264**: 620–624.

Martel, E.A. (1992). Radionuclide-induced evolution of DNA and the origin of life. *J. Mol. Evol.*, **35**: 346–355.

Mercer-Smith, J.A., Raudino, A. and Mauzerall, D.C. (1985). A model for the origin of photosynthesis III. The UV photochemistry of uroporphyrinogen. *Photochem. Photobiol.*, **42**: 239–244.

Mulkidjanian, A.Y. and Junge, W. (1997). On the origin of photosynthesis as inferred from sequence analysis. *Photosyn. Res.*, **51**: 27–42.

Muller, H.J. (1964). The relation of recombination to mutational advance. *Mutat. Res.*, **1**: 2–9.

Newcomb, T.G. and Loeb, L.A. (1998). Oxidative DNA damage and mutagenesis. In: J.A. Nickoloff and M.F. Hoekstra (eds) *DNA Damage and Repair, Volume 1: DNA Repair in Prokaryotes and Lower Eukaryotes*. New Jersey: Humana Press, pp. 65–84.

Nisbet, E.G., Cann, J.R. and Van Dover, C.E. (1995). Origins of photosynthesis. *Nature*, **373**: 479–480.

Ochs, C.A. and Eddy, L.P. (1998). Effects of UV-A (320 to 399 nanometers) on grazing pressure of a marine heterotrophic nanoflagellate on strains of the unicellular cyanobacteria *Synechococcus* spp. *Appl. Environ. Microbiol.*, **64**: 287–293.

Pierson, B.K. (1994). The emergence, diversification, and role of photosynthetic eubacteria. In: S. Bengtson (ed.) *Early Life On Earth. Nobel Symposium No. 84.* New York: Columbia University Press, pp. 161–180.

Pierson, B.K., Mitchell, H.K. and Ruff-Roberts, A.L. (1993). *Chloroflexus aurantiacus* and ultraviolet radiation: implications for archean shallow-water stromatolites. *Origins Life Evol. Biosphere*, **23**: 243–260.

Plotnick, R.E. (1980). Relationship between biological extinctions and geomagnetic reversals. *Geology*, **8**: 578–581.

Podlech, J. (1999). New insight into the source of biomolecular homochirality: an extraterrestrial origin for molecules of life? *Angew. Chem. Int. Ed.*, **38**: 477–478.

Proctor, M., Yeo, P. and Lack, A. (1996). *The Natural History of Pollination*. Portland, Oregon: Timber Press.

Rambler, M.B. and Margulis, L. (1980). Bacterial resistance to ultraviolet irradiation under anaerobiosis: implications for pre-Phanerozoic evolution. *Science*, **210**: 638–640.

Raup, D.M. (1985). Rise and fall of periodicity. *Nature*, **317**: 384–385.

Resnick, M.A. (1970). Sunlight-induced killing in *Saccharomyces cerevisiae*. *Nature*, **226**: 377–378.

Ridley, M. (1996). *Evolution*, second edition. Cambridge, Massachusetts: Blackwell Science Inc.

Rothschild, L.J. (1999). The influence of UV radiation on protistan evolution. *J. Euk. Microbiol.*, **46**: 548–555.

Rothschild, L.J. and Cockell, C.S. (1999). Radiation, microbial evolution and ecology, and its relevance to Mars missions. Mutation research/fundamental molecular mechanisms. *Mutagenesis*, **430**: 281–291.

Sagan, C. (1973). Ultraviolet selection pressure on the earliest organisms. *J. Theor. Biol.*, **39**: 195–200.

Sagan, C. and Pollack, J.B. (1974). Differential transmission of sunlight on Mars biological implications. *Icarus*, **21**: 490–495.

Salisbury, F.B. and Ross, C.W. (1992). *Plant Physiology*. Belmont, California: Wadsworth Publishing Company.

Sancar, A. and Tang, M.-S. (1993). Photobiology school. Nucleotide excision repair. *Photochem. Photobiol.*, **57**: 905–921.

Schwalm, P.A., Starrett, P.H. and McDiarmid, R.W. (1977). Infrared reflectance in leaf-sitting neotropical frogs. *Science*, **196**: 1225–1227.

Smith, G.R. (1998). DNA double-strand break repair and recombination in *Escherichia coli*. In: J.A. Nickoloff and M.F. Hoekstra (eds) *DNA Damage and Repair. Volume I: DNA Repair and Prokaryotes and Lower Eukaryotes*. Totowa, New Jersey: Humana Press, pp. 135–162.

Somers, D.E., Devlin, P.F. and Kay, S.A. (1998). Phytochromes and cryptochromes in the entrainment of the *Arabidopsis* circadian clock. *Science*, **282**: 1488–1490.

Thekaekara, M.P. (ed.) 1970). The solar constant and the solar spectrum measured from a research aircraft. NASA Technical Report NASA TR R-351. National Aeronautics and Space Administration, Washington, DC 20546.

Thresher, R.J., Vitaterna, M.H., Miyamoto, Y. *et al.*

(1998). Role of mouse cryptochrome blue-light photoreceptor in circadian photoresponses. *Science*, **282**: 1490–1494.

Ullrich, S.E. (1996). Does exposure to UV radiation induce a shift to a Th-2-like immune reaction? *Photochem. Photobiol.*, **2**: 254–258.

Vincent, W.F. and Roy, S. (1993). Solar ultraviolet-B radiation and aquatic primary production: damage, protection and recovery. *Environ. Rev.*, **1**: 1–12.

Walker, J.C.G., Klein, C., Schidlowski, M. *et al.* (1983). Environmental evolution of the Archean-early Proterozoic Earth. In: J.W. Schopf (ed.) *Earth's Earliest Biosphere*. Princeton, New Jersey: Princeton University Press, pp. 260–290.

Walter, M.R., Bauld, J., Des Marais, D.J. and Schopf, J.W. (1992). A general comparison of microbial mats and microbial stromatolites: bridging the gap between the modern and the fossil. In: J.W. Schopf and C. Klein (eds) *The Proterozoic Biosphere*. New York: Cambridge University Press, pp. 335–338.

Wynn-Williams, D.D. and Edwards, H.G.M. (2000). Proximal analysis of regolith habitats and protective biomolecules *in situ* by laser Raman spectroscopy: overview of terrestrial Antarctic habitats and Mars analogs. *Icarus*, **144**: 486–503.

7

Could life travel across interplanetary space? Panspermia revisited

Gerda Horneck

ABSTRACT

See Plate 13

More than a century ago H. Richter and later on S. Arrhenius formulated the theory of Panspermia which postulates that microscopic forms of life, for example spores, can be propagated in space driven by the radiation pressure of the Sun and thereby transport germs of life from one planet to another. Since its formulation this theory has been subjected to severe criticism, especially that it cannot be experimentally tested and that spores will not survive over long periods in the hostile environment of space. Recent discoveries have given support to revisit the theory of Panspermia, such as the Martian meteorites, the high UV resistance of microorganisms at the low temperature of deep space, and the recovery of viable spores after extended periods in space. During a hypothetical interplanetary transfer, the organisms have to cope with the following three major challenges: (i) the escape process; (ii) the long duration exposure to space; and (iii) the entering process.

A comprehensive experimental and theoretical study of the probabilities of microorganisms surviving the complex interplay of the parameters of space (e.g. vacuum, UV and ionizing radiation, temperature extremes) concluded that radiation-resistant microbes could survive a journey from one planet to another in our solar system if they were shielded by a layer of meteorite material.

7.1 Introduction

The history of life on Earth reaches back about 4 billion years. Fossil record gives evidence that life started as a 'microbial world' very early on the juvenile Earth, with a degree of certainty earlier than 3.5 billion years ago (Schopf, 1993) and probably earlier than 3.8 billion years ago (Schidlowski, 1988; Rosing, 1999). Because the oldest known microfossils are already structurally complex and stem probably from photoautotrophic organisms, the first appearance of life on Earth should be dated much earlier than this. Microbial prokaryotes dominated the Earth's biosphere during the first two billion years of life's history before the first unicellular mitotic eukaryotes appeared. Hence, microbial life has persisted on Earth almost since the cooling of the planet's crust.

The ingredients of life, such as the biogenic elements (carbon, hydrogen, oxygen, nitrogen, and phosphorus), organic molecules, and water, were either already present or synthesized on the primitive Earth (deDuve, 1991) or they were imported to the Earth via planetesimals, meteorites and comets which were abundant during the first billion years of Earth's existence (Horneck and Brack, 1992). During this early phase of heavy bombardment, which lasted until approximately 3.8 bil-

Evolution on Planet Earth
ISBN 0-12-598655-6

lion years ago, the impactors would, on the one hand, have been delivering the volatiles including water and organic precursors of life; on the other hand, if sufficiently large and fast, impactors would have eroded the atmosphere and certainly sterilized the Earth (Melosh and Vickery, 1989). This impact catastrophe scenario implies that, if life did already exist during this early period, it may have been extinguished several times as a consequence of such gigantic impacts until the end of the heavy bombardment (Oberbeck and Fogleman, 1990).

Furthermore, gigantic impacts lead to the ejection of a considerable amount of soil and rocks that are thrown up at high velocities, some fraction reaching escape velocity (Melosh, 1988). These ejecta leave the planet and orbit around the sun, usually for timescales of a few hundred thousand or several million years until they either impact another celestial body or are expelled out of the solar system. The question arises, whether such rock or soil ejecta could also be the vehicle for life to leave its planet of origin, or, in other words, whether spreading of life in the solar system via natural transfer of viable microbes in space is a feasible process. If so, this clearly could have had a profound effect on evolution, and would make the discovery of life elsewhere that much more difficult to interpret.

7.2　Panspermia revisited

The hypothesis of Panspermia postulates that microscopic forms of life, for example spores, can be propagated in space driven by the radiation pressure from the Sun thereby seeding life from one planet to another. This idea first appeared in 1865 (Richter, 1865) and was more precisely formulated later on by Arrhenius (1903). Also in 1865, Pasteur had experimentally disproved spontaneous generation of life. Hence, the generally accepted scenario at that time was that life is eternal and Panspermia is a likely process for life to be distributed throughout the universe.

However, already in 1871, Charles Darwin proposed an alternative pathway for the appearance of life on Earth. He first formulated a scenario of *in situ* production of complex organic compounds on Earth as prerequisites for the origin of life on Earth. 'If we could conceive in some warm little pond, with all sorts of ammonia and phosphoric salts, light, heat, electricity etc., present, that a protein compound was chemically formed' (Darwin, 1871). More than 50 years later, A.I. Oparin and J.B.S. Haldane independently developed a scenario for the origin of life from inanimate matter (Oparin, 1924; Haldane, 1928), but it took about 80 years until this idea of a 'warm little pond' was tested in an experiment when Stanley Miller and Harold Urey exposed a mixture of methane, ammonia, hydrogen, and water to electric discharges. Among the compounds formed, amino acids, the building blocks of proteins, were identified (Miller, 1953). Since then, a variety of possible atmospheres ranging from highly reducing (e.g. hydrogen, ammonia, methane, water) to nonreducing compositions (e.g. carbon dioxide, nitrogen, water) have been exposed to different energy sources to determine their potential for permitting abiotic organic synthesis. Nonreducing atmospheres resulted in very poor production of amino acids and other nitrogen-containing compounds, if they were formed at all. However, just such a non-reducing atmosphere is suggested to have prevailed on the early Earth. Therefore, *in situ* abiotic synthesis of amino acids on the early Earth seems to be a rather rare or even unlikely process. Today, the most favored conception is that the import of extraterrestrial organic compounds on Earth triggered life on Earth by evolving in liquid water (Horneck and Brack, 1992).

Since abiotic synthesis of organic compounds has been found to be a common process, either on the primitive Earth (e.g. in some protected niches of reducing gas mixtures) or elsewhere in the universe, Panspermia was no further considered a likely route to the appearance of life on Earth. The theory has been subjected to several criticisms, with arguments such as (i) it cannot be experimentally tested,

(ii) it shunts aside the question of the origin of life, and (iii) spores will not survive long-time exposure to the hostile environment of space, especially vacuum and radiation (reviewed in Horneck, 1995).

Although we do not know whether Panspermia is likely to have occurred in the history of the solar system, or whether it is feasible at all, with space technology a new tool is available for testing experimentally whether microorganisms are capable of surviving a hypothetical journey from one planet to another. Results from such experiments in space, which will be discussed later on, could give some answers to the criticism mentioned above. Additional support to revisit Panspermia comes from a variety of recent discoveries. These are: (i) the detection of meteorites, some of lunar and some of Martian origin (Kerr, 1987); (ii) the detection of organics and supposition of microbial fosssils in one of the Martian meteorites (McKay *et al.*, 1996); (iii) the probability of small particles of diameters between 0.5 µm and 1 cm (Moreno, 1988) or even boulder-sized rocks reaching escape velocities by the impact of large meteorites on a planet, for example on Earth (Melosh, 1988) or Mars (O'Keefe and Ahrens 1986; Vickery and Melosh 1987); (iv) the ability of bacterial spores to survive the shockwaves of a simulated meteorite impact to a certain extent (Horneck and Brack, 1992; Horneck *et al.*,

2001); (v) the high UV resistance of microorganisms at the low temperatures of deep space (Weber and Greenberg, 1985); (vi) the high survival of bacterial spores over extended periods in space (Horneck *et al.*, 1994); (vii) the paleogeochemical evidence of a very early appearance of life on Earth in the form of metabolically advanced microbial prokaryotic ecosystems leaving not more than approximately 0.4 Ga for the evolution of life from the simple precursor molecules to the level of a prokaryotic, photoautotrophic cell (Schidlowski, 1988); (viii) the biochemical evidence of a common ancestor for all life forms on Earth (Woese *et al.*, 1990); and (ix) the feasibility of artificial or directed transport by probes sent to other planets (Crick and Orgel, 1973).

In view of these observations and arguments, a critical revisit of the hypothesis of Panspermia will be undertaken. The hypothetical scenario of interplanetary transfer of life would involve the following steps: (i) escape process, i.e. removal to space of biological material which has survived being lifted from the surface to high altitudes; (ii) interim state in space, i.e. survival of the biological material over timescales comparable with the interplanetary or interstellar passage; (iii) entry process, i.e. non-destructive deposition of the biological material on another planet (Figure 7.1). In the following, results from experimentation in space will be presented to test the

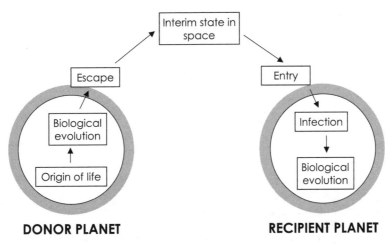

Figure 7.1 Hypothetical scenario of interplanetary transfer of life.

feasibility of step two, i.e. the impact of the space environment on seeding of life throughout the solar system.

7.3 Space environment of interest

The environment with which microorganisms in space have to cope is characterized by a high vacuum, an intense radiation climate of galactic and solar origin and extreme temperatures (Table 7.1). In free interplanetary space, pressures down to 10^{-14} Pa prevail. Within the vicinity of a body, the pressure may significantly increase due to outgassing. In a low Earth orbit of an altitude ≤ 500 km, pressure reaches 10^{-7} to 10^{-4} Pa. The major constituents of this environment are molecular oxygen and nitrogen as well as highly reactive oxygen and nitrogen atoms. In the vicinity of a spacecraft, the pressure further increases, depending on the

degree of outgassing of the spacecraft itself. In the Shuttle cargo bay, a pressure of 3×10^{-5} Pa was measured.

The planets in our solar system are continuously bombarded by charged particles coming from unknown sources outside our solar system and from the Sun (Figure 7.2). The galactic cosmic rays are composed of energetic particles which cover a broad spectrum of energy and mass values. About 98% are atomic nuclei and 2% electrons and positrons. The nuclear component consists of about 87% protons, about 12% α particles and about 1% nuclei of $Z > 2$, the so-called HZE particles. These nuclei are stripped off all their orbital electrons and have traveled for several million years through the galaxy before entering the solar system. When these charged particles enter the solar system, they interact with the outbound streaming of the solar wind and the interplanetary magnetic field which vary with the solar activity. In interplanetary space, the

Table 7.1 **Parameters of the environment of interplanetary space relevant for studies on microbial responses to space (modified from Horneck, 1999).**

Space parameter	Interplanetary space	Earth orbit	Simulation facility
Space vacuum Pressure (Pa)* Residual gas (part/cm^{-3})	!0^{-14} 1	10^{-7}–10^{-4} 10^4–10^5 H 10^4–10^6 He 10^3–10^6 N 10^3–10^7 O	10^{-7}–10^{-4} Different values[a]
Solar electromagnetic radiation Irradiance (W/m^2) Spectral range (nm)	Different values[b] Continuum	1380 Continuum	Different values[c] Different spectra[c]
Cosmic ionizing radiation Dose (Gy/a)	≤ 0.1[d]	400–10 000[e]	wide range[c]
Temperature (K)	>4[b]	wide range[b]	wide range[c]
Microgravity (g)	<10^{-6}	10^{-3}–10^{-6}	0–1000

* 1 Pa = 10^{-5} bar
[a] Depending on pumping system and requirements of the experimenter
[b] Depending on orientation and distance to Sun
[c] Depending on the radiation source and filter system
[d] Depending on shielding, highest values at mass shielding of 0.15 g/cm^2
[e] Depending on altitude and shielding, highest values at high altitudes and shield of 0.15 g/cm^2

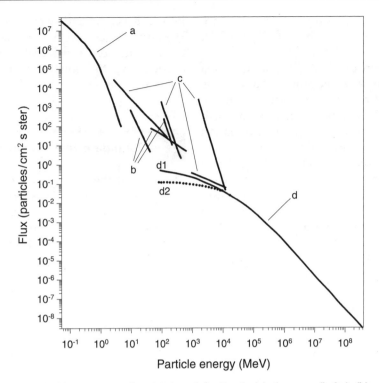

Figure 7.2 The radiation field in space in the vicinity of the Earth: (a) electrons (belts); (b) protons (belts); (c) solar flares; (d) heavy ions of galactic cosmic radiation; (d1) during solar minimum; (d2) during solar maximum.

annual radiation dose amounts to $\leq 0.1\,Gy/a$ depending on mass shielding with the highest dose at $0.15\,g/cm^2$ shielding due to built-up radiation. At the orbit of the terrestrial planets, a high solar activity is correlated with a low intensity of galactic cosmic rays and vice versa (Figure 7.2).

An additional source of particles impinging on the planets is the Sun itself. The flux of the solar particle radiation depends on the 11-year cycle of solar activity. It is composed mainly of protons with a smaller contribution of α particles and a few heavier ions. The solar wind protons are of relatively low energies. Sudden brightness in the solar disk indicates solar particle events that lead to the ejection of energetic plasma, produce magnetic storms and particulate radiation in the interplanetary space. These solar particle events release high fluxes of protons of considerably high energies and much lesser quantities of heavier ions and electrons (Figure 7.2). Their occurrence is linked to the

solar cycle with a higher probability at the end of a solar maximum.

In the vicinity of a planet, the flux of the cosmic ray nuclei is modulated by the magnetic field of the planet, if it exists. Among the terrestrial planets, the Earth is the only planet with a strong magnetosphere. This causes, on the one hand, the deflection of comic ray particles of low energies, and on the other hand the capture of solar wind light particles forming the radiation belts (Figure 7.2). Other than the Earth, no other terrestrial planetary body has radiation belts.

The spectrum of solar electromagnetic radiation spans over several orders of magnitude, from short wavelength X-rays to radio frequencies. At the distance of the Earth (1 AU), solar irradiance amounts to $1360\,W\,m^{-2}$, which is the solar constant. Of this radiation, 45% is attributed to the infrared fraction, 48% to the visible fraction and only 7% to the ultraviolet range. The extraterrestrial solar spectral UV irradiance has been measured during several space missions, such as Spacelab 1 and EURECA (Figure 7.3).

The temperature of a body in space depends on its position towards the Sun, and also on its surface, size and mass. It is determined by the absorption and emission of energy. In Earth orbit, the energy sources are the solar radiation ($1360\,W\,m^{-2}$), the Earth albedo ($480\,W\,m^{-2}$) and terrestrial radiation ($230\,W\,m^{-2}$). In Earth orbit, the temperature of a body can reach extreme values. In different space exposure experiments, the samples were subjected to temperatures in the range of 243 K to 318 K (Horneck, 1998).

Calculations have been made for the time it would take to travel from one planet of our solar system to another, for example from Mars to Earth by random motion. A mean time of several hundred thousands to millions of years has been estimated for boulder-sized rocks (Melosh, 1988), but only a few months for microscopic particles (Moreno, 1988). For an interstellar transport, such as from one solar system to another, even longer time periods up to 10 million years may be required. Weber and Greenberg (1985) suggest the random motion of interstellar clouds as a possible mechanism: as a solar system moves through a dense molecular cloud, microbes, for example, ejected by a meteorite impact, may be captured

into such a cloud and swept along with the gas. However, Mileikowsky *et al.* (submitted) have recently demonstrated that these time spans are too long to allow survival of any known microbial species.

7.4 Candidate test organisms

Throughout the history of life on Earth, there was probably never a need to adapt to conditions comparable to interplanetary space (unless life arose on Mars or elsewhere and was transported here). Therefore, perfectly suitable test systems adapted to the hostile conditions of space are not at hand. However, a variety of organisms exist that are adapted to grow or survive in extreme conditions of our biosphere. Some of them may be suitable candidates for studies on microorganisms in space. The rationale behind such studies is that if terrestrial organisms can survive the rigors of space, it is more likely that such transfer is a feasible process.

7.4.1 Bacterial spores

Several prokaryotic and eukaryotes microorganisms possess strategies of surviving unfavorable conditions in a kind of dormant state,

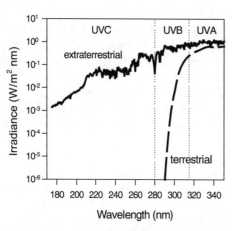
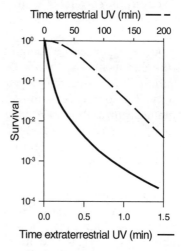

Figure 7.3 Spectrum and effects of extraterrestrial and terrestrial UV radiation on the survival of spores of *B. subtilis* (modified from Horneck and Brack, 1992).

and are capable of full metabolic activity if conditions become favorable again. Bacterial examples are *Bacillus* and *Clostridium* species, which are capable of producing dormant spores. In these spores, the DNA is extremely well protected against environmental stressors, such as desiccation, oxidizing agents, UV and ionizing radiation, low and high pH as well as temperature extremes. The high resistance of *Bacillus* spores is mainly due to two factors: (i) a dehydrated core enclosed in a thick protective envelop, the cortex and the spore coat layers (Figure 7.4), and (ii) the saturation of their DNA with small, acid-soluble proteins whose binding greatly alters the chemical and enzymatic reactivity of the DNA (Setlow, 1995). The space environment is governed by a high vacuum, temperature extremes and an intense radiation of solar and galactic origin. Yet, *Bacillus subtilis* spores have survived for extended periods in space (with the maximum duration tested to date of 6 years) (Horneck *et al.*, 1994). Isolates from Dominican amber suggest that *Bacillus* spores remain viable even for several millions of years (Cano and Borucki, 1995). Due to this exceptional resistance to environmental extremes, spores of *B. subtilis* have been subjected to intense studies on their survival in space (Horneck, 1993a, 1998).

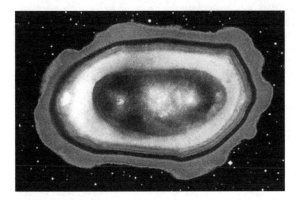

Figure 7.4 Electronmicrograph of a spore of *B. subtilis* showing the cytoplasmic area (spore core) and the surrounding layers, the membrane, the cortex, and the spore wall (kindly provided by S. Pankratz). The long axis of the whole spore is 1.2 μm, the core area 0.25 μm^2.

7.4.2 *Endolithic microbial communities*

Endolithic life forms, such as cyanobacteria, algae, fungi and lichens, are pioneers and survivors in extreme habitats. They represent a simple microbial ecosystem living inside rocks. Inside the rock is a favorable microclimate within a hostile macroclimate of low humidity, temperature extremes, and a high influx of solar radiation. Examples have been found in rocks of the dry valleys of Antarctica (Friedmann, 1982, Wynn-Williams, 1996) as well as in those from hot deserts. In addition, cyanobacteria have developed a variety of sun-protective mechanisms, such as the accumulation of UV-B-absorbing pigments and photorepair mechanisms for DNA and other functional molecules. Most lichens are extremely tolerant to desiccation. They occur in the driest and extremely cold regions on Earth and are known as extremists and pioneers (Kappen, 1973). Endolithic microbial communities from Antarctica will be studied in space on the platform EXPOSE which will be mounted on the external structure of the International Space Station (König and Horneck, 1997; Horneck *et al.*, 1999). The hypothesis to be tested states that a stratified endolithic habitat provides the extra protection against UV-B damage that enables cyanobacteria to survive conditions when their UV-B-protective pigment content alone in the free-living state is insufficient protection for metabolic activity and survival. These studies are of great significance for terrestrial and extraterrestrial biology, e.g. related to Mars (Wynn-Williams, 1996).

7.4.3 *Microorganisms in evaporites*

Osmophilic microorganisms, such as bacteria, cyanobacteria and algal assemblages, trapped in evaporite deposits show metabolic activity, although at a very slow rate (Rothschild *et al.*, 1994). These endoevaporitic microorganisms are capable of fixing carbon and nitrogen which allows them to survive while trapped in salt, and even to maintain all functions necessary for active life (Figure 7.5). Earlier space experiments have demonstrated the ability of osmophilic microbes to survive in the space

Figure 7.5 Evaporite comprising a layer of osmo-philic microorganisms (kindly provided by L. Rothschild).

environment (Mancinelli *et al.*, 1998). Endoevaporites will also be studied on the EXPOSE platform of the International Space Station (Table 7.2). The objectives are to determine the role of endogenous protection (e.g. by the carotenoid pigment) and exogenous protection (e.g. by salt crystals) against the deleterious effects of space, above all solar UV radiation and vacuum. Results of this study will be relevant to the potential of interplanetary transfer of viable microbes as well as the evolution of life on early Earth and possibly on Mars.

7.4.4 *Extremely radiation-resistant microorganisms*

Bacteria of the species *Deinococcus radiodurans* are the most radiation-resistant bacteria known

Table 7.2   Experiments of the Consortium ROSE to Study the Responses of Organisms to the Space Environment by Use of the EXPOSE Facility on Board of the International Space Station (Horneck *et al.*, 1999).

Experiment	Objective	Assay system	Investigator
ENDO	Impact of ozone depletion on microbial primary producers from sites under the 'ozone hole'	1. Endolithic microbial communities 2. Mats of cyanobacteria 3. Mats of algae	Wynn-Williams, D.D. UK
OSMO	Protection in evaporites against solar UV and anhydrobiosis	1. *Synechococcus* in and beneath gypsum-halite 2. *Haloarcula* (pigmented and non-pigmented) in and beneath NaCl 3. *Haloarcula* DNA in KCl	Mancinelli, R.L. USA
SPORES	Protection of spores by meteorite material against space (UV, vacuum, and ionizing radiation)	1. *B. subtilis* spores 2. Fungal spores 3. Lycopod spores All in or beneath meteorite material	Horneck, G. Germany
PHOTO	Main photoproducts in dry DNA and DNA from dry spores	1. DNA 2. Bacterial spores	Cadet, J. France
PUR	Sensitivity of the biologically effective UV radiation to ozone	1. T7 bacteriophage 2. Phage-DNA 3. Uracil	Rontó, G. Hungary
SUBTIL	Mutational spectra induced by space vacuum and solar UV	1. *B. subtilis* spores 2. Plasmid DNA	Munakata, N. Japan

to exist on Earth today. Although *D. radiodurans* is non-sporulating, it can go into a kind of dormancy under certain environmental adverse conditions, such as lack of food, desiccation or low temperatures (Moseley, 1983; Mattimore and Battista, 1996). During dormancy, the metabolism is inhibited or occurs only at a negligible rate. As a consequence, radiation damage to DNA accumulates. However, *D. radiodurans* possesses a very efficient DNA repair system which comes into action when dormancy ends. This very common soil microorganism was exposed to space during the European EURECA mission (Dose *et al.*, 1995). Because of their extraordinarily high radiation resistance cells of *D. radiodurans* are considered suitable for further space experiments.

7.4.5   *Anhydrobiotic higher organisms*

Although most studies under simulated space conditions have been performed on prokaryotic microorganisms or their cellular components or biomolecules, one should also consider higher organisms that have developed strategies of surviving complete dehydration. These organisms include the slime mold *Dictyostelium*, dry active baker's yeast, brine shrimp cysts of *Artemia salina*, and dry larvae and adults of several species of nematodes. In some of these organisms the survival of dehydration is related to the accumulation of polyols, in particular trehalose (Wormsley, 1981).

7.4.6   *Biomolecules*

To clarify the mechanisms of biological responses to the parameters of space, systematic studies on relevant biomolecules, such as DNA, protein or liposomes, provide complementary information to the studies on whole organisms.

7.4.7   *The ROSE consortium*

To study the Responses of Organisms to the Space Environment (ROSE), a consortium of scientists has formed to study photobiological processes in the simulated radiation climate of

planets and the probabilities and limitations for life to be distributed among the bodies of our solar system. These experiments will be performed in the EXPOSE facility of the International Space Station (König and Horneck, 1997). The experiments selected for the first flight are listed in Table 7.2. Heretofore, EXPOSE will provide opportunities for further studying exobiological questions regarding the potential pathways of the appearance of life on planets and the distribution of life beyond its planet of origin (Horneck *et al.*, 1999).

7.5   **Microbial responses to space environment**

7.5.1   *Effects of extraterrestrial solar UV radiation*

Solar UV radiation has been found to be the most deleterious factor of space, as tested with dried preparations of viruses, and of bacterial and fungal spores (Horneck, 1993a). The reason for this high biological efficiency is the highly energetic UV-C (190–280 nm) and vacuum UV radiation (<190 nm) that is directly absorbed by the DNA, as demonstrated by action spectroscopy in space (Figure 7.6) (Horneck *et al.*, 1995). This action spectrum follows the same wavelength dependence as the absorption spectrum of pure DNA (Ito and Ito, 1986). It is important to note that in addition to the well-known peak around 260 nm, the DNA absorption shows a second maximum towards shorter wavelengths. In the vacuum UV region, absorption of UV by the sugar phosphate backbone of the DNA becomes decisive. The full spectrum of extraterrestrial UV radiation kills unprotected spores of *B. subtilis* within seconds (Horneck and Brack, 1992). These harmful UV ranges do not reach the surface of the Earth because they are effectively absorbed by the Earth's atmosphere (Figure 7.3). Compared to the average UV radiation climate at the surface of the Earth, the extraterrestrial UV radiation is about three orders of magnitude more efficient in killing spores of *B. subtilis* (Horneck *et al.*, 1996).

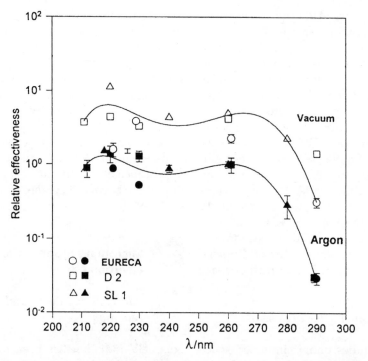

Figure 7.6 Action spectrum of extraterrestrial solar UV radiation for the inactivation of spores of *B. subtilis* after irradiation in space vacuum or in air or argon at atmospheric pressure (data from three space experiments; modified from Horneck *et al.*, 1995).

The photobiological and photochemical effects of extraterrestrial UV radiation are based on the production of specific photoproducts in the DNA that are highly mutagenic and lethal (reviewed in Cadet *et al.*, 1985). The most damaging photochemical lesions are thymine containing dimers: cyclobutadipyrimidines, the (6-4) pyrimidine–pyrimidone adducts and their Dewar valence isomer (reviewed in Cadet and Weinfeld, 1993). Vacuum UV has a specific mutagenic signature: compared to UV-C radiation, deletions become more frequent in inducing mutations as well as A:T → T:A transversions, whereas the frequency of G:C → C:G transversions become less frequent (Wehner and Horneck, 1995).

7.5.2 Effects of space vacuum

Because of its extreme dehydrating effect, space vacuum has been considered to be one of the factors that may prevent interplanetary transfer of life (Nussinov and Lysenko, 1983).

However, space experiments have shown that up to 70% of bacterial and fungal spores survive short-term (e.g. 10 days) exposure to space vacuum, even without any protection (Horneck, 1993a). The chances of survival in space are increased if the spores are embedded in chemical protectants such as sugars, or salt crystals, or if they are exposed in thick layers. For example, 30% of *B. subtilis* spores survived nearly 6 years of exposure to space vacuum, if embedded in salt crystals, whereas 80% survived in the presence of glucose (Horneck *et al.*, 1994). It has been suggested that sugars and polyalcohols stabilize the structure of the cellular macromolecules, especially during vacuum-induced dehydration, leading to increased rates of survival.

To determine the protective effects of different meteorite materials, 'artificial meteorites' were constructed by embedding *B. subtilis* spores in clay, meteorite dust or simulated Martian soil (Horneck *et al.*, 1995). Their survival has been determined after exposure to the space environment. Crystalline salt provided sufficient protection for osmophilic microbes in the vegetative state to survive at least 2 weeks in space (Mancinelli *et al.*, 1998). For example, a species of the cyanobacterium *Synechococcus*, which inhabits gypsum-halite crystals, was capable of nitrogen and carbon fixation and about 5% of a species of the extreme halophile *Haloarcula* survived after exposure to the space environment for 2 weeks in connection with a FOTON space flight.

The mechanisms of damage due to vacuum exposure are based on the extreme desiccation. If not protected by internal or external substances, cells in a vacuum experience dramatic changes in such important biomolecules as lipids, carbohydrates, proteins and nucleic acids. Upon desiccation the lipid membranes of cells undergo dramatic phase changes from planar bilayers to cylindrical bilayers (Cox, 1993). The carbohydrates, proteins and nucleic acids undergo so-called Maillard reactions, i.e. amino-carbonyl-reactions, to give products that become crosslinked, which eventually leads to irreversible polymerization of the biomolecules (Cox, 1993). Concomitant with these structural changes are functional changes including altered selective membrane permeability, inhibited or altered enzyme activity, decreased energy production, alteration of genetic information, etc.

Vacuum-induced damage to the genetic material of life, the DNA, is especially dramatic, because it may lead to death or mutation. An up to tenfold increased mutation rate over the spontaneous rate has been observed in spores of *B. subtilis* after exposure to space vacuum (Figure 7.7) (Horneck *et al.*, 1984). This mutagenic effect by vacuum is probably based on a unique molecular signature of tandem-double base changes at restricted sites in the DNA (Munakata *et al.*, 1997). In addition, DNA strand breaks have been observed to be induced by exposure to space vacuum (Dose *et al.*, 1992, 1995). Such damage would accumulate during long-term exposure to space

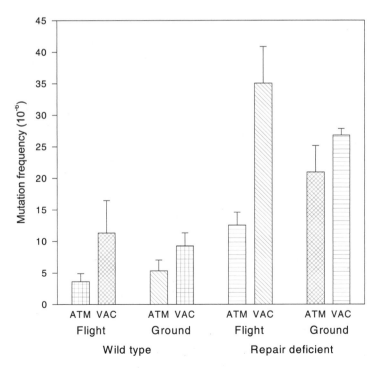

Figure 7.7 Exposure to space vacuum resulted in an increased mutation rate (histidine reversions) in spores of *B. subtilis*; results of an experiment during the EURECA space mission (Horneck *et al.*, 1995).

vacuum, because DNA repair is not active during anhydrobiosis. Survival ultimately depends on the efficiency of the repair systems after rehydration.

7.5.3 Combined action of solar UV and space vacuum

If spores are simultaneously exposed to solar UV radiation and space vacuum, they respond with an increased sensitivity to a broad spectrum of solar UV light (>170 nm) as well as to selected wavelengths (Horneck *et al.*, 1984; Horneck, 1993a) (Figure 7.6). Upon dehydration (e.g. in space vacuum) DNA undergoes substantial conformational changes, that is, from the fully hydrated B-conformation to the less hydrated A-conformation. This conversion in the physical structure leads to an altered DNA photochemistry, resulting in the observed increase in UV sensitivity of *B. subtilis* spores in space vacuum as compared to spores irradiated at atmospheric pressure. Several attempts were made to isolate and identify the photoproducts generated within the DNA of *B. subtilis* spores exposed to UV radiation in vacuum. Two thymine decomposition products, namely the *cis-syn* and *trans-syn* cyclobutadithymine (Thy<>Thy) as well as DNA-protein-crosslinking were found in addition to 5,6-dihydro-5(α-thyminyl) thymine (TDHT) (Horneck *et al.*, 1984; Lindberg and Horneck, 1991). From the efficiency of repair processes (photoenzymatic repair, spore photoproduct specific repair) it is concluded that photoproducts other than *cis-syn* Thy<>Thy and TDHT seem to be responsible for the UV supersensitivity of spores, if irradiated under vacuum conditions (Horneck, 1993a).

7.5.4 Effects of galactic cosmic rays

Among the ionizing components of radiation in space the heavy primaries, the so-called HZE particles, are the most biologically effective species (reviewed in Horneck, 1993b; Kiefer *et al.*, 1996). Because of their low flux (they contribute to approximately 1% of the flux of particulate radiation in space) methods have been developed to localize precisely the trajectory of an HZE particle relative to the biological object and to correlate the physical data of the particle relative to the observed biological effects along its path. In the Biostack method, visual track detectors are sandwiched between layers of biological objects in a resting state, for example *Bacillus subtilis* spores (Bücker and Horneck, 1975) (Figure 7.8). This method allows (i) localization of each HZE particle's trajectory in relation to the biological

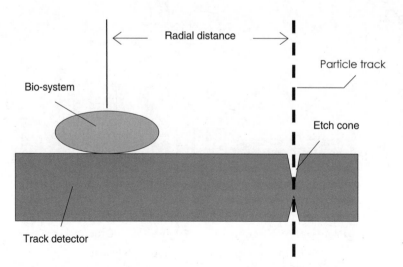

Figure 7.8 Biostack method to localize the effect of single particles (HZE particles) of cosmic radiation: biological layers are sandwiched between track detectors.

specimens; (ii) investigation of the responses of each biological individual hit separately, in regard to its radiation effects; (iii) the measurement of the impact parameter b (i.e. the distance between the particle track and the sensitive target); (iv) determination of the physical parameters (charge Z, energy E and linear energy transfer LET); and finally (v) correlation of the biological effects with each HZE particle parameter.

Bacterial spores having a cytoplasmic core with a geometrical cross-section of 0.2 to $0.3\,\mu m^2$ (Figure 7.4) are suitable test systems in HZE particle research, especially for sounding the biological effectiveness in dependence of the impact parameter for each particle's trajectory when applying the Biostack concept (Figure 7.8). For this purpose, spores of *B. subtilis* (wild type with regard to DNA repair) in a suspension of 0.001% aqueous polyvinyl acetate (PVA) solution were spread onto cellulose nitrate (CN) track detector sheets. After drying, a fixed contact of the spore monolayer with the track detector was achieved. These sheets, carrying the biological layer, were stacked between pure plastic detector sheets (Bücker and Horneck, 1975). The complete Biostack experiment package (about $10 \times 10 \times 10\,cm^3$ in size) accommodated in a hermetically sealed aluminum container was stowed inside the spacecraft (Apollo or Spacelab) and exposed to cosmic radiation. Total absorption by the spacecraft walls was approximately 2.4 g/ cm^2. The daily flux of HZE particles with LET $\geq 130\,keV/\mu m$ was 0.3–0.7 particles/cm^2.

After retrieval and disassembly, CN foils carrying a spore layer were etched (6 N NaOH, 30°C) only on the CN side resulting, along the trajectory of an HZE particle, in an etch cone at the side opposite to the spore layer. The procedure was followed by individual etching of each etch cone under microscopic control (pinpoint etching), until the tip of the etch cone approached the spore layer up to a few microns. By extrapolation of the etch cone, the target area in the spore layer was determined. The target area was defined as an area of a radius of $5\,\mu m$ off the point of intersection of an HZE particle with the spore layer. Spores of the target area were individu-

ally transferred onto nutrient agar by micromanipulation; germination, outgrowth and growth were followed until the microcolony was formed. The original position of each spore relative to the HZE particle's trajectory, i.e. the impact parameter, was obtained from micrographs and after breakthrough-etching of the etch cone (Schäfer *et al.*, 1977). The accuracy in determining the impact parameter – depending on the dip angle of the trajectory – was up to $0.2\,\mu m$, which is well below the size of a single spore. The physical parameters of the respective HZE particle, such as Z and LET, were obtained from the adjacent physical detector foils. Figure 7.9 shows the frequency of inactivated spores as a function of the impact parameter b. About 1000 individual spores were analyzed. Spores within $b \leq 0.25\,\mu m$ were inactivated by 73%. The frequency of inactivated spores dropped abruptly at $b > 0.25\,\mu m$. However, 15–30% of spores located within $0.25 < b < 3.8\,\mu m$ were still inactivated. Hence, spores were inactivated well beyond $1\,\mu m$, which distance would roughly correspond to the dimensions of a spore. At the distance of $1\,\mu m$, the mean δ-ray dose ranges between 0.1 Gy and 1 Gy, depending on the particle, and declines rapidly with increasing b (Facius *et al.*, 1978a). This value of 0.1 to 1 Gy is by several orders of magnitude below the D_{37} (dose reducing survival by e^{-1}) of electrons which amounts to 800 Gy. Therefore, the radial long-ranging biological effectiveness around the trajectory of an HZE particle (up to b = $3.8\,\mu m$) cannot merely be explained by the δ-ray dose. These results were largely confirmed by experiments at heavy ion accelerators (Horneck, 1993b).

Taking the results from the experiments in space as well as those obtained at accelerators, one can draw the following general conclusion:

i. The inactivation probability for spores, centrally hit, is always substantially less than one.
ii. The effective range of inactivation extends far beyond the range of impact parameter where inactivation of spores by δ-rays can be expected. This far-reaching effect is less pronounced for ions of low energies

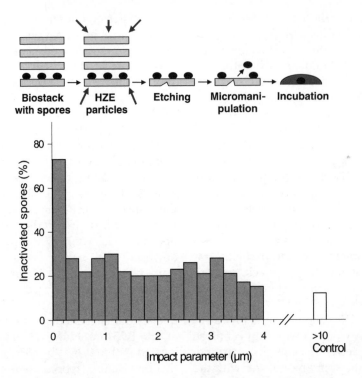

Figure 7.9 Results from Biostack experiments in space: inactivation probability of spores of *B. subtilis* as a function of the distance from the particles' trajectory (modified from Horneck, 1993b).

(1.4 MeV/u), a phenomenon which might reflect the 'thindown effect' at the end of the ion's path (Katz, 1986).

iii. The dependence of inactivated spores from impact parameter points to a superposition of two different inactivation mechanisms: a short-ranged component reaching up to about 1 μm may be traced back to the δ-ray dose and a long-ranged one that extends at least to somewhere between 4 and 5 μm off the particle's trajectory, for which additional mechanisms are conjectured, such as shockwaves, induced UV radiation, or thermophysical events (Facius *et al.*, 1978b).

These results of the Biostack experiments have shown that *B. subtilis* spores can survive even a central hit of an HZE particle of cosmic radiation. Such HZE particles of cosmic radiation are conjectured to set the ultimate limit on the survival of spores in space because they penetrate even thick shielding. A spore's maximum

time to escape a hit by an HZE particle (e.g. iron of LET >100 keV/μm) has been estimated to be several hundred thousands of years. This time span complies with estimates for boulder-sized rocks to travel from one planet of our solar system to another, e.g. from Mars to Earth, by random motion (Melosh, 1988).

Concerning shielding against radiation in space, a few micrometers of meteorite material are sufficient to give efficient protection against UV. If the material is without cracks, less than $0.5\,g/cm^2$ is required against the diffuse X-rays, and about $30\,g/cm^2$ against the solar particles. However, because the particles of cosmic radiation interact with the shielding material by creating secondary radiation, the dose rates go through a maximum with increasing shielding thickness. Based on experimental data from accelerator experiments with *B. subtilis* spores (Baltschukat and Horneck, 1991), an estimated density of the meteorite of $1.8\,g/cm^2$ (taken from data on Martian regolith) and a NASA model on an HZE transport code for cosmic radiation (Cuccinotta *et al.*, 1995), the dose rate has been calculated behind different shielding thickness: the dose rate reaches its maximum behind a shielding layer of about 10 cm, behind about 130 cm the value is the same as obtained without any shielding and only for higher shielding thickness the dose rate reduces significantly (Figure 7.10) (Mileikowsky *et al.*, 2000). The calculations also show that even after 25 million years in space, a substantial fraction of a spore population (10^{-6}) would survive the exposure to cosmic radiation if shielded by 2 to 3 m of meteorite material. The same surviving fraction would be reached after about 500 thousand years without any shielding, and after 350 thousand years after 1 m of shielding.

7.5.5 *Effects of long-term exposure to space*

During the LDEF mission, for the first time *B. subtilis* spores were exposed to the full environment of space for an extended period of time, i.e. for nearly 6 years, and their survival was determined after retrieval. The samples were separated from space only by a perforated aluminum dome, which allowed access of space vacuum, solar UV radiation and most of the components of cosmic radiation (Horneck *et al.*, 1994). Figure 7.11 shows that thousands of spores survived the space journey (from an initial sample size of 10^8 spores). All spores were exposed in multilayers and predried in the presence of glucose. The spore samples had turned from white into yellow during the mission, a phenomenon which is probably due to photochemical processes. This suggests that all spores in the upper layers were completely inactivated by the high flux of solar UV radiation. With time, they formed a protective crust which considerably attenuated the solar UV radiation for the spores located beneath this layer. Therefore the survivors probably originated from the innermost part of the samples. Of spores covered by an aluminum foil, which protected against UV radiation, nearly 100% survived the space journey (Figure 7.11).

To travel from one planet of our solar system to another, timescales of several hundreds thousands to millions of years have been estimated for rocks of sizes which would provide sufficient shielding against galactic cosmic radiation. For an interstellar transport, the following picture has been drawn: the clouds of

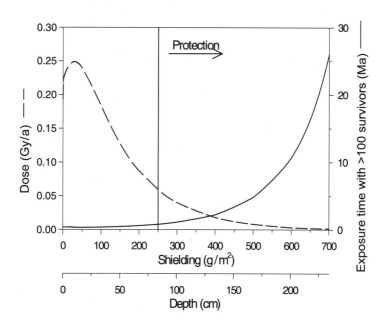

Figure 7.10 Shielding of bacterial spores against galactic cosmic radiation by meteorite material (data from Mileikowsky *et al.*, 2000).

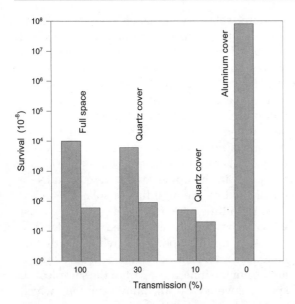

Figure 7.11 Survival of spores of *B. subtilis* after exposure to space for nearly 6 years (results of an experiment on LDEF) (modified from Horneck *et al.*, 1994; Horneck, 1998).

gaseous and particulate matter between the stars move in a random fashion at speeds of about 10 km/s. If a bacterial spore is captured in such a cloud, it will be swept along with the gas. Given the distance between neighboring stars of 0.1–1 pc, this corresponds to a passage time of 10^6–10^7 years. Thus, for interstellar transfer, spores would have to survive at least for 10^6–10^7 years in such a cloud (Weber and Greenberg, 1985). So far, results from exposure of spores to space up to several years are available (Horneck *et al.*, 1994). More data on long-term effects are required. Exposure platforms, such as the EXPOSE on the International Space Station, are especially suited for this kind of research. Such studies may eventually allow extrapolation to time spans, as required for interplanetary or interstellar transport of life.

7.6   Conclusion

Although it will be difficult to prove that life could be transported through our solar system, experiments in space estimate the chances of

resistant microbial forms to survive exposure to the complex matrix of the space environment. This will provide basic data on whether such travel is possible. Experiments in space, which were performed on free platforms since the Apollo era, becoming more sophisticated on the external platform of Spacelab and on free-flying satellites such as LDEF, EURECA, and FOTON, have provided some insight into responses of bacterial spores and other extremophile microorganisms to the parameters of space. The most interesting results are summarized as follows:

i. extraterrestrial solar UV radiation killed 99% of *B. subtilis* spores in 10 s;
ii. extraterrestrial solar UV radiation was a thousand times more efficient in killing spores than the UV radiation at the surface of the Earth;
iii. space vacuum increased the UV sensitivity of the spores;
iv. although spores survived extended periods of time in space vacuum (up to 6 years) genetic changes occurred, such as increased mutation rates;
v. spores could survive 6 years in space, especially if protected against solar UV radiation;
vi. in space, spores could escape a hit of a cosmic HZE particle (e.g. iron ion) for up to 1 million years;
vii. a meteorite layer of more than 2.5 m would effectively protect the spores against galactic cosmic radiation.

The data obtained so far on the responses of resistant microorganisms to the environment of space seem to support the supposition that space – although it is very hostile to terrestrial life – may not be an absolute barrier for cross-fertilization in the solar system. The various effects on the DNA of the space travelers might even serve evolutionary purposes. Hence, the transport of microbes inside rocks through the solar system might be a feasible process. If protected against solar UV and galactic cosmic radiation, spores may survive inside meteorites over extended periods of time.

Survival in space over extended periods of time is just one of the requirements to be met in a hypothetical scenario of transfer of life in the solar system (Figure 7.1). Several questions remain unanswered, such as:

i. Can spores survive the ejection process?
ii. Can spores survive the landing process?
iii. What are the most realistic test organisms for Panspermia?
iv. What kind of protection, chemical or physical, could be provided by the meteorite material or intrinsically by the organisms?

Relevant studies to tackle these questions will be performed in future studies in space and at ground-based facilities (Horneck, 1999; Horneck *et al.*, 2000). For future research on bacterial spores and other microorganisms in space, the European Space Agency ESA is developing the EXPOSE facility which will be attached to the balcony of the Columbus module of the International Space Station for 1.5 years. EXPOSE will support long-term *in situ* studies of microbes in artificial meteorites, as well as of microbial communities from special ecological niches, such as endolithic and endoevaporitic ecosystems (Horneck *et al.*, 1999). These experiments on the Responses of Organisms to the Space Environment (ROSE) include the study of photobiological processes in simulated radiation climates of planets (e.g. early Earth, early and present Mars, and the role of the ozone layer in protecting the biosphere from harmful UV-B radiation), as well as studies of the probabilities and limitations for life to be distributed beyond its planet of origin. Heretofore, the results from the EXPOSE experiments will eventually provide clues to a better understanding of the processes regulating the interactions of life with its environment.

References

Arrhenius, S. (1903). Die Verbreitung des Lebens im Weltenraum. *Die Umschau*, **7**: 481–485.

Baltschukat, K. and Horneck, G. (1991). Responses to accelerated heavy ions of spores of *Bacillus subtilis* of different repair capacity. *Radiat. Environm. Biophys.*, **30**: 87–103.

Bücker, H. and Horneck, G. (1975). Studies on the effects of cosmic HZE-particles in different biological systems in the Biostack experiments I and II, flown on board of Apollo 16 and 17. In: O.F. Nygaard, H.I. Adler and W.K. Sinclair (eds) *Radiation Research*. New York: Academic Press, pp. 1138–1151.

Cadet, J. and Weinfeld, M. (1993). Detecting DNA damage. *Anal. Chem.*, **65**: 675A–682A.

Cadet, J., Voituriez, L., Grand, A. *et al.* (1985). Photosensitized reactions of nucleic acids. *Biochimie*, **67**: 277.

Cano, R.J. and Borucki, M.K. (1995). Revival and identification of bacterial spores in 25 to 40 million-year-old Dominican amber. *Science*, **268**: 1060–1064.

Cox, C.S. (1993). Roles of water molecules in bacteria and viruses. *Origins Life Evol. Biosphere*, **23**: 29–36.

Crick, F.H.C. and Orgel, L.E. (1973). Directed panspermia. *Icarus*, **19**: 341–346.

Cucinotta, F.A., Wilson, J.W., Katz, R. *et al.* (1995). Track structure and radiation transport model for space radiobiology studies. *Adv. Space Res.*, **18**: (12)183–(12)194.

Darwin, C. (1871). Letter to Hocker, Febr. 1871.

DeDuve, C. (1991). *Blueprint for a Cell: The Nature and Origin of Life*. Burlington, North Carolina: Neil Patterson Publ.

Dose, K., Bieger-Dose, A., Labusch, M. and Gill, M. (1992). Survival in extreme dryness and DNA-single-strand breaks. *Adv. Space Res.*, **12**: (4)221–(4)229.

Dose, K., Bieger-Dose, A., Dillmann, R. *et al.* (1995). ERA-experiment 'space biochemistry'. *Adv. Space Res.*, **16**: (8)119–(8)129.

Facius, R., Bücker, H., Hildebrand, D. *et al.* (1978a). Radiobiological results from the *Bacillus subtilis* Biostack experiments within the Apollo and ASTP spaceflights. *Life Sci. Space Res.*, **16**: 151–156.

Facius, R., Bücker, H., Reitz, G. and Schäfer, M. (1978b). Radial dependence of biological response of spores of *Bacillus subtilis* around tracks of heavy ions. *Proc. 6th Symp. Microdosimetry, EUR 6064 DE-EN-FR*. London: Harwood Academic Press, pp. 977–986.

Friedmann, E.I. (1982). Endolithic microorganisms in the Antarctic cold desert. *Science*, **215**: 1045–1053.

Haldane, J.B.S. (1928). The origin of life. *Rationalist Annual*, **148**: 3.

Horneck, G. (1993a). Responses of *Bacillus subtilis* spores to space environment: results from experiments in space. *Origins Life Evol. Biosphere*, **23**: 37–52.

Horneck, G. (1993b). The Biostack concept and its application in space and at accelerators: studies on *Bacillus subtilis* spores. In: C.E. Swenberg, G. Horneck and E.G. Stassinopoulos (eds) *Biological Effects and Physics of Solar and Galactic Cosmic Radiation, Part A, NATO ASI Series A: Life Sciences 243A.* New York: Plenum, pp. 99–115.

Horneck, G. (1995). Exobiology, the study of the origin, evolution and distribution of life within the context of cosmic evolution: a review. *Planet. Space Sci.*, **43**: 189–217.

Horneck, G. (1998). Exobiological experiments in Earth orbit. *Adv. Space Res.*, **22**: (3)317–(3)326.

Horneck, G. (1999). Astrobiology studies of microbes in simulated interplanetary space. In: P. Ehrenfreund, C. Krafft, H. Kochan and V. Pirronello (eds) *Laboratory Astrophysics and Space Research.* Dordrecht: Kluwer, pp. 667–685.

Horneck, G. and Brack, A. (1992). Study of the origin, evolution and distribution of life with emphasis on exobiology experiments in Earth orbit. In: S.L. Bonting (ed.) *Advances in Space Biology and Medicine*, Vol. 2. Greenwich, CT: JAI Press, pp. 229–262.

Horneck, G., Bücker, H., Reitz, G. *et al.* (1984). Microorganisms in the Space Environment. Science, **225**: 226-228.

Horneck, G., Bücker, H., Reitz, G. (1994). Long-term survival of bacterial spores in space. *Adv. Space Res.*, **14**: (10)41–(10)45.

Horneck, G., Eschweiler, U., Reitz, G. *et al.* (1995). Biological responses to space: results of the experiment 'Exobiological Unit' of ERA on EURECA I. *Adv. Space Res.*, **16**: (8)105–(8)111.

Horneck, G., Rettberg, P., Rabbow, E. *et al.* (1996). Biological dosimetry of solar radiation for different simulated ozone column thickness. *J. Photochem. Photobiol. B: Biol.*, **32**: 189–196.

Horneck, G., Wynn-Williams, D.D., Mancinelli, R. *et al.* (1999). Biological experiments on the EXPOSE facility of the International Space Station. Proc. 2nd Europ. Symp. on Utilization of the Internat. Space Station, ESTEC, Noordwijk, ESA SP-433, pp. 459–468.

Horneck, G., Reitz, G., Rettberg, P. *et al.* (2000). A ground-based program for exobiological experiments on the International Space Station. *Planet. Space Sci.* 48: 507–513.

Horneck, G., Stöffler, D., Eschweiler, U. and Hornemann, U. (2001) Bacterial spores survive simulated meteorite impact. Icarus, **149**: 285–290.

Ito, A. and Ito, T. (1986). Absorption spectra of deox-

yribose, ribosephosphate, ATP and DNA by direct transmission measurements in the vacuum-UV (150–190 nm) and far-UV (190–260 nm) regions using synchrotron radiation as a light source. *Photochem. Photobiol.*, **44**: 355–358.

Kappen, L. (1973). Response to extreme environments. In: V. Ahmadjian and M.E. Hale (eds) *The Lichens III. 10.* New York: Academic Press, pp. 311–380.

Katz, R. (1986). Biological effects of heavy ions from the standpoint of target theory. *Adv. Space Res.*, **6**: (11)191–(11)198.

Kerr, R.A. (1987). Martian meteorites are arriving. *Science*, **237**: 721–723.

Kiefer, J., Kost, M. and Schenk-Meuser, K. (1996). Radiation biology. In: D. Moore, P. Bie and H. Oser (eds) *Biological and Medical Research in Space.* Berlin: Springer, pp. 300–367.

König, H. and Horneck, G. (1997). The Space Exposure Biology Assembly (SEBA). *ESA Microgravity News*, **10**: 8–12.

Lindberg, C. and Horneck, G. (1991). Action spectra for survival and spore photoproduct formation of *Bacillus subtilis* irradiated with short wavelength (200–300 nm) UV at atmospheric pressure and *in vacuo. J. Photochem. Photobiol. B: Biol.*, **11**: 69–80.

Mancinelli, R.L., White, M.R. and Rothschild, L.J. (1998). Biopan Survival I: exposure of the osmophiles *Synechococcus sp.* (Nageli) and *Haloarcula sp.* to the space environment. *Adv. Space Res.*, **22**: (3)327–(3)334.

Mattimore, V. and Battista, J.R. (1996). Radioresistance of *Deinococcus radiodurans*: functions necessary to survive ionizing radiation are also necessary to survive prolonged desiccation. *J. Bacteriol.*, **178**: 633–637.

McKay, D.S., Gibson, E.K., Thomas-Keprta, K.L. *et al.* (1996). Search for past life on Mars: possible relic biogenic activity in Martian meteorite ALH84001. *Science*, **273**: 924–930.

Melosh, H.J. (1988). The rocky road to Panspermia. *Nature*, **332**: 687–688.

Melosh, H.J. and Vickery, A.M. (1989). Impact erosion of the primordial atmosphere of Mars. *Nature*, **338**: 487–489.

Mileikowsky, C., Cucinotta, F.A., Wilson, J.W. *et al.* (2000). Natural transfer of viable microbes in space. Part 1. From Mars to Earth and Earth to Mars. *Icarus* 145: 391–427.

Mileikowsky, C., Cucinotta, F.A., Wilson, J.W. *et al.* (submitted). Natural transfer of viable microbes in space. Part 2. From a planet in one solar system to a planet in another solar system.

Miller, S.L. (1953). The production of amino acids

under possible primitive Earth conditions. *Science,* **117**: 528.

Moreno, M.A. (1988). Microorganism transport from Earth to Mars. *Nature,* **336**: 209.

Moseley, B.E.B. (1983). Photobiology and radiobiology of *Microccocus (Deinococcus) radiodurans.* In: K.C. Smith (ed.) *Photochemical and Photobiological Reviews,* Vol. 7. New York: Plenum, pp. 223–275.

Munakata, N., Saitou, M., Takahashi, N. *et al.* (1997). Induction of unique tandem-double change mutations in bacterial spores exposed to extreme dryness. *Mutation Research,* **390**: 189–195.

Nussinov, M.D. and Lysenko, S.V. (1983). Cosmic vacuum prevents Radiopanspermia. *Origins Life Evol. Biosphere,* **13**: 153–164.

Oberbeck, V.R. and Fogleman, G. (1990). Impact constrains on the environment for chemical evolution and the origin of life. *Origins Life Evol. Biosphere,* **20**: 181–195.

O'Keefe, J.D. and Ahrens, T.J. (1986). Oblique impact: a process for obtaining meteorite samples from other planets. *Science,* **234**: 346–349.

Oparin, A.I. (1924). *The Origin of Life* (in Russian). Moscow: Miscoviskiy Rabochii.

Richter, H. (1865). Zur Darwinschen Lehre. *Schmidts Jahrbuch Ges. Med.,* **126**: 243–249.

Rosing, M.T. (1999). 13C-Depleted carbon microparticles in >3700-Ma sea-floor sedimentary rocks from West Greenland. *Science,* **283**: 674–676.

Rothschild, L.J., Giver, L.J., White, M.R. and Mancinelli, R.L. (1994). Metabolic activity of microorganisms in evaporites. *J. Phycol.,* **30**: 431–438.

Schäfer, M., Bücker, H., Facius, R. and Hildebrand, D. (1977). High precision localization method for HZE particles. *Rad. Effects,* **34**: 129–130.

Schidlowski, M. (1988). A 3800 million-year isotopic record of life from carbon in sedimentary rocks. *Nature,* **333**: 313–318.

Schopf, J.W. (1993). Microfossils of the early Archean Apex chert: new evidence of the antiquity of life. *Science,* **260**: 640–646.

Setlow, P. (1995). Mechanisms for the prevention of damage to DNA in spores of *Bacillus* species, *Ann. Rev. Microbiol.,* **49**: 29–54.

Vickery, A.M. and Melosh, H.J. (1987). The large crater origin of SNC meteorites. *Science,* **237**: 738–743.

Weber, P. and Greenberg, J.M. (1985). Can spores survive in interstellar space? *Nature,* **316**: 403–407.

Wehner, J. and Horneck, G. (1995). Effects of vacuum-UV and UV-C radiation on dry *E. coli* plasmid pUC19: II. Mutational specificity at the *lacZ* gene. *J. Photochem. Photobiol. B: Biol.,* **30**: 171–177.

Woese, C.R., Kandler, O. and Wheelis, M.L. (1990). Towards a natural system of organisms: proposal for the domains Archaea, *Bacteria and Eucarya. Proc. Natl. Acad. Sci. USA,* **87**: 4576–4579.

Wormsley, C. (1981). Biochemical and physiological aspects of anhydrobiosis. *Comp. Biochem. Physiol.,* **70B**: 669.

Wynn-Williams, D.D. (1996). Response of pioneer microbial colonists to environmental change in Antarctica. *Microbial Ecology,* **31**: 177–188.

PART 3

Molecular evolution

8

Genome evolution and the impact of the physical environment

James A. Lake, Ravi Jain, Jonathan E. Moore and
Maria C. Rivera

ABSTRACT

Whole genome studies and increased knowledge of environments, especially of extreme environments, are providing a window into genome evolution and their interactions with the physical environment. The extensive amount of horizontal transfer between prokaryotes observed in genome studies was unthinkable in the pre-genome era. Clearly, horizontal gene transfer has contributed significantly to the evolution of genomes in both eukaryotes and prokaryotes, although possibly by very different mechanisms. The taxonomic breadth and extent of transfer has been so vast that, at least in prokaryotes, one can think of the operational gene component as a single global organism. Thus although prokaryotes have separate identities through their informational genes, they have global commonalities through their operational genes. The effective population size for the worldwide collection of operational genes is enormous and the potential for the creation of innovations is, and has been, correspondingly great.

8.1 Introduction

The possibility of analyzing complete genomes and an increased knowledge of environments, especially of extreme environments, has awakened interest in the influence of environments on genome evolution. Environmental change is a fact of life and it can be rapid if not instantaneous, at least on a geological timescale. Ice ages have come and gone in times far shorter than required to evolve new proteins, and global warming is happening within our lifetimes. Yet organisms may require a million years to evolve novel genes. To illustrate how slowly DNA evolves, consider the relationship between our closest animal relatives and us. Humans and chimpanzees have evolved separately for more than 4 million years. Yet we are still sufficiently similar that chimp blood can substitute for human in transfusions, and on average human and chimp DNA sequences are more than 99% identical!

How then do organisms evolve rapidly enough to respond to changes in their physical environment? New studies are suggesting that genome evolution is not so simple and that mechanisms for gene exchange between organisms may play a significant role in rapidly responding to changes in habitat. In prokaryotes, HGT can occur by transformation, conjugation, and transduction. In eukaryotes, mechanisms of HGT are not as completely understood, but include transformation, viral transfer, and endosymbiotic events.

This chapter provides an overview on these latest findings and on the role of horizontal gene transfer (HGT) in genome evolution.

Evolution on Planet Earth
ISBN 0-12-598655-6

8.2 Genomes and genome analyses

Analyses of complete genomes are profoundly changing our understanding of genome evolution. Before the first genomes were sequenced completely, there was nearly unanimous scientific agreement that prokaryotic genomes were evolving clonally, or approximately so. In other words, as generation after generation of bacteria divided, each bacterium would contain the DNA it inherited from its parent, except that an occasional DNA nucleotide might have mutated. Thus it was thought that the family tree derived from any one gene would look like the family tree from any other gene. It was also thought that if one constructed a tree from any bacterial gene, one would obtain the tree describing the history of past bacterial divisions. In other words, bacteria were thought to evolve clonally. Diploid eukaryotic cells with two copies of each gene per cell slightly complicated this picture, but they too were thought to be evolving as a clonal tree. Thus everyone felt comfortable that the family trees were calculated from gene trees. In particular, ribosomal RNA (rRNAs) genes were favored, since rRNA was easy to sequence and it was assumed trees calculated from rRNA would probably be like those calculated from any other genes.

Because so much attention was focused on the approximately clonal evolution of rRNA, few noticed that horizontal gene transfer may have had an important role to play in genome evolution. Recently, studies of the evolution of life, based on complete genomes, have revealed the flaws in the old, clonal view. Scientific perceptions have only now substantially shifted and favor a significant role for horizontal transfer in genome evolution.

Three amazing new findings, based on analyses of whole genomes, have engendered appreciation for this new paradigm. First, horizontal gene transfer is now generally recognized to be rampant among genomes (rampant at least on a geological timescale). Second, not all genes are equally likely to be horizontally transferred. Informational genes (involved in transcription, translation, and related processes) are rarely transferred, while operational genes (involved in amino acid biosynthesis, and numerous other housekeeping activities) are readily transferred. Third, biological and physical factors appear to have altered horizontal gene transfer. These include intracellular structural constraints between proteins (the 'complexity hypothesis'), interactions among organisms (eukaryotic endosymbionts), and interactions with the physical environment.

8.3 Horizontal gene transfer appears to be rampant

The simple clonal view began to crumble a decade or more ago. When diverse genes from proteins were used to reconstruct eukaryotic and prokaryotic phylogenies, the protein-derived trees often differed substantially from the traditional rRNA-based tree of life. Furthermore, the individual protein genes frequently disagreed with each other.

Early on, these results began to suggest a chimeric origin for nuclear-coded eukaryotic genes. In the mid 1980s, it was noted that the evolution of lipids was inconsistent with the evolution of other molecules in eukaryotes, suggesting a chimeric eukaryotic origin (Lake *et al.*, 1982). This view was reinforced when Zillig, Palm and Klenk (Zillig *et al.*, 1991), using trees based on DNA-dependent RNA polymerase genes, found additional evidence for chimeric eukaryotic origins, and Sogin (1991), noting major differences between rRNA and protein gene trees, proposed a novel type of chimeric eukaryotic origin. Golding and Gupta (1995), utilizing heat shock protein (hsp70) gene sequences, interpreted their results as being consistent with earlier chimeric proposals (Lake *et al.*, 1982; Rivera and Lake, 1992). Once complete genomes were available, the pace of discovery accelerated, as highlighted in recent work from the laboratories of R. Doolittle (1998), F. Doolittle (Brown and Doolittle, 1999), Golding (Ribeiro and Golding, 1998), and ourselves (Rivera *et al.*, 1998).

8.4 Informational and operational genes have different chances of horizontal transfer

In the late 1990s, Koonin and coworkers (Koonin *et al.*, 1997) analyzed relationships between functional groups of proteins. They observed that genes for translation, transcription, replication, and protein secretion in *Methanococcus*, a hyperthermophilic methanogen, were more similar to those of eukaryotes, whereas *Methanococcus* genes for metabolic enzymes and cell wall biosynthesis were more similar to those of eubacteria. Although they incorrectly interpreted the data to imply that the *Methanococcus* genome was a chimera, with genes for translation and transcription coming from *Saccharomyces*, and genes for intermediary metabolism coming from eubacteria, nevertheless they obtained the first genome-wide data suggesting chimeric origins for eukaryotes.

Since our lab had generated the earliest chimeric proposal, we were eager to test theories for chimeric origins. In theory, complete genome sequences can provide the data to test the available chimeric theories. Using genomes from the proteobacterium *Escherichia coli*, from the cyanobacterium *Synechocystis*, from the methanogen *Methanococcus*, and from the eukaryote *Saccharomyces*, Rivera *et al.* (1998) found that genes fell into two functional super-classes: informational genes (those involved in transcription, translation, and related processes) and operational genes (those involved in cellular housekeeping such as genes for amino acid biosynthesis, nucleotide biosynthesis, cell envelopes, lipid synthesis, regulation, etc.). Surprisingly, the two gene classes appear to have different paths of inheritance in both eukaryotes and prokaryotes.

8.5 In eukaryotes, informational and operational genes are descended from different prokaryotes

Perhaps one of the strangest findings of early whole genome analyses was that informational and operational genes have had different pathways of inheritance. In experiments in our lab, we identified a set of approximately 500 orthologous genes from yeast, *E. coli*, a cyanobacterium, and a methanogen, and aligned each set of orthologs. Then we calculated phylogenetic trees for each of these genes and for each set asked which prokaryote branches next to the eukaryote, that is, which prokaryote gave each gene to the eukaryote. From these data we found a startling fact – the two classes of genes had different prokaryotic origins. Eukaryotes obtained their informational genes almost exclusively from an organism on the *Methanococcus* side of the tree (perhaps a hyperthermophile). In contrast, the operational genes of eukaryotes were obtained primarily (70%) from the eubacterial side of the tree (40% from *Escherichia*, and 30% from cyanobacteria). Thus these results indicated that the genes of the eukaryote are indeed a chimera, but they are a special type of chimera, one in which operational genes have come from many sources (although predominantly from eubacteria), but informational genes from a single source. These results are shown schematically in Figure 8.1.

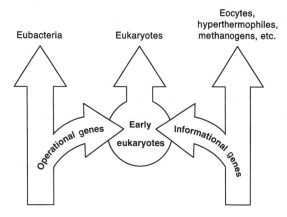

Figure 8.1 A stylized illustration of the prokaryotic origins of eukaryotic nuclear genes. The informational genes (translation, transcription, etc.) are principally derived from eocytes, hyperthermophiles, methanogens, etc., most likely from the eocyte prokaryotes (Tourasse and Gouy, 1999; Galtier *et al.*, 1999). The operational genes have come from multiple sources, including organisms related to the purple bacteria and the cyanobacteria. Understanding the chimeric origins of eukaryotes is an especially active scientific field.

8.6 How might operational genes have found their way into the eukaryotic nucleus: the role of interactions between organisms

Unlike prokaryotes where the mechanisms of conjugation, transformation, and transduction (viral transfer) offer ready explanations for the preponderance of observed horizontal gene transfer, eukaryotes use only one of these three methods to obtain foreign genes, transduction (viral transfer). But eukaryotes have another mechanism not found in prokaryotes, namely the ability to transfer genes from their endosymbionts to the nucleus via intracellular horizontal gene transfer (Palmer, 2000). Indeed, eukaryotes may themselves be highly modified fusions of several prokaryotic genomes. Here we explore how these processes may explain the results of whole genome analysis.

Recall from Figure 8.1 that most operational genes have come from eubacteria (*E. coli*, and cyanobacteria), and that informational genes came from methanogens. The genes didn't come from precisely these organisms, but these were simply the closest representatives for which whole genomes were then available. Thus the informational genes may not have come from a methanogen, but based on other data (Rivera and Lake, 1992; Hashimoto and Hasegawa, 1996; Baldauf *et al.*, 1996; Galtier *et al.*, 1999), the informational genes may have come from a group of hyperthermophilic sulfur-metabolizing prokaryotes, the eocytes, only distantly related to the methanogens.

This is what would be expected if eukaryotes were formed by the fusion of several prokaryotic cells. Scientists already have good evidence for two endosymbiotic fusions. These are the two eukaryotic organelles, the mitochondrion and the chloroplast. Phylogenetic analyses of mitochondrial DNA indicate that it is definitely derived from a eubacterium, probably from a proteobacterium, and chloroplast DNA is very likely derived from a cyanobacterium (reviewed in Gray and Spencer, 1996). Although mitochondrial and chloroplast chromosomes contain fewer genes than the organisms from which

they were derived, not all the excess genes have been lost since many have been simply transferred from the organelle to the nucleus. Thus we have had, in essence, intracellular horizontal gene transfer. It is clear that eukaryotes have had considerable intracellular HGT, but whether eukaryotes have experienced much extracellular HGT, that is, from an outside prokaryotic donor, is not clear.

The source of the informational genes in eukaryotes is more speculative. But endosymbioses of eocyte prokaryotes, and of methanogens, have been proposed as major events in eukaryotic evolution. Martin and Muller (1998) and Moreira and Lopez-Garcia (1998) have proposed models in which methanogens participate in the formation of the first mitochondria, and we (Lake, 1983) have proposed a model in which an eocyte participates in the formation of the nucleus. The model for the origin of the nucleus provides a nice illustration of how the physical environment may have been a major driving force in the creation of the first eukaryotic cell.

8.7 How interactions with the physical environment may have driven the origin of the nucleus: a model

Organisms do not evolve in a vacuum. Most prokaryotes live in habitats teeming with life. Consequently their interactions with their neighbors and with their physical environments influence their evolution. Organisms living together occasionally develop symbiotic relationships. In a few rare cases this has led to endosymbioses, the intracellular capture of former symbionts. In the following section, we explore how such relationships may have created the nucleus, and how this process may have been driven by a need to obtain scarce resources, in this instance hydrogen and sulfur. These ideas have been influenced considerably by the hydrogen hypothesis of Martin and Muller (1998).

In eukaryotes, both the mitochondrion and the chloroplast are surrounded by double membranes, characters that are thought to be vestiges of their endosymbiotic origins. Less well known is the fact that the eukaryotic nucleus is also enveloped by a system of double membranes similar to those found surrounding the mitochondrion and chloroplast (Lake *et al.*, 1982). It has not been obvious what the selective advantages could have been that would favor a symbiotic association of eocyte and eubacterial prokaryotes.

Searcy (1992), in a thoughtful paper, summarized potential roles of sulfurous compounds in symbioses, and Martin and Muller (1998) and Moreira and Lopez-Garcia (1998) both proposed explicit hypotheses for the role of H_2, CO_2, and CH_4 in the origin of mitochondria. In the following we outline a hypothesis that uses reasonable metabolic steps to show how the nucleus and its content of eocyte-related genes could have been acquired.

Any hypothesis for the origin of the nucleus needs to be both metabolically reasonable and provide a selective advantage for both partners of the symbiosis. In the following, we show how H_2 and sulfurous compounds can be used as electron donors and acceptors, in a process which brings together a photosynthetic eubacterium and a hydrogen metabolizing eocyte as symbionts, in a mutually beneficial relationship.

Purple sulfur bacteria and eocytes are naturally complementary endosymbiotic partners, since the metabolic waste products of one match the metabolic needs of the other, and vice versa. The purple sulfur bacteria (Figure 8.2, upper left panel) are unicellular photoautotrophs capable of growing anaerobically in the light using CO_2; some also use acetate (OA_c^-) as a carbon source. Their ATP is provided by cyclic photophosphorylation and reducing power is provided by H_2S which is oxidized anaerobically to S_i and to H_2SO_4. In most purple sulfur bacteria, sulfur is deposited within the cell, but in some (*Ectothiorhodospira*) it is excreted where it would be available for sulfur recycling by a symbiotic partner. Anaerobic eocyte prokaryotes typically generate energy by the reduction of sulfur with hydrogen, forming H_2S. Many can be grown on a range of carbon sources whereas others are

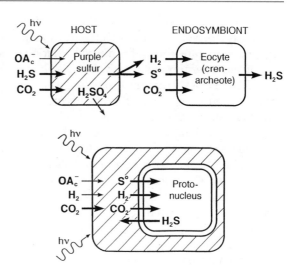

Figure 8.2 A model for the formation of the first eukaryotic cell from a eubacterial host cell and an eocyte endosymbiont, using the hydrogen hypothesis mechanism.

chemolithoautotrophs (Stetter, 1996) (Figure 8.2, upper right panel).

An endosymbiotic association between a purple bacterial host and an anaerobic eocyte symbiont, based on hydrogen and sulfur recycling, could provide advantages for both partners, since the metabolic waste products of one are nourishment for the other, and vice versa. As illustrated at the bottom of Figure 8.2, the host would have a reliable supply of H_2S, provided by the endosymbiont, and the symbiont a reliable supply of S_i and H_2, provided by the host. Sulfur–hydrogen recycling would thus decouple ATP generation by photophosphorylation from H_2S availability, thereby nearly eliminating the need for sulfur, and would require only sufficient H_2 to replace that which escapes by diffusion. Although the endosymbiont would be completely surrounded by the host, its metabolic needs could be supplied by the host without any evolutionary innovations since hydrogen and sulfur are host waste products, and CO_2 import was initially established in the free living host. Furthermore, once the host and endosymbiont relationship was established, if they were removed from a concentrated source of sulfur, both would be irreversibly dependent upon each other.

We, of course, don't know if this model is historically correct, but it is metabolically correct, and illustrates how the mutual relationship of eubacterium with eocyte could have great selective advantages. Proposing such a detailed model by explicitly specifying the host and endosymbiont organisms, also has the advantage that it can be tested when complete genomes of the appropriate host and endosymbiont organisms become available. It is indeed possible that one of the greatest events in biological history, the origin of the eukaryotic nucleus, occurred when two disparate prokaryotes joined together to form an organism, that would be less dependent on its physical environment.

8.8 Although prokaryotes lack organelles, their genomes provide evidence of massive horizontal transfer

Prokaryotes, like eukaryotes, contain both informational and operational genes. But because prokaryotes lack organelles, the basis for separation of the gene types was dif-

ficult to ascertain. In a study using six complete prokaryotic genomes, Jain *et al.* (1999) showed that operational genes are frequently horizontally transferred in prokaryotes, at least on a billion year timescale, whereas informational genes are rarely exchanged (Figure 8.3). They showed this by calculating gene trees for all proteins in six different genomes, then, selecting only those trees with the strongest statistical support, they counted how many horizontal transfers occurred for each operational and informational tree. Yet the same studies showed that there are no measurable differences in the rate of gene evolution (nucleotide substitution rates) between operational and informational genes. These results, and others, caused a revision of the traditional understanding of horizontal gene transfer, which was previously thought to be a rare phenomenon affecting a few genes. Instead, we now find that horizontal transfer strongly affects operational genes (containing some two thirds of the genes in prokaryotic genomes). Furthermore, these studies have given us a preliminary understanding that one of the major distinctions between operational and informational genes is due to the complexity of their interactions.

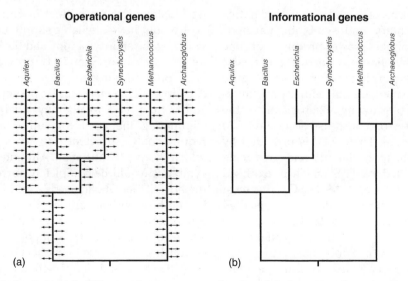

Figure 8.3 A fundamental difference between operational and informational genes. Operational genes, shown in (a), are more likely to be exchanged between organisms by horizontal gene transfer. Informational genes, shown in (b), are more likely to evolve clonally.

8.9 How interactions between proteins may have driven prokaryotic evolution: the complexity hypothesis

Horizontal gene transfer is not an abstract theoretical process. At least some of the basic facts about horizontal transfer in prokaryotes are known. The probability that a specific gene will be successfully transferred to a new host depends upon the specific mechanistic details of transformation, transduction, and conjugation (Syvanen and Kado, 1998); it also depends upon the types of nucleic acids which are being transferred (single stranded, double stranded, linear, circular, etc.) (Day, 1998); it depends upon such variables as the distribution of integrases in organisms (Hall, 1998); and it depends upon the probability that a gene once transferred will function successfully in its new cellular environment. It is this last factor that seems to explain many of the differences between the rates of HGT between operational and informational genes.

The complexity hypothesis attributes differences in rates of HGT to the probability that a transferred gene will function successfully, once it has been incorporated within a prokaryotic genome and is based on an obvious distinction between operational and informational gene types. Most informational genes, particularly the translational and transcriptional apparatuses, are large, complex systems, whereas most operational genes are members of small assemblies of a few gene products. This difference makes it much easier for operational genes, compared to informational genes, to function successfully in a new cellular environment. Two examples will help to illustrate this point.

Translation in *Escherichia* requires the coordinated and complex interactions of at least 100 gene products. The ribosome assembly map (Mizushima and Nomura, 1970) shown in Figure 8.4a summarizes the principal assembly interactions observed for the *Escherichia* small ribosomal subunit. In practice, the assembly map represents only part of the interactions of this complex (other small subunit interactions, not shown, include those with initiation factors, elongation factors, termination factors, 86 tRNAs, numerous messenger RNAs, and the large ribosomal subunit containing 31 proteins, and two ribosomal rRNAs). According to the assembly map, on average a small subunit protein interacts during assembly with four to five other ribosomal gene products.

In contrast, many operational proteins interact with fewer gene products. In the thioredoxin–thioredoxin reductase complex, Figure 8.4b, for example, each protein interacts with just one other gene product.

To understand how complexity can influence the probability of successful horizontal transfer, consider the fate of a foreign gene that has been integrated into a host chromosome. For simplicity, assume that the gene product will function provided that it can make the necessary bonding interactions with its

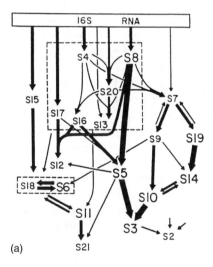

(a)

(b)

$$Th \rightleftharpoons ThR$$

Figure 8.4 Examples of the complexity of gene product interactions in informational genes (a) and operational genes (b). The assembly map of the *E. coli* small ribosomal subunit is shown in (a) as an illustration of the high complexity that is frequently present in the translational apparatus. The thioredoxin (Th) and thioredoxin reductase (ThR) complex is shown in (b) as an example of the reduced complexity present in some operational genes.

neighbors. According to this simplified model, the probability that a transferred thioredoxin could successfully interact with thioredoxin reductase would be 0.25, while the probability that a transferred S5 could be assembled into a small subunit is 0.25^6 (=0.00024) or about 1000 times less. Thus the probability of a successful horizontal transfer is predicted to be higher for proteins that make fewer interactions with host proteins (operational genes) and lower for proteins that make many interactions with host proteins (informational genes).

8.10 The physical environment as a possible barrier to horizontal gene transfer

Many of the observed differences between the horizontal transfer of informational and operational genes can be explained by complexity, but there remain other unexplained differences. These difference are likely due to other factors that affect HGT, perhaps directly related to the transfer process itself. The physical environment itself is one likely candidate for directly affecting HGT.

Accordingly, it now becomes important to consider just how much the physical environment may affect horizontal gene transfer. How has the environment affected horizontal transfer? For example, are genes preferentially transferred within habitats? Would we expect to find that bacteria living in hot springs exchange genes with only other high temperature organisms, or can they donate and receive genes from organisms living at diverse temperatures? How does temperature affect HGT? Can organisms in hot springs transfer genes from organisms living nearby in the soil, and vice versa? Can prokaryotes in the warmest part of a spring receive genes from nearby neighbors occupying niches in cold portions of the spring? If pH varies throughout a thermal spring, can high pH organisms receive genes from those that live at low pH, and vice versa? We can even consider the potential effects of internal constraints upon horizontal gene transfer. For example, it is well known that the GC composition of the DNA is

fairly constant across organisms, and those parts of genomes having different GC compositions are thought to have come from foreign DNA. Pathogenicity islands (defined by atypical DNA compositions) are thought to delineate DNAs that have recently been incorporated into a cell (Ochman *et al.*, 1996). It would be interesting to find out if the GC composition of DNA is a significant barrier to the horizontal transfer of DNA.

Finally, we would like to consider the effect of metabolisms on the horizontal transfer of DNA among organisms. Do carbon heterotrophs, for example, horizontally exchange DNA with other heterotrophs, or is DNA freely exchanged between heterotrophs and autotrophs. In order to test ideas such as these in future studies, methods are being developed to quantify the effect of environmental variables upon horizontal transfer. In analogy with the selection coefficient, s, one can define horizontal transfer coefficients, hx, for each environmental variable, x. Like selection coefficients, transfer coefficients, the hx, can vary between +1.0 and −1.0. A value of $hx = 0$ would indicate that horizontal transfer between organisms is not affected by parameter x. Positive values of hx would imply that horizontal transfers are more probable between organisms having similar x values, and negative hx values would imply that horizontal transfers are less probable between organisms with similar x values. Thus a value of $h_{temp} = +0.1$ would indicate a preference for organisms living at similar temperatures to exchange genes.

Figure 8.5 provides an example, using the high temperature limits (shown at the tips of the tree) of bacteria A, B, C, etc., to illustrate how transfer coefficients may be estimated in the future using this theory. First, it is well known that by using parsimony one can estimate temperatures (or any other evolving parameter) at internal positions within a phylogenetic tree using parsimony arguments, provided that the values at the tips of the branches are known. Suppose that we know that gene 'X' is transferred from organism G to organism C, and that at the time of transfer organism G (the donor) has an upper growth temperature limit of 85°C and organism C (the

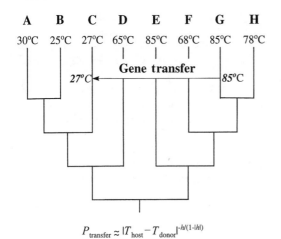

$$P_{transfer} \approx |T_{host} - T_{donor}|^{-h/(1-|h|)}$$

Figure 8.5 An example of a proposed genome-wide test of the effect of environmental parameters on horizontal gene transfer (HGT), illustrating how the coefficient of horizontal transfer might be defined. Studies of multiple genomes promise to yield important information on how the environment affects HGT.

recipient) has an upper temperature limit of 27°C. The probability of transfer is given by the equation $P_{transfer} = |T_{Host} - T_{Donor}|^{(-h/(1-|h|))}$, where P is the probability that gene X will be transferred, T_{Host} and T_{Donor}, are the upper temperature limits of the host and donor, respectively, and h is the horizontal transfer coefficient for the upper temperature limits. In the case that $h = 0$, all transfers would be unaffected by temperature. If h were positive $P_{transfer}$ would be the largest when the temperatures of host and donor were similar, and if h were negative $P_{transfer}$ would be greatest when the temperatures of host and donor are most dissimilar. Thus the coefficient of horizontal transfer would have just the desired properties. Using these definitions, and the enormous numbers of genes that are available, it appears feasible to estimate transfer coefficients directly from whole genome comparisons. We hope that in the coming years we may begin to appreciate some of the fundamental parameters quantifying the influence of the physical environment upon horizontal gene transfer and genome evolution.

References

Baldauf, S.L., Palmer, J.D. and Doolittle, W.F. (1996). The root of the universal tree and the origin of eukaryotes based on elongation factor phylogeny. *PNAS.* **93**: 7749–7754.

Brown, J.R. and Doolittle, W.F. (1999). Gene descent, duplication, and horizontal transfer and the evolution of glutamyl- and glutaminyl-tRNA synthetases. *J. Mol. Evol.,* **49**: 485–495.

Day, M. (1998). In M. Syvanen and C.I. Kado (eds) *Horizontal Gene Transfer.* London: Chapman and Hall, pp. 144–167.

Doolittle, R.F. (1998). Microbial genomes opened up. *Nature,.* **392**: 6674.

Galtier, N., Tourasse, N. and Gouy, M. (1999). A non-hyperthermophilic common ancestor to extant life forms. *Science,* **283**: 139–141.

Golding, G.B. and Gupta, R.S. (1995). Protein-based phylogenies support a chimeric origin for the eukaryotic genome. *Mol. Biol. Evol.,* **12**: 1–6.

Gray, M.W. and Spencer, D.F. (1996). Organellar evolution. In *Evolution of Microbial Life* (Society for General Microbiology Symposium 54) (Roberts, D.M., Sharp, P., Alderson, G., Collins, M. eds), pp. 109–126, Cambridge University Press, UK.

Hall, R.M. (1998). In M. Syvanen and C.I. Kado (eds) *Horizontal Gene Transfer.* London: Chapman and Hall, pp. 53–62.

Hashimoto, T. and Hasegawa, M. (1996). Origin and early evolution of eukaryotes inferred from the amino acid sequences of translation elongation factors 1-alpha/Tu and 2/G. *Adv. Biophys.,* **32**: 73–120.

Jain, R., Rivera, M.C. and Lake, J.A. (1999). Horizontal gene transfer among genomes: the complexity hypothesis. *PNAS,* **96**: 3801–3806.

Koonin, E.V., Mushegian, A.R., Galperin, M.Y. and Walker, D.R. (1997). Comparison of archaeal and bacterial genomes: computer analysis of protein sequences predicts novel functions and suggests a chimeric origin for the archaea. *Mol. Micro.,* **25**: 619–637.

Lake, J.A. (1983). Ribosome evolution: the structural bases of protein synthesis in archaebacteria, eubacteria, and eukaryotes. *Prog. Nucleic Acid Res. Mol. Biol.,* **30**: 163–194.

Lake, J.A., Henderson, E., Clark, M.W. and Matheson, A.T. (1982). Mapping evolution with ribosome structure: intralineage constancy and interlineage variation. *PNAS,* **79**: 5948–5952.

Martin, W. and Muller, M. (1998). The hydrogen hypothesis for the first eukaryote. *Nature*, **392**: 37–41.

Mizushima, S. and Nomura, M. (1970). Assembly mapping of 30S ribosomal proteins from E. coli. *Nature*, **226**: 1214–1218.

Moreira, D. and Lopes-Garcia, P. (1998). Symbiosis between methanogenic archaea and delta-Proteobacteria as the origin of eukaryotes: the syntrophic hypothesis. *J. Mol. Evol.*, **47**: 517–530.

Ochman, H., Soncini, F.C., Solomon, F. and Groisman, E.A. (1996). Identification of a pathogenicity island required for Salmonella survival in host cells. *PNAS*, **93**: 7800–7804.

Palmer, J.D. (2000). A single birth of all plastids? *Nature*, **405**: 32–33.

Ribeiro, S. and Golding, G.B. (1998). The mosaic nature of the eukaryotic nucleus. *Mol. Biol. Evol.*, **15**: 779–788.

Rivera, M.C. and Lake, J.A. (1992). Evidence that eukaryotes and eocyte prokaryotes are immediate relatives. *Science*, **257**: 74–76.

Rivera, M.C., Jain, R., Moore, J.E. and Lake, J.A. (1998). Genomic evidence for two functionally distinct gene classes. *PNAS*, **95**: 6239–6244.

Searcy, D. (1992). Origins of mitochondria and chloroplasts from sulfur-based symbioses. In H. Hartman and K. Matsuno (eds) *The Origin and Evolution of the Cell*. World Scientific Press, pp. 47–48.

Sogin, M.L. (1991). Early evolution and the origin of eukaryotes. *Curr. Opin. Genet. Dev.*, **1**: 457–463.

Stetter, K.O. (1996). Hyperthermophilic procaryotes. *GEMS Micro. Rev.*, **18**: 1491–1458.

Syvanen, M. and Kado, C.I. (eds) 1998). *Horizontal Gene Transfer*. London: Chapman and Hall.

Tourasse, N.J., Gouy, M. (1999) Accounting for evolutionary rate variation among sequence sites consistently changes universal phylogenies deduced from rRNA and protein-coding genes. *Mol. Phylogenet. Evol.* **13**: 159–168.

Zillig, W., Palm, P., Klenk, H.P. *et al.* (1991). Phylogeny of DNA-dependent RNA polymerases: testimony for the origin of eukaryotes. In F. Rodriguez-Valera (ed.) *General and Applied Aspects of Halophilic Microorganisms*. New York: Plenum Press, pp. 321–332.

PART 4

Gravity

9

The impact of gravity on life

Emily R. Morey-Holton

ABSTRACT

See Plate 14

Gravity is a well-known, but little understood, physical force. Its intensity and direction have been constant throughout evolutionary history on Earth, making it difficult to understand what role, if any, this vector force may have on life as we know it. Only since the launch of Sputnik in October of 1957 has life left the planet Earth and ventured into space, so that we could begin to investigate what happens to life with minimal gravity. To date, we only have fascinating snapshots of life in space. Completion of the International Space Station should allow long-duration studies over multiple generations in multiple species.

This chapter explores four questions: What is gravity? What happens to life when gravity changes? Is gravity necessary for life as we know it? Did gravity play a role in evolution of life on Earth? Life from the cellular level through adult humans exposed to spaceflight is briefly examined and examples from spaceflight and ground-based experiments are discussed. The conclusion from these studies suggests that gravity is necessary for life as we know it, and that 'gravity shapes life'.

9.1 Introduction

Gravity has been constant throughout the history of Earth. This simple fact masks the complexity of gravity as an evolutionary force. Gravity is a vector, i.e. a force that has magnitude and direction at each point in space. Gravitational loading is directional toward the center of the Earth. Gravitational loading acts on all masses at the Earth's surface and defines the weight of each object. Weight is the product of the object's mass times the force of gravity, which on Earth is equal to 1 G. Weight drives many chemical, biological, and ecological processes on Earth. Altering weight changes these processes. Given these facts, one should not be surprised that changes in gravity could alter life, as we know it. If gravity causes changes to biology, then gravity, *per se*, must be a major physical environmental force shaping life on Earth.

Life evolved from the sea. Neutrally buoyant aquatic species still have gravity acting upon them, but the uniform pressure around them and internal organs, such as the swim bladder, tend to counterbalance the intensity of the gravitational signal. However, some aquatic species appear to use gravity as a directional cue. When life evolved from the sea, it likely experienced gravitational loading for the first time. Land species changing their orientation with respect to the gravity vector or increasing in height probably began to develop adaptive mechanisms for coping with directional changes and for moving fluids and structures against this load. Due to gravity, the force necessary to lift an object above the surface of the Earth increases with the distance that an object is lifted. Birds had to solve the lift/drag problem related to air density and gravity before they could fly and had to evolve a musculoskeletal system that could provide adequate thrust. As species on land increased in size, they required support structures appropriate for the loads imposed. Species that crawled along the ground didn't need the same mechanisms for countering gravity's effects as those species alternating between horizontal and vertical positions. The latter species required more complex systems for balance or gravity sensing, fluid regulation, and locomotion. So, gravity, though constant, may have played a major role in evolution as species crawled from the sea and began to populate the land (see also Rayner, this volume, Chapter 10).

By altering gravity, we are able to investigate those biological systems that were developed to detect or oppose this unique force. Decreasing gravity on Earth for more than several seconds is impossible with existing technology. Until Sputnik was launched in October 1957, we had little opportunity to study how lowering this physical force influenced life. By decreasing gravity through spaceflight, we are beginning to understand that not only gravity, but also the physical changes that occur in the absence of gravity, may have profound effects on evolution of species and their ecologies. By going into space, we can gain a better understanding of how gravity shaped life on Earth. This chapter attempts to provide answers to four questions:

- What is gravity?
- What happens to life when gravity changes?
- Is gravity necessary for life as we know it?
- Does gravity play a role in evolution? If so, what role might it play?

I am privileged to share recent research results from investigators whom I personally thank for allowing me to present their data, suggesting that gravity has been and continues to be a major player in the evolution of species.

9.2 What is gravity?

In 1665/1666, Sir Isaac Newton first developed the universal law of gravitation and the laws of motion, which form the basis for our understanding of planetary motion and spaceflight

(Guillen, 1995). The universal law of gravitation states that the attractive force between any two bodies is given by:

$$F_g = G_u \frac{Mm}{d^2} \qquad (9.1)$$

where M (of Earth) and m (of any object) are the masses of the two attracting bodies, d is the distance between their centers of mass and G_u is the universal gravitational constant $(6.67 \times 10^{-8} \, \mathrm{cm^3/g \cdot s^2})$ (Pace, 1977). In other words, the force of gravity is directly proportional to the product of the masses and inversely proportional to the square of the distance between them. Thus, each time the distance between the center of two masses doubles, the force is cut to one quarter of the previous value. Microgravity $(10^{-6} \, \mathrm{G})$ requires a significant distance between the two masses (~ 1000 earth radii or $6.37 \times 10^6 \, \mathrm{km}$). Low Earth orbit is only about 300 km above Earth. How, then, can we state that microgravity is found in low Earth orbit? The next paragraph suggests an answer to this apparent discrepancy.

A force is defined as equal to the mass of an object times its acceleration (i.e. $F = ma$). Equation (9.1) can be rewritten as:

$$a = G_u \frac{M}{d^2} \qquad (9.2)$$

Thus, an object of any mass at the surface of the Earth accelerates toward the center of the Earth at approximately $9.8 \, \mathrm{m/sec^2}$. This gravitational acceleration is 1 G. A spacecraft in orbit above Earth moves at a constant velocity in a straight trajectory (Figure 9.1). Earth's gravitational acceleration at that vehicle's center of mass alters the direction of the spacecraft from a straight path into a circular orbit normal to the gravitational vector via centripetal acceleration. Centrifugal force, the apparent force in a rotating system, deflects masses radially outward from the axis of rotation and is equal and opposite centripetal acceleration per unit mass. Thus, a spacecraft in a circular orbit above Earth is in 'free' fall around Earth. Centrifugal force counterbalances centripetal acceleration causing momentary resultant gravitational forces that range between 10^{-3} and $10^{-6} \, \mathrm{G}$ even though gravity *per se* is

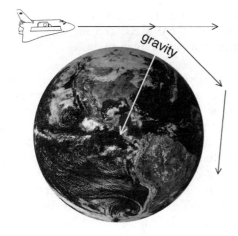

Figure 9.1 Spacecraft move in a straight line at a given acceleration rate. Rather than flying into space in a straight trajectory, forces acting on the craft cause it to 'free fall' and circle the Earth. The force of gravity in low Earth orbit is only reduced about 10% from that on Earth.

reduced only about 10% at the altitude of low Earth orbit (Klaus, 2001).

Gravity is one of the four fundamental physical forces of nature. The other three are the nuclear strong and weak forces, and electromagnetic forces. Given the intensity of the forces (adapted from http://learn.lincoln.ac.nz/ phsc103 / lectures / intro / 4_forces_ of_ physics.htm)

Nuclear strong force	10^{40}
Electromagnetic force	10^{38}
Nuclear weak force	10^{26}
Gravitational force	10^{1}

one quickly sees that gravitational force is far weaker than other forces. How could such a weak force affect all living systems? A brief description of the various forces may help in understanding this apparent discrepancy. The strength of a force depends on the distance over which it is acting. The strong force holds together protons and neutrons in the nucleus of an atom and is effective over a relatively short distance. Electromagnetic force (emf) is the force between charged particles; whether the force is attractive or repulsive is determined by the charges between interacting particles. The strength of the force

drops with the inverse of the distance between charges. The weak force is effective over an incredibly small distance and can be pictured as the force that causes the decaying processes of unstable nuclear particles through time. Gravitational force is the weakest of the four fundamental physical forces of nature. Similar to emf, this force gets smaller as the objects get further apart. Yet, you feel the force of gravity and not emf, because an object at rest on Earth is pressed against Earth's surface by the force of gravity so that continuous loading is imposed upon the object. In orbit around Earth, objects have mass but almost no weight because the acceleration due to gravity is balanced by the centrifugal acceleration that keeps the object in orbit.

9.3   What happens to life when gravity changes?

Gravitational acceleration has been constant throughout the ~4 billion years of biological evolution on Earth. Gravity interacts with other environmental factors to produce today's Earth; for example, gravity is responsible for giving weight to objects on Earth so gravity is necessary for rain to fall, for water to drain, for heat to dissipate (i.e. convective force), for air and water to separate, etc. In addition to gravity's influence on the environment, it is probably a major contributor to biological changes as species evolve from water to land. To counteract gravity, new land species would need to develop systems for fluid flow and regulation, postural stability, structural support and locomotion to function and thrive in a 1 G terrestrial environment. How will terrestrial biota transported beyond Earth evolve in different gravity regimes? How much gravity is required to maintain life, as we know it? Is the Moon's gravity (1/6 G) sufficient for stimulating gravity thresholds while the lower gravity levels in space (10^{-6} G) are not? Could life evolving in space successfully return to Earth? Could Earth-based life readily evolve on planets larger than Earth with a higher gravity field? The ability to evolve under increased gravity appears related to size. Single cells and nematodes withstand 10^5 G for brief periods, young plants easily cope for 10 minutes at 30–40 G without noticeable structural changes, rats withstand 15 G for 10 minutes while 20 G is lethal, and humans are capable of tolerating only 4–5 G for 10 minutes. Gravitational levels, like other physical environmental factors, appear to determine the boundaries for life.

Microbes are less gravity sensitive than larger species and should have less difficulty transiting between planets and different gravity levels than humans. Complex spacecraft are required to transport and maintain humans off Earth while microbes survive outside spacecraft with minimal protection. Microbes fit into many ecological niches and began to evolve as soon as an environment is hospitable to their life form. Complex life forms require complex ecosystems for survival and evolution, suggesting that prototypes of Earth's ecology may have to be included, at least initially, to allow survival and evolution of these complex forms on other planets. Thus, the ability to thrive beyond Earth may be determined, at least initially, by the size of organisms and their environmental requirements.

To fully appreciate the effects of altered gravity on biological species, multiple generations must be studied at that gravity level. Subtle biological changes due to altered gravity are difficult to define over a single generation. Acute changes can be studied in less than one generation; the duration of most altered gravity experiments. Spaceflight studies in vertebrates suggest that gravity plays an incredibly important role in the development of these organisms even though studies have been limited to a small portion of the life cycle of the animal. Most spacecraft return specimens to Earth after several days to several weeks in space and most samples are collected following, rather than during, flight. Postflight data are confounded by a recovery period superimposed upon the spaceflight. Minimal spacecraft flights have had the time and the facilities for collecting inflight samples. Thus, predicting evolutionary changes from these meager, yet extremely important, data is a monumental challenge. The time

required to initiate a second generation for species that NASA is considering for the International Space Station is listed in Table 9.1.

We seldom are exposed to gravity levels other than 1 G for any length of time on Earth making it very difficult to grasp the subtleties of altered gravity. Thus, we have developed an evolutionary '1 G mentality'. '1 G mentality' means that we use gravity in our daily life without even thinking about it and have difficulty comprehending life without gravity. In fact, we subconsciously design hardware and habitats for a 1 G, rather than an altered G, environment.

According to NASA, approximately 40% of equipment flown in space for the first time does not work, often due to heat build-up from lack of convection, air bubbles impeding fluid flow, or habitats based on designs more appropriate for Earth. In space, animals can use all sides of their cages and aren't limited to the floor as on Earth suggesting that housing standards appropriate for specimens on Earth may not be as appropriate for specimens in space. Understanding and appreciating the differences between Earth's physical environment and the spaceflight environment is critical if one is to provide a habitat that will keep organisms healthy and happy in altered gravity. To answer the question 'Can terrestrial life be sustained and thrive beyond Earth?' we need to understand the importance of gravity to living systems and to appreciate the role of gravity during evolution on Earth. When we can readily transition our thought processes between multiple gravity levels, we finally will be able to design space hardware and habitats that take advantage of, rather than depend on, the ambient gravity level.

Dr Maurice Averner, NASA Program Manager for Fundamental Biology, has likened gravity levels to light levels; that is, microgravity and 1 G can be compared to no light vs light. Profound differences occur in dark vs light environments and subtle changes occur as the light level increases above ambient. Gravity levels may be similar, i.e. more striking changes occur when

Table 9.1 **Species doubling times vary greatly between species. Invertebrates ranging from bacteria through insects can go through multiple generations during the 90d crew rotation planned for the International Space Station, while vertebrates have a minimal doubling time of approximately two months.**

SPECIES	APPROXIMATE DOUBLING TIME
Escherichia coli (Bacteria):	0.01d (16 min)
Yeast:	0.07d (100 min)
Protozoa (Euglena in the dark):	0.5d
Paramecium:	0.75d
Eukaryotic cells in culture:	1d
C. Elegans:	4d (on plates) or 8d (in suspension culture)
Arabidopsis (Plant):	25d (light dependent)
Drosophila:	13d (at 25C)
Rodent:	63d (2 mo)
Zebrafish:	90d (3 mo)
Quail/chicken:	90d (3 mo)
Xenopus (Frog):	152d (diploid, 5 mo) or 730d (pseudo tetraploid, 2 yr)
Human:	5380d (15 yr)

gravity is turned on or off with more subtle difference as the gravity level increases above 1 G.

The science of gravitational biology took a giant step forward with the advent of the space program. It provided the first opportunity to examine living organisms in gravity environments lower than could be sustained on Earth. Organisms ranging in complexity from single cells through humans are responsive to Earth's gravity; thus, these organisms most likely would be affected by a lack of gravity. Our knowledge of the biological consequences of decreased gravity (i.e. spaceflight) has increased significantly since 1957, yet we only have snapshots of biological changes in multiple species. This chapter will focus primarily on altered gravity responses of cells in culture, ecosystems, vertebrate development, and adult humans.

9.3.1 *Cells*

Physics predicts that altered gravity will not cause any changes in cells because gravity is extremely weak compared with other physical forces acting on or within cells (Brown, 1991). Yet, cellular changes have been reported. Are the physical scientists wrong or are there other previously unconsidered factors at work on cells when gravity changes? Purely physical mechanisms for gravitational responses probably can be eliminated (Hemmersbach *et al.*, 1999). Yet, cells appear to respond to changes in the environment (Klaus *et al.*, 1997) and to have evolved structures that interact directly with the outside environment to sense the environmental loads placed upon them (Hemmersbach *et al.*, 1999; Ingber, 1998).

The bacterium *E. coli* has flown experimentally in culture seven times aboard the space shuttle (Klaus *et al.*, 1997). During spaceflight, *E. coli* exhibited a shortened lag phase, an increased duration of exponential growth, and an approximate doubling of final cell population density compared to ground controls. These differences may be related to the lack of convective fluid mixing and sedimentation, processes that require gravity. During exponential growth in minimal gravity, the more uniform distribution of suspended cells may initially increase nutrient availability compared to the 1 G-sedimenting cells that concentrate on the container bottom away from available nutrients remaining in solution. Also, local toxic by-products could become concentrated on the bottom of the 1 G container with cells in increased proximity to each other. Such a process could limit cell growth. Thus, changes in *E. coli* and possibly other cells during spaceflight may be related to alterations in the microenvironment surrounding non-motile cells. If true, then the extracellular environment plays a critical role in evolution of single cells through controlling nutrients and waste. This response to the extracellular environment suggests that intracellular gravity sensors are not essential for cells to elicit a gravitational response. Earlier predictions that microgravity could not affect cells were focused on the physical inability of gravity, an extremely weak intracellular force, to elicit an immediate or 'direct' response from organisms of such small mass. Rather than a 'direct' response, reduced gravity more likely initiates a cascade of events – the altered physical force leads to an altered chemical environment, which in turn gives rise to an altered physiological response. Modeling cell behavior predicts how cells evolve in different physical environments, including Earth, by including gravity as an integral part of the equations; hence, changes in sedimentation, convection, nutrient availability, and waste removal with altered gravity can be predicted.

Hammond, Kaysen, and colleagues (Kaysen *et al.*, 1999) cultured renal cells under different conditions. They concluded that differentiation of renal cells in culture most likely requires three simultaneous conditions: low shear and low turbulence, three-dimensional configura-

tion of the cell mass (i.e. free-floating), and co-spatial arrangement of different cell types and substrates. They have cultured human renal cells in rotating-wall vessels and in centrifuged bags on Earth, and in stationary bags flown aboard the shuttle (Hammond *et al.*, 1999). Controls for all experiments were simultaneous, ground-based, bag cultures. All cultures contained liquid medium and the bags were made of material that was non-adherent for cells. A plethora of changes in steady-state level of mRNA expression occurred in space-flown human cells (1632 of 10 000 genes or 16.3%) compared to the Earth-based bag cultures. These patterns were unrelated to the changes in gene expression found in rotating-wall vessel experiments. Shear stress response elements and genes for heat shock proteins showed no change in steady-state gene expression in the flight culture. Specific transcription factors underwent large changes during flight (full data set at http://www.tmc.tulane.edu/astrobiology/microarray). In the rotating-wall vessel, 914 genes or 9% changed expression. In the centrifuge, increasing gravity to 3 G caused only four genes to change expression greater than threefold. In addition to the unique changes in gene expression noted during flight, structural changes in the cultured rat kidney cells also occurred. Far more microvilli were formed in renal cells grown in space or in the rotating-wall vessel than in the 1 G static bag culture or during centrifugation (Hammond *et al.*, 2000). These studies suggest that renal cells flown in space have unique patterns of gene expression unrelated to the best Earth-based model of spaceflight (i.e. rotating-wall vessel), and that the ability to form a three-dimensional, free-floating structure in culture appears critical to induce tissue-specific, differentiated features in renal cells.

The data from bacterial and renal cells suggest that spaceflight may affect cells via their external environment and that differentiation of renal tissue may be enhanced during spaceflight. Such studies are demonstrating how physical factors, specifically gravity, regulate expression of specific genes, creating an organism specific for that environment. Thus, some cells and tissues may show greater differentiation of specific features while others may show the reverse. In fact, the timing of gene expression may be beneficial or detrimental to downstream effects and, hence, alter the final protein product and, ultimately, the organism. Evolution is more likely to cause changes through altered gene expression rather than through genomic modifications as the latter are more likely spontaneous mutations. Data are indicating that gravity may actually be a critical environmental factor in determining the differentiation and maturation of cells on Earth.

Early results with cultured cells from the musculoskeletal system suggest that spaceflight induces a variety of responses. Delayed differentiation and changes in the cytoskeleton, nuclear morphology, and gene expression have been reported for bone cells (Hughes-Fulford and Lewis, 1996; Landis, 1999). Dr Herman Vandenburgh has flown fused myoblasts (i.e. muscle fibers) to investigate the effects of microgravity on cultured muscle fibers. He found that flight muscle organoids were 10–20% thinner (i.e. atrophied) compared with ground controls due to decreases in protein synthesis rather than increases in protein degradation (Vandenburgh *et al.*, 1999). Interestingly, atrophy of the isolated muscle fibers in culture was very similar to the amount of muscle atrophy reported in flight animals. These preliminary data from bone and muscle cells suggest that spaceflight affects adherent cells and tissues even when isolated from systemic factors and that the physical environment might direct the ultimate development of cells, organs, and tissues.

Changes in the physical environment surrounding cells, *in vivo* or *in vitro*, can lead indirectly to changes within the cell. Little is known about if or how individual cells sense mechanical signals or how they transduce those signals into a biochemical response. A cellular mechanosensing system might initiate changes in numerous signaling pathways. Such a system has been found in cells that attach to an extracellular matrix (i.e. the cell substratum) and the cellular components are beginning to be defined. These cellular interactions probably suppress or amplify signals

generated by gravitational loading. We now know that the extracellular matrix to which cells attach contains adhesive proteins that bind to regulatory proteins that traverse the cell membrane. These transmembrane regulatory proteins (e.g. integrins), in turn, connect to the cytoskeleton and the cytoskeleton ultimately connects to the cell nucleus. Given these connections, activation of the regulatory proteins in the cell membrane can lead directly to regulation of gene expression, thereby eliminating the need for a solely intracellular gravity sensor. Living cells may be hardwired to respond immediately to external mechanical stresses. Exciting research on the interaction of the cell cytoskeleton with membrane components and the extracellular matrix is shedding light on possible 'force sensors' at the cellular level that might be essential for the differentiation process (Ingber, 1997, 1998; Globus *et al.*, 1998; Schwuchow and Sack, 1994; Wayne *et al.*, 1992). Ingber has applied to cells the concept of 'tensegrity' (i.e. tensional integrity), a tension-dependent form of cellular architecture that organizes the cytoskeleton and stabilizes cellular form (Ingber, 1999). This architecture may be the cellular system that initiates a response to mechanical loading as a result of stress-dependent changes in structure and may have been a key factor in the origin of cellular life (Ingber, 2000).

Definition of the cellular connections that might sense and transduce mechanical signals into a biochemical response may also shed light on the events initiating cell maturation. As a cell matures, it stops dividing and begins to express characteristics of a mature cell type. If a cell does not mature, it will continue to divide – the definition of a cancer cell. The maturation process may be triggered by multiple factors, including loads placed on the extracellular matrix during different phases of development.

With exciting new molecular tools in hand and the development of facilities for increasing gravity on Earth (i.e. centrifugation) or decreasing gravity on space platforms, great strides will be made in understanding the influence of gravity in living systems at the cellular level within the next decade.

In summary, the local environment around cells may be altered in space. Such changes may affect cellular metabolism and steady-state gene expression may change. Potential adaptive systems in eukaryotic cells include force coupling through the cellular skeleton, ion channels, and other load-sensitive cellular structures that might alter cellular signaling. Further investigation into cellular changes at multiple gravity levels is required. Research may show that cellular architecture in eukaryotic cells evolved to oppose loading or amplify directional cues. Thus, physical changes in the aqueous medium surrounding cells in culture and cellular structures that oppose or respond to mechanical loads may provide cells with the ability to respond to gravity.

9.3.2 Ecosystems

Algal mats and protists are fascinating! If orientation and stratification are weight-dependent, then the microgravity of space could significantly alter interactions between organisms. How these organisms would fare without gravity is unknown, but changes from Earth-based mats would be predicted. For example, microbes migrating during the day using gravity as an environmental cue to minimize exposure to solar radiation would not be able to migrate and could have greater radiation damage. Such damage would tend to select species with radiation resistance. Protists appear to detect gravity at about 0.1 G. Thus, the general principle of mechanoreceptors in metazoa is represented in unicellular organisms (Hemmersbach et al., 1999) suggesting that the ability to detect and use Earth's gravity must have occurred very early in the evolutionary history. In fact, the Hemmersbach et al. (1999) review article suggests that these organisms have evolved structures, such as mechano- or stretch-sensitive ion channels, cytoskeletal elements and second messengers, to amplify the gravity signal rather than evolving intracellular gravity sensors. Interestingly, the genes for light sensing and gravity sensing occur very early in evolution. Some organisms seem to be able to use either or both as directional cues. Space, with very low gravity levels,

provides a unique laboratory to sort out the importance of light in a relatively gravity-free environment. Or, perhaps these physical environmental factors have created redundant biological sensing systems.

Decreased gravity causes very complex changes in the environment. For example, gaseous boundary layers build up due to lack of convective mixing in the atmosphere and these boundary layers expose plants and simple ecosystems to stratified environments not present on Earth. Soil substrates in space have a different shape, do not pack like Earth soils, and wet in a very different way than on Earth. In space, water does not drain through soil, as that process requires gravity. Such changes make the management of simple ecosystems in space rather difficult. Plants may be the most difficult species to evolve efficiently in space, as they must adapt simultaneously to two environments (above and below the ground) that change during spaceflight (Musgrave et al., 1997). Atmospheric issues include pollination and mixing of gases. During spaceflight, insects, gravity, and wind may be missing in the plant habitat. In addition, the gaseous environment above the ground may stratify due to lack of convective mixing and expose plant shoots to boundary layers that are new and novel. Such stratification may be responsible for the uneven ripening of seedpods noted during flight, i.e. ripening begins at the tip and is not uniform. Levels of potentially dwarfing compounds (e.g. ethylene and CO_2) might cling to the plants creating shorter plants rather than the taller ones that one might expect with lower gravity levels. Not only will plant shoots have to adapt to novel environments in flight, but also water management for the roots may be problematic.

Soil issues during spaceflight focus on root-zone management. The team from Utah State University has grappled with water-management issues while studying evolution of plant systems in space by attempting to grow multiple generations of wheat on the Russian space station MIR (Jones and Or, 1999). Their results show that we have a lot to learn before we can achieve successful and repeatable plant growth under reduced gravity. Root media wet differently in space. On Earth, water drains vertically through a soil column where each particle is held in contact by gravity. In space, the wetting front is free to move in all directions. The wetting front bulges out to reach a neighbor particle that may be floating a small distance away, 'gulps' the particle, and then opens an air space on the other side of it. Where particles are touching, water wicks along the particle surfaces partially filling the channels around the soil particulates. This tends to trap air between the particles rather than forming the saturated slurries that one finds on Earth. In a partially hydrated system, water wicks between particles as it does on Earth, but in all directions rather than just flowing 'down'. This wicking, due to capillary attraction, inhibits exchange of nutrients and, by bridging between the particle contact points, can suffocate plants at lower water contents than would be observed on Earth. Soil-based systems in space can be managed to resemble hydroponic (i.e. water-based) systems on Earth. It is impossible to simulate space-substrate/water-content conditions using the same soil substrate on Earth. In space, water forced into a substrate does not drain and can fill up to 90% of the soil matrix. Until the water is wicked out of the substrate, the flooded volume will stay in the substrate, creating an oxygen-free zone that is not usable by most plants. Proper root-zone management in space is an active process that requires sensors for continuous monitoring of both water and oxygen content in the soil matrix during spaceflight. Such monitoring is essential for effectively managing plant systems and for learning how the decreased gravity experienced during spaceflight alters the environment. Understanding and controlling the environment is often a prelude to survival and adaptation in unique environmental niches.

Many challenges remain for plant growth and crop management in space, including understanding of boundary layers above the ground and water and oxygen management in the root media. All physical environmental factors must be considered when predicting the evolutionary fate of a species in a unique environment.

9.3.3 *Vertebrate development*

Studies on Earth and in space suggest that gravity has shaped life. Studies with tadpoles, birds, and rats on Earth and in space are shedding light on the importance of gravity to animal systems. Unlike plants, no vertebrate has completed a life cycle in space. In fact, humans have spent about 1% of their life cycle in space, and rats have spent about 2% of a single generation. Building habits for multiple generations of complex species is a challenge not only in space, but also on Earth.

9.3.3.1 *Amphibian development*

The most elegant and definitive developmental biology experiment in space used the amphibian as a model (Souza *et al.*, 1995, Figure 9.2). This experiment had an on-board 1G centrifuge control. On Earth, the fertilized frog egg rotates upon sperm penetration, and this rotation is thought to be essential for normal development. Upon fertilization, the egg begins to divide and form the embryo that, after an appropriate time, emerges from the jelly-like egg as a tadpole.

Female frogs were sent into space and induced to shed eggs that were artificially inseminated. The eggs did not rotate and yet, surprisingly, the tadpoles emerged and appeared normal. After return to Earth within 2–3 days of hatching, the tadpoles metamorphosed and matured into normal frogs.

Development appeared normal during spaceflight, yet some morphological changes in embryos and tadpoles occurred. The embryo had a thicker blastocoel roof that should have created abnormalities in the tadpole, but no deformations appeared, suggesting plasticity of the embryo. The flight tadpoles did not inflate their lungs until they returned to Earth. The lungs appeared normal by the time the tadpoles were 10 days old. If the lungs didn't inflate and the animals remained in space, then would the gills remain as the tadpoles metamorphosed into frogs? If the gills resorbed without inflated lungs, would the defect be lethal? Why didn't the lungs inflate? We don't know the answers to these

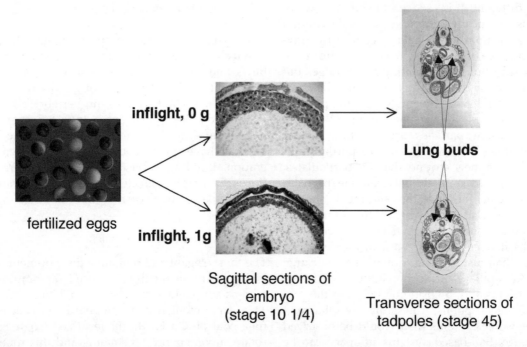

inflight, 0 g

Lung buds

fertilized eggs

inflight, 1g

Sagittal sections of
embryo
(stage 10 1/4)

Transverse sections of
tadpoles (stage 45)

Figure 9.2 Fertilized frog eggs on Earth rotate upon fertilization and this rotation was hypothesized to be required for normal development. In space, the eggs appeared to develop normally whether or not they rotated. Several interesting abnormalities were noted in histological sections taken from the embryos and tadpoles. In the embryos, an extra cell layer was noted in the embryo and yet no gross abnormalities were found in the tadpoles. In the tadpole, the lung buds did not appear to inflate. The lack of inflation of the lungs might be cause for concern for frogs, but the tadpoles were returned to Earth when they were less than four days of age, i.e. before they metamorphosed to frogs. Pictures courtesy of Kenneth Souza, NASA Ames Research Center.

questions, but we do know that air bubbles were present in the tadpole aquatic habitat on orbit. Possibly, lack of directional cues or increased surface tension between the air/ water interface interfering with penetration of the air bubbles may be involved in this interesting observation.

This developmental study produced multiple important findings. It showed that vertebrates can be induced to ovulate in space and that rotation of fertilized eggs is not required for normal development in space. The flight-induced changes, including a thicker blastocoel roof with more cell layers and uninflated lungs, appeared correctable in this experimental paradigm. In conclusion, the vertebrate embryo is very adaptive and the system is plastic, yet the long-term fate of the animal throughout its life in space remains unknown.

9.3.3.2 Quail

Adult quail on MIR adapted quickly to the space environment. They learned to soar with minimal wing flapping and held onto their perch for stability when eating rather than being propelled backwards when they pecked their food (i.e. for every action there is an equal and opposite reaction). Fertilized quail eggs appeared to undergo normal embryogenesis in space, but serious problems occurred after hatching (Jones, 1992). When a cosmonaut took a hatchling from its habitat, the chick appeared content as long as it was held. But once released, the bird first flapped its wings for orientation and began to spin like a ballerina, then kicked its legs causing it to tumble – it became a spinning ball. The cosmonaut noted that the chick would fix its eyes on the cosmonaut while trying to orient in space. When placed in their habitat, the chicks had difficulty flying to their perch to eat, and, unlike the adults, had difficulty grasping the perch for stability when eating. The hatchlings ate normally only when fed by the crew and, thus, did not survive.

9.3.3.3 Rat development

The force of gravity may influence events underlying the postnatal development of motor function in rats, similar to those noted in hatchling quail. Such effects most likely depend on the age of the animal, duration of the altered gravitational loading, and the specific motor function.

Walton (1998) reported differences in righting-reflex and locomotion in neonatal rats when the musculoskeletal system did not bear weight. Walton's data suggest that there are critical development periods during which biomechanical loading of limbs is essential to give cues to nerves. Without the cues, brain development and limb innervation may not occur normally and animals may develop an abnormal walking behavior. At the Final Results Symposium for the 17-day Neurolab Shuttle Mission, Dr Walton suggested that neonatal rats flown in space exhibited altered loco-motor behavioral development that persisted for the 1-month recovery period and that righting-reflex strategies were still abnormal 5 months after return to Earth. Dr Danny Riley showed delayed development of certain nerve connections to muscles in these neonates. The connections returned to normal after return to Earth, yet fibers in hindlimb muscles did not reach normal size even after a month back on Earth. The Riley team found similar results in neonates that were not allowed to bear weight on their hindlimbs on Earth (Huckstorf *et al.*, 2000). The data suggest that biomechanical loading of limbs during early development may be essential for innervation of muscles and development of normal muscle fiber size.

These vertebrate studies suggest that embryonic development in frogs and birds proceeds normally in space, although unexplained changes occur during embryogenesis and early development. In birds and rats, biomechanical loading may be required for Earth-like development and innervation of certain structures. We are learning that habitats in which early development occurs on orbit may have to be very different from Earth cages. Without gravity, rat and bird neonates float freely. Without a surface to crawl against, the animals thrash about and their health may degrade if the housing provided is too large. In space the animals can use all three dimensions of their habitat rather than the two dimensions available in Earth habitats. Perhaps space habitats should be sized to the individuals, suggesting that more confining habitats may be appropriate for neonates until they are able to grasp and walk. Only after development of appropriate motor function should cage size be expanded. Cages that accommodate all stages in the life of vertebrates are critical if we are to understand the influence of gravity on development of vertebrate systems in a free-fall environment. Interestingly, evolutionary development with increased gravity may also require special habitats so that the pups are not crushed beneath the dam and can still obtain essential nutrients. If the habitat for a particular species is not compatible with survival throughout life, then evolution of that species will not occur in that environment.

9.3.4 Adult humans

Early predictions of the response of humans to spaceflight assumed that space adaptation would be analogous to human disease processes rather than to normal physiology. Through studies of bed-rested healthy adults and medical examinations of crews returning from space, we now recognize the adaptive nature of the human responses to spaceflight or its ground-based models. We are also aware of the necessity to minimize the flight-induced changes so that crews maintain their Earth-readiness and avoid injury on landing. Lack of gravitational loading affects multiple physiological systems, especially fluid flow, balance, and support structures that are particularly vulnerable to change or injury during re-entry and renewed exposure to gravitational forces. To minimize these changes, most crew members exercise extensively during long-duration flight. Although many physiological systems appear to be affected by spaceflight, only the cardiovascular, vestibular, and musculoskeletal systems are covered in this chapter.

9.3.4.1   Cardiovascular system

To understand how the human cardiovascular system adapts to gravitational loading, it is helpful to think about the system as the body's 'plumbing', which consists of the 'pump' (heart), 'pipes' (blood vessels), and 'control system' (nerves, hormones, and local factors). The cardiovascular system is designed for a 1 G environment. When crews go into space, strange things happen. Spaceflight causes a fluid shift from the legs toward the head, producing a puffy face and bird-like legs. The fluid shift increases the amount of blood in the chest region, causing the heart and fluid-volume sensors in the neck to detect an increase in fluid volume. The increased chest fluid initially increases heart size (i.e. amount of blood), but regulatory mechanisms quickly kick in and return the fluid to an appropriate, lower level. The loss of fluid results in a reduced plasma or blood volume. To keep blood thin, the decrease in plasma volume triggers a destruction of newly synthesized, immature, red blood cells, probably by a mechanism of programmed cell death or apoptosis (Alfrey *et al.*, 1996). The shift of fluids to the upper body and the distended facial veins noted in astronauts suggest that central venous pressure should increase. Surprisingly, it decreases, suggesting that our concepts of pressure and volume regulation need revision (Buckey *et al.*, 1996). These changes are appropriate for the spaceflight environment. However, upon return to Earth, many crew members have difficulty standing, usually due to the rush of blood to the feet that can cause fainting (Buckey *et al.*, 1996). This readaptation to Earth's gravitational force following spaceflight could pose a problem if crews are expected to stand and function normally immediately after landing on any planetary body.

9.3.4.2   The vestibular system

The vestibular system is our guidance system that controls eye movements, posture, and balance. Its main purpose is to create a stable platform for the eyes so that we can orient to the vertical – up is up and down is down. Deep within our inner ear is the vestibular organ with thousands of tiny hair cells. Resting atop these hair cells are microscopic crystals that move and bend the hair cells, sending information to the central nervous system for the reflex control of eye movements, posture, and balance. In space, the eyes send signals that confuse the brain because the visual references that we rely on for stability are missing (Merfeld, 1996; Merfeld *et al.*, 1996; Oman *et al.*, 1996). These mismatched sensory inputs may be one cause of 'Space Adaptation Syndrome' (SAS), an adaptive process that often involves nausea and can lead to vomiting. Another possible cause of SAS is sensor adaptation to a novel gravitational environment to increase the gain of sensory cells, possibly by increasing the number of synapses (Ross and Tomko, 1998). After several days in space, crews begin to function effortlessly, signaling that adaptation is complete. Crews initially rely on touch, sight, and muscle sensors for orientation (Young *et al.*, 1996). As soon as they switch to an internal alignment and use the feet to signal down, they are able to function normally. Upon return to Earth, the brain is confused once again as gravity is now available for orientation. This confusion creates postural instability that is compounded with the cardiovascular difficulty in standing. Also, reflexes associated with posture and balance are slowed even on short-duration missions. With long-duration flights, changes in reflexes, visual perception, and eye/hand coordination may become major issues for re-entry and readaptation to Earth.

9.3.4.3   The musculoskeletal system

The musculoskeletal system provides the magic of movement. This system is very responsive to changes in load. In fact, exercise is necessary to maintain muscle and bone mass on Earth. Without gravitational load, muscles and bones associated with posture and weight-bearing become weaker. The intensive exercises performed by crews are not able to counteract the loss of bone/muscle mass and strength because exercising in space without gravity does not produce the same level of mechanical loading possible on Earth. With the fluid shifts and decreased bone loading, calcium is lost from bone and calcium excretion increases. The higher calcium load presented to the kidneys is of concern for potential kidney-stone formation. Our bodies tend to conserve calcium; during spaceflight, the amount of mineral in some bones, including the head, may increase to offset losses from other sites. Bone and muscle are lost *only* in the legs, back, and neck indicating that the musculoskeletal changes are site-specific – loss does not occur throughout the entire body. Bone loss primarily occurs at sites in weight-bearing bones where muscles (that are also losing mass) attach to that bone. The muscles that help maintain posture are most severely affected and change phenotype. This new phenotype resembles skeletal muscle that fatigues more readily. Muscles in the jaw may change function during spaceflight because on Earth the jaw opens with gravity and people have to work to keep their mouth shut. Upon return to Earth, reduced muscle strength and power, and even pain, occur. Following extended spaceflight missions, certain muscles and bones might be weaker and fracture more easily. Thus, re-entry from space is similar to returning from a long boat trip on a rough sea, but space adds the additional complexity of fluid redistribution and muscle weakness in addition to the dizziness.

In summary, the changes in humans are appropriate adaptations to the space environment. They are not life threatening for at least 1 year, which is the longest that humans have been in space. The adaptations are functional (see Fregly and Blatteis (1996) and Sulzman (1996) – Results of SLS1 and SLS2). That's the good news. The bad news is that adaptation to space creates problems upon returning to Earth. Difficulty standing, dizziness, and muscle weakness present problems after landing. Appropriate countermeasures must be devel-

oped. Crew members exercise in space to minimize the difficulties of re-entry. We ethically cannot request that they stop exercising. To learn about adaptation of mammals that do not exercise in space, we use appropriate animal models.

9.4 Is gravity necessary for life as we know it?

Life most likely will look and, perhaps, move quite differently after many generations in space. We have learned that life is 'plastic' and changes with the environment; it adapts at least transiently to changes in gravity. The microenvironments of spaceflight require more study so that we will understand how to use them effectively. We certainly have a lot to learn about the complexity of biological responses to altered gravity. Data to date suggest that certain biological structures have evolved to sense and oppose biomechanical loads, and those structures occur at the cellular as well as at the organismal level. Certainly, the Earth-tuned physiological systems of vertebrates change following acute exposure to space; what will happen over multiple generations is speculative. The 'functional hypothesis' theory suggests 'use it or lose it'. If this theory holds over multiple generations in space, then gravity-dependent structures may ultimately disappear or assume a very different appearance. Based on the studies described in this chapter, gravity most likely is essential for life, as we know it.

9.5 Does gravity play a role in evolution?

Gravity affects the environment. Its attractive force gives weight to mass and weight is required for many ecological processes on Earth. Sprinkled throughout this chapter are examples and suggestions of the importance of gravity to life, as we know it. Particularly important is the apparent evolutionary development of unique biological structures that amplify the force of gravity and specific gravity sensors that are required for orientation, balance, and movement in a gravity environment. Will these structures and sensors change with gravity levels less than 1 G? Only extended time in space with multiple generations will begin to answer this question.

One might predict that plants would grow taller without gravity. Yet, the boundary layers produced by a lack of gravity might concentrate growth-inhibitory or ageing factors around the plants, thereby causing them to dwarf; increased gravity might facilitate the dispersal of such factors and actually lead to taller plants. If plants on Earth are fine-tuned to a 1 G environment, then they might not function as well at either increased or decreased gravity levels.

Ecologies, such as algal mats, that stratify by weight on Earth might tend to form as three-dimensional communities without gravity. If the hierarchical structure achieved by stratification is essential for survival or fitness, then the communities would either become extinct or change their fitness leading to evolved characteristics appropriate for the new environment.

Gravity level is important in development of load-bearing structures. The scaling effect of gravity is well known: the percentage of body mass relegated to structural support is proportional to the size of a land animal (e.g. 20 g mouse $= \sim 5\%$, 70 kg human $= \sim 14\%$, and 7000 kg elephant $= \sim 27\%$). The scaling effect in land animals would likely change in space and could result in a static scale comparable to marine mammals on Earth ($\sim 15\%$ of mass as supporting tissues over a wide range of weights). However, increasing gravity would require altered support structures as scaling up existing structures without any modification in geometry would ultimately lead to failure.

Load-bearing limbs, so important on Earth, are less necessary in space. Human legs not only get in the way during spaceflight but also are involved in the fluid shifts that occur early in flight. Whether legs would disappear over time without gravity (perhaps similar to the extraterrestrial ET) or become more like grasping talons is unknown. Unlike evolution

in a decreased gravity environment, higher gravity levels may lead to a different posture and a bipedal stance might become unusual with most species possibly existing as quadrupeds or even hexipeds. Larger species might become extinct at higher gravity levels unless these animals quickly adjust for brain-blood flow and placement of internal organs.

To 'fall down' probably requires a certain gravity level and the reflexes related to posture and equilibrium at 1 G are sluggish following spaceflight. Would such reflexes be innervated in species evolving in lower gravity fields? If the reflexes that keep us from falling down in a gravity environment do not develop, then would species evolving in a lower gravity field be able to move when placed in a higher gravity field?

Would the evolutionary response of biological systems be linear, logarithmic, or degraded at gravity levels other than 1 G? Some biological systems (e.g. metabolic rate which is proportional to body weight) increase with increased size on Earth. On the other hand, some life systems may have adapted to be maximally efficient at 1 G and degrade with changes in gravity (e.g. body temperature). Life as we know it is extremely adaptable and usually fits form to function during evolution in a hospitable environment. Humans readily adapt to a lower gravity regime aboard spacecraft yet require an extensive 'recovery' period when returning to Earth from space voyages suggesting that initial adaptation to a lower gravity environment might be easier than adapting to a higher gravity environment.

A fascinating suggestion that gravity might play a role in evolution comes from snakes (Figure 9.3). On Earth, snakes have evolved in different environments. For example, tree snakes spend their days crawling up and down trees and exist in an environment where they must cope with gravity. Land snakes spend most of their life in a horizontal position. Sea snakes are neutrally buoyant and spend their life swimming within their habitats. In other words, the orientation of the snake to the direction of the gravity force differs depending upon habitat, without a concomitant alteration in magnitude of gravity.

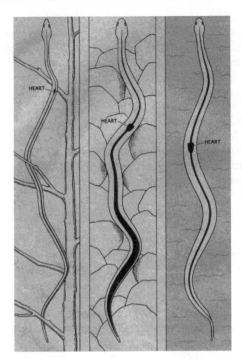

Tree Land Sea

Figure 9.3 The location of the heart in different species of snakes indicates a potential effect of gravity on organ position. The tree snake continuously changes its orientation with respect to the direction of gravity as it climbs up and down trees. Its heart is located closer to the head compared to either the land snake or the sea snake. The heart in the sea snake is approximately in the middle of its body. Pictures courtesy of Dr Harvey Lillywhite, University of FL, Gainesville, and published with the permission of Nelson Prentiss (Lillywhite, H.B. (1988) Snakes, blood circulation and gravity. *Scientific American*, **256**: 92–98).

Lillywhite (1988) noticed that the heart of the tree snake was closest to the brain, suggesting that it might be more gravity tolerant than the other snakes as it did not have to carry blood over as great a distance from the heart to the brain. He centrifuged the animals and found that the sea snake had the least gravity tolerance (i.e. fainting with increased gravity), the tree snake had the most, and the land snake was intermediate (Lillywhite *et al.*, 1997). Changes in heart position, likely related to gravity, most certainly happened over evolu-

tionary, rather than single-generation, time-scales. These studies suggest that changes in orientation of a species with respect to the direction of a gravitational force, without an alteration in the magnitude of gravity, may play a role in the evolution of that species on Earth. Gravity may determine the location and size of internal organs such as the heart.

So, what might evolving species at a higher or lower gravity field look like? Form follows function and as function changes, so will form. How much change and what form organisms and ecologies will assume over time in altered gravity is currently unknown. Increasing gravity within a survivable range will probably not cause dramatic differences in evolving Earth-like species while a gravity-free environment will likely produce significant changes in ecologies and species. ET may be a good example of a species evolving at a lower gravity field with the rotund body, duck-like flappers for feet, minimal legs, long thin arms and fingers, and a large head, large eyes, and minimal hair.

To quote ET, 'Love your planet', meaning that you are a product of your physical environment. We will begin to understand the influence of gravity on evolution of species only after prolonged exposure to different gravity levels. Today, the role of gravity in evolution remains speculative. But one certainly can say that *gravity shapes life*!

Acknowledgements

I am most grateful to a number of investigators who have spent many hours discussing various aspects of their experiments, sharing their unpublished observations with me, and critiquing the manuscript. In particular, I would like to acknowledge Malcolm Cohen, James Kaysen, Cary Mitchell, Lynn Rothschild, and Ken Souza (NASA Ames Research Center), Gail Bingham (Utah State University), Millie Hughes-Fulford (UC-San Francisco and VAMC), David Klaus (University of Colorado/BioServe), Tim

Hammond (Tulane University), Tim Jones (University of Missouri/Columbia), Danny Riley (Medical College of Wisconsin), and Kerry Walton (NYU Medical School).

Web addresses

E. coli: http://www.colorado.edu/ASEN/asen5016/
http://www.colorado.edu/engineering/
Fundamental forces of nature: http://learn.lincoln.ac.nz/phsc103/lectures/intro/4_forces_of_physics.htm
Human kidney cells: http://www.tmc.tulane.edu/astrobiology/microarray
Astrobiology/Life Sciences: http://space.arc.nasa.gov/
http://astrobiology.arc.nasa.gov
http://lifesci.arc.nasa.gov

References

Alfrey, C.P., Udden, M.M., Leach-Huntoon, C. *et al.* (1996). Control of red blood cell mass in spaceflight. *Journal of Applied Physiology*, **81**: 98–104.

Buckey, J.C., Jr, Gaffney, F.A., Lane, L.D. *et al.* (1996). Central venous pressure in space. *Journal of Applied Physiology*, **81**: 19–25.

Brown, A.H. (1991). From gravity and the organism to gravity and the cell. *ASGSB Bull.*, **4**: 7–18.

Fregly, M.J. and Blatteis, C.M. (eds) (1996). *Environmental Physiology: Handbook of Physiology*, Section 4, Chapters 29–40. New York: Oxford University Press, pp. 631–970.

Globus, R.K., Doty, S.B., Lull, J.C. *et al.* (1998). Fibronectin is a survival factor for differentiated osteoblasts. *Journal of Cell Science*, **111**: 1385–1393.

Guillen, M. (1995). *Five Equations that Changed the World*. Chapter 2: Apples and oranges. New York: MJF Books, pp. 9–63.

Hammond, T.G., Lewis, F.C., Goodwin, T.J. *et al.* (1999). Gene expression in space. *Nature Medicine*, **5**: 359.

Hammond, T.G., Benes, E., O'Reilly, K.C. *et al.* (2000). Mechanical culture conditions effect gene expression: gravity-induced changes on the space shuttle. *Physiological Genomics*, **3**: 163–173.

Hemmersbach, R., Volkmann, D. and Häder, D.-P. (1999). Graviorientation in protists and plants. *Journal of Plant Physiology*, **154**: 1–15.

Huckstorf, B.L., Slocum, G.R., Bain, J.L. *et al.* (2000). Effects of hindlimb unloading on neuromuscular development of neonatal rats. *Developmental Brain Research* **7**: 169–178.

Hughes-Fulford, M. and Lewis, M. (1996). Effects of microgravity on osteoblast growth activation. *Experimental Cell Research*, **224**: 103–109.

Ingber, D.E. (1997). Tensegrity: the architectural basis of cellular mechanotransduction. *Annual Review of Physiology*, **59**: 575–59.

Ingber, D.E. (1998). The architecture of life. *Scientific American*, **278**: 48–57.

Ingber, D.E. (1999). How cells (might) sense microgravity. *FASEB Journal*, **13**: S3–S15.

Ingber, D.E. (2000). The origin of cellular life. *BioEssays*, **22**: 1160–1170.

Jones, S.B. and Or, D. (1999). Microgravity effects on water flow and distribution in unsaturated porous media: analysis of flight experiments. *Water Resources Research*, **35**: 929–942.

Jones, T.A. (1992). Gravity and the ontogeny of animals. *The Physiologist*, **35**: S77–S79.

Kaysen, J.H., Campbell, W.C., Majewski, R.R. *et al.* (1999). Select *de novo* gene and protein expression during renal epithelial cell culture in rotating wall vessels is shear stress dependent. *Journal of Membrane Biology*, **168**: 77–89.

Klaus, D., Simski, S., Todd, P. and Stodieck, L. (1997). Investigation of space flight effects on E. coli and a proposed model of underlying physical mechanisms. *Microbiology*, **143**: 449–455.

Klaus, D.M. (2001) Clinostats and bioreactors. *Gravitational and Space Biology Bulletin*, **14**: 55–64.

Landis, W.J. (1999). Collagen–mineral interaction in vertebrate calcification. *Gravitational and Space Biology Bulletin*, **12**: 15–26.

Lillywhite, H.B. (1988). Snakes, blood circulation and gravity. *Scientific American*, **256**: 92–98.

Lillywhite, H.B., Ballard, R.E., Hargens, A.R. and Rosenberg, H.I. (1997). Cardiovascular responses of snakes to hypergravity. *Gravitational and Space Biology Bulletin*, **10**: 145–152.

Merfeld, D.M. (1996). Effect of spaceflight on ability to sense and control roll tilt: human neurovestibular studies on SLS-2. *Journal of Applied Physiology*, **81**: 50–57.

Merfeld, D.M., Polutchko, K.A. and Schultz, K. (1996). Perceptual responses to linear acceleration after spaceflight: human neurovestibular studies on SLS-2. *Journal of Applied Physiology*, **81**: 58–68.

Musgrave, M.E., Kuang, A. and Porterfield, D.M. (1997). Plant reproduction in spaceflight environments. *Gravitational and Space Biology Bulletin*, **10**: 83–90.

Oman, C.M., Pouliot, C.F. and Natapoff, A. (1996). Horizontal angular VOR changes in orbital and parabolic flight: human neurovestibular studies on SLS-2. *Journal of Applied Physiology*, **81**: 69–81.

Pace, N. (1977). Weightlessness: a matter of gravity. *The New England Journal of Medicine*, **297**: 32–37.

Ross, M.D. and Tomko, D.L. (1998). Effect of gravity on vestibular neural development. *Brain Research Reviews*, **28**: 44–51.

Schwuchow, J. and Sack, F.D. (1994). Microtubules restrict plastid sedimentation in protonemata of the moss Ceratodon. *Cell Motility and Cytoskeleton*, **29**: 366–374.

Souza, K.A., Black, S.D. and Wassersug, R.J. (1995). Amphibian development in the virtual absence of gravity. *Proceedings of the National Academy of Sciences*, **92**: 1975–1978.

Sulzman, F.M. (1996). Life sciences Space Mission, Overview. *Journal of Applied Physiology*, **81**: 3–6.

Vandenburgh, H., Chromiak, J., Shansky, J. *et al.* (1999). Space travel directly induces skeletal muscle atrophy. *FASEB Journal*, **13**: 1031–1038.

Walton, K. (1998). Postnatal development under conditions of simulated weightlessness and space flight. *Brain Research Reviews*, **28**: 25–34.

Wayne, R., Staves, M.P. and Leopold, A.C. (1992). The contribution of the extracellular matrix to gravisensing in characean cells. *Journal of Cell Science*, **101**: 611–623.

Young, L.R., Mendoza, J.C., Groleau, N. and Wojcik, P.W. (1996). Tactile influences on astronaut visual spatial orientation: human neurovestibular studies on SLS-2. *Journal of Applied Physiology*, **81**: 44–49.

10

Gravity, the atmosphere and the evolution of animal locomotion

Jeremy M.V. Rayner

ABSTRACT

See Plates 15–17

Morphology, locomotion physiology and movement behaviour in animals have evolved to be closely tuned to such physical features of their environment as gravity, fluid density, viscosity and pressure. When exposed to environments with different physical properties animals can adopt unusual locomotion patterns: human lunar astronauts adopt a 'loping' gait because their leg length is too short for a normal walk in the lower gravity. The wing dimensions and body size of flying animals are strongly influenced by the magnitude of gravity and the density and viscosity of the surrounding air. Many extant animals – alcid birds which use their wings both to swim and fly are a good example – have evolved locomotor systems able to work in media with widely differing physical properties, and many modern birds migrate at great altitudes and experience a wide range of air densities.

Simple biomechanical models predicting how optimal locomotor design of animals with existing Bauplans may be determined by properties of their physical environmental are combined with estimates of palaeogravity and the palaeoatmosphere to explore historical diversity in morphology in response to physical environmental change. Relatively high levels of atmospheric oxygen and accordingly high air density and low viscosity during the Carboniferous and early Permian may have triggered the Amniote radiations and the evolution of flight and gigantism in flying insects, but birds and pterosaurs evolved flight when air density was low. Low density and high kinematic viscosity of air may have encouraged high aspect ratio in pterosaurs.

Gravity is difficult to estimate with any confidence, but estimates of the effect of extreme models are still insufficient for it to have played a major role in the evolution of gigantism in flying vertebrates. Large size in pterosaurs and Miocene teratorn birds must be explained by autotrophic or physiological factors. While they are unlikely to have been important in the origin of flying vertebrates, high oxygen levels may have influenced the physiology of the archosaur and mammal precursors of flying birds and bats.

10.1 Introduction

Animal movement is a process of interchange of energy between an animal and its environment. The physical forces associated with loco-motion biomechanics and the energy flows required to produce them are determined by the morphology of the organism and the way it moves, and also by the physical properties of the environment. In the past atmospheric and geophysical conditions differed from those of today, and in which the design and activity of extant animals is assessed. Thus there is potential for changes in the properties of the environment over geophysical time to interact with macroevolutionary processes. It has been proposed that dramatic variations in oxygen levels have been responsible for major diversification events and the origin of novel locomotor modes such as flight (e.g. Graham *et al.*, 1995, 1997; Dudley, 1998, 2000; Thomas, 1997), and extreme palaeoatmospheric conditions have been invoked to explain such phenomena as gigantism in flying animals (Harlé and Harlé, 1911).

Here I take a broad approach to palaeobiomechanics by attempting to determine the extent to which the changes in the physical properties of the atmosphere and environment have altered evolutionary constraints on the locomotor systems of vertebrates. The methodological approach is to apply simple scaling models to quantify the limits to possibility in two particular problems. First is to test the hypothesis that changes in the geophysical properties of the Earth and its atmosphere could have been responsible for major macroevolutionary events, primarily in the radiation of the tetrapods. Second, I explore the locomotor capabilities of terran animals when hypothetically transposed to other planets, and then consider the likely Baupläne of organisms evolving on those other planets to fulfil locomotor roles comparable to those which have appeared on Earth. I consider primarily the biomechanical problems of flying and of cursorial locomotion, arguing that physical factors relating to forces and accelerations in flight are likely to form dominant constraints (*cf.* Rayner, 1995, 1996). I give less attention to physiological or autotrophic aspects of locomotion, although these are potentially more important than biomechanical factors and must not be ignored.

The list of examples of macroevolutionary events which could be potentially related to changes in gravity or atmosphere is long, and ranges from the Cambrian explosion ($\sim$ 520–500 My BP), through radiations onto land ($\sim$ 350–300 My BP), the Permian and Triassic Amniote radiations ($\sim$ 300–250 My BP) to the early Tertiary radiations of higher tetrapods ($\sim$ 65 My BP). Gravity may be associated with instances of gigantism, in plants (with development of gymnosperm and angiosperm trees), in Paleozoic insects ($\sim$ 300 My BP) and in Jurassic sauropod (and other) dinosaurs ($\sim$ 200 My BP), and more recently (and probably less probably) in Mesozoic mammals (*Indricotherium*, $\sim$ 35 My BP). Both gravity and physical properties of the air dominate flight, and may have influenced the evolution of the four lineages of true flying animals (see, e.g., Dudley, 1998, 2000), and the instances of gigantism in Cretaceous pterodactyls (100–70 My BP) and in Miocene birds ($\sim$ 10 My BP). Here, I explore the extent to which any of these may be possible.

10.1.1 *Scaling and similarity methods*

Scaling and the closely related similarity arguments are widely used in biology (see, e.g., Schmidt-Nielsen, 1972, 1985; McMahon, 1973; Alexander, 1982, 1992) and in physics (see, e.g., Barrow and Tipler, 1986) to model and to quantify functional interdependence of critical physical quantities. The approach often involves broad assumptions and (sometimes) generous approximations, but is powerful for clarifying the role of important adaptive trends, as for instance of changes in size. Scaling models are used frequently by biomechanicists to predict such quantities as the speeds and sizes of moving organisms, and to identify physically based constraints on evolution and adaptation. A powerful approach is to identify physical quantities or ratios (dimensionless numbers) which remain constant or have a predictable effect over a broad range of sizes. Thus, Alexander (1976, 1982, 1989a, b; Alexander and Jayes, 1983) has used the Froude number (the ratio of kinetic to potential energy) to show that moving cursors should change gait from walk to trot at a consistent speed related to body morphology. In a somewhat different approach, Barrow and Tipler (1986) and Hokkanen (1986) have used scaling and similarity arguments to determine the physical and biomechanical factors which set the largest size for terrestrial animals. Comparable arguments can be applied to the scaling of flight performance and to maximum size in aerial animals (Pennycuick, 1975; Rayner, 1988, 1995, 1996).

Arguments of this kind are used here to quantify the effect of changes in gravity and in atmospheric physics on locomotion biomechanics.

In developing these arguments I adopt the common hypothesis of *isometric* or *geometric* similarity: this means that all animals are assumed to have the same shape and uniform density. This is a convenient assumption. In practice design across related taxa of different sizes rarely varies isometrically, and deviations from isometry in any lineage can be considered as an average of the adaptive responses to conflicting ecological pressures and physical constraints (see, e.g., Rayner, 1985, 1995). These deviations are generally relatively small, and should not distort these arguments.

10.1.2 Palaeogeophysical changes in the Phanerozoic

10.1.2.1 Hypothetical changes in gravity

Limited and conflicting information is available about gravity over geological time, and the potential influence of varying gravity has been largely neglected in studies of evolutionary morphology.

There is considerable geomorphological evidence that can be taken to imply that the Earth has expanded rapidly over geological time, possibly as a result of radioactive heating (Jordan, 1971; Carey, 1976). In the light of modern plate tectonics some of this evidence is open to alternative explanation, but the topic is currently undergoing something of a revival (Carey, 1983, 1996). Carey (1976) estimated a 40% change in the radius of the Earth over the past 3×10^8 years, and links this with alternative extreme hypotheses as to whether the mass of the Earth or the density of the Earth has remained constant. I follow the speculations of Pennycuick (1987) by quantifying the potential change in gravity according to Carey's model of expansion. Under Carey's alternate extreme assumptions (Figure 10.1) the potential change in gravity between the early Phanerozoic and the present is considerable. If Earth mass has remained constant, then gravity has decreased sharply, and at the start of the Phanerozoic could have been up to a factor of 3 higher than today. If Earth density is assumed constant, then early Phanerozoic gravity would have been only around 55% of present (note that this assumption also requires a resolution to the apparently intractable prob-

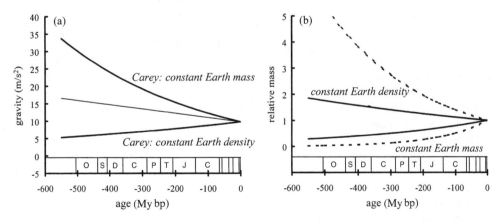

Figure 10.1 Hypothetical and speculative extremes of palaeogravity (a), and consequences (b), for the size of terrestrial cursors. Curves in (a) are computed following the hypotheses of Carey (1976), with alternative assumptions of constant Earth density and constant Earth mass. Direct measurements of palaeogravity (Stewart, 1978) suggest gravitational acceleration at the start of the Jurassic (~200 MY BP) was between 10 and 15 m/s², and has never been less than present. In (b), solid lines correspond to maximum size set by posture and compressive stresses in limb bones, and dashed lines to expected variation in size for organisms moving at constant speed and Froude number.

lem that Earth's mass would have risen by a factor of around 5 in 5×10^8 years).

Measurements of palaeogravity are few. From chemical properties of kimberlite deposits Stewart (1978) estimates that gravity has declined slowly, and has never been less than its present value; gravity in the early Phanerozoic was estimated as between 1 and 2 times present. Stewart interprets this as inconsistent with the rapidly expanding Earth hypothesis.

With these reservations, estimates of gravity based on Carey's hypotheses are used here. The purpose of this exercise is not to argue that gravity has changed, but to determine bounds on palaeogravity to identify whether hypothetical influences on macroevolutionary events or on the evolution of gigantism are possible.

On relativistic grounds Dirac (1937) argued that the universal constant of gravitation varied linearly over astrophysical time. Although now considered unlikely, this too may have had an appreciable effect on terrestrial gravity over geological time. It is not considered here. It should be noted that the consequences of both Earth expansion (according to Carey's model) and of shift in gravitational constant should be measurable by contemporary physical techniques. They have not been identified. Any significant variation in gravity during the Phanerozoic should be regarded as speculative.

10.1.2.2 *Atmospheric variation*

By comparison with the lack of information about gravity, knowledge of the palaeoatmosphere and of palaeoclimatological conditions is far more developed; predictions of atmosphere composition have received far greater acceptance, and palaeoclimate models are becoming increasingly sophisticated. The two physical properties of air, which have a dominant influence on the performance of aerial organisms, are density and viscosity. The values of both are dominated by the largest components of gases in air, that is by nitrogen and oxygen. Sufficient information is now available to estimate density and viscosity through the Phanerozoic with some degree of confidence.

Berner (1990, 1993) and Berner and Canfield (1989) have modelled carbon cycling and bionic effects to estimate CO_2 and O_2 levels through the Phanerozoic (Figure 10.2a). Hart (1978) and Holland (1984) have determined that nitrogen partial pressure had reached almost its present level by the start of the Phanerozoic and since has remained static (see also Canfield and Teske, 1966). Reliable estimates of mean global surface temperature are available from oxygen isotope measurements for the Cenozoic and the latter part of the Cretaceous (Buchardt, 1977; Anderson, 1990) and for the Paleozoic (Popp *et al.*, 1986). The Mesozoic was generally warmer than the present and terrestrial ice masses were absent (Budyko *et al.*, 1987; Crowley and North, 1991), and I have obtained a sufficiently reliable estimate of temperature for the present purpose by interpolation.

By combining Berner's estimates of gas concentrations, these estimates of temperature, the physical properties of gases (Braker and Mossman, 1971; Perry *et al.*, 1984) and standard formulae (Reid, 1977) it is possible to estimate the history of surface atmospheric density and viscosity (Figure 10.2b, c). Density broadly follows the mass of oxygen in the atmosphere (Figure 10.2a), with a peak at the end of the Carboniferous. Although oxygen levels were high through the Cretaceous and into the Tertiary, this is not reflected in air density because of the high temperatures during that time. Over the range of variation of pressure, temperature and gas composition in the Phanerozoic, the dynamic viscosity of air varies little, and the kinematic viscosity (which is the ratio of dynamic viscosity to density) therefore varies approximately as the inverse of density.

The confidence ranges of Berner's estimates of oxygen and carbon dioxide levels are broad. I have used the means of his time series (following Graham *et al.*, 1995, 1997). Because of the additional assumptions involved about temperature and gas partial pressures I have not attempted to estimate bounds to density and viscosity. The curves in Figure 10.2b and c are likely to indicate the major trends in variation of atmosphere geophysics, but should not be interpreted as accurate predictions of the actual values.

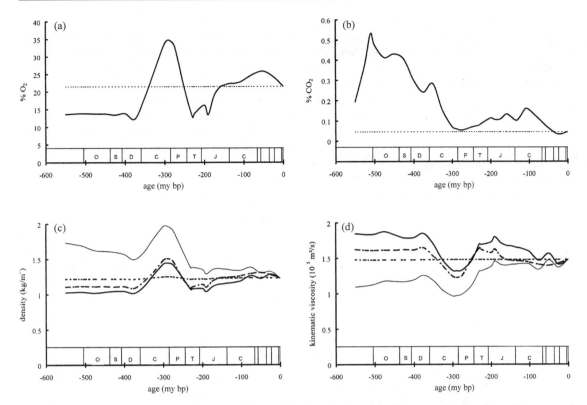

Figure 10.2 Models of the palaeoatmosphere. (a) Mean estimates of oxygen levels, from Berner and Canfield (1989). (b) Mean estimates of carbon dioxide levels, from Berner (1990, 1993). In (a) and (b) dotted lines show present levels. (c) Estimated air density. (d) Estimated kinematic viscosity of air. Carbon dioxide level has a weaker effect on air properties than does oxygen level. In (c) and (d), the dashed lines are based on estimated air composition with present-day pressure, gravity and temperature; neither density nor viscosity vary materially from present-day values. The chained lines show the effect of incorporating pressure variations, on the assumption that nitrogen partial pressure has remained constant. The heavy solid lines show the effect of incorporating variation in estimated mean sea-level temperatures (see text). These are considered to be the most realistic estimates of palaeoatmosphere properties. Air properties are sensitive to gravity: the light solid lines incorporate an estimate of decreasing gravity (light solid line in Figure 10.1a): in this case air density was always higher, and air viscosity always lower, than present-day values.

10.2 Effects of gravity and physical properties of air on locomotion biomechanics

10.2.1 Terrestrial locomotion and gravity

10.2.1.1 Size and posture

Animals moving on the ground must support their body weight against gravity, and move their limbs to propel themselves. The limbs will experience a mixture of compressive and buckling stresses. As the largest sizes are approached, limbs tend to become columnar in posture reflecting the dominance of compressive forces associated with the animal's weight (Alexander, 1989; Benton, 1990; Hokkanen, 1986), and in mammals and dinosaurs there is a systematic change in posture with size (Alexander, 1992). Compared to weight, smaller animals can tolerate larger bending and buckling moments during locomotion than can the largest animals (Fariña *et al.*, 1997). The risk of fracture by these bending or buckling moments can be reduced by increasing the cross-sectional size of the limb bones relative to bone length as size increases, but at the very largest sizes diameter will increase too much to be practical, and at the same time limb weight and the size of muscles required to move the limb will become impossibly large. These factors set the maximum size of terrestrial animals (Hokkanen, 1986).

Compressive forces appear to have had the dominant influence on evolution of limb morphology in the largest terrestrial cursors. These forces are directly proportional to weight, and therefore – in animals of equal mass and uniform shape – to the gravitational acceleration *g*. As a result, gravity may be expected to have a large influence on evolution of the locomotion system. Predicted maximum body masses are inversely proportional to gravity, and therefore in a low-gravity environment animals may be predicted to evolve larger size. There have been two marked episodes of gigantism in cursorial tetrapods, the first in sauropod dinosaurs in the Mesozoic, and the second in Miocene perissodactlys. Sauropods reached masses estimated at up to 140 tonnes (maxima around 80 t can be considered reliable; Alexander, 1998), and *Indricotherium* may have had a mass up to around 20 t, *c.* 35 My BP (but see Fortelius and Kappelman, 1993; Alexander, 1998: recent estimates of size for this species are more conservative). These may be compared with masses up to 8–10 t in living elephants. Can these sizes be explained by variations in palaeogravity?

It seems unlikely that gravity can form the full explanation. The estimated gravity at the start of the Jurassic ranges from a factor of 0.7 (assuming constant Earth density) to 1.6 (constant Earth mass) of present gravity; these are extreme, and as explained above contentious, estimates. A conservative estimate (Stewart, 1978) would place the factor between 1 and 1.5 times present gravity. The largest cursorial animals should therefore be within at most ±40% of the present-day maximal mass (Figure 10.1b), and if gravity were higher than present the effect would be to *reduce* the maximum size attainable. For larger maximum sizes to be explicable by Earth expansion, palaeogravity must have been well below pres-

ent values, and this is unrealistic. Gravity variation is unable to explain the order of magnitude difference in mass between the largest sauropods and present-day elephants.

By the Miocene period any difference in geophysical properties from the present would have been relatively small, and cannot be invoked to explain gigantism in *Indricotherium* or its relatives.

10.2.1.2 Froude numbers and gait transitions

The non-dimensional Froude number *F* (Appendix 10.1, p. 180) characterizes the speed at which an animal changes gait, and because its effect on gait is determined mainly by contact between the foot and the ground it should be invariant in different environments and with different morphologies. It may be used to predict locomotor strategies and morphologies in different gravities.

Two questions may be posed: how should we expect size to vary if an animal evolves in different gravities, and how should an animal move if transplanted into a different environment? To answer the first, assume that locomotion speed is constant. This may be because of factors such as control and sensing the environment, or of minimizing the risk of injury. Then invariance of *F* implies that length *L* should be inversely proportional to *g*. If design is isometric, so that lengths vary as the cube root of body mass, then body mass *M* varies as

$$M \propto g^{-3} \qquad (10.1)$$

With lower gravity, as, for example, with Carey's hypothesized gravity change and constant Earth density, the constraint on gait transition modelled by Froude number implies that either gait changed at much lower speeds, or that animals were larger (or, at least, had longer limbs). The maximum allowable factor of about 2 in mass for the early Jurassic (Figure 10.1b) remains insufficient to explain large size in sauropods. Increased palaeogravity (Stewart, 1978) would suggest that Froude number was a sharp constraint on maximum size. Possibly therefore sauropods only walked, and it is unlikely they could change gait to a trot. They would therefore not be subject to limits on size associated with faster running gaits and elastic energy storage, would not experience large bending moments on the limb bones, and this could have encouraged a columnar stance, and therefore possibly indirectly have facilitated large size.

The Froude number constraint becomes far more important for animals evolving in, or transported to, alien environments. This is considered below.

10.2.2 *Flight*

Theoretical methods for predicting flight performance from morphology by scaling arguments are well developed, and can readily be modified to predict the effects of changing palaeogeophysical factors on flight performance (primarily speed or power consumption) or on optimum morphology; the relevant derivations are set out in Appendix 10.2 (p. 180). The key constraints are to balance wing size and flight speed against the need to support the weight, which is the single largest force acting in flight, and which dominates the solution in both fixed wing or flapping flight.

This analysis incorporates a considerable number of potential variables all of which interact: morphology is defined as is normal in studies of animal flight (e.g. Pennycuick, 1975; Rayner, 1988) by body mass M, wingspan b and wing area S, and flight performance can be measured by speed V, power P and cost of transport C, among other quantities. Flying animals experience a number of constraints, the most critical of which are generating sufficient lift, which is quantified by lift coefficient C_L and being able to produce sufficient power from the flight muscles (and ultimately from food extracted from the ecosystem).

Optimum speed or power for any animal depends on morphology (see Appendix 10.2), and therefore selection for flight performance has implications for evolution of morphology. I have identified three possible trajectories through the morphology/performance space which could indicate how flying animals might adapt or evolve, namely, flying at a constant speed and lift coefficient, flying at the

optimum speed and constant lift coefficient, and flying at optimal morphology; this last assumes that the selective pressures which determine optimal morphology are the same in the different palaeoatmospheres or different planets that I consider.

10.2.2.1 *Constant speed and C_L*

Under this assumption, small changes in density or gravity may have a marked effect on size, but air viscosity v plays no role (M varies as $v^{1/4}$ (A2.8); Figure 10.3a). Density changes have an appreciable effect only in the Carboniferous and Permian, when oxygen levels were high, and through much of the Mesozoic optimum sizes were somewhat smaller than now. There is no evidence to support atmospheric change as a mechanism for vertebrate gigantism.

10.2.2.2 *Optimum speed and C_L*

Viscosity now has a marked effect on size, overlaying that of density (A2.12). The low oxygen and high CO_2 levels of the Mesozoic could have led to a possible slight size increase, but this would have been insufficient to explain the largest sizes of pterosaurs by comparison with extant animals.

The distinction between constant speed and C_L and optimum speed and C_L is important; it relates to assumptions about speed, and how animals may select speed. Larger sizes can be possible if much faster flights were selected, and this might result indirectly from other environmental factors such as wind or atmospheric circulation. The two optima should be seen as extremes setting bounds on the adaptation of body size for flying animals.

10.2.2.3 *Optimum morphology*

If morphology evolves in response to the same atmospheric pressures as today, then it is not necessarily the case that animals should always have been the same size or shape as extant species. The extent of variation in wing design has tended to decrease over time (Figure 10.4). The Paleozoic oxygen pulse has the effect of permitting slightly larger animals to fly with

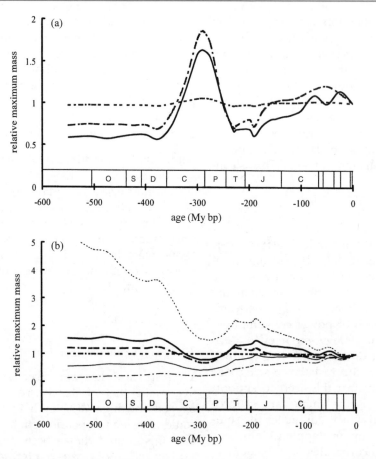

Figure 10.3 Predicted variation in maximum body mass for flying animals, relative to the maximum mass of extant animals experiencing similar selective pressures, assuming constant lift coefficient, and (a) constant flight speed (from equation A2.2) or (b) flight at optimum flight speed (equation A2.12). Heavy lines show the effect of the three atmospheric models of Figures 10.2c and d: constant pressure; varying pressure and constant temperature; varying pressure and temperature. In (b), faint lines add extremes of gravity (Figure 10.1) to the varying pressure and temperature model. Note the inversion in predicted mass between the two models, which results from the influence of viscosity on drag in model (b). During the Carboniferous and Permian the maximum possible sizes for flying animals able to fly at all were relatively large compared to the present (this is consistent with gigantism in Carboniferous insects), but these animals could have flown rather slowly. Migrants during this period (if they existed) would have been rather smaller than they are today.

rather smaller or thinner wings, and this may well be related to the appearance of pterygote insects and to insect gigantism. By the Mesozoic the scope for variation in morphology is much smaller, and again does not seem to encourage either the origin of vertebrate fliers, or the evolution of giant pterosaurs. There seems to be no scope to invoke atmospheric effects in the evolution of bats (*pace* Dudley, 1998, 2000).

10.2.3 *Swimming*

Morphology and locomotion patterns of underwater swimming animals are determined by the interaction of buoyancy and weight, and by the hydrodynamic drag (and to some extent lift) on their bodies. For a neutrally buoyant swimmer any variation in gravity should have no direct effect since the weights of the organism and of the displaced water would be in constant pro-

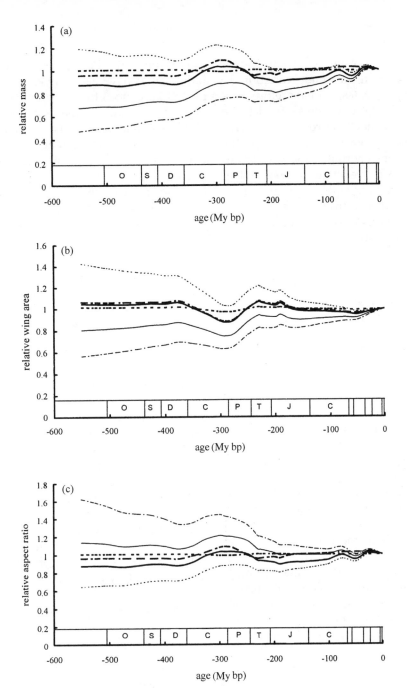

Figure 10.4 Optimum morphology ((a) body mass, (b) relative wing area S^* and (c) aspect ratio) for flying animals, computed from equations (A2.13)–(A2.17). This calculation shows how sizes and shapes might vary relative to those of flying animals today, assuming that morphologies were optimum for ecological niche in a way comparable to those today, and that selective pressures associated with different flight niches were comparable to those determining the structure and ecomorphology of present communities. Models of atmospheric density and viscosity are as in Figures 10.2c and d and 10.3. The effects of the extreme range of gravity are also shown as fainter lines (see Figure 10.3 for legend). The range of variation admitted is not great owing to balancing of the effects of changing viscosity and density. Animals should have been larger, with less rounded wings, during the oxygen pulse of the Carboniferous and Permian. The optimum size during the later part of the Mesozoic was slightly smaller than at present, but with larger and more pointed wings as density was relatively low and kinematic viscosity high.

portion. For a non-buoyant organism gravity changes would result in small variation in climbing or sinking rates, and therefore possibly more intense selective pressures for the evolution of buoyancy control organs. This may – speculatively – have been a factor in the radiations of aquatic chordates and vertebrates in the Ordovician to Devonian periods, when (according to Carey's hypotheses) gravity may have differed markedly from present; it may also have implications for the evolution of the cyanobacteria gas vacuole.

<h3>10.3 Palaeogeophysics and macroevolution</h3>

10.3.1 Hypothetical changes in gravity

Carey's models allow us to set bounds to changes in surface gravity over geological time. If gravity variation alone was to have been responsible for gigantic body masses in sauropod dinosaurs, gravity in the early Jurassic would have to have been as little as one tenth of present-day values. Variations in gravity of this magnitude are inconceivable; it is not possible to identify processes permitting the Earth's mass to increase by accretion to anything approaching the extent required. If, for the moment, we accept the hypothesis that the Earth has expanded while its mass has remained approximately constant, then in the past the maximum sizes of cursorial animals would have been lower than they are today. This is inconsistent with the rich fossil record of large forms, quite apart from the tendency to large size in several dinosaur clades, and indeed the existence of these giant forms becomes in itself compelling evidence for the absence of giant forms. These factors can indeed be taken as indirect evidence that during the evolution of tetrapods gravity has changed little if at all.

Broadly similar considerations apply also to the evolution of giant pterosaurs in the Cretaceous, and to giant birds in the Miocene. Gravity changes are unable to account for the observed size ranges. It appears unlikely, at least through the Phanerozoic, that macroevolution could have been driven or triggered, or indeed influenced to any great extent, by gravitational changes. It remains possible, although highly speculative, that the evolution of teleost swim bladders and of other mechanisms for buoyancy control in swimming fish was associated with varied levels of gravity and the need to control sinking and climbing rates. Other biological or autecological processes must be invoked to explain instances of gigantism in tetrapods.

10.3.2 Atmospheric changes

Gravity changes can largely be discounted, if only for their inherent implausibility, but changes in the physical properties of air have greater potential to have been a real effect on animals. Here I concentrate on the effects on locomotion aerodynamics, but there could also have been major physiological consequences, high oxygen levels possibly encouraging the Cambrian radiations (Thomas, 1997), the evolution of lungs and buoyancy strategies in different fish lineages, and the transition to land in the Devonian (Graham *et al.*, 1995, 1998) through their effects on respiration and diffusion, and the exchange of energy with the environment. The major biomechanical consequences of atmospheric change probably come later, with the origin of pterygote insects and of flying vertebrates.

10.3.3 Locomotor macroevolution

10.3.3.1 Origin of flight

The origin of flight in the four separate groups of organisms capable of true flapping has attracted continuous discussion, considerable controversy, and little agreement. This is not the appropriate place to review these debates (for aerodynamic aspects see Rayner, 2001a). Dudley (1998, 2000) argued that changes in oxygen concentration may have triggered the origins of insects, birds and pterosaurs, but had to push the origin of flying archosaurs into the late Permian or early Triassic when it might have been triggered by the Paleozoic oxygen pulse. Equally, he proposed that bats may have originated in the late Cretaceous, when oxygen levels were elevated. This proposal is entirely plausible for insects, which underwent their first radiations coincident with the Paleozoic oxygen pulse; it is, however, optimistic for vertebrates, for there is no evidence for such early origins of any of these groups, and the present analysis implies that the benefits for fliers would have been modest. The probable close relationship of pterosaurs to dinosaurs implies a mid-Triassic origin for the group at the earliest, at a time when oxygen levels, and maximum masses for flying animals, were falling; the origin of birds among derived theropod dinosaurs points to an origin of avian flight in the early Jurassic, between 180 and 150 My BP. However, the effect of low oxygen through air viscosity to favour high aspect ratio during the Triassic and Jurassic (Figure 10.4c) may have aided the flight performance, particularly in take-off, of the incipient vertebrate fliers; this tentative hypothesis, which is the opposite of that proposed by Dudley, deserves further investigation.

10.3.3.2 Gigantism in insects, pterosaurs and birds

At various periods there have been episodes in which flying animals have reached large sizes, compared to extant animals, including Carboniferous insects (Kukulova-Peck, 1985; Wootton and Ellington, 1991), several lineages of Cretaceous pterodactyloid pterosaurs (Wellnhofer, 1978, 1980, 1991; Padian and Rayner, 1992; Frey and Martill, 1996; Rayner, 1996; Alexander, 1998), and several lineages of Miocene birds (Halstead and Middleton, 1976; Campbell and Tonni, 1980, 1981, 1983; Rayner, 1996; Alexander, 1998). All have reached maximum sizes considerably larger than comparable modern species. The maximum size of modern flying birds is of the order of 12–15 kg, but pterosaurs and Miocene teratorns (related to extant condors) may have reached masses up to 60–80 kg (Rayner, 1996), and some estimates place them even higher (Alexander, 1998). Harlé and Harlé (1911) invoked high atmospheric pressure (not, as sometimes

quoted, air densities) to explain giant insects and pterosaurs. The present analysis supports this hypothesis to some extent, at least for insects. During the Carboniferous and Permian maximum mass of flying animals may have increased by up to a factor of about 1.6 (Figure 10.4); optimum wing size could have been smaller than today, and for comparable aerodynamic efficiency wings may have been rather less narrow; wings of equivalent aspect ratio to modern insects would therefore have permitted even larger sizes. However, bounds on mass decrease rapidly into the Mesozoic, and there is no reason to expect large insects during the Jurassic such as those found in the Solnhofen limestones.

The situation in pterosaurs is less clear. Atmospheric factors do not indicate large size during the Mesozoic, and it must therefore be concluded that gigantism in Cretaceous pterodactyloids did not result from changes in atmospheric gas composition. The relatively high aspect ratio of most pterosaurs (Padian and Rayner, 1992; Hazelhurst and Rayner, 1992), which compared to modern birds and bats – a trend which was not confined to larger species – has been interpreted as a response to trophic factors or to preservational bias (Wellnhofer 1991; Hazelhurst and Rayner, 1992), but the present analysis suggests that aspect ratio up to 50% greater than modern values could have been aerodynamically optimum during the period of low air density and high kinematic viscosity in the Jurassic and into the Cretaceous. If this hypothesis is correct, then similar pressures might have affected Mesozoic birds. Few species from the limited fossil record are sufficiently completely preserved for wing outline to be reconstructed, but circumstantial support is provided by the wing shape of *Confuciusornis*, from the lower Cretaceous of China, which had a high aspect ratio compared to modern birds of comparable size (Peters and Qiang, 1999; Rayner, 2001a).

10.3.3.3 What sets the maximum sizes of organisms?

Animals are unlikely to reach the absolute maximum sizes set by biomechanical factors.

Even allowing for appropriate safety factors, this implies too panglossian a universe. Adaptation must not only fit organisms for their environment, but also for likely fluctuations in that environment. Seen narrowly from a biomechanical or a physiological point of view this might imply suboptimal design, but this does not necessarily make organisms less fit. The biomechanical approach sets aside the critical role of autotrophic factors, and in particular of the environment and community, in determining morphology.

There are several consequences of this. First, animals may not reach the mechanically possible maximum sizes because the largest possible forms may not be able to guarantee sufficient to eat, or may be vulnerable to predation or to environmental fluctuations, or may, for example, be unable to raise young comparable with their own sizes. Second, the ecological factors which constrain size in this way may be very labile over geological time, as the physical and geographical environment change, and as the structure of the community of which the organisms are a part adapt. There is every reason to expect that these changes may respond non-linearly to forcing geophysical factors.

A plausible application is that absolute maximum sizes vary, if perhaps weakly, with geophysical changes, as I have explored in this chapter. In some circumstances, which currently are not understood, ecosystems may be structured with relatively diverse small forms, which are limited primarily by autotrophic factors, while in other circumstances the ecosystem may be structured so that the largest forms may be much bigger, but there are fewer of them and diversity is low. These widely differing communities may represent separate self-organizing stable states, and the nature and stability of these states may be driven by external factors such as climate, or the progression of tectonic movements. Thus, large birds during the Miocene of South America may have been a unique response to the absence of a cordilleran mountain range and therefore to trade winds over the continental land mass (Campbell and Tonni, 1981, 1983). Comparable conditions may have permitted gigantism in pterosaurs or in

sauropod dinosaurs, but we know insufficient of the Mesozoic environment or of pterosaur palaeogeography to suggest what the causes might have been. What this does mean is that, in any lineage, explaining the palaeobiology of the largest organisms is not simply a matter of identifying biomechanical or physiological constants and the evolutionary mechanism for subverting them, but is a far broader question covering many aspects of the adaptation and environment of that species.

10.4 Exozoology

10.4.1 Locomotion of terran animals on other planets

The ability of a terran organism to survive on a different planet is a fundamental question which must be solved for effective interplanetary travel. Bodies in the solar system vary widely in surface gravity and in atmosphere density and viscosity (Table 10.1), and some environments may pose substantial challenges to terran organisms transplanted into them.

Cursorial movement is dominated by gravity, and Froude number can be used to predict walking speeds. Extensive experimental studies have investigated locomotion patterns for humans under conditions of reduced gravity, to simulate conditions on the Moon or on Mars (Cavagna *et al.*, 1998; Donelan and Kram, 1997; Farley and McMahon, 1992; He *et al.*, 1991; Kram *et al.*, 1997; Newman *et al.* (1994; Wickman and Luna, 1996), and most indicate that the responses in locomotion speeds and energetics are broadly as predicted by Froude number, that is by the balance between potential and kinetic energies at the walk/trot transition. Where experimental results deviate from this prediction, it is because in low gravity walking gaits with bent legs may be more efficient (Donelan and Kram, 1997; Kram *et al.*, 1997), and because when gravity is low the weight may be insufficient to stretch tendons during faster gaits.

The NASA Apollo project has provided extensive information on locomotion patterns

Table 10.1 **Surface gravity and atmospheric conditions for planets and major satellites of the Solar System. Gravity, surface conditions and atmosphere composition (not shown) from Lewis (1997), Barnet (1998) and Lodders and Fegley (1998).**

Planet	Gravitational acceleration (m/s^2)	Surface temperature (K)	Surface air density (kg/m^3)	Dynamic viscosity (10^5 Pa s)	Kinematic viscosity (10^5 m^2/s)
Mercury	3.70		–	–	–
Venus	8.87	735	64.8	3.47	0.053
Earth	9.81	288	1.20	1.81	1.48
Moon	1.62		–	–	–
Mars	3.75	214	0.0155	1.07	6.49
Jupiter	25.05	165	0.173	0.74	4.48
Io	1.80		–	–	–
Europa	1.31		–	–	–
Ganymede	1.43		–	–	–
Callisto	1.24		–	–	–
Saturn	10.55	135	0.196	0.57	3.10
Titan	13.58	94	5.55	0.79	0.14
Uranus	8.89	76	0.365	0.45	1.08
Neptune	11.21	72	0.438	0.45	1.04

in sustained low gravity on the Moon. Lunar gravity is about one sixth of the gravity on Earth, and is the lowest gravity level on the accessible bodies in the solar system (Table 10.1). This poses particular problems, because the expected speed of the walk/trot transition is only 41% of the speed on Earth. Much of the problem experienced by astronauts lies in acclimatization to the impossibility of walking quickly: records of the Apollo landings (NASA, 1999a, b) include many observations on walking strategies, and there was considerable inter-individual variation. At relatively fast speeds a hopping or loping gait was popular (Letko *et al.*, 1966). As yet there are no observations of other cursors in lunar gravity: the acclimatization of a terrestrial quadruped to lunar life will be of considerable interest!

These effects of gravity on locomotion are not the only consequences of gravity on spaceflight and the colonization of other planets. Extensive studies have focused on problems of development and of bone composition, maintenance and growth. The effect on balance and indirectly on swimming in frog tadpoles was mentioned above. These aspects are beyond the scope of the present problem, but have the potential to have more far-reaching consequences than locomotion alone.

Flight is much less likely to pose a serious practical problem, since there seems little purpose beyond sentiment for introducing terrestrial birds to a terra-formed Mars. Limited but inconclusive studies on flight behaviour of pigeons in reduced gravity during parabolic flight (Oosterveld and Greven, 1975) are more indicative of unfamiliarity and reorientation than of aerodynamics. A theoretical model of human flapping flight reached conclusions similar to those in Appendix 10.2 (Collins and Graham, 1994). However, a possible application of the equations in Appendix 10.2 is in setting design parameters for small-scale aerofoil-based flying machines, possibly using flapping flight, for use on other planets, should these even be possible.

10.4.2 Exobaupläne

A very different type of speculation is posed by the possible course of evolution on other plan-

ets. Science fiction writers tend to speculate that organisms in alien environments evolve to fit niches broadly similar to those on Earth. Thus, we might expect to find swimming, running and flying forms in widely differing physical environments. Is this plausible? What would these forms look like if selection acted similarly?

The arguments developed here give some quantitative answers to these questions. Presuming that problems of respiration and thermoregulation in the different atmosphere can be solved, and also that exobiomaterials are similar in physical properties to those

Table 10.2 **Locomotion of cursorial animals on the surface of planets and major satellites of the Solar System. Maximum sizes for animals evolving on the planet, based on Froude number (columns 1a and b) and limb bone strengths, and compressive strengths (column 2), and predicted factorial change in speed of walk–trot transition for terran animals (column 3). Size (= body length), mass and speeds are shown as factors relative to values on Earth. Columns 1 and 2 assume isometric design. Bodies marked * do not have a solid surface.**

Planet	1a Size for walking at the same speed $\propto g^{-1}$	1b Mass for walking at the same speed $\propto g^{-3}$	2 Maximum mass $\propto g^{-1}$	3 Walking speed for terrestrial animal $\propto g^{-\frac{1}{2}}$
Mercury	2.7	19	2.7	0.61
Venus	1.1	1.4	1.1	0.95
Earth	1	1	1	1
Moon	6.1	222	6.1	0.41
Mars	2.6	18	2.6	0.62
Jupiter*	0.4	0.06	0.4	1.60
Io	5.5	162	5.5	0.43
Europa	7.5	420	7.5	0.37
Ganymede	6.9	323	6.9	0.38
Callisto	7.9	495	7.9	0.36
Saturn*	0.9	0.80	0.9	1.04
Titan	0.7	0.38	0.7	1.18
Uranus*	1.1	1.3	1.1	0.95
Neptune*	0.9	0.7	0.9	1.07

based on carbon, then cursorial animals are unlikely to differ by more than an order of magnitude from equivalent terran forms (Table 10.2). They can be expected to become rather larger on worlds with low gravity: a lunar organism might be around six times heavier than its cousins on Earth; if it retained the same shape it would still have to walk slowly, but this constraint could be circumvented by the evolution of very long legs. A similar, but less extreme, response can be expected on Mars. Planets with significantly higher gravity than Earth do not have a solid surface.

The situation with flying animals is very different, since in addition to variation in gravity other worlds have dramatically different atmospheres (Table 10.1). For sustained flight, animals would generally be slightly smaller than on Earth (Table 10.3, col. 4), although they would have to fly very fast to generate sufficient lift (Table 10.3, cols 1 and 3). It is in opti-

mal morphology that animals may differ dramatically from what might be familiar on Earth. An aerodynamically optimum flying animal in the atmospheres of Mars or Saturn would have extremely long and thin wings, perhaps demanding mechanical competence beyond any known biomaterial, but on Titan flight should be possible with very short and thin wings. Flying organisms *might* be able to evolve in these environments, but they would appear very strange. It is unlikely that they could operate by aerofoil action, or that the adaptations of terran birds or bats would give any guide to their body form. Barrow and Tipler (1986) describe the *anthropic cosmological principle*, which expresses the idea that unique features of terrestrial physics have restricted the origin of human life to Earth (or closely similar planets in other solar systems). Evidently much the same (the *Ornithopic cosmological principle*) applies to the possibility of flying organisms.

Table 10.3   **Constraints on flying in the lower atmosphere of planets and major satellites of the Solar System. Columns 1 and 2 show the flight speed and power for flight at minimum cost speed for animals of similar morphology, from equations (A2.8) and (A2.9), relative to performance on Earth. Columns 3 and 4 are the two models equations (A2.2) and (A2.12) for maximum body mass compared to that on Earth under different assumptions about flight speed. Columns 5–7 are the optimum sizes and shapes computed by equations (A2.13)–(A2.17), relative to values on Earth.**

Planet	1. Relative speed	2. Relative power	3. Maximum mass flying at constant speed	4. Mass to maintain optimal flight speed	5. Relative mass	6. Relative wing area S^*	7. Relative aspect ratio
Venus	0.16	0.1	2×10^5	0.075	4.2	0.03	3.8
Earth	1	1	1	1	1	1	1
Mars	8.6	2.0	3.6×10^{-5}	0.64	5.3	52	2.0
Jupiter	3.0	10	1.7×10^{-4}	0.51	0.35	12.7	0.9
Saturn	2.6	2.7	3.3×10^{-3}	0.65	1.1	7.0	1.2
Titan	0.6	0.6	35	0.03	3.6	0.47	5.0
Uranus	2.0	1.4	3.5×10^{-2}	0.18	2.8	4.3	2.5
Neptune	1.9	1.8	3×10^{-2}	0.15	2.1	4.1	2.4

10.5 Conclusion

The analyses in this chapter should be regarded as somewhat speculative, and some of the specific numerical predictions should be treated with caution. However, the arguments I have developed are valuable in indicating that the scope for explanation of macroevolutionary events in locomotion through changes in the physical properties of the environment is limited.

1. Atmospheric changes have been critical triggers for or facilitators of some macroevolutionary events.
2. Density and viscosity changes are unlikely to explain most episodes of gigantism, but might have contributed to improved flapping flight performance in evolving insects, and possibly in proto-birds and -pterosaurs.
3. Increased aspect ratio in pterosaurs may have been a response to low density and high viscosity during the Mesozoic.
4. Gravity changes – admittedly controversial – potentially have had a large effect on terrestrial life, but cannot on their own explain the extent of vertebrate gigantism.
5. Cursorial organisms should be able to move on the other terrestrial planets with speeds of the same order of magnitude as on Earth.
6. Cursorial organisms evolving in lower gravity can be significantly larger than on Earth.
7. Terrestrial flying animals should expect serious problems adapting to the atmospheres of most other planets.
8. Animals evolving flight on other planets may look very strange to human eyes: birds will look like birds only on Earth.

References

Alexander, R.McN. (1976). Estimates of speeds of dinosaurs. *Nature, Lond.*, **261**: 129–130.

Alexander, R.McN. (1982). Size, shape, and structure for running and flight. In: C.R. Taylor, K. Johansen and L. Bolis (eds) *A Companion to Animal Physiology*. Cambridge University Press, pp. 309–324.

Alexander, R.McN. (1989a). Optimization and gaits in the locomotion of vertebrates. *Physiol. Rev.*, **69**: 1199–1227.

Alexander, R.McN. (1989b). *The Dynamics of Dinosaurs and Other Extinct Vertebrates*. New York: Columbia University Press.

Alexander, R.McN. (1992). *Exploring Biomechanics*. New York: W.H. Freeman.

Alexander, R.McN. (1998). All-time giants: the largest animals and their problems. *Palaeontology*, **41**: 1231–1245.

Alexander, R.McN. and Jayes, A.S. (1983). A dynamic similarity hypothesis for the gaits of quadrupedal mammals. *J. Zool., Lond.*, **201**: 135–152.

Anderson, T.F. (1990). Temperature from oxygen isotope ratios. In: D.E.G. Briggs and P.R. Crowther (eds) *Palaeobiology, a Synthesis*. Oxford: Blackwell, pp. 403–406.

Barnet, C. (1998). Standard planetary information, formulae and constants. http://atmos.nm-su.edu/jsdap.encyclopediawork.html.

Barrow, J.D., Tipler, F.J. (1986). *The Anthropic Cosmological Principle*. Oxford University Press.

Benton, M.J. (1990). Evolution of large size In: D.E.G. Briggs and P.R. Crowther (eds) *Palaeobiology, a Synthesis*. Oxford: Blackwell, pp. 147–152.

Berner, R.A. (1990). Atmospheric carbon dioxide levels over Phanerozoic time. *Science*, **249**: 1382–1386.

Berner, R.A. (1993). Paleozoic atmospheric CO_2: importance of solar radiation and plant evolution. *Science*, **261**: 68–70.

Berner, R.A. and Canfield, D.E. (1989). A new model for atmospheric oxygen over Phanerozoic time. *Am. J. Sci.*, **289**: 333–361.

Braker, W. and Mossman, A.L. (1971). *Matheson Gas Data Book* (fifth edition). Milwaukee: Matheson Gas Products.

Buchardt, B. (1977). Oxygen isotope palaeotemperatures from the Tertiary period in the North Sea area. *Nature, Lond.*, **275**: 121–123.

Budyko, M.I., Ronov, A.B. and Yanshin, A.L. (1987). *History of the Earth's Atmosphere*. Berlin: Springer Verlag.

Campbell, K.E. and Tonni, E.P. (1980). A new genus of teratorn from the Huayquerian of Argentina (Aves: Teratornithidae). *Contr. Sci. Nat. Hist. Mus. Los Angeles Co.*, **330**: 59–68.

Campbell, K.E. and Tonni, E.P. (1981). Preliminary observations on the paleobiology and evolution of teratorns (Aves: Teratornithidae*). J. Vert. Paleont.*, **1**: 265–272.

Campbell, K.E. and Tonni, E.P. (1983). Size and locomotion in teratorns (Aves: Teratornithidae). *Auk*, **100**: 390–403.

Canfield, D.E. and Teske, A. (1996). Late Proterozoic rise in atmospheric oxygen concentration inferred from phylogenetic and sulphur-isotope studies. *Nature*, **382**: 127–132.

Carey, S.W. (1976). *The Expanding Earth*. Amsterdam: Elsevier.

Carey, S.W. (ed.) (1983). *The Expanding Earth: a Symposim*. University of Tasmania Department of Geology. http://www.geocities.com/capecanaveral/launchpad/8098/symposium/symposium.htm.

Carey, S.W. (1996). *Earth, Universe, Cosmos*. University of Tasmania Department of Geology. http://www.geocities.com/CapeCanaveral/Galaxy/9981/.

Cavagna G.A., Willems P.A. and Heglund N.C. (1998). Walking on Mars. *Nature, Lond.*, **393**: 636.

Chai, P. (1997). Hummingbird hovering energetics during moult of primary flight feathers. *J. Exp. Biol.*, **200**: 1527–1536.

Chai, P. and Dudley, R. (1995). Limits to vertebrate locomotor energetics suggested by hummingbirds hovering in heliox. *Nature, Lond.*, **377**: 722–724.

Chai, P. and Dudley, R. (1996). Limits to flight energetics of hummingbirds hovering in hypodense and hypoxic gas mixtures. *J. Exp. Biol.*, **199**: 2285–2295.

Chai, P. and Millard, D. (1997). Flight and size constraints: hovering performance of large hummingbirds under maximal loading. *J. Exp. Biol.*, **200**: 2757–2763.

Chai, P., Chen, J.S.C. and Dudley, R. (1997). Transient hovering performance of hummingbirds under conditions of maximal loading. *J. Exp. Biol.*, **200**: 921–929.

Chai, P., Harrykissoon, R. and Dudley, R. (1996). Hummingbird hovering performance in hyperoxic heliox: effects of body mass and size. *J. Exp. Biol.*, **199**: 2745–2755.

Collins, P.Q. and Graham, J.M.R. (1994). Human flapping wing flight under reduced gravity. *Aero. J.*, **98**: 177–184.

Crowley, T.J., North, G.R. (1991). *Paleoclimatology*. New York: Oxford University Press.

Denny, M.W. (1993). *Air and Water*. Princeton University Press.

Dirac, P.A.M. (1937). The cosmological constants. *Nature, Lond.*, **139**: 323.

Donelan, J.M. and Kram, R. (1997). The effect of reduced gravity on the kinematics of human walking: a test of the dynamic similarity hypothesis for locomotion. *J. Exp. Biol.*, **200**: 3193–3201.

Dudley, R. (1995). Extraordinary flight performance of orchid bees (Apidae: Euglossini) hovering in heliox (80% He/20% O_2). *J. Exp. Biol.*, **198**: 1065–1070.

Dudley, R. (1998). Atmospheric oxygen, giant paleozoic insects and the evolution of aerial locomotor performance. *J. Exp. Biol.*, **201**: 1043–1050.

Dudley, R. (2000). The evolutionary physiology of animal flight: paleobiological and present perspectives. *A. Rev. Physiol.*, **62**: 135–155.

Dudley, R. and Chai, P. (1996). Animal flight mechanics in physically variable gas mixtures. *J. Exp. Biol.*, **199**: 1881–1885.

Fariña, R.A., Vizcaíno, S.F. and Blanco, R.E. (1997). Scaling of the indicator of athletic capability in fossil and extant land tetrapods. *J. Theor. Biol.*, **185**: 441–446.

Farley, C.T. and McMahon, T.A. (1992). Energetics of walking and running: insights from simulated reduced-gravity experiments. *J. Appl. Physiol.*, **73**: 2709–2712.

Feinsinger, P., Colwell, R.K., Terborgh, J. and Chaplin, S.B. (1979). Elevation and the morphology, flight energetics, and foraging ecology of tropical hummingbirds. *Am. Nat.*, **113**: 481–497.

Fortelius, M. and Kappelman, J. (1993). The largest land mammal ever imagined. *Zool. J. Linn. Soc.*, **107**: 85–101.

Frey, E. and Martill, D.M. (1996). A reappraisal of *Arambourgiana*: the world's largest flying animal. *N. Jb. Geol. Paläont. Abh.*, **199**: 221–247.

Graham, J.B., Aguilar, N., Dudley, R. and Gans, C. (1997). The late Paleozoic atmosphere and the ecological and evolutionary physiology of tetrapods. In: S.S. Sumida and K.L.M. Martin (eds) *Amniote Origins: Completing the Transition to Land*. New York: Academic Press, pp. 141–167.

Graham, J.B., Dudley, R., Aguilar, N.M. and Gans, C. (1995). Implications of the late Paleozoic oxygen pulse for physiology and evolution. *Nature, Lond.*, **375**: 117–120.

Halstead, B. and Middleton, J. (1976). Fossil vertebrates of Nigeria. Part I. *Nigerian Field*, **41**: 55–63.

Harlé, E. and Harlé, A. (1911). Le vol des grands reptiles et insectes disparus semble indiquer une pression atmospherique élevée. *Bull. Soc. Géol. Fr.*, **4**: 117–121.

Hart, M.H. (1978). The evolution of the atmosphere of the Earth, *Icarus*, **33**: 23–39.

Hazlehurst, G. and Rayner, J.M.V. (1992). Flight

characteristics of Jurassic and Triassic Pterosauria: an appraisal based on wing shape. *Paleobiology*, **18**: 447–463.

He, J.P., Kram, R. and McMahon, T.A. (1991). Mechanics of running under simulated low gravity. *J. Appl. Physiol.*, **71**: 863–870.

Hokkanen, J.E.I. (1986). The size of the largest land animal. *J. Theor. Biol.*, **118**: 491–499.

Holland, H.D. (1984). *The Chemical Evolution of the Atmnosphere and Oceans*. Princeton, NJ: Princeton University Press.

Jordan, P. (1971). *The Expanding Earth*. Oxford: Pergamon Press.

Kram, R., Domingo, A. and Ferris, D.P. (1997). Effect of reduced gravity on the preferred walk–run transition speed. *J. Exp. Biol.*, **200**: 821–826.

Kukulova-Peck, J. (1985). Ephemeroid wing venation based upon new gigantic Carboniferous mayflies and basic morphology, physiology and metamorphosis of pterygote insects (Insecta, Ephemerida). *Can. J. Zool.*, **63**: 933–955.

Letko, W., Spady, A.A. and Hewes, D.E. (1966). Problems of man's adaptation to the lunar environment. In: *Second Symposium on the Role of the Vestibular Organs in Space Exploration*, pp. 25–32. NASA **SP-115**.

Lewis, J.S. (1997). *Physics and Chemistry of the Solar System*. New York: Academic Press.

Lockwood, R., Swaddle, J.P. and Rayner, J.M.V. (1998). Avian wingtip shape reconsidered: wingtip shape indices and morphological adaptations for migration. *J. Avian. Biol.*, **29**: 273–292.

Lodders, K. and Flegley, B. (1998). *The Planetary Scientist's Companion*. Oxford University Press.

Maybury, W.J. (2000). *The Aerodynamics of Bird Bodies*. PhD thesis, University of Bristol.

McMahon, T. (1973). Size and shape in biology. *Science*, **179**: 1201–1204.

NASA (1999a). Apollo 11 lunar surface journal: mobility and photography. http://www.hq.nasa.gov/office/pao/History/alsj/a11/a11.mobility.htm

NASA (1999b). Apollo 17 lunar surface journal: Geology Station 5 at Camelot Crater. http://www.hq.nasa.gov/office/pao/History/alsj/a17/a17.sta5.htm

Newman, D.J., Alexander, H.L. and Webbon, B.W. (1994). Energetics and mechanics for partial gravity locomotion. *Aviat. Space Env. Med.*, **65**: 815–823.

Norberg, U.M. and Rayner, J.M.V. (1987). Ecological morphology and flight in bats (Mammalia, Chiroptera): wing adaptations, flight performance, foraging strategies and echolocation. *Phil. Trans. R. Soc. Lond. B*, **316**: 335–427.

Oosterveld, W.J. and Greven, A.J. (1975). Flight behaviour of pigeons during weightlessness. *Acta oto-lar.*, **79**: 233–241.

Padian, K. and Rayner, J.M.V. (1992). The wings of pterosaurs. *Am. J. Sci.*, **293-A**: 91–166.

Pennycuick, C.J. (1975). Mechanics of flight. In: D.S. Farner and J.R. King (eds) *Avian Biology*, vol. 5. London: Academic Press, pp. 1–75.

Pennycuick, C.J. (1978). Fifteen testable predictions about bird flight. *Oikos*, **30**: 165–176.

Pennycuick, C.J. (1987). Cost of transport and performance number, on Earth and other planets. In: P. Dejours, L. Bolis, C.R. Taylor and E.R. Weibl (eds) *Comparative Physiology: Life in Water and on Land*. Berlin: Springer-Verlag, pp. 371–386.

Perry, R.H., Green, D.W. and Maloney, J.O. (eds) (1984). *Perry's Chemical Engineer's Handbook* (sixth edition). New York: McGraw-Hill.

Peters, D.S. and Qiang, J. (1999). Mußte *Confuciusornis* klettern? *J. Orn.*, **140**: 41–50.

Popp, B.N., Anderson, T.F. and Sandberg, P.A. (1986). Brachiopods as indicators of original isotopic compositions in some Paleozoic limestones. *Bull. Geol. Soc. Am.*, **97**: 1262–1269.

Rayner, J.M.V. (1985). Linear relations in biomechanics: the statistics of scaling functions. *J. Zool., Lond. A*, **206**: 415–439.

Rayner, J.M.V. (1988). Form and function in avian flight. *Curr. Orn.*, **5**: 1–77.

Rayner, J.M.V. (1993). On aerodynamics and the energetics of vertebrate flapping flight. In: A.Y. Cheer and C.P. van Dam (eds) *Fluid Dynamics in Biology*, pp. 351–400. Contemporary Mathematics No. 141: xii, 1–586. American Mathematical Society, Providence.

Rayner, J.M.V. (1995). Flight mechanics and constraints on flight performance. *Israel J. Zool.*, **41**: 321–342.

Rayner, J.M.V. (1996). Biomechanical constraints on size in flying vertebrates. In: P.J. Miller (ed.) *Miniature Vertebrates, Symp. Zool. Soc. Lond.*, **69**: 83–109.

Rayner, J.M.V. (1999). Estimating power curves for flying vertebrates. *J. Exp. Biol.*, **202**: 3449–3461.

Rayner, J.M.V. (2001a). On the origin and evolution of flapping flight aerodynamics in birds. In: J. Gauthier (ed.) *New Perspectives on the Origin and Early Evolution of Birds*. Spec. Publs Yale Peabody Mus., pp. 363–385. New Haven: Peabody Museum of Natural History, Yale University.

Rayner, J.M.V. (2001b). Mathematical modelling of the avian power curve. *Math. Meths. Appl. Sci.* **24**: 1485–1514.

Reid, R.C. (1977). *The Properties of Gases and Liquids* (third edition). New York: McGraw-Hill.

Schmidt-Nielsen, K. (1972). Locomotion: energy cost of swimming, flying, and running. *Science*, **177**: 222–228.

Schmidt-Nielsen, K. (1985). *Scaling: Why is Animal Size so Important?* Cambridge University Press.

Stewart, A.D. (1978). Limits to palaeogravity since the late Precambrian. *Nature, Lond.*, **271**: 153–155.

Thomas, A.L.R. (1997). The breath of life – did increased oxygen levels trigger the Cambrian explosion? *TREE*, **12**: 44–45.

Vogel, S. (1994). *Life in Moving Fluids: the Physical Biology of Flow*. Princeton University Press.

Wellnhofer, P. (1978). *Handbuch der Paläoherpetologie. Teil 19. Pterosauria*. Stuttgart: Gustav Fischer.

Wellnhofer, P. (1980). *Flugsaurier*. Neue Brehm-Bücherei, **534**. Wittenburg-Lutherstadt: A. Ziemsen.

Wellnhofer, P. (1991). *The Illustrated Encylopedia of Pterosaurs*. London: Salamander Books.

Wickman, L.A. and Luna, B. (1996). Locomotion while load-carrying in reduced gravities. *Aviat. Space Environ. Med.*, **67**: 940–946.

Wootton, R.J. and Ellington, C.P. (1991). Biomechanics and the origin of insect flight. In: J.M.V. Rayner and C.P. Ellington (eds) *Biomechanics in Evolution*. Cambridge University Press, pp. 98–112.

10.1 APPENDIX Froude number and the scaling of terrestrial locomotion

The scaling of locomotion dynamics is determined by the interaction between propulsive and supportive forces. In mammals in the slowest gait, the walk, the limbs tend to be straight while they are supporting the weight. As a result, the position of the centre of mass oscillates vertically during a stride, and there is an interchange between potential and kinetic energy. The limb/body system functions as an inverted pendulum. Alexander (1982, 1992; Alexander and Jayes, 1983) has shown how this can be modelled by the Froude number

$$F = \frac{V^2}{Lg} \qquad (A1.1)$$

where V is the speed of travel, L is the length of the limb, and g is the gravitational acceleration. At the mid-point of a step the centre of mass is at its highest position, and is being accelerated upwards by a centripetal force. If this force exceeds the weight then it becomes impossible to keep the foot on the ground. The condition for this is

$$\frac{MV^2}{L} < Mg \qquad (A1.2)$$

or $F < 1$. F can be interpreted as proportional to the ratio between the mean kinetic ($\frac{1}{2}MV^2$) and mean potential (MgL) energies. For larger Froude numbers the animal cannot continue to walk, and must change gait to a run or trot. Typically, in these paces the legs are bent and elastic energy is stored in the limbs or the trunk. In fact animals change gait at Froude numbers appreciably less than unity, because as a value of 1 is approached the contact force on the ground reduces and it becomes increasingly difficult to maintain sufficient friction to keep the foot still during the step. Alexander and Jayes (1983) and Alexander (1992) have shown that mammals consistently change from walk to trot at a uniform Froude number of around 0.5. Since this constraint is determined by friction, it should apply to walking animals in all environments provided the substrate has comparable properties.

10.2 APPENDIX Scaling of animal flight performance

Lift generation by the aerofoil sets maximum size

In flight, the lift force L generated by aerofoil action on the wings must *at least* support the weight Mg. That is

$$L = \tfrac{1}{2}\rho S C_L V^2 \le Mg \qquad (A2.1)$$

where ρ is air density, V is flight speed, S is wing area and C_L is the lift coefficient. Lift coefficient is a non-dimensional measure of the aerodynamic performance of the wings. It is the ratio of the pressures on the upper and lower surfaces of the wing, and is independent of viscosity. Lift coefficient is expected to be invariant, and within wide bounds should be independent of atmosphere geophysics. Aerodynamic constraints set a maximum lift coefficient (in steady fixed-wing flight, this is somewhat less than 2), and this in turn sets a maximum to weight support and a minimum to flight speed.

Assume now (1) that lift coefficient is invariant, (2) that design is isometric ($S \propto M^{2/3}$), and (3) that flight speed is held constant. Then by eliminating V, S and C_L from equation (A2.1) we find

$$M \le M_{max} \propto \rho^3/g^3 \qquad (A2.2)$$

as a constraint on maximum mass. This is independent of air viscosity. Maximum linear size should scale as ρ/g.

Flight speed might plausibly be constant under changing conditions, since speed is intimately related to sensory performance and to the risk of collision, and to typical wind speeds. A model derived below will show somewhat different size scaling of mass under alternative assumptions about speed.

Scaling of flapping flight performance

Flight performance is normally described by the variation in the mechanical power P required to fly as a function of flight speed, morphology, and – in the present context – of air geophysics. Performance is expressed as the *power curve*, which can be modelled as

$$P = \alpha \frac{M^2 g^2}{\rho b^2 V} + \beta \rho S C_D V^3 \qquad (A2.3)$$

(Rayner, 2001b), where the terms represent respectively the work done against induced (=vortex) and friction (=body parasite + wing profile) drags (Figure A2.1). Here α and β are numerical constants, $2b$ is wingspan, and C_D is the drag coefficient of the body and wings. Relatively little is known about drag of animal bodies, and published values are often little better than general estimates. For typical bodies in the size range of birds, C_D is expected to vary as

$$C_D = C_{D0} Re^{-\frac{1}{2}} \qquad (A2.4)$$

where Reynolds number

$$Re = \frac{Vb}{\nu} \qquad (A2.5)$$

expresses the relative importance of viscous and inertial forces (see, e.g., Denny, 1993; Vogel, 1994). (ν is the kinematic viscosity of

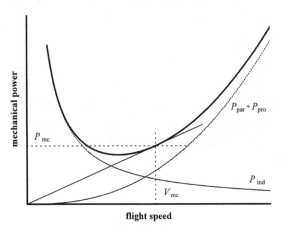

Figure A2.1 The mechanical power curve for flight (equation A2.3) is U-shaped, since induced power P_{ind} falls but friction drag (wing profile + body parasite $P_{par} + P_{pro}$) rises as speed increases. The speed V_{mc} at which cost of transport $C = P/MgV$ is minimum characterizes flight performance; this is the speed where the tangent from the origin meets the curve, and is usually understood to predict the speed at which a bird should migrate in still air (Pennycuick, 1975).

air.) Recent measurements of drag for starling bodies (Maybury, 2000) have measured close to the predicted $Re^{-1/2}$ behaviour.

Strictly, equation (A2.3) applies to fixed wing (i.e. gliding) flight, and should be modified to take account of flapping flight. For the present purposes this simple model is sufficient to capture the functional response of performance and morphology to the atmosphere. For discussion of power curves, including their history and aerodynamic background, see Rayner (1993) and Rayner (1999, 2001b).

A valuable measure of locomotion performance is the *Cost of Transport* (Schmidt-Nielsen, 1974), defined as

$$C = \frac{P}{MgV} \qquad (A2.6)$$

or the energy required to transport unit weight of the animal through unit distance. (This is the same quantity as Pennycuick's (1987) 'performance number'.) A migrant bird in still air should opt to fly at the speed which minimizes cost of transport (Figure A2.1). Selection for sustained long-distance locomotion such as migration is associated with the long, thin and pointed wings which minimize cost of transport and mechanical power at this speed (Norberg and Rayner, 1987; Rayner, 1988; Lockwood *et al.*, 1998).

The dependence of sustained flight performance on morphological parameters M, b and S and on geophysical parameters ρ, ν and g can be expressed by evaluating the *minimum cost speed* as the speed V_{mc} which satisfies

$$\left. \frac{\partial}{\partial V} \frac{P}{MgV} \right|_{V=V_{mc}} = 0 \qquad (A2.7)$$

namely

$$V_{mc} \propto \left[\frac{g^4}{\rho^4 \nu} \frac{M^4}{b^3 S^2} \right]^{\frac{1}{7}} \qquad (A2.8)$$

and by determining the corresponding power, cost of transport and lift coefficient at that speed

$$P_{mc} = P(V_{mc}) \propto \left[\frac{g^{10} \nu}{\rho^3} \frac{M^{10} S^2}{b^{11}} \right]^{\frac{1}{7}} \qquad (A2.9)$$

$$C_{\mathrm{mc}} = \frac{P_{\mathrm{mc}}}{MgV_{\mathrm{mc}}} \propto \left[\frac{\rho v^2}{g} \frac{S^4}{Mb^8} \right]^{\frac{1}{7}} \qquad (A2.10)$$

and
$$C_{\mathrm{L,mc}} \propto \left[\frac{\rho v^2}{g} \frac{b^6}{MS^3} \right]^{\frac{1}{7}} \qquad (A2.11)$$

from equation (A2.1). Similar formulae were derived by Pennycuick (1978) to model the effect of air density on climbing performance, but his analysis was incomplete in neglecting height-related changes in viscosity. The formulae above, which incorporate viscosity, are novel.

It should be noted that this model applies to aerial flight in which gravity and weight are the dominant forces; it is not appropriate in very low gravity even in gases, for aerodynamic drag then becomes comparable in magnitude to weight.

10.2.3 Characteristic speed and lift determines maximum body mass

The argument in equation (A2.2) derived a maximum limit to size from lift coefficient on the assumption that speed remained constant, as might be the case where selection favoured control of speed for prey capture or for take-off or landing. What happens if selection favours long-distance flight, and therefore flight at the optimum minimum cost speed?

Assume as before (1) that lift coefficient is invariant and (2) that design is isometric ($S \propto M^{2/3}$), but now let flight speed vary in proportion to V_{mc}. Then, maximum body mass will be constrained as

$$M \leq M_{\mathrm{max}} \propto \rho v^2 / g \qquad (A2.12)$$

from equation (A2.11). Compared to equation (A2.2) there is a weaker relationship between ρ/g and body mass, and this relationship is tempered by kinematic viscosity, which in air tends to be approximately inversely proportional to ρ. As a result the predicted constraint on body size under this constraint is somewhat different.

10.2.4 Flight performance determines body size and wing shape

The arguments above assumed isometric design, that is that all dimensions vary in proportion, and therefore wing shape is constant and wing size can be determined directly from body mass. This is not necessarily the case. Although flying animals are close to isometry on average, selection acts on wing morphology to optimize different measures of performance within a range of constraints. Selection to favour flight speed or flight economy (i.e. power or cost of transport, depending on the context) within overall constraints on lift generation modelled by lift coefficient C_{L} has given rise to a range of morphologies (Rayner, 1988, 1996). On the assumption that the processes of selection in different atmospheres acted similarly to those acting on recent organisms, it is possible to predict how optimum morphology might vary in response to atmosphere by keeping speed, power and lift coefficient constant in equations (A2.8), (A2.9) and (A2.11). By algebraic rearrangement of these equations we find the following functional relationships for flying animal morphology:

body mass	$M \propto \rho^{-\frac{1}{2}} g^{-1} v^{-1}$	(A2.13)
wingspan	$b \propto \rho^{-1} v^{-1}$	(A2.14)
wing area	$S \propto \rho^{-\frac{3}{2}} v^{-1}$	(A2.15)
aspect ratio	$A = \frac{4b^2}{S} \propto \rho^{-\frac{1}{2}} v^{-1}$	(A2.16)
wing loading	$N = \frac{Mg}{S} \propto \rho^{1}$	(A2.17)
relative wing span	$b^* = b/M^{\frac{1}{3}} \propto \rho^{-\frac{5}{6}} g^{\frac{1}{3}} v^{-\frac{2}{3}}$	(A2.18)
relative wing area	$S^* = S/M^{\frac{2}{3}} \propto \rho^{-\frac{7}{6}} g^{\frac{2}{3}} v^{-\frac{1}{3}}$	(A2.19)

Aspect ratio is a valuable non-dimensional measure of wing shape: higher values correspond to longer, thinner wings. Wing loading is not non-dimensional, but is closely associated with flight speed (cf. equation A2.1) and normally increases with body mass and depends on gravity. For the present purposes the relative wing area S^* is more useful.

Interestingly, body mass is affected by gravity, but aspect ratio and wing loading are not; all parameters are affected by density and viscosity. Wings should become smaller in response to rising density or viscosity because a lower area is needed to support the weight. Few data are available to test these models directly, but they provide a formal functional explanation for Rensch's rule, that the size of organisms and the relative size of their appendages tend to increase with altitude. Hummingbirds living at high altitude tend to have larger and longer wings than species from lower levels (Feinsinger *et al.*, 1979), as predicted by equations (A2.18) and (A2.19).

PART 5

Temperature

11

Evolution and low temperatures

Andrew Clarke

ABSTRACT

See Plates 18 and 19

Life as we know it depends absolutely on solution chemistry in a highly unusual solvent (water), a complex range of chemicals based on a few simple small atoms, and relatively small values of free energy. The unusual physical properties of water are greatly influenced by hydrogen bonding, and these properties in turn affect almost all aspects of organismal existence. The boundary conditions for life on Earth are set by the existence of liquid water but the physical challenges set by high and low temperatures are very different. At high temperatures cells are faced with the disruptive effect of the high thermal energy, and at temperatures above about 55°C (roughly half the range of temperatures over which life is found) effectively only microbes exist. At low temperatures organisms face the risk of cellular water freezing, and must deal with the low thermal energy which reduces the rate of many physiological processes.

Freezing of intracellular water is usually fatal, and many organisms have evolved antifreezes to maintain water in the liquid state at low subzero temperatures. There are also widespread, though not universal, adjustments to protein structure to maintain cellular architecture and physiological rates. There is an unresolved debate as to whether evolution itself proceeds more slowly at lower temperatures. Consideration of the possibility of life elsewhere suggests that ammonia is the only likely alternative to water as a solvent for life.

11.1 Introduction

Although the thermal range of an individual species is often relatively narrow, life as a whole exists over a surprisingly wide range of temperatures. The boundaries of this range are, by definition, extreme environments and they are of considerable interest to biologists concerned with determining those processes that set limits to the existence of life. The physiological challenges set by high and low temperatures are, however, different. At low temperatures organisms have to face the possibility of water freezing and the very low

thermal energy of the environment, with the consequent reduced rate of many physical processes. At high temperatures cells are faced with the disruptive effect of the very high thermal energy on cellular architecture, and the need to maintain metabolic control.

These differing thermal challenges are exemplified in the distribution of organisms in relation to temperature (Rothschild and Mancinelli, 2001). Although a wide range of taxa exist at low temperatures, once habitat temperature exceeds about 50°C essentially only microbes are found. The highest aquatic temperatures at which vertebrates can complete their life cycle is about 40°C (fish living

Evolution on Planet Earth
ISBN 0-12-598655-6

in hot springs at a variety of locations: Reite *et al.*, 1974; Gerking *et al.*, 1979; Gerking and Lee, 1983). Terrestrial organisms living in hot environments utilize a variety of mechanisms for regulating their body temperature, and for vertebrates this rarely exceeds 45°C. It is striking that for roughly half the temperature-range over which life can be found on Earth, metazoans appear not to exist.

11.1.1 *Water as the basis of life*

The envelope of physical conditions for life on Earth is in many ways remarkably wide: environmental pH can range from <1 to 10, pressure from <1 to >500 atmospheres, and salinity from zero (freshwater) to saturation in saline lakes (Rothschild and Mancinelli, 2001). Life as we know it does, however, depend absolutely on the presence of liquid water, and the existence of the liquid state sets the boundary conditions for life itself. Although some forms of life are capable of withstanding the formation of ice within the body, or of dehydrating almost completely, these states reflect extreme examples of resistance to environmental challenges. The organisms are unable to grow or reproduce while frozen or dehydrated, and must return to the hydrated state to complete their life cycles. Within these constraints life can be found across the whole temperature range at which liquid water is found on Earth (Table 11.1).

The absolute dependence of life on liquid water in the cell can be linked to two fundamental factors. The first is that water is a very common compound on Earth (indeed also in the universe at large: Mason, 1992): aquatic environments dominate the surface of the Earth, and life evolved in water. The second factor is that the physical properties of water are highly unusual.

It has long been recognized (Henderson, 1913) that most, if not all, of the physical properties of water are very different from those that would be predicted on the basis of the properties of the homologous compounds of the other Group VIA elements (H_2S, H_2Se, H_2Te) (Table 11.2). This anomalous behaviour is primarily the result of the structure of the water molecule and its propensity to form hydrogen bonds (Wald, 1964; Tanford, 1980; Pain, 1982; Franks, 1985).

Thus the unusually high specific heat capacity, together with the unusual latent heats of fusion (low) and evaporation (high), are critical in determining the thermal environment of Earth. The high specific heat means that aquatic environments change temperature only slowly, and the latent heat of evaporation sets the upper limit for oceanic temperatures under current salinities and atmospheric composition. Organisms also change temperature relatively slowly; were the specific heat capacity of water more typical then metabolic heat would raise body temperature very rapidly. Furthermore, the high thermal conductivity of water allows organisms to equilibrate temperature very quickly within tissues and cells, thereby avoiding convective forces which would otherwise damage internal cellular architecture. This high thermal conductivity incidentally makes it very difficult for a marine organism with gills to maintain a temperature differential with that environment, for gills exchange heat even more effectively than they do gases.

Of particular importance to physiology is the relatively inert nature of water as a compound, coupled with its very high dielectric permittivity. The latter is a key factor in rendering water such a powerful solvent, particularly for ionic compounds. Water is often referred to colloquially as the universal solvent, and the range of compounds it dissolves far exceeds that of any other known liquid. Equally important, however, are the compounds that water either does not dissolve, or does so only sparingly; the hydrophobic interactions of many lipids are critical in structuring the membranes which act as an outer boundary for the cell and as internal architecture (Tanford, 1980). Water is also intimately involved in the maintenance of other key cellular macromolecular structures such as proteins and nucleic acids.

The recognition of the central importance of water to life goes back to the Ancient Greeks: Thales viewed water as the origin of everything, and water was included as one of the four fundamental elements by Empedocles

Table 11.1 **Temperature and life.**	
Temperature	*Comment*
77 K (−196°C)	Boiling point of nitrogen. Temperature of successful cryopreservation of many microbes (bacteria, unicellular algae, yeasts) and cell lines.
209 K (−64°C)	Minimum successful undercooling temperature yet recorded for terrestrial arthropods in the wild (three species of willow gall insects from interior Alaska, where minimum extended winter temperatures experienced are typically −50°C and may reach −55°C: Miller and Werner, 1987).
233 K (−40°C)	Lower limit for presence of undercooled water at 1 atmosphere. General lower limit of survival for many taxa. Winter minimum isotherm corresponds roughly to high latitude limit of tree growth.
255 K (−18°C)	Lowest recorded temperature for active aquatic microbial communities (hypersaline lakes in east Antarctica: Deep Lake, Vestfold Hills, Lake Hunazoko, Skarvs Nes: Kerry *et al.*, 1977; Tominage and Fukui, 1981; Ferris and Burton, 1988).
263 K (−10°C)	Approximate lower limit for ice-associated marine organisms (typically unicellular algae and associated microfauna in brine channels).
271 K (−2°C)	Lower limit for typical marine organisms (seawater of salinity 35 freezes at −1.86°C).
308 K (35°C)	General upper limit for typical marine organisms (maximum bulk seawater temperature set by latent heat of evaporation, though solar heating of shallow waters can produce locally higher temperatures).
313 K (40°C)	General upper limit for freshwater metazoans (hot springs).
383 K (110°C)	Approximate upper limit for microbial life associated with hydrothermal vents or geothermal environments.

and Aristotle. Although invocation of a strong anthropic view of the role of water in life as we know it here on Earth might lead to an unduly restrictive view of the possibility of alternative physiologies elsewhere, the central role of water in the biosphere is often underestimated.

Not only does water act as solvent, but there are very few metabolic processes in which water does not play a part as either reactant or product (Franks, 1985). The four major classes of chemical reactions in biology are oxidation, reduction, condensation and hydrolysis, and water plays a key role in each. At the broadest level, the whole biogeochemical cycle of the planet may be viewed as a balance between photosynthesis, which splits water in order to reduce carbon, thereby liberating oxygen, and respiration, which reduces that oxygen back to water. To a first (and somewhat crude) approximation, the mass of free oxygen in the biosphere may be taken as an indication of the mass of reduced carbon fixed in organic matter. Such calculations do, however, ignore the microbial biomass, which, it has been speculated, may be found in geothermal waters (Gold, 1992, 1999). In the absence of hard data it is difficult to estimate the true importance of this geothermal community to current biogeo-

Table 11.2   Physical properties of liquid water. Where these vary with temperature data are provided generally for 273 K (0°C).

Property	Value	Comment
Equilibrium freezing point	273 K (0° C)	At 1 atmosphere pressure (0.1 MPa); anomalously high.
Boiling point	373 K (100° C)	At 1 atmosphere pressure (0.1 MPa); high.
Specific heat capacity	4.75 kJ $(g\ K)^{-1}$ 75.3 J $(mol\ K)^{-1}$	Unusually high; varies with temperature, with minimum value at 35° C.
Dielectric constant (permittivity)	78.5 (at 298 K)	One of the highest values known for a liquid. Renders ionic compounds very soluble.
Latent heat of vaporization	2257 J g^{-1} (at 373 K)	Very high; allows for rapid cooling by evaporation and thereby regulates upper bulk temperature of aquatic habitats.
Latent heat of fusion	334 J g^{-1}	Low.
Thermal conductivity	0.561 W m^{-1} K^{-1} (at 273 K)	High; allows equalization of temperatures within cells and tissues without convection damage.
Surface tension	75.7 mN m^{-1} (at 273 K)	High, important for fluid motion in plants.
Dynamic viscosity	1.792 mPa s (at 273 K)	Relatively low, enabling use of water as efficient transport medium.

chemical cycling, but it is undeniably important in providing insight into physiological pathways that may have preceded the photosynthetic fixation of carbon which dominates the surface biosphere today.

The physical basis for the unusual behaviour of water lies in the structure of the water molecule itself. The water molecule is essentially a sphere bearing two positive charges (from the hydrogen atoms) and two negative charges from the lone pair of electrons in the electronic structure (Franks, 1985). With two proton-donor and two proton-acceptor sites the water molecule thus has the ability to form four hydrogen bonds, the spatial distribution of which gives rise to a three-dimensional network. Thus, each oxygen atom is surrounded by four other oxygen atoms, with a hydrogen atom placed on each O-O axis (Figure 11.1). This structure has been confirmed for ice by X-ray diffraction studies. The structure of liquid water is less well understood but is

believed to be similar, with the number of hydrogen bonds reducing with increasing temperature (and hence molecular motion) (Taylor, 1987).

The hydrogen bond is a weak interaction, but the density of these bonds in water is sufficient to produce the unusual physical properties of water. The importance of these hydrogen bonds to physiological processes is illustrated by the effect of replacing hydrogen with deuterium. The greater molecular mass of deuterium has a small, and predictable, effect on physical properties such as melting point or boiling point. There are also solvent kinetic effects for chemical reactions performed in D_2O. For life, however, the disturbance of the hydrogen bonding structure is sufficient to make deuterium highly toxic, and often lethal at substitutions approaching 40% (Franks, 1985).

The unusual physical properties of water include many that are critical to life, and define the temperature range over which organisms

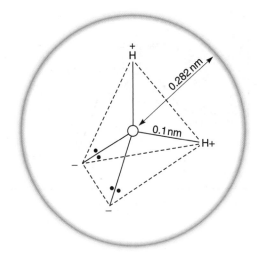

Figure 11.1 The four point charge model of the water molecule. The oxygen atom is placed at the centre of a regular tetrahedron, the vertices of which are occupied by two positive (hydrogen atoms) and two negative (lone electron pairs) charges. The O-H distances are 0.1 nm, and the distant of closest approach of two molecules (van der Waals radius) is 0.282 nm. Redrawn from Franks (1985).

can live (Table 11.1). Within this range, however, temperature also has an important role in influencing the rate of physiological processes involved in life.

11.1.2 *Temperature and physiology*

Life on Earth involves two fundamental processes; indeed the combination of these two processes may be taken to define life itself. They are the transfer of information from one generation to the next, and the manipulation of free energy and elemental resources to ensure that this transfer of information can take place.

The central role of information in life processes was highlighted by the eminent physicist Erwin Schrödinger in a small but highly influential book (Shrödinger, 1944). Schrödinger also recognized that the small error rate in information transfer between generations was particularly critical, for this variation forms the essential raw material for evolution. We now understand that the processes involved in generating this variation

are considerably more complex and extensive than envisaged by Schrödinger. Some of these processes are independent of temperature, for example quantum effects and mutation induced by natural radiation. Others, and the repair processes involved in eliminating many mutations, are physiological in nature and hence will be affected by temperature. There are also genes that influence the rate of mutation elsewhere in the genome.

All processes involving a change in free energy are affected by temperature, which thus has a profound effect on physiology. The theoretical understanding for this relationship is founded in statistical thermodynamics, which predicts a broadly exponential relationship between reaction rate and temperature in simple chemical systems. Part of this sensitivity of reaction rate to temperature is related to the marked variation with temperature of the physical properties of water itself, and part to variations in kinetic energy.

The frequency distribution of molecular kinetic energies can be calculated for any given temperature from the Maxwell–Boltzmann equation. The change in mean kinetic energy of substrate molecules with increasing temperature is, however, too small to explain the increase in reaction rate commonly observed. The critical insight came from Svante Arrhenius, who introduced the concept of an intermediate transition state between the initial reactants and final products. For a typical biological reaction this might represent the formation of an enzyme/substrate complex (though more complex possibilities involving ligands and cofactors are frequent). Since this intermediate state exists at a higher energy level, only a small fraction of the substrate molecules have sufficient energy to achieve the transition state and hence proceed to completion of the reaction (Figure 11.2). Arrhenius showed that the consequence of this intermediate transition state would be a much greater thermal sensitivity of reaction rate to temperature than would have been predicted on the statistical distribution of substrate energy values alone. The relationship is effectively exponential and is described by the Arrhenius equation (for

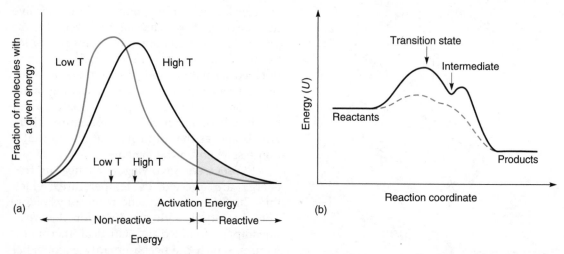

Figure 11.2 (a) The frequency distribution of energy levels for two populations of substrate molecules, as described by the Maxwell–Boltzmann distribution function for two temperatures. The veil line at high energy levels marks the threshold energy for achieving the activation energy which allows the reaction to proceed. Note that the threshold energy shown here provides only the enthalpic contribution to the free energy of activation. The fraction of substrate molecules exceeding this threshold increases and is greater at the higher temperature, the overall relationship with temperature being approximately exponential, as described by the Arrhenius equation. (b) The temporal pathway of a physiological reaction. The line represents the changes in free energy during the course of a hypothetical reaction. The free energy content is plotted on the abscissa, while the coordinate is a measure of the extent of the reaction. The free energy of activation (G‡) required for the reaction to proceed is provided by two components, one enthalpic (H‡; see (a) and one entropic (S‡).

fuller discussion see Clarke, 1983; Hochachka and Somero, 1984; Cossins and Bowler, 1987).

Interestingly, the relationship between temperature and many aspects of whole-organism physiology is also broadly exponential when examined within individuals, or between individuals from a given population or species. In very simple terms this is often expressed as the van t'Hoff rule, which is that the rate of a physiological reaction often approximately doubles for an increase in temperature of 10 K (usually formalized as $Q_{10} = 2$, where Q_{10} is the ratio of the rates of a process measured at two temperatures differing by 10 K). For many physiological reactions in unstressed organisms, Q_{10} typically falls in the range 2 to 3 (Clarke, 1983). The most parsimonious explanation for this widespread observation is that organismal physiology is tightly integrated and highly regulated, and reflects the basic underlying physics very directly. It may, however, also reflect an inherent property of a complex sys-

tem. It is important to recognize that any system exhibiting Arrhenius behaviour has a Q_{10} which varies with temperature, although this effect is small at physiological temperatures (Clarke, 1983).

11.2   Life and low temperature

Despite the profound sensitivity of life processes to temperature, organisms can be found across the entire temperature range at which water remains liquid (Table 11.1). The limits to this range clearly set severe evolutionary challenges to life, but the nature of these changes differs at the two extremes. At the higher end there is plenty of kinetic energy in the system and the challenge is essentially one of maintaining structural integrity and metabolic balance. At low temperatures the problems are primarily the potential for freezing and the low cellular energy content. That these differences are important is shown by the fact that only microbial organisms are found with cellular temperatures above about 50°C.

Low temperatures for life on Earth may be taken as those below about 5°C, and in this range the laws of physics set three distinct challenges to physiology:

1. Freezing (change of state).
2. The maintenance of macromolecular structure and physical integrity.
3. The maintenance of physiological rate in the face of reduced system enthalpy

The adaptations of organisms which have evolved in response to these challenges, can be classified into two categories (Cossins and Bowler, 1987):

1. *Resistance adaptations.* These allow the organism to withstand a particular environmental challenge, which would otherwise prevent life completely. In many cases the duration of the stress is shorter than the life span of the organism, which then waits until the stress eases before resuming normal activity. In some cases, however, the physiological challenge is continuous, and the resistance adaptation needs to be present throughout the life span of the individual.
2. *Capacity adaptations.* These allow life processes to proceed at the required rate despite the physical nature of the environment.

Organisms living at low temperatures typically exhibit both types of evolutionary response: the threat of tissue water freezing induces a resistance adaptation, whereas the low temperature requires capacity adaptations to maintain a functional physiology.

11.2.1 A classic resistance adaptation: avoidance of freezing in marine teleost fish

Physiological processes require liquid water. It therefore follows that these processes will cease should cellular water freeze; indeed freezing of intracellular water is almost invariably lethal. The only exception to this rule reported to date, outside the cryopreservation

laboratory, is the nematode *Panagrolaimus davidi* which has been shown to withstand freezing of all body water (Wharton and Ferns, 1995).

Nevertheless many organisms live at temperatures below the equilibrium freezing point of their body fluids, and to do so requires specific adaptations. The precise physiological challenge and the evolutionary response differ between the terrestrial and marine environments. This is largely to do with the much greater specific heat capacity of water $(4.75 \, kJ \, g^{-1} K^{-1})$ compared with air $(0.23 \, kJ \, g^{-1} K^{-1})$. As a result the thermal environment of the sea is much more stable than on land, and it is consequently easier to discern the evolutionary signal from the noise of short-term acute or acclimatory responses.

Life evolved in the sea and most organisms have a body fluid osmotic strength (but not ionic composition) similar to seawater. For most marine organisms, freezing is therefore not a major problem: if the sea remains fluid then so do they. Important exceptions are intertidal organisms exposed to low air temperatures during low tides in winter, and teleost fish. Teleost fish are unusual in having dilute blood, roughly half the osmotic strength of seawater. This dilute blood probably reflects an early phase of evolutionary history in fresh waters, when a more dilute blood would reduce osmoregulatory costs. It does, however, mean that teleost fish swimming in polar seas are living in an environment significantly below the equilibrium freezing point of their blood. Since teleost fish need to drink continuously to offset the loss of water driven by the osmotic gradient between their body fluids and the surrounding seawater, they are also continually exposing themselves to the risk of inoculating freezing from ingested ice nuclei.

Teleost fish living in polar waters avoid freezing by a whole suite of anatomical, biochemical and physiological adaptations. The key adaptation is, however, the production of protein-based antifreeze molecules. The nature of a teleost antifreeze was first elucidated by DeVries in notothenioid fish from Antarctica (DeVries and Wohlschlag, 1969). The notothenioid antifreeze is a glycoprotein based on a repeating tripeptide unit backbone (alanine-alanine-threonine), with a disaccharide (galactose-N-galactosamine) attached to each threonine residue. Differing numbers of repeats of the AAT tripeptide unit give rise to antifreezes of differing molecular mass and physiological function (DeVries, 1982, 1983, 1988).

Studies of purified antifreeze glycoprotein (AFGP) have shown that it acts non-colligatively, probably by binding to specific faces of growing ice nuclei and preventing them from reaching sufficient size to achieve thermodynamic stability and thereby inoculate bulk freezing (DeVries, 1988). This is a dynamic process and a relatively low concentration of antifreeze thus provides sufficient protection (DeVries, 1971). Notothenioid fish also have aglomerular kidneys, which prevent loss of AFGP through filtration into the urine (Eastman, 1993). The genes for antifreeze glycopeptides in notothenioid fish have an unusual structure. While most protein genes encode one protein molecule per gene, notothenioid AFGP genes encode a polyprotein precursor containing many AFGP molecules (46 in *Notothenia coriiceps* and 41 in *Dissostichus mawsoni*: Hsiao *et al.*, 1990; Chen *et al.*, 1997a, b). Recent cladistic work has indicated that antifreeze evolved only once in notothenioid fishes, at the base of the radiation (Eastman, 1993; Clarke and Johnston, 1996). This evolutionary event occurred after the splitting of the bovichthid clade, all of whose representatives lack antifreeze and live outside the Southern Ocean. The genes for antifreeze appear not to be expressed in notothenioids which have secondarily colonized warmer waters to the north of Antarctica, nor in at least one species which lives close to the Antarctic continent but at depths where contact with external ice crystals never occurs. Molecular evidence indicates a surprising evolutionary origin for the notothenioid antifreeze glycoprotein, namely a pancreatic trypsinogen gene (Chen *et al.*, 1997a).

Notothenioids are the dominant teleost fishes on the continental shelf of Antarctica and they represent a major evolutionary radiation (Eastman and Clarke, 1998). It has recently become clear that there has also been a signifi-

cant radiation of an entirely unrelated group of fishes, the lipariids, in the deeper waters of the Antarctic continental slope (Eastman and Clarke, 1998), but we know very little about the physiology of freezing protection in these fish.

The waters in which many Antarctic fishes live is permanently below the equilibrium freezing point of their body fluids, and they therefore need antifreeze throughout their life span. Counteracting the threat of freezing thus represents a resistance adaptation whose energetic costs must be borne continuously. In contrast many fish living on the fringes of the Arctic basin are only exposed to low temperatures in winter. Many of these fish synthesize antifreezes seasonally (Lin, 1979, 1983), and this is a more typical example of a resistance adaptation, where the duration is shorter than the organism's lifetime. The energetic cost of the adaptation may, however, be high and thus have an effect on the ecology of the organism. Arctic fish also contrast with the Antarctic notothenioids in that many use peptide, rather than glycopeptide, antifreeze molecules. This suggests that evolution of antifreeze protection is not a particularly difficult evolutionary problem. Indeed gadoids (true cods) have evolved a glycoprotein antifreeze identical to that of notothenioids, but although the parent molecule has yet to be identified the codon structure shows that it was clearly different from that in notothenioids (Chen *et al.*, 1997b).

11.2.2 *Freezing avoidance on land*

The thermal environment on land is far more variable than in the sea, and the evolutionary response has been correspondingly complex. Seasonal variation in air temperature can be very large in polar and alpine regions, with winter minima below −50°C. The need to avoid freezing is thus not an exclusively polar phenomenon; night or winter temperatures exposing plants and terrestrial invertebrates to the possibility of freezing are extremely widespread geographically (Grout, 1987).

Much of the work on freezing avoidance in terrestrial animals has centred on arthropods, and especially insects, although it is likely that the principles will prove to be general for terrestrial invertebrates. As early as Salt (1966) it was recognized that insects utilized one of two broad strategies in response to potentially freezing by undercooling temperatures. These were either to avoid freezing (species in which freezing of any body fluids would be lethal), or to tolerate freezing of extracellular water. The latter group, typically the minority, utilize ice nucleating proteins to induce freezing at a relatively high subzero temperature. Water in the cells, however, remains fluid and metabolism continues, albeit at a low level. Freezing of extracellular water during winter is also known from a small number of frogs, turtles and one snake (Storey and Storey, 1996).

Among terrestrial invertebrates it would appear that the more frequent strategy is avoidance of freezing, though this simple binary classification into species tolerant or intolerant of freezing is somewhat oversimplified. The ecology of freezing avoidance is rendered complex by a range of other physiological factors including chill injury and death at temperatures above the nucleation point (Bale, 1993).

Species intolerant of freezing generally (but not universally) produce two classes of protective compound: antifreezes (usually called thermal hysteresis proteins by insect physiologists) and cryoprotectants. The first insect thermal hysteresis protein was isolated from larvae of the tenebrionid beetle *Meracantha contracta* (Duman, 1977a) and was shown to be synthesized seasonally, with high levels present only in winter (Duman, 1977b). Insect antifreezes have so far all proved to be proteins, similar but not identical to some fish protein antifreezes; to date no glycoprotein antifreezes have been reported from terrestrial invertebrates. The functional role of thermal hysteresis proteins in insects appears to be similar to that in polar fish, which is to prevent the growth of ice nuclei to the size where they might initiate bulk freezing. Protein antifreezes isolated from the spruce budworm *Choristoneura fumefurana* and the yellow mealworm beetle *Tenebrio molitor* have recently been shown to be many times more effective than fish antifreezes (Graham *et al.*, 1997; Tyshenko *et al.*, 1997). This is presumably related to the much wider range of subzero temperatures over which the insect antifreezes have to be effective in order to maintain organism viability. Interestingly thermal hysteris proteins (antifreezes) are also found in insects which are tolerant of freezing. Here their role appears to be prevention of recrystallization, so as to avoid tissue damage from an increase in ice crystal size during the frozen state (Knight and Duman, 1986).

Many arthropods exposed to freezing stress in polar and alpine regions also produce one or more of a range of cryoprotectants. These are typical poly-hydroxy compounds (polyols), such as glycerol, glucose, trehalose, mannitol and sorbitol. In some cases (for example, many Antarctic terrestrial microinvertebrates) the concentration of these cryoprotectants is sufficient to provide significant colligative depression of the equilibrium freezing point. Some species utilize just one compound, others a complex suite (although it is not clear that the minor components have a true cryoprotectant role). It is likely that this varied pattern reflects a variety of physiological functions. Although the concentration of some compounds undoubtedly provides some freezing protection through colligative depression of the equilibrium freezing point, it is likely that the primary role of some of these compounds is to minimize damage to cellular structure. Particularly important in this regard is trehalose, a sugar which has been shown to be important in the maintenance of membrane integrity during anhydrobiosis.

Cellular dehydration is a significant problem for organisms whose extracellular fluids freeze, for the increased osmolarity of the unfrozen residual water will be very high and will tend to pull water from the cell. Many unicellular organisms living in hypersaline environments or tolerant of drought are often incidentally tolerant of freezing (Clarke and Leeson, 1985) and the integrity of the cell membrane plays an important role in this resistance. It is possible that there are important physiological parallels and evolutionary convergences between drought resistance and tolerance of freezing. These include the synthesis of stress (or chaperone) proteins and similar polyhydroxy osmolytes (Bartels and Nelson, 1994; Joset *et al.*, 1996). Recent work on cold-hardiness in arthropods has also emphasized important parallels between cold tolerance and drought tolerance.

Whereas terrestrial animals are mobile and can seek out particular areas that minimize environmental stresses (for example, overwintering sites), plants are sessile and must be able to withstand whatever challenges the environment sets them. Plants typically initiate freezing in xylem and extracellular fluids at a relatively high subzero temperature, using a variety of ice nucleating agents (Griffith and Antikainen, 1996). Freezing of intracellular water appears to be invariably lethal in plants and antifreeze activity has been identified in a diverse range of plant taxa (Urrutia *et al.*, 1992; Duman and Olsen, 1993). The formation of extracellular ice withdraws water from the cells, and successful freezing avoidance involves tolerance to cellular dehydration and associated stresses (Burke *et al.*, 1976; George *et al.*, 1982; George and Burke, 1984). Again there is a clear functional link between the responses to freezing and drought.

11.2.3 *Capacity adaptation: physiological function at low temperature*

All processes involving a change in free energy are affected by temperature. In the absence of any evolutionary compensation physiological processes will thus proceed at quite different rates in polar and tropical ectotherms. The typical temperature dependency of physiological processes would suggest that in a representative tropical fish or marine invertebrate living at 30°C these would proceed at between 27 ($Q_{10} = 3$) and 8 ($Q_{10} = 2$) times the rate in a polar species. For many life processes this difference would pose severe ecological difficulties.

One mechanism for circumventing the influence of environmental temperature on physiology is to maintain a constant internal body temperature. Not only does this prevent environmentally driven changes in physiological rates but it also allows for an evolutionary fine-tuning of physiology to a particular temperature. Internal body temperature is maintained at a set point, typically in the range 35°C to 40°C, utilizing cellular metabolism as the primary source of heat and a series of mechanisms either to retain or dissipate that heat depending on external environmental temperature. This mechanism (endothermy, or homeothermy in the earlier literature) appears to have evolved independently in mammals and birds. Although endothermy is energetically expensive (Hemmingsen, 1960; Calder, 1984), it has allowed mammals and birds to occupy a far wider range of thermal habitats than reptiles or amphibians.

The set point tends to be higher in birds than in mammals (Louw, 1993) but the interesting evolutionary question is why endotherms maintain a body temperature in the range 35–40°C (as against, say, 25°C or 50°C). A theme common to many organisms with a need to generate very high muscle power output is a warm body temperature; this is true not only of mammals and birds, but also some ectotherms. A number of lizards, predatory fish such as tuna, and many active insects all maintain their muscles at relatively warm temperatures during activity. Perhaps the set point for endotherms represents a balance between the energetic costs of maintaining body temperature and the benefits to be gained from enhanced muscle power output. One intriguing suggestion is that the set point is adjusted to the temperature at which the specific heat of water is minimal (Paul, 1986). This would minimize the amount of energy needed to change body temperature (but it would also maximize the sensitivity of body temperature to thermal input or loss). The effect is small, but it may reflect once again the exquisite and subtle link between physiology and the physical properties of water.

For polar endotherms the key adaptations are insulating layers of fat (blubber, most notably in marine mammals) together with thicker fur in terrestrial mammals or feathers in birds. These are associated with a complex suite of anatomical, physiological and behavioural mechanisms for reducing heat loss to the environment. All other organisms living in polar regions (microbes, plants, invertebrates, fish) are ectotherms. Their body temperature is determined primarily by the ambient environmental temperature (albeit modified in some cases by incident radiation), and their physiology is subject to the rate-depressing effects of the low polar temperatures.

The simple observation that a rich and diverse marine fauna can be found from the tropics to the poles indicates that some form of compensation has evolved to offset these direct temperature effects. Even if not perfect in theoretical terms, it is clearly highly effective in ecological and evolutionary terms. The existence of an arctic terrestrial flora and invertebrate fauna carries the same implication, though here the relatively low diversity compared with many tropical habitats also reveals the impact of recent climatic history (as does the extremely impoverished Antarctic terrestrial flora and fauna).

11.2.4 Compensation

These general considerations lead directly to a straightforward definition of temperature compensation, which is:

the maintenance of physiological rate in the face of temperature change.

This is shown conceptually in Figure 11.3.

In the past few decades many physiological processes have been examined to determine the extent to which such compensation has evolved, and the mechanisms by which this has been achieved. It is now clear that some processes have evolved very good compensation for temperature, others have evolved only partial compensation, and some almost none at all (for a recent review see Clarke and Crame, 1997; for fuller discussions see Hochachka and Somero, 1984; Cossins and Bowler, 1987). This immediately poses the evolutionary questions of why this variation exists, and what the consequences might be.

11.2.5 A conceptual framework for temperature compensation

In simple evolutionary terms, an organism can be viewed as taking energy and material from

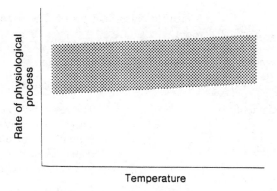

Figure 11.3 A conceptual model of compensation as homeostasis; evolution has modified cellular or organismal physiology such that the same process has the capacity to operate at broadly the same speed in organisms living at different temperatures. Note that the compensation plot has both bandwidth and slope. The bandwidth indicates that at any given environmental temperature there may be a range of rates for a given physiological process in different organisms, perhaps associated with ecology or phylogenetic history. Perfect compensation across the ecological temperature range would be indicated by a slope of zero (that is, the same rate at all temperatures). The slight slope shown here is to indicate that underlying physical or physiological constraints may mean that compensation is not perfect (reproduced from Clarke, 1991; Clarke and Crame, 1997).

its environment and manipulating these to maximize its fitness (that is, to maximize the relative representation of its genes in the next generation). Since this process is generally fuelled by the oxidation of foodstuffs, with consequent reduction of oxygen to water, the rate and efficiency of energetic pathways are critical to fitness.

An organism contains a vast number of different enzymes, involved in a complex web of metabolic interactions. This complexity can, however, be reduced conceptually to a small number of key processes, namely, gene expression, catabolism (production of energy, reducing power and intermediary compounds), work (utilization of energy) and homeostasis (Table 11.3). Homeostasis is, of course, a form of work but it is convenient to distinguish it because it encompasses those processes necessary to keep the organism alive in the absence

Table 11.3 **A simple classification of the major cellular energetic processes (from Clarke and Crame, 1997).**

Process	Description
Gene expression	The regulated transcription and translation of genetic information.
Catabolism	The regulated oxidation of reserves or exogenous foodstuffs to provide a supply of high-energy phosphate, reducing power and a suite of intermediary compounds,[a] together with the production of water, CO_2, heat and nitrogenous waste (typically ammonia and urea) as by-products.
Work	Biosynthesis of macromolecules for turnover, growth, and mechanical work (including locomotor activity).
Homeostasis	Chemical work associated with the active regulation of the intracellular environment within boundary conditions set by physiological viability. This includes ion exchange, macromolecular turnover, and removal of waste products. In endotherms it also involves heat production (heat produced by ectotherms is largely lost to the environment).

[a] The major intermediary compounds are DHAP (dihydroxyacetone phosphate), PGA (phosphoglyceric acid), G6P (glucose-6-phosphate), R5P (ribulose-5-phosphate), E4P (erythrose-4-phosphate), PEP (phosphenol pyruvate), pyruvate, acetyl CoA, succinyl CoA and oxaloacetate. The major high energy phosphate is usually ATP (adenosine triphosphate) and the major sources of reducing power are nicotinamide adenine dinucleotide (NADH) and its phosphate (NADPH).

of any other physiological processes such as growth, locomotion or reproduction. It is therefore the major component of basal (maintenance) metabolic rate, and is likely to be subject to selection pressures different from those influencing active metabolism or the production of new tissue.

Despite the fundamental importance of gene expression, surprisingly little is known about temperature compensation in this process. Although there are a number of well-documented examples of individual genes being switched on or off by changes in temperature, essentially nothing appears to be known about the impact of temperature on the processes involved in routine transcription, message transport and processing, or translation. As with all enzyme-mediated processes these activities will slow down when an individual organism is cooled to lower temperatures, and at a sufficiently low temperature the cell cycle ceases. It is not known what organisms have been able to do about this on an evolutionary timescale, or even whether it is ever an energetic or evolutionary problem.

Catabolic processes have the main function of generating ATP, reduced pyridine nucleotides and intermediary compounds for biosynthesis. Anabolic processes utilize these for growth and reproduction, and homeostatic processes use them in the maintenance of organism integrity. All of these involve enzymes, and the major evolutionary possibilities for eliciting compensation for the rate-depressing effects of a low cellular temperature have long been recognized to be essentially threefold (Hochachka and Somero, 1984). They are the quantitative strategy (more enzymes), the qualitative strategy (different enzymes, working more effectively at low temperatures) and the modulation strategy (modification of the enzyme micro-environment) (Table 11.4). These strategies are not mutually exclusive, and all appear to be involved in evolutionary adaptation to low temperature.

Table 11.4 **The major categories of compensatory response to a lowered cellular temperature (Clarke and Crame, 1997, from Hochachka and Somero, 1984).**

Response	Description
Quantitative strategy	Increase the number of enzyme molecules to offset the reduced catalytic efficiency of an individual enzyme molecule at lower temperatures.
Qualitative strategy	A change in the type of enzyme(s) present. This could include a shift in the balance of allozymes with differing kinetic properties.
Modulation strategy	Modulation in the activity of pre-existing enzymes. This could include changes in enzyme-substrate binding kinetics with temperature, changes in cofactor binding kinetics, or modification of the immediate enzyme environment (membrane or cytosol).

Three general conclusions have emerged from comparative studies of enzyme homologues isolated from marine organisms adapted over evolutionary time for living at different temperatures. These are:

1. Enzymes from organisms with lower body temperature tend to have lower free energies of activation. These changes are, however, typically quite small and appear to contribute relatively little to the observed higher activity of enzymes isolated from cold water species (Hochachka and Somero, 1984).
2. Enzymes from organisms with a lower body temperature tend to have lower thermal stabilities (when comparisons are undertaken at the same experimental incubation temperature), although thermal denaturation of these enzymes generally occurs only at temperatures well above those at which the organisms live.
3. There is a large shift in the relative contributions of enthalpy and entropy to the overall free energy of activation, with a greater entropic contribution at lower temperatures. This is related to the lower thermal stability, and the mechanism appears to be changes in primary structure (amino acid sequence) to produce a more flexible molecule.

The shift in relative contributions to the free energy of activation from enthalpy and entropy is particularly important, for it explains why effective compensation for the effects of low cellular temperature can be achieved with only a small change in the overall free energy of activation (see Figure 11.2). For recent discussions of the role of enzyme number and type in evolutionary temperature compensation see Moerland (1995), Clarke (1997) and Pörtner *et al.* (1997). The molecular aspects of these evolutionary adjustments are beyond the scope of this review; for a recent review with particular reference to microtubule assembly see Detrich (1998).

The extent to which organisms offset the effects of low temperature through modifications to the enzyme microenvironment has received slightly less attention (Morris and Clarke, 1987). There is clear evidence for an adjustment to the viscosity of the cellular membrane (which affects the activity of membrane-bound enzymes), and some evidence for regulation of the internal cellular milieu (see Cossins, 1994; Pörtner *et al.*, 1997 for recent reviews).

11.2.6 *Costs and constraints*

The available evidence thus suggests that evolutionary adaptation to low temperature involves a wide array of molecular and physiological adjustments. These are accompanied by ultrastructural and anatomical changes in a complex suite of adaptations (Tyler and Sidell, 1984; Egginton and Sidell, 1989). While laboratory measurements can reveal the efficiency of these compensatory adaptations, their energetic costs, and any ecological constraints they impose, are also of evolutionary importance.

One important measure of the costs comes from resting metabolism. This is the best practical measure of basal or maintenance metabolism, and it reflects the energetic costs of staying alive independent of the costs of growth, reproduction or activity. It has now been established that, at least in marine organisms, these basal costs are significantly lower in polar than tropical ectotherms (recent data for fish are shown in Figure 11.4; data for invertebrates are discussed by Clarke, 1983, 1991).

The lower rate of resting metabolism at lower body temperatures in ectotherms might be taken to represent a simple thermodynamic rate-limitation. In fact it does not; rather it reflects the overall lower costs of maintaining a suite of cold-adapted enzymes and associated anatomical structures. This is a somewhat subtle argument, presented in detail elsewhere (Clarke, 1987). The essential point is that the effect of a low temperature over evolutionary

time is not to drive down respiration mechanistically (as it undoubtedly does in short-term experiments in the laboratory), but to reduce the overall costs of existence. Oxygen consumption is a measure of the instantaneous demand for ATP; it represents the power consumption of the organism at the time of measurement. For polar organisms the low cellular temperatures have allowed for protein turnover, osmoregulation, and other maintenance processes to be scaled down, with a consequent reduction in the power demand for resting metabolism.

In ecological and evolutionary terms, however, it is also important to know whether this adaptation to low temperature brings with it any constraints. At present this is unclear but two areas which have attracted attention recently have been the extent to which cold-adapted marine organisms are obligate stenotherms, and whether the low cost of basal metabolism places an upper bound on maximal aerobic scope (in other words, whether adaptation to low temperature necessarily prevents the evolution of particularly active or energy-intensive lifestyles). One area of investigation which might throw valuable light on these questions would be the energetics of terrestrial polar organisms, which are presented with an extra environmental challenge in the wide daily and/or seasonal variation, superimposed on the low mean temperature.

11.3   Evolution at low temperatures

The current steep latitudinal cline in temperature from the tropics to the poles is not typical of Earth history as a whole. Current evidence suggests that there have been just four major periods of glaciation during the history of life (including the Precambrian glaciation); at other times the thermal environment of the Earth has been far more uniform. This poses the question as to whether adaptation to low temperatures is something special, or simply a particular facet of temperature adaptation in general. This in turn raises the far more difficult question of the temperature at which life itself

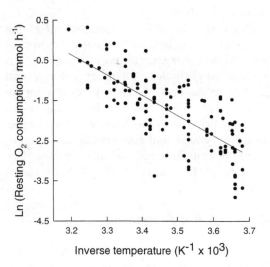

Figure 11.4 Relationship between resting oxygen consumption (mmol h^{-1}) and body temperature for 69 species of marine teleost fish. Data corrected to a body size of 50 g wet mass assuming a mass exponent of 0.79, with Arrhenius statistical model fitted. Reproduced from Clarke and Johnston (1999).

evolved. Although recent phylogenetic work points to an ancestry in anoxic and perhaps hot environments, other analyses have suggested temperatures more typical of today's warmer aquatic environments (Galtier *et al.*, 1999).

Metazoan life since the Cambrian explosion, however, appears to have existed largely within the current thermal envelope of about 50 K (with a wider range for terrestrial organisms in some circumstances allowed through resistance adaptations and resting stages). It is likely therefore that important insights into the relationship between temperature and evolution will come from a detailed study of the most recent large-scale climatic change on Earth, namely the development of the present glaciation during the Cenozoic. The climatic and faunal history of Gondwana thus has enormous importance in helping us to unravel general questions concerning the relationship between evolution and temperature.

11.3.1 *Evolutionary adaptation to low temperature during the Cenozoic*

The Cenozoic era opened with global temperatures generally warm and evenly distributed. Early Cenozoic marine temperatures appear to have been generally warm, although there is some evidence for terrestrial glaciation even in the Eocene. This was followed by a sharp cooling of the Southern Ocean in the middle Miocene (from about 17 to 14 Ma BP), probably associated with the opening of the Drake Passage and the onset of the Antarctic Circumpolar Current. From this point the general trend has been for the Southern Ocean to continue cooling, albeit with brief temporary periods of warming, to present polar temperatures (Clarke and Crame, 1992).

This naturally raises the long-standing and much debated question of the extent to which the present polar biotas evolved *in situ*, or have invaded polar regions from lower latitudes. Evidence from the Southern Ocean marine fauna suggests strongly that the current fauna has evolved *in situ*, tracking the gradual cooling since the Miocene (Lipps and Hickman, 1982; Clarke and Crame, 1989, 1992).

There appears to have been a rich and diverse shallow water marine fauna around Gondwana in the late Cretaceous and early Tertiary. Unfortunately the fossil record of the Antarctic marine fauna is poor and its history through the Cenozoic therefore largely unknown. Most major groups of organisms appear to have survived the onset of glaciation and cooling, although two notable exceptions are the large decapods and many groups of fish (both teleosts and chondrichthyes) (summarized by Clarke and Crame, 1989, 1992; Eastman and Clarke, 1998).

The evolution of the Southern Ocean fish fauna is particularly interesting in that subsequent to an extinction event in the Cenozoic, two groups have undergone a striking adaptive radiation: notothenioids on the continental shelf and liparids on the deeper continental slope (Eastman and Clarke, 1998). Several lines of evidence now point to a diversification of the notothenioid clade about the time of the mid-Miocene cooling (Bargelloni *et al.*, 2000), and cladistic analysis indicates that the evolution of the glycoprotein antifreeze appears to have been a single event at about this time (Eastman, 1993; Clarke and Johnston, 1996; Eastman and Clarke, 1998; Cheng, 1998). These exceptions notwithstanding, the bulk of the present Southern Ocean marine fauna appears to have survived the onset of cooling and glaciation. This will have required a continuous evolution of the physiology to track the change in temperature.

Although we might regard the dominant feature of the climatic evolution of the Southern Ocean to be the long-term cooling, this is only one facet of temperature variation in the sea. Environmental temperature varies over a wide range of temporal scales, from seconds to millennia. This variability is not strictly fractal in that the variance is not independent of temporal scale, and there are also significant differences in variance structure between marine and terrestrial environments that may have important evolutionary implications (Steele, 1985).

The shortest timescales of temperature variation may represent environmental noise, but they can still set the organism a physiological challenge. Recently, increasingly sophisticated models have been used to examine the response of organisms to environmental change (often temperature) and these have shown that at moderate rates of change and with the inclusion of meaningful phenotypic variance, traits can evolve to track that change, albeit with a lag between mean population performance and the optimum environmental value (Lynch and Lande, 1993).

The present cold-water marine fauna of the Antarctic will thus have evolved from warm-water ancestors. The necessary physiological evolution may not necessarily have been accompanied by morphological change. Indeed until the advent of molecular techniques, evolutionary history could only be reconstructed because morphological change was present but conveniently slow. This raises the point that morphological stasis may hide significant evolutionary change in physiology. Indeed given that environmental change is the norm, organisms are faced with the need to evolve continuously in a physiological analogue to the competitive Red Queen (Clarke, 1993b). In physiological terms at least, evolutionary stasis simply does not exist. This can be shown very clearly in that although the ancestry of many living Southern Ocean molluscs can be traced back to warmer water relatives, the polar species are unable to survive even quite small elevations in temperature (e.g. the bivalves *Limopsis* and *Laternula*: Pörtner *et al.*, 1999a, b).

11.3.2 *Extinction at low temperature*

Climate change has long been recognized as an important factor in evolutionary history. In particular a decrease in temperature has often been invoked to explain extinction in both marine and terrestrial organisms. Although there are some convincing associations between regional extinctions of warm-water faunas and cooling in the sea (e.g. Stanley, 1986), the history of the Southern Ocean marine fauna shows that extinction is not an inevitable consequence of climatic cooling (Clarke and Crame, 1989, 1992).

The problem with temperature as a direct causal agent of extinction is that in many cases the mean rate of change is far too slow. Typical rates of climatic change in Earth's history are slower by orders of magnitude than rates which living organisms appear to tolerate without problem. Although there is increasing evidence for rapid shifts in climate at certain periods of recent Earth history, it is likely that the proximate causes of extinction are ecological, possibly related to the emergent properties of complex ecological systems (see discussion of these points in Clarke, 1993b and Jablonski, this volume, Chapter 13).

11.3.3 *Does temperature affect the rate of evolution itself?*

The raw material for evolutionary change is mutation. These mutations arise through a number of mechanisms, some of which may be influenced by temperature. Whereas quantum effects and ionization from cosmic rays or local radiation are independent of temperature, at least over the range of interest to physiologists, other sources of point mutations may be correlated with temperature. One which has been discussed recently is the effect of metabolic rate (Rand, 1994). Oxidative metabolism produces free oxygen radicals which can cause point mutations in the genome, and this has led to the suggestion that the rate of mutation might be higher in endotherms than in many ectotherms (Martin and Palumbi, 1993). Since the metabolic rate of ectotherms generally increases with environmental temperature (Clarke, 1983; Clarke and Johnston, 1999), this argument could be extended to predict a lower rate of mutation in polar ectotherms compared with related tropical taxa (Clarke and Johnston, 1996). Given that mitochondria are the site of oxidative metabolism, it might also be expected that any effect would be more obvious in the mitochondrial than in the nuclear genome. There are almost no data to test this prediction, although a recent study of crustaceans suggests that there is currently no evidence for slower rates of evolution in Antarctic compared with tropical taxa (Held, 2001).

A second potentially important influence on overall evolutionary rate is generation time. The precise mechanisms involved are complex and the overall effect of generation time encompasses a suite of factors including meiosis, recombination and natural selection on the adult organism. Overall, however, since generation timescales with body mass (Calder, 1984) generally vary inversely with temperature, the potential exists for evolution to proceed more slowly at lower temperatures (Rohde, 1992).

The measurement of evolutionary rate is very difficult, and two recent studies have come to differing conclusions. Crame and Clarke (1997) looked at speciation rates in a variety of clades of gastropod and could detect no difference in median evolutionary rate. The model underlying this measure assumes an exponential increase in extant species with time, but underestimates actual speciation rate because it can take no account of lineages that become extinct. An alternative approach is to examine the age structure of whole faunas. Flessa and Jablonski (1996) utilized this technique and found a tendency for tropical marine faunas to show greater turnover than cooler water faunas. These comparisons are, however, beset with difficulties of sampling and both these studies are based on species established primarily on morphological grounds. They thus hide a considerable amount of underlying evolutionary change, including adaptation to temperature.

Overall there appears to be no convincing evidence as yet that evolution proceeds significantly more slowly in the cold. Indeed there is evidence as yet that some taxa are actively speciating at present (Clarke and Crame, 1989), and the presence of two recent radiations of teleost fish in the Southern Ocean suggests that low temperature is no bar to innovative evolution (Eastman and Clarke, 1998).

11.4 A final comment: life elsewhere

Life as we know it depends absolutely on solution chemistry in an unusual solvent (water), a complex range of chemicals based on a few simple, small, atoms (predominantly H, N, C, O, P and S) and relatively small values of free energy. Because of the intimate involvement of water in life processes, and because water appears to be a common compound in the universe, it is often assumed that life elsewhere, if it exists, will probably be based on liquid water and carbon.

Alternative physiologies may exist, however, and these could conceivably extend the temperature range over which life is possible. One commonly discussed possibility is life based around silicon, principally on the grounds that it is abundant and capable of complex chemistry. Silicon compounds, however, do not form hydrogen bonds and the large size of the silicon atom tends to make its compounds solid and insoluble at moderate temperatures.

Speculation has also centred on solvents other than water which might provide the basis for life, particularly those based on elements in the second period of the periodic table which are relatively small and capable of forming stable compounds with hydrogen. These compounds include methane (CH_4), ammonia (NH_3) and hydrogen fluoride (HF) (see, e.g., Clement, 1990).

Methane has the disadvantage that it is nonpolar and hence a poor solvent for the polar compounds we might suspect to be necessary for the complex interactions required for life. Ammonia is more promising in that it is polar, and liquid over a reasonable range of temperature. HF is also highly polar and exists as a liquid over a wide range of temperatures at low pressure. It is, however, extremely reactive and thermodynamically unstable in the presence of oxygen. Since oxygen is an extremely common element in the universe, HF would seem an unlikely candidate for a basis of life.

Extrapolation from our experience of life on Earth would thus suggest that the two solvents most likely to act as a basis for life are water and ammonia. Since ammonia is liquid at lower temperatures than water, a life chemistry based around ammonia would probably operate over a lower range of temperatures than here on Earth. The lower free energies involved, and the intimate involvement as reactant in almost all physiological processes that we can currently observe, would mean that any such form of life would likely be very different in nature, temporal and spatial scale from that we recognize on Earth.

Acknowledgements

I thank Lynn Rothschild and Adrian Lister for the invitation to present this chapter, and NASA, the Linnean Society and the Centre for Ecology and Evolution, University College London, for supporting my attendance. My research is supported by the British Antarctic Survey (Natural Environment Research Council), and I thank Lloyd Peck, Alistair Crame, Bill Block, Bruce Sidell and Ian Johnston for many fruitful discussions of topics covered in this review. I would also like to thank David Wynn-Williams for much helpful input on low-temperature lakes, and Felix Franks for first introducing me to the peculiarities of water and its relationship to all aspects of life on Earth.

References

Bale, J.S. (1993). Classes of insect cold-hardiness. *Functional Ecology*, **7**: 751–753.

Bargelloni, L., Stefanio, M., Lorenzo, Z. and Patarnello, T. (2000). Mitochondrial phylogeny of notothenioids: a molecular approach to Antarctic fish evolution and biogeography. *Systematic Biology*, **49**: 114–129.

Bartels, D. and Nelson, D. (1994). Approaches to improve stress tolerance using molecular genetics. *Plant, Cell and Environment*, **17**: 659–667.

Burke, M.J., Gusta, L.V., Quamme, H.A. *et al.* (1976). Freezing and injury in plants. *Annual Review of Plant Physiology*, **27**: 507–528.

Calder, W.A. (1984). *Size, Function and Life History*. Cambridge: Harvard University Press.

Chen, L., DeVries, A.L. and Cheng, C.-H.C. (1997a). Evolution of antifreeze glycoprotein gene from a trypsinogen gene in Antarctic notothenioid fish. *Proceedings of the National Academy of Sciences of the USA*, **94**: 3811–3816.

Chen, L., DeVries, A.L. and Cheng, C.-H.C. (1997b). Convergent evolution of antifreeze glycoproteins in Antarctic notothenioid fish and Arctic cod. *Proceedings of the National Academy of Sciences of the USA*, **94**: 3817–3822.

Cheng, C.-H.C. (1998). Origin and mechanism of evolution of antifreeze glycoproteins in polar fishes. In G. di Prisco, E. Pisano and A. Clarke (eds) *Fishes of Antarctica: A Biological Overview*. Berlin: Springer-Verlag, pp. 311–328.

Clarke, A. (1983). Life in cold water: the physiological ecology of polar marine organisms. *Oceanography and Marine Biology: An Annual Review*, **21**: 341–453.

Clarke, A. (1987). Temperature, latitude and reproductive effort. *Marine Ecology Progress Series*, **38**: 89–99.

Clarke, A. (1991). What is cold adaptation, and how should we measure it? *American Zoologist*, **31**: 81–92.

Clarke, A. (1993a) Seasonal acclimatisation and latitudinal compensation in metabolism: do they exist? *Functional Ecology*, **7**: 139–149.

Clarke, A. (1993b). Temperature and extinction in the sea: a physiologist's view. *Paleobiology*, **19**: 499–518.

Clarke, A. (1998). Temperature and energetics: an introduction to cold ocean physiology. In: H.O. Pörtner and R.C. Playle (eds) *Cold Ocean Physiology. Society for Experimental Biology Seminar Series 66*. Cambridge: Cambridge University Press, pp. 3–30.

Clarke, A. and Crame, J.A. (1989). The origin of the Southern Ocean marine fauna. In: J.A. Crame (ed.) *Origins and Evolution of the Antarctic Biota. (Geological Society of London Special Publication No. 47)*. London: The Geological Society, pp. 253–268.

Clarke, A. and Crame, J.A. (1992). The Southern Ocean benthic fauna and climate change: a historical perspective. *Philosophical Transactions of the Royal Society of London, Series B*, **338**: 299–309.

Clarke, A. and Crame, J.A. (1997). Diversity, latitude and time: patterns in the shallow sea. In: R.F.G. Ormond, J.D. Gage and M.V. Angel (eds) *Marine Biodiversity: Causes and Consequences*. Cambridge: Cambridge University Press, pp. 122–147.

Clarke, A. and Johnston, I.A. (1996). Evolution and adaptive radiation of Antarctic fishes. *Trends in Ecology and Evolution*, **11**: 212–218.

Clarke, A. and Johnston, N.M. (1999). Scaling of metabolic rate with body mass and temperature in teleost fish. *Journal of Animal Ecology*, **68**: 893–905.

Clarke, K.J. and Leeson, E.A. (1985). Plasmalemma structure in freezing tolerant unicellular algae. *Protoplasma*, **129**: 120–126.

Clement, H. (1990). Alternative life designs. In: B. Bova and B. Preiss (eds) *First Contact: The Search for Extraterrestrial Intelligence*. London: Headline Book Publishing, pp. 54–70.

Cossins, A.R. (1994). Homeoviscous adaptation of biological membranes and its functional significance. In: A.R. Cossins (ed.) *Temperature Adaptation of Biological Membranes*. London: Portland Press, pp. 63–76.

Cossins, A.R. and Bowler, K. (1987). *Temperature Biology of Animals*. Chapman and Hall, London, 339pp.

Crame, J.A. and Clarke, A. (1997). The historical component of taxonomic diversity gradients. In: R.F.G. Ormond, J.D. Gage and M.V. Angel (eds) *Marine Biodiversity: Causes and Consequences*. Cambridge: Cambridge University Press, pp. 258–273.

Detrich, H.W. (1998). Molecular adaptation of microtubules and microtubule motors from Antarctic fish. In: G. di Prisco, E. Pisano and A. Clarke (eds) *Fishes of Antarctica: A Biological Overview*. Berlin: Springer-Verlag, pp. 139–149.

DeVries, A.L. (1971). Freezing resistance in fishes. In: W.S. Hoar and D.J. Randall (eds) *Fish Physiology, Vol. 6*. London: Academic Press, pp. 157–190.

DeVries, A.L. (1982). Biological antifreeze agents in coldwater fishes. *Comparative Biochemistry and Physiology*, **73A**: 627–640.

DeVries, A.L. (1983). Antifreeze peptides and glycopeptides in cold-water fishes. *Annual Review of Physiology*, **45**: 245–260.

DeVries, A.L. (1988). The role of antifreeze glycopeptides and peptides in the freezing avoidance of Antarctic fishes. *Comparative Biochemistry and Physiology*, **90B**: 611–621.

DeVries, A.L. and Wohlschlag, D.E. (1969). Freezing resistance in some Antarctic fishes. *Science*, **163**: 1073–1075.

Duman, J.G. (1977a). The role of macromolecular antifreeze in the darkling beetle, *Meracantha contracta*. *Journal of Comparative Physiology*, **115**: 279–286.

Duman, J.G. (1977b). Variations in macromolecular antifreeze levels in larvae of the darkling beetle, *Meracantha contracta*. *Journal of Experimental Zoology*, **201**: 85–92.

Duman, J.G. and Olsen, T.M. (1993). Thermal hysteresis protein activity in bacteria, fungi, and phylogenetically diverse plants. *Cryobiology*, **30**: 322–328.

Eastman, J.T. (1993). *Antarctic Fish Biology: Evolution in a Unique Environment*. San Diego: Academic Press.

Eastman, J.T. and Clarke, A. (1998). A comparison of adaptive radiations of Antarctic fish with those of non-Antarctic fish. In: G. di Prisco, E. Pisano and A. Clarke (eds) *Fishes of Antarctica: A Biological Overview*. Berlin: Springer-Verlag, pp. 3–26.

Egginton, S. and Sidell, B.D. (1989). Thermal acclimation induces adaptive changes in subcellular structure of fish skeletal muscle. *American Journal of Physiology*, **256**: R1–9.

Ferris, J.M. and Burton, H.R. (1988). The annual cycle of heat content and mechanical stability of hypersaline Deep Lake, Vestfold Hills, Antarctica. In: J.M. Ferris, H.R. Burton, G.W. Johnstone and I.A.E Bayly (eds) *Biology of the Vestfold Hills, Antarctica. (Developments in Hydrobiology 34)*. Kluwer Academic Publishers, Dordecht. *Hydrobiologia*, **165**: 115–128.

Flessa, K.W. and Jablonski, D. (1996). Geography of evolutionary turnover. In: D. Jablonski, D.H. Erwin and J. Lipps (eds) *Evolutionary Paleobiology*. Chicago: Chicago University Press, pp. 176–197.

Franks, F. (1985). *Biophysics and Biochemistry at Low Temperatures*. Cambridge: Cambridge University Press, 210pp.

Galtier, N., Tourasse, N. and Gony, M. (1999). A nonhyperthermophilic common ancestor to extant life forms. *Science*, **283**: 220–221.

George, M.F. and Burke, M.J. (1984). Supercooling of tissue water to extreme low temperatures in overwintering plants. *Trends in Biochemical Sciences*, **9**: 211–214.

George, M.F., Becwar, M.R. and Burke, M.J. (1982). Freezing avoidance by deep undercooling of tissue water in winter-hardy plants. *Cryobiology*, **19**: 628–639.

Gerking, S.D. and Lee, R.M. (1983). Thermal limits for growth and reproduction in the desert pupfish *Cyprinodon n. nevadensis*. *Physiological Zoology*, **56**: 1–9.

Gerking, S.D., Lee, R.M. and Shrode, J.B. (1979). Effects of generation-long temperature acclimation on reproductive performance of the desert pupfish, *Cyprinodon n. nevadensis*. *Physiological Zoology*, **52**: 113–121.

Gold, T. (1992). The deep, hot biosphere. *Proceedings of the National Academy of Sciences of the USA*, **89**: 6045–6049.

Gold, T. (1999). *The Deep Hot Biosphere*. Berlin: Springer-Verlag, 235pp.

Graham, L.A., Liou, Y.-C., Walker, V.K. and Davies,

P.L. (1997). Hyperactive antifreeze protein from beetles. *Nature*, **388**: 727–728.

Griffith, M. and Antikainen, M. (1996). Extracellular ice formation in freezing-tolerant plants. In: P.L. Steponkus (ed.) *Advances in Low-temperature Biology, Vol. 3*. Greenwich, Connecticut: JAI Press, pp. 107–140.

Grout, B.W.W. (1987). Higher plants at freezing temperatures. In: B.W.W. Grout and G.J. Morris (eds) *The Effects of Low Temperatures on Biological Systems*. London: Edward Arnold, pp. 293–314.

Held, C. (2001). Polar submergence in Antarctic isopods and falsification of the molecular slow-down hypothesis in cold waters. *Polar Biology* **24**: 497–501.

Hemmingsen, A.M. (1960). Energy metabolism as related to body size and respiratory surfaces, and its evolution. *Reports of the Steno Memorial Hospital and Nordinsk Insulin Laboratorium*, **9**: 6–110.

Henderson, L.J. (1913). *The Fitness of the Environment*. Macmillan, New York (or Beacon Press, Boston).

Hochachka, P.W. and Somero, G.N. (1984). *Biochemical Adaptation*. Princeton University Press, New Jersey, 537pp.

Hsiao, K., Cheng, C.C., Fernandes, I.E. *et al.* (1990). An antifreeze glycopeptide gene from the Antarctic cod *Norothenia coriiceps neglecta* encodes a polyprotein of high peptide copy number. *Proceedings of the National Academy of Sciences of the USA*, **87**: 9265–9269.

Joset, F., Jeanjean, R. and Hagemann, M. (1996). Dynamics of the response of cyanobacteria to salt stress: deciphering the molecular events. *Physiologia Plantarum*, **96**: 738–744.

Kerry, K.R., Grace, D.R., Williams, R. and Burton, H.R. (1977). Studies on some saline lakes of the Vestfold Hills, Antarctica. In: G.A. Llano (ed.) *Adaptations within Antarctic Ecosystems*. Smithsonian Institution, Washington DC, pp. 839–858.

Knight, C.A. and Duman, J.G. (1986). Inhibition of recrystallisation of ice by insect thermal hysteresis proteins: a possible cryoprotective role. *Cryobiology*, **23**: 256–262.

Lin, Y. (1979). Environmental regulation of gene expression. *Journal of Biological Chemistry*, **254**: 1422–1426.

Lin, Y. (1983). Regulation of the seasonal biosynthesis of antifreeze peptides in cold-adaptation fish. In: A.R. Cossins and P. Sheterline (eds) *Cellular Acclimatisation for Environmental Change*. Cambridge: Cambridge University Press, pp. 217–226.

Lipps, J.H. and Hickman, C.S. (1982). Origin, age and evolution of Antarctic and deep-sea faunas. In: W.G. Ernst and J.G. Morin (eds) *The Environment of the Deep Sea (Rubey Volume II)*. Englewood Cliffs, New Jersey: Prentice Hall, pp. 325–356.

Louw, G.N. (1993). *Physiological Animal Ecology*. Harlow: Longman, 299pp.

Lynch, M. and Lande, R. (1993). Evolution and extinction in response to environmental change. In: P.M. Kareiva, J.G. Kingsolver and R.B. Huey (eds) *Biotic Interactions and Global Change*. Sunderland, MA: Sinaeur Associates, pp. 234–250.

Martin, A.P. and Palumbi, S.R. (1993). Body size, metabolic rate, generation time, and the colecular clock. *Proceedings of the National Academy of Sciences of the USA*, **90**: 4087–4091.

Mason, S.F. (1992). *Chemical Evolution: Origins of the Elements, Molecules, and Living Systems*. Oxford: Oxford University Press, 317pp.

Miller, L.K. and Werner, R. (1987). Extreme super-cooling as an overwintering strategy in three species of willow gall insects from interior Alaska. *Oikos*, **49**: 253–260.

Moerland, T.S. (1995). Temperature: enzyme and organelle. In: P.W. Hochachka and T.P. Mommsen (eds) *Biochemistry and Molecular Biology of Fishes, Vol. 5*. Amsterdam: Elsevier, pp. 57–71.

Morris, G.J. and Clarke, A. (1987). Cells at low temperatures. In: B.W.W. Grout and G.J. Morris (eds) *The Effects of Low Temperatures on Biological Systems*. London: Edward Arnold, pp. 72–119.

Pain, R.H. (1982). Molecular hydration and biological function. In: F. Franks (ed.) *Biophysics of Water*. Chichester: John Wiley, pp. 3–14.

Paul, J. (1986). Body temperature and the specific heat of water. *Nature*, **323**: 300 only.

Pörtner, H.O., Hardewig, I., Sartoris, F.J. and van Dijlz, P.L.M. (1997). Energetic aspects of cold adaptation: critical temperatures in metabolic, ionic and acid-base regulation? In: H.O. Pörtner and R.C. Playle (eds) *Cold Ocean Physiology. Society for Experimental Biology Seminar, Series 66*. Cambridge: Cambridge University Press, pp. 88–120.

Pörtner, H.O., Peck, L.S., Zielinski, S. and Conway, L.Z. (1999a). Intracellular pH and energy balance in the highly stenothermal Antarctic bivalve *Limopsis marionensis* as a function of ambient temperature. *Polar Biology*, **22**: 17–30.

Pörtner, H.O., Hardewig, I. and Peck, L.S. (1999b). Mitochondrial function and critical temperature in the Antarctic bivalve, *Laternula elliptica*.

Comparative Biochemistry and Physiology, Part A, **124**: 179–189.

Rand, D.M. (1994). Thermal habit, metabolic rate and the evolution of mitochondrial DNA. *Trends in Ecology and Evolution,* **9**: 125–131.

Reite, O.B., Maloiy, G.M.O. and Aasehaug, B. (1974). pH, salinity and temperature tolerance of Lake Magadi tilapia. *Nature,* **247**: 315 only.

Rohde, K. (1992). Latitudinal gradients in species-diversity – the search for the primary cause. *Oikos,* **65**: 514–527.

Rothschild, L.J. and Mancinelli, R.L. (2001). Life in extreme environments. *Nature (London),* **409**: 1092–1101.

Salt, R.W. (1966). Relation between time of freezing and temperature in supercooled larvae of *Cephus cinctus* Novt. *Canadian Journal of Zoology,* **44**: 947–952.

Schrödinger, E. (1944). *What is Life?* Cambridge: Cambridge University Press, 96pp.

Stanley, S.M. (1986). Anatomy of a recent regional mass extinction: Plio-Pleistocene decimation of the western Atlantic bivalve fauna. *Palaios,* **1**: 17–36.

Steele, J.H. (1985). A comparison of terrestrial and marine ecological systems. *Nature,* **313**: 355–358.

Storey, K.B. and Storey, J.M. (1996). Natural freezing survival in animals. *Annual Review of Ecology and Systematics,* **27**: 365–386.

Tanford, C. (1980). *The Hydrophobic Effect.* New York: John Wiley.

Taylor, M.J. (1987). Physio-chemical principles in low temperature biology. In: B.W.W. Grout and G.J. Morris (eds) *The Effects of Low Temperatures on Biological Systems.* London: Edward Arnold, pp. 3–71.

Tominage, H. and Fukui, F. (1981). Saline lakes at Syowa Oasis, Antarctica. *Hydrobiologia,* **82**: 375–389.

Tyler, S. and Sidell, B.D. (1984). Changes in mito-chondrial distribution and diffusion distances in goldfish muscle upon acclimation to warm and cold temperatures. *Journal of Experimental Zoology,* **232**: 1–9.

Tyshenko, M.G., Doncet, D., Davies, P.L. and Walker, V.K. (1997). The antifreeze potential of the spruce budworm thermal hysteresis protein. *Nature Biotechnology,* **15**: 887–890.

Urrutia, M.E., Duman, J.G. and Knight, C.A. (1992). Plant thermal hysteresis proteins. *Biochimica Biophysica Acta,* **1121**: 199–206.

Wald, G. (1964). The origins of life. *Proceedings of the National Academy of Sciences of the United States,* **52**: 595–611.

Wharton, D.A. and Ferns, D.J. (1995). Survival of intracellular freezing by the Antarctic nematode *Panagrolaimus davidi. Journal of Experimental Biology,* **198**: 1381–1387.

12

Temperature, tectonics, and evolution

Geerat J. Vermeij

ABSTRACT

See Plates 20–22

Temperature – the average kinetic energy of molecules of a substance – affects nearly every aspect of life and its evolution. As temperature rises from near freezing to about 30°C, rates of biological processes such as photosynthesis, metabolism, suspension-feeding, and locomotion in water increase, whereas costs associated with water transport and the formation of hard skeletons decrease. Most importantly from an evolutionary point of view, a rise in temperature is accompanied by an increase in the scope of adaptation – the range of possible or attainable phenotypes – by reducing physical limitations on the many functions carried out by organisms whose body temperatures are close to those of their surroundings. As a result, warming and the spread of warm conditions should stimulate evolution by providing opportunity. This should be especially so for energy-intensive modes of life and for innovations requiring a high energy investment.

The direct effects of temperature are amplified by the complex interplay amongst temperature, the availability of essential nutrients, and the geological processes that affect the character and composition of the Earth's crust, oceans, and atmosphere. Although some processes by which raw materials become available to organisms are associated with or amplified by cool conditions (e.g. physical erosion, formation and spread of dust, and upwelling of nutrient-rich water), others lead to or operate faster at higher temperatures and in the aggregate are more important (e.g. hydrothermal activity, carbon dioxide-releasing volcanism, chemical weathering, decomposition, and direct consumption of organisms by other organisms). Higher temperatures and broader warm belts should, on balance, make resources more accessible to living things, lead to a rise in primary productivity, increase concentrations of oxygen in the atmosphere, and expand the range of metabolically feasible phenotypes.

Geological evidence indicates that periods of warming and of poleward expansion of warm conditions coincide with rising sea levels, increased rates of production of crust, massive eruptions of volcanoes beneath the sea surface, the release of surface-warming greenhouse gases such as carbon dioxide and methane, and the breakup and subsequent dispersal of continental land masses. Equatorial temperatures during the Paleozoic era may have been significantly lower than those of the subsequent Mesozoic and Cenozoic eras. The Vendian to Early Cambrian, Late Cambrian to Middle Ordovician, Silurian to Early Carboniferous, Jurassic to Late Cretaceous, and several shorter intervals of the Cenozoic (early Eocene, late Oligocene to early Miocene, and early Pliocene) were time intervals when tectonic activity and thermal conditions were most favorable to evolution. These are also the times of frequent and dramatic evolutionary innovation and of diversification. Not all warm intervals coincide with innovation, nor are all major evolutionary events (especially those on land) associated with continental rifting and other manifestations of warming. Important questions remain about how temperature, oxygen, nutrients, and their controlling geological processes affect biological processes and evolutionary opportunity.

Evolution on Planet Earth
ISBN 0-12-598655-6

12.1 **Introduction**

Biological evolution is an economic process in which the entities of life – all organizational units from genes to species – change through time according to the circumstances in which they live and which they helped to create. It is a process governed by opportunities, challenges, and limitations. Living things and their parts metabolize and interact; they compete for nutrients, whose availability and conversion to fitness are affected by life itself as well as by the physics and chemistry of the Earth (Vermeij, 1999).

Every activity and every adaptation of living things requires energy. As a practical matter, the amount of energy available to an individual and convertible to biological work is limited, so that energy devoted to one activity of adaptation is unavailable to another. The relation of life to energy is therefore fundamental to any understanding of ecology or evolution (Van Valen, 1976). Temperature – the average kinetic energy of molecules of a substance – is a simple and convenient measure of energy in the environment. Physiologists and biochemists have accumulated an enormous amount of information on the biological effects of temperature, but evolutionary biologists have been slow to grasp the historical implications of these findings.

In this chapter, I first review briefly the numerous direct effects temperature has on living systems and their environments. Then, I shall argue that episodes in Earth history during which temperatures rose or high-temperature zones spread are highly favorable to evolutionary innovation and diversification. This is so not only because increases in temperature make many adaptations catalytically more attainable, but also because global increases in temperature are accompanied by higher oxygen levels in the atmosphere, as well as by a greater availability of nutrients and higher primary productivity. Finally, I caution against atomizing the environment into discrete physical and biological factors, and argue instead that feedbacks between life and its environment demand a holistic approach to the problem of cause and effect in evolution.

12.2 **The direct effects of temperature**

12.2.1 *Temperature, physiology, and adaptive scope*

Almost every physical and biological process depends on temperature. As the average kinetic energy of molecules rises, so do the diffusion coefficients of carbon dioxide (CO_2) and oxygen in water, the dynamic viscosity of air, the rate of nerve conduction, water-vapor content of air, and rates of chemical reaction. Rising temperatures are also accompanied by decreases in the dynamic viscosity of water, concentrations of dissolved CO_2 and oxygen in water, density of air, and solubility of the important skeletal mineral calcium carbonate (see Hochachka and Somero, 1973; Clarke, 1983, 1993; Denny, 1993; Podolsky, 1994).

These relationships have far-reaching biological consequences. For example, viscosity-related costs of swimming and suspension-feeding decrease as temperature rises. So do the costs of calcification. On the other hand, the average basal metabolic rate of ectothermic organisms (those whose body temperature is indistinguishable from that of the environment) rises with an increase in environmental temperature, at least up to some 'optimal' temperature. The effects of temperature can be partially overcome by a variety of quantitative and qualitative changes in enzymes, whose function is to lower the activation energy of chemical reactions; but such compensation requires energy and is rarely complete (Hohachka and Somero, 1973; Holeton, 1974).

These general relationships between biological activities and temperature conceal an important and usually overlooked phenomenon. Although it is true that average metabolic rate increases as temperature rises, a more precise statement is that the variance in metabolic rates increases. At low temperatures, ectothermic organisms have more or less uniformly low basal metabolic rates; whereas at higher temperatures, many organisms have higher rates but some maintain very low metabolism. For example, brachiopods with extremely low metabolic rates occur in tropical reefs in the same thermal regime as do bivalved molluscs, whose basal rates are three to ten times higher. Averages, in other words, are deceiving. The range of evolutionarily attainable states among ectotherms is smaller in the cold than under warm conditions. Just as importantly, the potential difference in performance (biological function) and therefore the degree of potential specialization are much greater at high than at low temperatures (Vermeij, 1978, 1987). For example, the difference between potential maximum growth rates of land plants at low temperatures is so small that specialization to the rapidly growing vine habit as compared to that of slow-growing understory shrubs is unlikely to be great. In tropical regions or in temperate climates with hot, moist summers, however, this difference is large enough for such specializations to become feasible. Similarly, the range in shell thicknesses and strengths among tropical marine molluscs is huge compared to that in cold-water temperate and polar molluscs. In cold conditions, limitations on calcification place restrictions on the range of attainable phenotypes of shells (Graus, 1974).

The effect of rising temperatures is therefore to increase the scope of adaptation – the range of permissible or attainable phenotypes – by decreasing the energy of activation and by reducing limitations on the rates of life processes by which organisms compete for resources. It should therefore stimulate evolution, especially the establishment of innovations requiring a lot of energy (Vermeij, 1978, 1987, 1995). Consistent with this expectation is Jablonski's (1993) finding that a disproportionate number of innovations, as represented by the first appearances of order level clades of marine animals in the fossil record of the past 250 million years, began in the tropics.

Perhaps as a direct consequence of the relationship between adaptive scope and temperature, the intensity and selective importance of interactions between species and their predators are greater at higher temperatures, and tend to increase from polar and temperate latitudes to the tropics (Vermeij, 1978). In other words, the degree of expression or specialization of antipredatory traits increases toward the tropics, as does the percentage of adult individuals being eaten by predators. These relationships have been empirically demonstrated in many groups, including molluscs and their predators (shell-breakers, shell-enterers, and shell-drillers) (Vermeij, 1978, 1993, and references therein), ants and their spider predators (Jeanne, 1979), and plants and leaf-eating insects (Coley and Barone, 1996).

The spread of warm conditions to higher latitudes amplifies the permissive effects of higher temperatures at the population level. As their environment expands, warm-adapted species increase in population size. As a consequence, many small isolates that potentially act as founders for new species are able to persist long enough to become established and to expand themselves. Population expansion in the presence of sufficient resources allows for a certain evolutionary 'freedom': the usual adaptive constraints and compromises imposed by populational stability or decline are lifted, allowing more variations to be tested by selection (Vermeij, 1978, 1987, 1995).

The stimulatory effects of higher temperatures reverse when temperatures rise above 30°C. Water temperatures above this limit harm the photosynthesizing symbionts in corals, and inhibit many physiological functions in a wide variety of aquatic organisms. Complex enzymes lose function and become denatured above specific threshold temperatures. Overheating places strict upper size limits on leaves of land plants and on terrestrial animals, and may restrict the activities of many animals during daylight hours (Denny, 1993; Heinrich, 1993; McElwain *et al.*, 1999). Although many prokaryotes can operate at very much higher temperatures, the stimulatory effects of increasing temperature that I emphasize in this chapter occur mainly up to 30°C.

12.2.2 *Temperature, nutrients, and productivity*

Although temperature affects all life processes, it cannot be the only factor influencing evolutionary potential. Primary producers must have access to inorganic elements in order to convert inorganic carbon to organic compounds, on which all other living things that do not make their own food depend. The quantity of available nutrients, access to raw materials, and rate of supply of essential elements all affect the size of an individual's energy budget and therefore the range of adaptive possibilities in the pattern of energy allocation. A critical aspect of the role of temperature in evolution is therefore how temperature affects, directly and indirectly, the availability of, and access by organisms to, essential nutrients. I shall argue that, on balance, an increase in temperature brings about an increase in the supply of phosphorus and other nutrients, and therefore indirectly promotes global primary production and the scope of adaptation.

The ultimately limiting nutrient in most of the world ocean and in many terrestrial environments is phosphorus (Melillo *et al.*, 1993; DeAngelis, 1992; Tyrrell, 1999). Although such other elements as nitrogen, iron, silicon, zinc, and potassium act as limiting nutrients in the short term, it is phosphorus which regulates the rate of production in the long run (for discussion see Tyrrell, 1999).

Inorganic phosphorus enters the biosphere as runoff in rivers, dust from erosion and volcanic eruptions, and mantle-derived minerals through volcanic and hydrothermal activity. According to Froelich and colleagues (1982), river runoff accounts for more than 92% of the phosphorus supply globally. Dust accounts for about 0.5% of the phosphorus today, but could have been much more important at times of more intense volcanism or erosion. Hydrothermal sources at deep-sea vents account for about 7% of the phosphorus in the ocean today, much of which is unavailable to organisms, either because it is bound with metals or because it is incorporated into recently erupted basalts when the latter come into contact with seawater (Froelich *et al.*, 1982). The hydrothermal component would have been much more important at times of intense submarine volcanism, especially if such activity were associated with large-scale convection of warm bottom waters, allowing the mantle-derived phosphorus to mix with seawater far from the hot source. Any geological process

that enhances runoff, the formation and spread of dust, and the liberation and connection of mantle-derived phosphorus in the ocean should therefore increase the rate of supply of phosphorus, and therefore stimulate global productivity. How does temperature affect these processes, and how do the mechanisms controlling the availability of phosphorus affect temperature?

Phosphorus, iron, silicon, and many other elements at the Earth's surface do not occur in a gas phase in the atmosphere, and become available to organisms only through the breakdown of rock or through recycling within the biosphere. Two processes are involved in the breakdown of rock: physical erosion and chemical weathering. Physical erosion – the breakage of rock into smaller particles – is an episodic process that is most intense in steep mountainous terrain, in dry unvegetated deserts, and on land scoured by ice in glaciers. High rates of erosion, including the formation and spread of dust, should therefore be associated with the presence of alpine and high-latitude glaciers, as well as with large expanses of uplifted land and of desert (Capo and DePaolo, 1990; Edmond, 1992; Raymo and Ruddiman, 1992; François *et al.*, 1993; Rea, 1994; Filippelli, 1997). These circumstances imply widespread cool, dry climates. Dust, which also enters the atmosphere through volcanic eruptions, may cause further cooling by intercepting and scattering sunlight (Sigurdsson, 1990). A high dust load of the atmosphere during glacial intervals of the Pleistocene was associated with high rates of productivity by phytoplankton in the ocean, probably because settling dust provides iron and other nutrients (see Sancetta, 1992; Paytan *et al.*, 1996). Moreover, dust blown from deserts and volcanoes also prevents loss of nutrients in soils on land. Experiments and comparative studies by Chadwick and colleagues (1999) in Hawaii show that dust provides enough nutrients for trees to maintain reasonable productivity despite the loss of soluble nutrients through solution in aging forest soils.

Chemical weathering – the transformation of silicate and carbonate minerals in rock into soluble ions and clay minerals through microbial activity and by the respiration and acid secretion of plant roots – occurs mainly in young soils, those formed recently as the result of erosion and subsequent deposition of freshly exposed rock particles (Stallard, 1988, 1992; Chadwick *et al.*, 1999). Like other life processes, chemical weathering speeds up as temperature rises. The release of nutrients should be especially rapid in warm regions with high rainfall. In Hawaii, most of the more soluble ions in forest soils are released by chemical weathering during the first 150 000 years of a soil's existence. As a soil ages, its nutrient content declines, because more and more of the remaining nutrients are bound up in insoluble compounds unavailable to organisms (Chadwick *et al.*, 1999).

Forms of nutrient recycling that involve decomposition, predation, and herbivory are also temperature-dependent. In fact, direct herbivory – the consumption of photosynthesizing tissues by animals – occurs at higher rates in the tropics than in cooler zones, and provides a faster means of distribution of nutrients through an ecosystem than does decomposition of dead plant material (DeAngelis, 1992).

An important physical mechanism by which nutrients are recycled to the marine photic zone is upwelling. When organisms swimming or floating in the water die, their remains sink into waters below the photic zone, with the result that the nutrients these organisms contain are no longer available to the primary producers in surface waters. Winds and currents, however, often transport surface waters horizontally by advection, creating conditions in which the surface waters are replaced by nutrient-rich waters from below. This upwelling, which supports high rates of production, is associated with cooler than normal surface conditions, because the upwelled water tends to be colder than the surface water it replaces. The zones of highest productivity in today's ocean occur in a narrow equatorial zone in the Pacific Ocean, at mid latitudes on the western coasts of the Americas and Africa, and in the Antarctic region. There are also important upwelling sites in the tropical eastern Pacific, the north coast of South America, the northwestern Indian Ocean, and the Indonesian

and Philippine archipelagos. Upwelling is most effective in stimulating primary production where chemically weathered nutrients enter the ocean and become available for recycling (Walsh, 1988; Dugdale and Wilkerson, 1998; Smetacek, 1998; Johnson *et al.*, 1999).

Mantle-derived phosphorus and other elements, notably iron, reach the Earth's surface through volcanic eruptions and hydrothermal activity at deep-sea vents (Baker *et al.*, 1987; Garzanti, 1993; Coffin and Eldholm, 1994; Coale *et al.*, 1996; Wells *et al.*, 1999). Iron limits productivity in large parts of the world ocean, so that its addition leads to short-term and in some cases perhaps to long-term increases in primary productivity by phytoplankton (Martin and Fitzwater, 1988; Martin *et al.*, 1994; Sullivan *et al.*, 1993; Coale *et al.*, 1996; Watson, 1997; Takeda, 1998; Johnson *et al.*, 1999). The tectonic input of essential nutrients like phosphorus and iron is accompanied by the release of large amounts of carbon dioxide (CO_2), a potent greenhouse gas. Increases in CO_2 also enhance primary production. Doubling the atmospheric concentration of CO_2 means that plants transpire only half as much water vapor to fix a given amount of carbon (Farquhar, 1997; Chaloner and McElwain, 1997). This effect, which is strongest when initial concentrations of CO_2 are low (in the range of modern preindustrial levels, 280 parts per million per volume of air), therefore allows rapidly growing plants to fix carbon when CO_2 accumulates in the atmosphere (DeLucia *et al.*, 1999). Riebessel and colleagues (1993) described a similar effect in rapidly growing marine phytoplankters. In short, increased volcanic and hydrothermal activity stimulate primary production not only through the release of nutrients at a higher rate, but also because of the increased liberation of CO_2, which in turn causes temperatures to rise and which makes primary production by plants more efficient.

From this discussion, I conclude that some non-biological processes by which nutrients become available to organisms – extensive uplift, physical erosion, glaciation, the formation and spread of dust, and upwelling – are associated with or amplified by cool conditions; whereas volcanism, hydrothermal activity, and the biological processes of chemical weathering, decomposition, and the direct consumption of organisms lead to or operate faster at higher temperatures. In today's world, and probably throughout much of Earth history, organisms play and have played decisive roles in many if not most chemical cycles on the Earth's surface. As a result, biological processes that liberate or recycle nutrients may be more important in controlling the availability of limiting elements than are non-biological processes.

I therefore suggest that a rise in temperature is associated globally with increases in the rate at which nutrients become available and with greater access by organisms to these nutrients. Higher temperatures should, on balance, lead to a rise in primary productivity and to a wider range of attainable phenotypes.

Several lines of evidence support this hypothesis. The warm tropics supply a disproportionate quantity of nutrients to the world ocean. Stallard (1992), for example, points out that the humid tropics supply 65% of the dissolved silica and 38% of the ions delivered by rivers to the sea, even though they occupy only 25% of the Earth's land surface (see also Berner, 1992; Treguer *et al.*, 1995). Tropical rain forests cover only 14% of the world's land area, but contribute 38% of terrestrial primary production (Melillo *et al.*, 1993). Moreover, high productivity under cold conditions does not demonstrably result in unusual evolutionary innovation or diversity. Antarctic waters are exceptionally productive, but their biota is phenotypically as well as taxonomically limited. Finally, there is no evidence that high productivity in the cool tropical Pacific during glacial intervals of the Pleistocene promoted evolution. In fact, the Pleistocene is a conspicuously quiet time in terms of speciation and innovation (see Bennett, 1997).

12.2.3 Temperature and oxygen

The effects of temperature do not stop there. I think it is possible that rising temperatures, insofar as they are accompanied by an increase in primary productivity, may lead to enrichment of the atmosphere and photic zone of the ocean with oxygen. Globally, photosynthesis and other processes that add oxygen to the atmosphere are approximately balanced in the modern biosphere by respiration and other processes that remove oxygen. The photic zone, where photosynthesis takes place, comprises only a small fraction of the environment in the biosphere where oxygen is consumed. Even within the photic zone, areas of low productivity such as the mid-latitude ocean gyres are net sinks of oxygen (Duarte and Agustí, 1998). Geographic enlargement of highly productive environments, where there is net production of oxygen, therefore has the potential to raise oxygen concentrations in the atmosphere. Warming, as well as the spread of warm conditions, would lead to oxygen enrichment if the rates of photosynthesis and other oxygen-producing processes go up more rapidly with temperature than do rates of oxygen consumption. The expansion of humid lowland forests and of shallow coastal or inland seas capable of supporting high planktonic as well as benthic productivity by plants should be especially favorable.

One effect that causes or amplifies this possible link between temperature and oxygen is the burial of organic matter and sulfide, which are the reduced forms of carbon and sulfur respectively. Most of the organic matter produced by plants is eventually oxidized by animals, decomposers, and the plants themselves. Because this material is being recycled in the biosphere, there is no net accumulation or loss of oxygen. In the modern biosphere, about 0.2 to 0.3% of organic matter escapes oxidation by becoming buried in sediments, where it is unavailable to most life forms until geological processes or human activities such as mining coal and extracting oil once again expose it to oxidation (Van Valen, 1976; Des Marais *et al.*, 1992; D'Hondt *et al.*, 1998). When the quantity of organic matter being buried increases relative to the rate of oxygen consumption, the concentration of oxygen in the atmosphere goes up (Schopf, 1980; Lovelock and Whitfield, 1982; Berner and Canfield, 1989; Des Marais, 1994; Canfield and Teske, 1996; Landis *et al.*, 1996). In the ocean, high rates of burial of organic matter occur in areas where sediments accumulate rapidly (Schopf, 1980; Berner and Raiswell, 1983) and where the sediments and the waters above them contain little or no oxygen. Rapid sedimentation characterizes coastal seas receiving runoff through rivers draining ice-scoured or mountainous land masses. Oxygen-poor conditions in areas where sediments accumulate arise either because water is stagnant, preventing water that has become depleted of oxygen by respiration from being replaced with oxygen-rich water, or because the water is warm. Respiration is faster and oxygen concentrations are lower in warm than in cold water. Warm water is therefore more apt to become depleted of oxygen and therefore to allow sinking organic particles to escape oxidation (Berry and Wilde, 1978; Hallam, 1981; Hoffman *et al.*, 1991). Intense erosion and runoff together with warm water therefore stimulate primary production as well as the accumulation of organic matter, and should therefore favor the accumulation of oxygen in the atmosphere and in the upper layers of the ocean. The burial of sulfide amplifies the effects of burial of organic carbon (Paytan *et al.*, 1998) and is under similar controls (Berner and Raiswell, 1983; Canfield and Teske, 1996).

Like an increase in temperature, a rise in atmospheric oxygen concentration greatly expands the range of evolutionary possibilities for aerobically metabolizing organisms. Because respiration in most organisms is limited by diffusion of oxygen across membranes, a higher concentration of oxygen makes possible higher metabolic rates and larger body sizes. Moreover, environments previously free of oxygen, such as deep layers of sediment, may become oxygenated enough to allow the establishment of aerobically metabolizing life (Runnegar, 1982; Graham *et al.*, 1995; Dudley, 1998; Gans *et al.*, 1999). The increased density of air resulting from the accumulation of oxygen may have made possible the Mid-Carboniferous evolution of flight in insects (Graham *et al.*, 1995). One possible adverse consequence of rising oxygen levels is the experimental finding by Beerling and colleagues (1998) that the rate of primary produc-

tion in two woody plant species exposed to 35% oxygen in the air is about 29% lower than that of control plants grown at the modern oxygen concentration of 21%. This effect therefore represents a potential negative feedback as temperatures rise, but the total production and biomass of land plants at higher temperatures may compensate for the per-capita decline in productivity.

12.3 Temperature and its geological controls

Although the direct effects of high temperatures on physiology, populations, and chemical cycles in the biosphere have important implications for evolutionary potential, as discussed in the preceding section, the full story of the role of temperature in evolutionary history can be told only when we have taken full account of the geological factors that control temperature on the Earth's surface. Surface temperatures in the photic zone, that part of the biosphere where primary production of organic carbon through photosynthesis is taking place, are affected by the movement of materials and heat among the atmosphere, water, soil, crust, and mantle. These movements and relationships, which involve a host of positive and negative feedbacks, are far from being understood completely (for a review see Raven, 1998). Instead of dealing with this subject in all its complexity, I shall try to show that the geological processes and conditions which bring about an increase in temperature or a broadening of warm belts magnify the effects of the higher temperatures themselves on evolutionary opportunity.

One mechanism that can trigger global warming is the addition of such greenhouse gases as carbon dioxide (CO_2), water vapor (H_2O), and methane (CH_4) to the atmosphere, where these gases trap heat (see Crowley and North, 1991). The mantle is the principal source of new CO_2 for the biosphere. Volcanic eruptions, underwater hydrothermal activity, and the metamorphism of carbonate rocks as the latter sink and are heated in the mantle, liber-

ate CO_2 into the ocean and atmosphere. These
mantle sources become increasingly important
during times when continental blocks break
apart and then separate along mid-ocean
ridges (see McLean, 1985; Larson, 1991b;
Kaiho and Saito, 1994). As heat rises up
through the mantle into the crust, the mid-
ocean ridges displace large amounts of seawater onto low-lying parts of adjacent land
masses, with the result that sea level rises
(see also Arthur *et al.*, 1985; Nance *et al.*, 1986;
Veevers, 1990). In these shallow coastal or
inland seas, sediments collect rapidly and productivity is high. Meanwhile, the open ocean
receives relatively little sediment and may
become relatively unproductive (Fischer and
Arthur, 1977; Hallock and Schlager, 1986;
Garzanti, 1993).

An increase in the luminosity of the Sun may
provide a second mechanism by which the
Earth's average surface temperature may have
increased. The Sun's output of radiation has
increased by about 25% since the Early
Archean (3.8 Ga) and by about 4.6% since
550 Ma, just before the beginning of the
Phanerozoic (Berner, 1993, 1994; Lenton, 1998).
In the absence of plants, which by chemically
weathering rock and transpiring water vapor
have kept temperatures down, the Earth
would be some 18°C warmer now than at
3.8 Ga (Lenton, 1998). Although Schwartzman
and colleagues (1993) and Schwartzman (1998)
argue that the Earth's surface has cooled in
stages as larger and more deeply rooted land
plants have evolved, I know of no compelling
arguments in favor of the idea that plants have
completely canceled the solar effect. In fact,
some paleontological evidence (see below)
implies that surface temperatures, and tropical
temperatures in particular, may have risen
through the Phanerozoic. In any case, the
increase in luminosity must be considered as a
potentially important cause of surface heating.

The distribution of heat around the globe is
influenced by patterns of circulation in the
atmosphere and ocean. These patterns, in
turn, depend on climate as well as on the distribution and sizes of land masses and oceans.
Brass and his colleagues (1982) suggested that
zones of warm climate expanded poleward

when deep ocean water sinks in regions of
high evaporation at middle and low latitudes
and then flows away from the equatorial belt.
The formation of such warm, deep water is
most likely in shallow coastal or inland seas
at times when sea level is high. If polar regions
are ice-free, either because the poles are located
in or near an ocean (Crowley *et al.*, 1987) or
because warm currents deliver large amounts
of water to the far north and south, plant cover
can reinforce the warm high-latitude conditions by reducing the amount of heat reflected
back into the atmosphere from unvegetated or
ice-covered land (Otto-Bliesner and Upchurch,
1997). Another mechanism for transporting
heat to polar regions is the diversion of currents. Equatorial currents, which flow east to
west or west to east, are deflected northward
and southward when land masses interrupt the
flow (Crowley and North, 1991). Changes in
atmospheric circulation, which are as yet
poorly understood, may also distribute equatorial heat to a wider latitudinal band.

12.4   The geological evidence

In order to see how well the history of life con-
forms with the expectation that warm conditions brought on by rifting-associated
addition of CO_2 to the atmosphere triggered
evolutionary innovation, it is important to
review some of the geological evidence pertaining to the nature, timing, and consequences
of tectonic events of the past. In this section, I
provide a thumbnail sketch of what we know
about crust production, seafloor spreading, sea
levels, volcanism, mountain-building, gas composition of the atmosphere, and temperature
through time.

12.4.1   Crust production and sea level

A synthesis of data by Kaiho and Saito (1994)
indicates that maximum rates of production
of crust during the past 125 My occurred
during the mid-Cretaceous (Cenomanian),
Late Paleocene, and Late Oligocene to early
Miocene. From 120 to 80 Ma during the

Cretaceous, the rate of crust production may have risen by 50 to 75% (Larson, 1991a), and the rate of seafloor spreading increased five- to tenfold from 1 to 2 cm per million years to more than 10 cm per year (Landis *et al.*, 1996). High rates of seafloor spreading also characterized the Middle Jurassic (Callovian to Oxfordian, 165 to 160 Ma), when they may have been 2.8 times higher than during the early Cretaceous (Sheridan, 1997; see also Veevers, 1989).

Increases in sea level coinciding with or following these episodes of rapid crust production and seafloor spreading were dramatic. During the Jurassic, as break-up and rifting of the supercontinent Pangaea were well under way, sea levels rose by about 100 m to a peak in the Late Jurassic (Kimmeridgian) (Hallam, 1992). From the Early Aptian to the Turonian stages (120 to 90 Ma), Cretaceous sea levels rose by at least 125 m to a maximum height of some 250 m above present levels. At peak high stands, the sea covered 20% of the continental area and about 77% of the Earth's surface, as compared to about 71% today (Thierstein, 1989; Hallam, 1992; Haq *et al.*, 1987; Crowley and North, 1991; Larson, 1991b). From the latest Cretaceous onward, sea levels show a general downward trend, but significant rises occurred in the early Paleocene, early Eocene, early Oligocene, early to middle Miocene, and early Pliocene (Haq *et al.*, 1987; Hallam, 1992). Only the Pliocene rise in sea level does not correspond with or follow a known episode of enhanced seafloor spreading, although there was substantial volcanic activity and uplift near New Guinea during the late Miocene to Pliocene (8 to 3 Ma), as well as mountain-building in Asia and South America (see Raymo and Ruddiman, 1992; Filippelli, 1997; Wells *et al.*, 1999). Early Pliocene sea levels (4.5 to 4.4 Ma) were 20 to 35 m higher than those today (Dowsett *et al.*, 1992; Krantz, 1991), whereas in the mid-Pliocene (4.0 to 3.2 Ma) sea levels reached a maximum of some 35 to 40 m above present level. During the Pleistocene, as in other glacial intervals of the Phanerozoic, sea levels fluctuated greatly in association with the expansion and contraction of glaciers at high latitudes.

The history of sea level in the Paleozoic is marked by a rise in the Early Paleozoic and a general fall late in the era (Hallam, 1992). Notable rises in sea level occurred during the Cambrian, Ordovician, Devonian, and Early Carboniferous, though this general trend was frequently interrupted by episodes of lower sea level, such as at the end of the Early Cambrian, the Tremadoc–Arenig boundary in the Early Ordovician, and the Late Ordovician. A general trend toward lower sea levels began during the Namurian stage of the Middle Carboniferous, and culminated in low stands during the Late Permian (Hallam, 1992).

12.4.2 *Volcanism*

The history of major volcanism is being pieced together by those who study flood-basalt eruptions on the continents and beneath the oceans. Full appreciation of the magnitude, timing, duration, number, and consequences of these episodes has come only in the past 15 years, and a great deal remains to be learned about volcanism in Earth history, especially for the Paleozoic and earlier times. Already, however, it has become clear that the consequences of continental flood-basalt volcanism are very different from those of massive eruptions of lava beneath the sea.

The end-Permian, end-Triassic, and end-Cretaceous mass extinctions coincide with, and may be causally linked to, huge flood-basalt eruptions on continental land masses. This connection has been reasonably well established for the end-Permian Siberian Traps of Russia (Campbell *et al.*, 1992; Renne and Basu, 1991; Renne *et al.*, 1995; Bowring *et al.*, 1998), circum-Atlantic flood-basalts erupted at the end of the Triassic (Marzoli *et al.*, 1999; Olsen, 1999), and the Deccan Traps of India at the end of the Cretaceous (Courtillot *et al.*, 1986; McLean, 1985; Officer and Drake, 1985; Rampino and Caldeira, 1993). Extinctions of smaller magnitude at the Paleocene–Eocene boundary and near the end of the early Miocene may be associated with circum-Caribbean eruptions (Bralower *et al.*, 1997) and the Columbia Plateau flood-basalts (Hooper, 1990; Rampino and Stothers, 1988;

Rampino and Caldeira, 1993) respectively, with the North Atlantic flood-basalts perhaps also playing a role at the end of the Paleocene (Rea *et al.*, 1990; Koch *et al.*, 1992; Zachos *et al.*, 1993). All of these large and relatively brief eruptive episodes, lasting not more than 1 to 2 million years at their peak, may have introduced vast quantities of sulfur-rich particles and dust into the atmosphere, bringing about cooling and other temporary disruptions to primary production as well as a longer-term warming (Rampino *et al.*, 1979; Sigurdsson, 1990; Vogelmann *et al.*, 1992; Erwin and Vogel, 1992; Rampino and Self, 1992; Zielinski *et al.*, 1997).

Submarine volcanism resulting in the formation of large oceanic plateaus appear to have had a stimulatory rather than a disruptive effect on the biosphere (Vermeij, 1995). The release of CO_2 during these eruptions, which were often extremely large, may have been accompanied by relatively minor amounts of volcanic rust. Huge submarine volcanic eruptions are known or inferred for the Norian stage of the Late Triassic (Richards *et al.*, 1991), the latest Jurassic to early Cretaceous (Renne *et al.*, 1992; Sager and Han, 1993), the Aptian and Albian stages of the early to middle Cretaceous (Arthur *et al.*, 1985; Larson, 1991a; Tarduno *et al.*, 1991; Bralower *et al.*, 1994), the Turonian to Coniacian stages of the early late Cretaceous (Tarduno *et al.*, 1998), the end-Cretaceous (Coffin and Eldholm, 1993, 1994), the end-Paleocene (Rea *et al.*, 1990; Koch *et al.*, 1992; Zachos *et al.*, 1993), and the early Oligocene (Hoffmann *et al.*, 1997).

The Paleozoic history of volcanism remains largely unknown thanks to the tendency for volcanic terrains to erode rapidly and to leave little trace of their existence. Extensive submarine volcanism associated with continental rifting has been proposed for the Early Cambrian (Brasier, 1992) and the Late Devonian to Early Carboniferous (Racki, 1998). Continental flood-basalts are known for the Middle Permian (Garzanti, 1993).

12.4.3 *Uplift*

Strontium-isotopic ratios in carbonate rocks have been used as indicators of the extent of chemical weathering from the continents. The lighter isotope, ^{86}Sr, is more common in mantle-derived rock, introduced into the biosphere by hydrothermal activity, whereas the heavier isotope, ^{87}Sr, is preferentially liberated from continental rocks. High $^{87}Sr/^{86}Sr$ ratios are thought to indicate intense chemical weathering, and to be associated with large areas of uplifted or mountainous continental terrain. Maxima in this ratio are known for the Middle to earliest Late Cambrian (Montañez *et al.*, 1996), the Late Devonian, and the Late Cenozoic (see also Capo and DePaolo, 1990; Richter *et al.*, 1992; Ingram *et al.*, 1994; Kadko *et al.*, 1995). They correspond well with other evidence of extensive areas of high continental elevation (Hay *et al.*, 1988; Raymo and Ruddiman, 1992) and with continental collisions.

12.4.4 *Carbon dioxide*

Estimates of the concentration of CO_2 in the atmosphere, the role of volcanism in raising these concentrations, and relationships between temperature and atmospheric levels of CO_2 remain highly controversial topics. Calculations of temperatures and gas concentrations based on models of the carbon cycle are highly sensitive to estimates of the rate of basalt production (and therefore the rate of addition of CO_2 from the mantle to the atmosphere), the duration of major episodes of volcanism, and assumptions about how heat is transported in the atmosphere and oceans. Moreover, greenhouse gases other than or in addition to CO_2 may be important in controlling temperature. Many other assumptions surround the use of various geochemical indicators of the CO_2 content of seawater and air. The problem with these indicators, as well as with the models, is that there are too many variables influencing the sizes of sources and sinks, rates of formation and loss, concentrations of gases, and isotopic compositions of preserved minerals and organic matter. Not surprisingly, therefore, wide disagreements exist about the magnitude (if not the direction) of the effects of great volcanic eruptions on the ocean and atmosphere.

Past concentrations of CO_2 in the atmosphere can be expressed as multiples of the modern pre-industrial value of 280 parts per million per volume of air. Estimates of this multiple for the middle Cretaceous (Late Aptian to Cenomanian) range from 1.0 (Liu and Schmitt, 1996), 5 to 7 (Crowley, 1991a), and 3.4 to 14.7 (Coffin and Eldholm, 1994; Sellwood *et al.*, 1994; Landis *et al.*, 1996). Geochemical markers indicate a sudden sharp decrease in atmospheric concentrations at the Cenomanian–Turonian boundary, which is marked by an extinction event as well as by widespread burial of organic carbon (Kuypers *et al.*, 1999). Atmospheric concentrations of CO_2 following the end-Paleocene eruptions were somewhere between 1.5 and 6 times modern values, depending on which model or geochemical proxy is used (Zachos *et al.*, 1993; Liu and Schmitt, 1996). Although Early Eocene concentrations of CO_2 in the atmosphere may have been high, evidence from stomatal indices of tropical-forest leaves and especially from boron-isotopic ratios of open-ocean foraminifers in the Pacific indicate that concentrations by the Middle Eocene (43 Ma) had dropped to levels 1.3 to 2 or perhaps 3 times modern pre-industrial levels (McElwain, 1998; Pearson and Palmer, 1999). Data for the late Paleogene and Neogene generally indicate low levels of CO_2. Using an alkenone-based proxy, Pagani *et al.* (1999a, b) suggest a drop in CO_2 concentration from the late Oligocene to the early Miocene, followed by more or less uniformly low levels at or below modern pre-industrial concentrations until the early late Miocene, about 9 Ma (see also Flower, 1999). This Miocene interval of stability encompasses both an episode of warmth, when subtropical conditions extended very far north and south, and the middle Miocene cooling event at 14.5 to 14.1 Ma. The inferences of Pagani and colleagues conflict with Liu and Schmitt's analysis, which is based on the abundance of cesium relative to other elements in ocean water, another geochemical proxy for CO_2. Their work indicates a CO_2 concentration 1.5 times modern values at 17 Ma, when warm conditions were latitudinally extensive and at the peak eruption of the

Columbia Plateau flood-basalts. The low CO_2 levels postulated for much of the Miocene by Pagani and colleagues (1999a, b) are also at odds with inferences from the distribution of C_3 and C_4 photosynthetic pathways in land plants. C_4 plants are capable of photosynthesis at very low concentrations of CO_2, and begin to outcompete C_4 plants, which do better at higher CO_2 concentrations and lower temperatures, when the CO_2 level falls below 500 parts per million per volume of air, or 1.8 times modern values (see Cerling *et al.*, 1997).

C_4 plants existed 15 Ma during the middle Miocene, but they expanded significantly (as recorded by carbon-isotopic ratios in the tooth enamel of herbivorous mammals) during late Miocene to early Pliocene time (9 to 4 Ma) (Morgan *et al.*, 1994; Cerling *et al.*, 1997). If CO_2 is the dominant control on the distribution of C_4 plants – and this is a questionable conjecture (Morgan *et al.*, 1994; Pagani *et al.*, 1999b) – then the Miocene before about 9 Ma should have been characterized by carbon dioxide levels 1.8 times or more those prevailing in the Holocene. For the early Pliocene, a combination of isotopic data and densities of stomata in tree leaves implies that CO_2 levels in the relatively warm early half of the epoch were not more than 35% higher than modern pre-industrial values (van der Burgh *et al.*, 1993; Raymo *et al.*, 1996). Concentrations may have reached a low (180 parts per million) during the last glacial interval of the Pleistocene (Farquhar, 1997; Street-Perrot *et al.*, 1997; Petit *et al.*, 1999).

The general consensus for the Paleozoic era is that atmospheric concentrations of CO_2 were very high until the Late Silurian, and then fell very rapidly as vascular plants evolved and grew in stature (Holland, 1984; Berner, 1993, 1994, 1997; Mora *et al.*, 1996; Retallack, 1997). Land plants show a marked increase in the number of stomata relative to other surface cells in photosynthesizing stems and leaves, implying a reduction in CO_2 from Devonian to Carboniferous time (see Chaloner and McElwain, 1997, for a review). Just how high CO_2 levels were during the pre-Silurian Paleozoic remains uncertain. Arguing that chemical weathering by the newly evolved land plants was the chief mechanism by which CO_2 concentrations fell during the middle Paleozoic, various authors (Berner, 1993, 1994; Mora *et al.*, 1996; Algeo and Scheckler, 1998; Lenton, 1998) postulated Early Paleozoic concentrations 10 or more times higher than modern values. It is possible, however, that such levels are exaggerated. Although vascular plants weather soils more rapidly and more deeply than do mosses and lichens (Moulton and Berner, 1998), substantial chemical weathering due to microbial and cryptogamic communities was occurring long before the evolution of vascular plants (see, e.g., Jackson and Keller, 1970; Schwartzman, 1993; Horodyski and Knauth, 1994; Yapp and Poths, 1994). CO_2 concentrations were probably high before the Devonian, but the fall from Early to Middle Paleozoic times may have been much less dramatic than the current consensus indicates.

Inferences about oxygen concentrations in the atmosphere are even more indirect and problematic than those for CO_2. Data from stable oxygen isotopes, the occurrence of oxidized minerals, and the record of aerobic life itself nevertheless indicate the broad outlines of the history of oxygen in the atmosphere (Holland, 1984; Des Marais *et al.*, 1992; Des Marais, 1994; Landis *et al.*, 1996; Canfield, 1998). Briefly, major episodes of oxygen enrichment occurred during the Early Proterozoic (2.2 to 1.9 Ga), the Late Neoproterozoic (900 to 600 Ma) (see also Towe, 1970), the Devonian to Carboniferous interval of the Middle Paleozoic (see also Graham *et al.*, 1995; Dudley, 1998; Gans *et al.*, 1999), and the Jurassic and Cretaceous periods of the Mesozoic (see also Berner and Landis, 1988). Whereas Middle Devonian levels of oxygen were still perhaps only 15% of the volume of air (as compared to 21% today), those during the Carboniferous and Cretaceous may have been as high as 35%. Levels approaching this high level may have characterized much of the Cretaceous except for the last 5 My of that period, when there may have been a steep decline (Landis *et al.*, 1996). Oxygen may also have been more plentiful in the Eocene and Miocene atmosphere than it is today, although I know of no reliable estimates of its concentration during the Cenozoic.

12.4.5 Temperature

It is likely that the surface temperature of the tropical oceans as well as the latitudinal extent of the equatorial zone has varied through time, but the precise relationships among tropical temperatures, breadth of warm-climate zones, heat transport from equatorial to polar regions, and concentrations of greenhouse gases in the atmosphere are not well understood. Temporal trends must in any case be evaluated against spatial variations observable in the modern tropics. Subtropical or tropical conditions extend for 65° of latitude or more in the western Pacific and western Atlantic Oceans, but only about 30° in the eastern Pacific and eastern Atlantic. Sea-surface temperatures are warmer in the western Pacific than elsewhere in the tropics. It is therefore an oversimplification to think of the tropics as uniform at any single time, or to infer tropical conditions for particular time intervals on the basis of a few highly localized observations.

Inferences from the oxygen-isotopic record of foraminifers indicate that tropical sea-surface temperatures during the mid-Cretaceous were higher than those prevailing today. For the Aptian stage, Coffin and Eldholm (1994) suggest that the tropical ocean was anywhere from 2.8 to an unlikely 12°C above modern temperatures. The Late Albian and Cenomanian stages were probably not more than 2°C warmer in the tropical ocean than today (Sellwood *et al.*, 1994), but temperatures of 30 to 32°C (at the high end of modern western Pacific values) may have prevailed during this interval (Norris and Wilson, 1998). Isotopic data indicate a general warming, with occasional reversals, from the Aptian to the early Turonian, and a subsequent general cooling from the late Turonian to the end of the Cretaceous (Clarke and Jenkyns, 1999). Tropical surface temperatures during the Late Paleocene, Eocene, and Oligocene may have been slightly lower than today (Adams *et al.*, 1990; Crowley, 1991a; Zachos *et al.*, 1994).

Various indicators point to the poleward extension of tropical conditions during intervals of the Mesozoic and Cenozoic eras. Carbonate-dominated deposits on continental shelves, accumulating during warm times, occur over 65° of latitude in today's oceans. They extended over about 85° of latitude during the Aptian stage of the early Cretaceous, 75° during the Campanian stage of the Late Cretaceous, and 80° during the Late Eocene (Hay *et al.*, 1988). Particularly extensive poleward transport of heat, associated with warm polar climates, is indicated for the Turonian–Coniacian and Late Maastrichtian stages of the Late Cretaceous (Upchurch and Wolfe, 1987; D'Hondt and Arthur, 1996; Herman and Spicer, 1996; Otto-Bliesner and Upchurch, 1997; Li and Keller, 1998; Tarduno *et al.*, 1998), the early to middle Eocene (Adams *et al.*, 1990; Crowley, 1991a), the first half of the Miocene (Kafanov and Volvenko, 1997), and the early Pliocene (Crowley, 1991b; Raymo *et al.*, 1996; see also Hudson and Anderson, 1989, for a general review of temperature through time). During the Late Maastrichtian, Late Paleocene to Middle Eocene, and early Pliocene at least, latitudinally extensive warm conditions may have been accompanied by tropical sea-surface temperatures that were slightly below those prevailing in the modern ocean (Dowsett and Poore, 1991; Zachos *et al.*, 1994; D'Hondt and Arthur, 1996).

Paleozoic conditions are again problematic. Isotopically derived estimates, variously massaged with corrections, imply that tropical Devonian temperatures in North America were around 25°C, but estimates are accompanied by considerable uncertainty (Gao, 1993). Late Ordovician brachiopod shells contain isotopic signatures indicating temperatures of somewhere between 13 and 20°C in the North American equatorial belt (Railsback *et al.*, 1989). The coal forests of the equatorial latitudes of Europe and North America during the Carboniferous flourished under aseasonal conditions (Chaloner and McElwain, 1997), but whether the tropics were as warm then as they are today is doubtful (see Beerling *et al.*, 1998). The presence of reefs during the Ordovician to Devonian has long been taken to imply tropical conditions similar to those prevailing in modern regions where reefs are common (see, e.g., Crowley and North, 1991). Reefs, like rain forests, can exist at tempera-

tures at or even below 20°C. I suspect strongly that the Middle Paleozoic tropics were generally cooler than those of the Late Cenozoic. Relatively low temperatures are compatible with slow decay and widespread coal formation in Carboniferous coal swamp forests, as well as with the Mesozoic and Cenozoic geographical and habitat restriction of many shallow-water Paleozoic clades (brachiopods, crinoids, hexactinellid sponges, etc.) to cool, deep ocean waters (see Vermeij, 1987).

12.5 The record of realized evolutionary opportunity

How does the record of physical history of the atmosphere, ocean, and crust mesh with the evolutionary history of innovation and diversification? Do increases in temperature and in the width of warm zones indeed lead to a greater scope of adaptation and therefore to an expanded evolutionary realm?

Uncertainties in the timing of both physical and biological events precludes a complete answer to this question. Two strategies are available for finding the answers. First, we can identify all intervals of warming or of tropical expansion and ask whether they encompass a disproportionate number of evolutionary innovations or episodes of diversification in comparison to intervals of cooling or tropical contraction. Second, we can compile a list of innovations or diversifications and ask if they cluster in intervals or warming or poleward expansions of tropical zones. To do this properly even with the available evidence would require a book-length treatment. I shall therefore confine myself to a bare-bones outline of this important topic.

There is general agreement that the early Earth was warm, so warm in fact that life may not have been possible. The earliest forms of life either evolved in, or at least tolerated, conditions of very high temperatures, in the range of 60 to 70°C (Schwartzman *et al.*, 1993; Schwartzman, 1998; Nisbet and Fowler, 1996). The complex partnerships exemplified by eukaryotic and multicellular organisms may not have become possible until temperatures no longer routinely exceeded 50°C (Schwartzman, 1998). Glaciations, possibly of global extent, occurred as early as the early Proterozoic, near 2.2 Ga (Evans *et al.*, 1997), and in the Late Neoproterozoic (Kaufman *et al.*, 1997), implying that a temperature regime broadly comparable to that of the Phanerozoic eon was established very early.

The time of origin of animals and their various body plans remains a highly controversial topic, but present evidence does not conflict with the possibility that it occurred during Late Neoproterozoic time, perhaps 850 or more likely 700 Ma or a little later (Vermeij, 1996; Lynch, 1999; Valentine *et al.*, 1999). This timing would correspond broadly to continental rifting and to an increase in oxygen concentration in the atmosphere (see also Kadko *et al.*, 1995; Knoll, 1996; Canfield, 1998). Other crown-group metazoans, however, arose during the Cambrian (Knoll and Carroll, 1999).

The mineralization, antipredatory evolution, deeper burrowing by animals, and increase in body sizes of living things began in the latest Proterozoic (about 600 Ma) and culminated in the Cambrian explosion, dated by Valentine and colleagues (1999) at 530 to 520 Ma. Brasier (1992) and Pelechaty (1996) link this time interval with rifting and submarine volcanism. Other indicators point to an increase in chemical weathering, productivity, and sea level (see also Cook, 1992; Hallam, 1992; Richter *et al.*, 1992; Montañez *et al.*, 1996). The latest Neoproterozoic to Early Cambrian interval may have uniquely combined an initially low sea level with high concentrations of CO_2, rising oxygen levels and temperatures, and a high nutrient supply.

The evolutionary opportunities manifested by the large-scale diversification of marine life from the Late Cambrian to at least the Middle Ordovician (Miller and Foote, 1996) coincide with generally rising sea levels and with continental rifting (Hallam, 1992, 1994). Although strontium-isotopic ratios indicate a globally reduced intensity of chemical weathering after the earliest Late Cambrian (Richter *et al.*, 1992; Montañez *et al.*, 1996), extensive volcanism may have released plentiful nutrients during much

of the Cambrian–Ordovician interval. Miller and Mao (1995) argued that the highest marine diversity during the Ordovician was to be found in volcanically active regions.

The major Mid-Paleozoic (Silurian to Early Carboniferous) events – evolution in the sea of predators and antipredatory defenses, the rapid diversification of life on land – are correlated with rises in sea level and, beginning in the Early Devonian, with increases in oxygen and decreases in CO_2 concentrations in the atmosphere. Although Hallam (1994) recorded various continental collisions and episodes of mountain-building beginning in the Late Ordovician, culminating in the assembly of the supercontinent Pangaea in the Permian, there were also continental rifting and hydrothermal activity during the Late Devonian and Early Carboniferous (Racki, 1998). The increases in temperature, widening of warm belts, rising oxygen, and probable increases in primary productivity associated with this rifting and hydrothermal release of nutrients may have provided the evolutionary opportunities manifested in the Early Carboniferous establishment of tetrapod vertebrates on land and the Middle Carboniferous (Namurian) evolution of flight in insects.

The Late Paleozoic (Late Carboniferous and Permian) was a time of few notable innovations in the sea but important evolutionary events on land. Raymond and her colleagues (1990) noted high rates of origination among tropical brachiopod clades during the late Middle Carboniferous (Namurian B) immediately following equatorial warming and the beginning of southern-hemisphere glaciation during Namurian A. There was a sharp rise in the consumption of living land plants by arthropods and tetrapods during the latest Carboniferous (Stephanian stage) (see Shear, 1991; Labandeira, 1998). This jump is more or less contemporaneous with a rise in CO_2 inferred by Cleal and colleagues (1999) from the stomatal index of seed-fern fronds, and with the poleward spread of relatively warm conditions. Bakker (1980) noted that the earliest tetrapod herbivores are from warm, low latitudes, where they presumably evolved, and that tetrapods penetrated to higher latitudes during the Early Permian. The Early Permian witnessed further increases in herbivory (Beck and Labandeira, 1998) as well as increased metabolic rates among tetrapods, culminating in the Early Triassic in endothermy in some groups (Bakker, 1980; Hillenius, 1994). It remains unknown what if any physical circumstances triggered or favored these trends.

Some authors view the end-Permian catastrophe as having provided unprecedented evolutionary opportunities, but the full realization of this evolutionary potential came only after the Triassic. Biotic events in the Jurassic and Cretaceous periods – evolution of deep-burrowing animals, new groups of predators, antipredatory defenses, mineralized single-celled planktonic organisms, flowering plants, and social insects – took place against a backdrop of generally rising sea levels, broad warm zones, high CO_2 and rising oxygen concentrations, massive submarine volcanism, and tectonic separation of large continental blocks (Vermeij, 1987, 1995). One of the times of most rapid change in the plankton and in the spread of flowering plants was the Aptian–Albian interval of the early Cretaceous (Kemper, 1982; Crane et al., 1995), when the largest known episode of submarine volcanism took place, with attendant warming worldwide.

Details of timing of tectonic and biological events are for the most part unavailable for the Middle Paleozoic. Raymond and colleagues (1990), however, noted high rates of origination of tropical brachiopod clades during the Middle Carboniferous (Namurian B), immediately following equatorial warming and the beginning of southern-hemisphere glaciation during Namurian A. The Namurian was also a time when flying insects first appeared.

The Middle Carboniferous to Permian, or Late Paleozoic, was a time of few noteworthy innovations among marine organisms, but several important evolutionary events took place on land. During the latest Carboniferous (Stephanian), there was a sharp rise in the consumption of land plants by arthropods and vertebrates (Shear, 1991; Labandeira, 1998). The Early Permian witnessed further increases in herbivory (Beck and Labandeira, 1998), as well as increased metabolic rates among tetra-

pod vertebrates, culminating in the evolution of endothermy in the Early Triassic. The evolution of insect flight in the Middle Carboniferous has been interpreted as being made possible by a rise in atmospheric oxygen (Graham et al., 1995; Dudley, 1998; Gans et al., 1999). The vertebrate increases in metabolism and the rise of herbivory, however, significantly postdate the Devonian to Early Carboniferous rise in oxygen. Bakker (1980) noted that the earliest tetrapod herbivores are from low latitudes, and that it was only during the Permian when tetrapods penetrated to more seasonal climates at higher latitudes. It is also interesting that the jump in herbivory during the latest Carboniferous is more or less contemporaneous with a rise in CO_2, as inferred from data by Cleal et al. (1999) on the stomatal indices of seed-fern fronds, and a poleward spread of warm conditions.

The end-Permian catastrophe may have provided unprecedented evolutionary opportunities, but the full realization of those opportunities came only after the Triassic. The Jurassic and Cretaceous diversification, evolution of burrowers, escalation between predators and prey, rise of flowering plants and social insects, development of mineralized phytoplankters and photosynthesizing zooplankters, and many other biological innovations (Vermeij, 1977, 1987) took place against a backdrop of generally rising sea levels, broad warm zones, high carbon dioxide and increasing oxygen concentrations, and tectonic separation of major continental blocks. There were extraordinary episodes of submarine volcanism during both the Jurassic and Cretaceous periods (for a review see Vermeij, 1987, 1995). Again, the detailed temporal correspondence between geological and biological events remains to be worked out. It is interesting, however, that Kemper (1982) emphasized the Aptian–Albian as a time of profound change in the marine plankton without referring to the great Aptian volcanic outpourings beneath the sea, and that the modern angiosperm-dominated vegetation was first in evidence in the Late Albian (see, e.g., Crane et al., 1995).

Both biologically and tectonically, the Cenozoic era is to the later Mesozoic as the

Middle to Late Paleozoic is to the Early Paleozoic. The Cenozoic consolidation of innovations and trends begun during the Mesozoic involved further escalation between marine predators and prey, the rise of social insects, extensive diversification on land and in the sea, and the emergence late in the era of intelligent animals (Vermeij, 1987). These and other biological events are concentrated during the early to middle Eocene and the late Oligocene to early Miocene, with some additional diversification and the evolution of hominids taking place during the early Pliocene. All these intervals are characterized by broad warm belts. There was relatively rapid seafloor spreading in the early Eocene and late Oligocene to early Miocene. Warming during the late Oligocene to Miocene and the early Pliocene was accompanied by diversification of marine life worldwide, but not by the predicted higher concentrations of CO_2 such as are found during the early Eocene. In correspondence with this history of atmospheric CO_2, submarine volcanism was important during the Cenozoic only near the end of the Paleocene and earliest Eocene. Continental rifting occurred during some later intervals such as the Oligocene as well (in the Rift Valley of Africa and the Red Sea, for example), but the dominant tectonics of the Cenozoic involved collisions rather than the dispersal of land masses (Hallam, 1994). The Pleistocene, an interval of conspicuously little evolution (Bennett, 1997), is marked by an alternation between glacial episodes, when temperatures worldwide were low and CO_2 concentrations reached perhaps unprecedented lows, and globally warmer interglacials during which CO_2 levels climbed (see Broecker, 1996; Colinvaux et al., 1996; Lee and Slowey, 1999; McCulloch et al., 1999; Petit et al., 1999).

This record, preliminary and incomplete as it is, seems to indicate that episodes of diversification and evolutionary innovation broadly coincide with intervals of high temperatures or broad tropical belts, or both. Submarine volcanism and its attendant tectonic phenomena, including the release of large amounts of CO_2 into the atmosphere, seem to be associated with the most dramatic evolutionary events in the sea (e.g. during the Cretaceous), but do not

account for all instances of temperature rise, nor are all episodes of diversification and innovation triggered by such volcanism. The most dramatic evolutionary changes on the dry land sometimes occurred when CO_2 concentrations and sea levels were high (Middle Paleozoic and Middle Mesozoic), but during the Late Paleozoic they may have taken place in (and perhaps caused) an atmosphere of increasingly scarce CO_2.

Not every episode of warming precipitated evolutionary ferment. This point is well illustrated by the periodic global warming characteristic of the Pleistocene interglacials, and by the warming that took place during the evolutionarily quiescent Early Triassic, after the end-Permian extinctions (Retallack et al., 1996). A major challenge for future investigators will be to understand what circumstances in addition to warming transform evolutionary potential into evolutionary reality.

One interesting possibility that merits serious consideration is that equatorial zones since early Phanerozoic time have become warmer. Patterns of habitat and geographical restriction show that clades found in shallow equatorial waters during the Paleozoic and sometimes the early Mesozoic have since become confined to cooler waters, either at greater depths in the ocean or at higher latitudes (Vermeij, 1987). This restriction undoubtedly has a biological component in that new, energy-intensive competitors and predators originating in warm regions have collectively pushed earlier incumbents to more marginal environments (Vermeij, 1987), but it is also consistent with long-term tropical warming. Thermal conservatism is very widespread among plants and animals, meaning that modern survivors live in temperature regimes not very different from those of their shallow-water tropical Paleozoic ancestors.

My suggestion that major Phanerozoic evolutionary events are triggered by processes that cause temperatures to rise, and that the equatorial zone has generally warmed over the course of the Phanerozoic, conflicts with proposals that increased plant cover has cooled the Earth in steps. According to the cooling scenario, the successive evolution of photosynthesizing microbes during the Archean, eukaryotes during the Proterozoic, vascular plants during the Middle Paleozoic, and flowering plants during the Middle Mesozoic brought about stepwise surface cooling of the Earth through increased rates of water-vapor transpiration and chemical weathering (Schwartzman *et al.*, 1993; Schwartzman, 1998; Lenton, 1998). Without the increased cover and activity of land plants, the increase in luminosity of the Sun by 25% since the early Archean (3.8 Ga) would have caused the surface of the Earth to warm by an estimated 18°C (see Lenton, 1998). Although plants have undoubtedly modified Earth's environment greatly, as indeed have other forms of life, I am not convinced that they have entirely erased the heating effect of the increasing solar luminosity. Schwartzman and colleagues (1993) supported their cooling scenario by arguing that glaciation may be only a Late Cenozoic phenomenon. Most authors, however, agree that there is good evidence for glaciation during the Early and Late Proterozoic and the Late Paleozoic (see, e.g., Crowley and North, 1991; Evans *et al.*, 1997; Kaufman *et al.*, 1997). Another possibility is that tropical temperatures increased even as the average temperature of the Earth's surface declined. In this connection it is important to remember that evolution is a local process dependent on local, not globally average, conditions.

12.6 Conclusion and questions

How important is temperature as a factor facilitating evolution? The answer to this question is neither short nor simple. Temperature, nutrients, and oxygen levels all profoundly affect evolution, but their effects cannot, and probably should not, be separated. Feedbacks causally interconnect all three of these factors, as well as the biological controls on evolution, such as competition, predation, migration among ecosystems and among geographical regions, symbioses, exploitation of environments in which nutrients were previously protected from recycling by organisms, and a host of other phenomena. Cause and effect are inextricably intertwined in evolution, and in historical processes generally. Temperature must play a fundamental role in evolution simply because it affects every biological, chemical, and physical process that controls the supply, demand, and availability of resources. However, the mechanisms by which temperatures are raised or lowered must be understood if the role of temperature in evolution is to be evaluated properly. A simple increase in temperature will not suffice to increase evolutionary opportunity or to expand adaptive options. Similarly, an increase in the supply of nutrients by itself is insufficient as an evolutionary stimulus. Instead, the causal linkage between these two factors, as well as with the concentration and availability of oxygen, provides the mechanism by which physical factors affect fitness and opportunity. This linkage, in turn, arises in part from the tectonic control of conditions in the atmosphere, ocean, crust, and biosphere.

Many questions about physical triggers in evolution remain unanswered. Is the potential for evolution increased by rising temperature or by expansions of warm zones, or both? Do times of extraordinary warmth coincide with times of broad warm zones, or are they separate? Are there thresholds in temperature, gas concentration, and sea level above which increases in these quantities would have little effect? In general, should we be more concerned with absolute values or with trends? Are slow rises in temperature or sea level as effective as rapid rises? Were evolutionary events on land occurring simultaneously with those in the sea? As a contribution that paleobiology can make to our understanding of how humans are changing the planet, answers to these and other questions raised in this chapter will become central targets of biological and evolutionary research during this century.

Acknowledgements

This chapter is a contribution from the Center for Population Biology, University of California at Davis. Support from NSF Grant 97-06749 is gratefully acknowledged.

References

Adams, C.J., Lee, D.E. and Rosen, B.R. (1990). Conflicting isotopic and biotic evidence for tropical sea-surface temperatures during the Tertiary. *Palaeogeography, Palaeoclimatology, Palaeoecology*, 77: 289–313.

Algeo, T.J. and Scheckler, S.E. (1998). Terrestrial-marine teleconnections in the Devonian: links between the evolution of land plants, weathering processes, and marine anoxic events. *Philosophical Transactions of the Royal Society of London B*, 353: 113–130.

Arthur, M.A., Dean, W.E. and Schlanger, S.O. (1985). Variations in the global carbon cycle during the Cretaceous related to climate, volcanism, and changes in atmospheric CO_2. *Geophysical Monographs*, 32: 504–529.

Baker, E.T., Massoth, G.J. and Feely, R.A. (1987). Cataclysmic hydrothermal venting on the Juan de Fuca Ridge. *Nature*, 329: 149–151.

Bakker, R.T. (1980). Dinosaur heresy – dinosaur renaissance: why we need endothermic archosaurs for a comprehensive evolution of bioenergetic evolution. In: R.D.K. Thomas and E.C. Olson (eds) *A Cold Look at the Warm-blooded Dinosaurs*. Boulder, Colorado: Westview Press, pp. 351–462.

Beck, A.L. and Labandeira, C.C. (1998). Early Permian insect folivory on a gigantopterid-dominated riparian flora from north-central Texas. *Palaeogeography, Palaeoclimatology, Palaeoecology*, 142: 139–173.

Beerling, D.J., Woodward, F.I., Lomas, M.R. *et al.* (1998). Influence of Carboniferous palaeoatmospheres on plant function: an experimental and modeling assessment. *Philosophical Transactions of the Royal Society of London B*, 353: 131–140.

Bennett, K.D. (1997). *Evolution and Ecology: The Pace of Life. Cambridge*: Cambridge: Cambridge University Press.

Berner, R.A. (1992). Weathering, plants, and the long-term carbon cycle. *Geochimica et Cosmochimica Acta*, 56: 3225–3231.

Berner, R.A. (1993). Paleozoic atmospheric CO_2: importance of solar radiation and plant evolution. *Science*, 261: 68–70.

Berner, R.A. (1994). 3GEOCARB II: a revised model of atmospheric CO_2 over Phanerozoic time. *American Journal of Science*, 294: 56–91.

Berner, R.A. (1997). The rise of plants and their effect on weathering and atmospheric CO_2. *Science*, 276: 544–546.

Berner, R.A. and Canfield, D. (1989). A new model for atmospheric oxygen over Phanerozoic time. *American Journal of Science*, 289: 333–361.

Berner, R.A. and Landis, G.P. (1988). Gas bubbles in fossil amber as possible indicators of the major gas composition of ancient air. *Science*, 239: 1406–1409.

Berner, R.A. and Raiswell, R. (1983). Burial of organic carbon and pyrite sulfur in sediments over Phanerozoic time: a new theory. *Geochimica et Cosmochimica Acta*, 47: 855–862.

Berry, W.B.N. and Wilde, P. (1978). Progressive ventilation of the oceans: an explanation for the distribution of the Lower Paleozoic black shales. *American Journal of Science*, 278: 257–275.

Bowring, S.A., Erwin, D.H., Jin, Y.G. *et al.* (1998). U/Pb zircon geochronology and tempo of the end-Permian mass extinction. *Science*, 280: 1039–1045.

Bralower, T.J., Arthur, M.A., Leckie, R.M. *et al.* (1994). Timing and paleoceanography of oceanic dysoxia/anoxia in the Late Barremian to Early Aptian (Early Cretaceous). *Palaios*, 9: 335–369.

Bralower, T.J., Thomas, D.J., Zachos, J.C. *et al.* (1997). High-resolution records of the Late Paleocene thermal maximum and circum-Caribbean volcanism: is there a causal link? *Geology*, 25: 963–966.

Brasier, M.D. (1992). Paleoceanography and changes in the biological cycling of phosphorus across the Precambrian–Cambrian boundary. In: J.H. Lipps and P.W. Signor (eds) *Origin and Early Evolution of the Metazoa*. New York: Plenum, pp. 483–523.

Brass, G.W., Saltzman, E., Sloan, J.L. II *et al.* (1982). Ocean circulation, plate tectonics, and climate. In W.H. Berger and J.C. Crowell (eds) *Climate in Earth History*. Washington: National Academy Press, pp. 76–82

Broecker, W. (1996). Glacial climate in the tropics. *Science*, 272: 1902–1903.

Burgh, J. van der, Visscher, H., Dilcher, D.R. and Kürschner, W.M. (1993). Paleoatmospheric signatures in Neogene fossil leaves. *Science*, 260: 1788–1790.

Campbell, I.H., Czamanske, G.K., Fedorenko, V.A. *et al.* (1992). Synchronism of the Siberian Traps and the Permian–Triassic boundary. *Science*, 258: 1760–1763.

Canfield, D.E. (1998). A new model of Proterozoic ocean chemistry. *Nature*, **396**: 450–453.

Canfield, D.E. and Teske, A. (1996). Late Proterozoic rise in atmospheric oxygen concentration inferred from phylogenetic and sulfur-isotope studies. *Nature*, **382**: 127–132.

Capo, R.C. and DePaolo, D.J. (1990). Seawater strontium isotopic variations from 2.5 million years ago to the present. *Science*, **249**: 51–55.

Cerling, T.E., Harris, J.M., McFadden, B.J. *et al.* (1997). Global vegetation change through the Miocene/Pliocene boundary. *Nature*, **388**: 153–158.

Chadwick, O.A., Derry, L.A., Vitousek, P.M *et al.* (1999). Changing sources of nutrients during four million years of ecosystem development. *Nature*, **397**: 491–497.

Chaloner, W.G. and McElwain, J. (1997). The fossil plant record and global climatic change. *Review of Paleobotany and Palynology*, **95**: 73–82.

Clarke, A. (1983). Life in cold water: the physiological ecology of polar marine ectotherms. *Oceanography and Marine Biology Annual Review*, **21**: 341–453.

Clarke, A. (1993). Temperature and extinction in the sea: a physiologist's view. *Paleobiology*, **19**: 499–518.

Clarke, L.J. and Jenkyns, H.C. (1999). New oxygen isotope evidence for long-term Cretaceous climate change in the southern hemisphere. *Geology*, **26**: 699–702.

Cleal, C.J., James, R.M. and Zodrow, E.L. (1999). Variation in stomatal density in the Late Carboniferous gymnosperm frond *Neuropteris ovata*. *Palaios*, **14**: 180–185.

Coale, K.H., Fitzwater, S.E., Gordon, R.M. *et al.* (1996). Control of community growth and export production by upwelled iron in the equatorial Pacific Ocean. *Nature*, **379**: 621–624.

Coffin, M.F. and Eldholm, O. (1993). Scratching the surface: estimating the dimensions of large igneous provinces. *Geology*, **21**: 515–519.

Coffin, M.F. and Eldholm, O. (1994). Large igneous provinces: crustal structure, dimensions, and external consequences. *Reviews of Geophysics*, **32**: 1–36.

Coley, P.D. and Barone, J.A. (1996). Herbivory and plant defenses in tropical forests. *Annual Review of Ecology and Systematics*, **27**: 305–335.

Colinvaux, P.A., Oliveira, P.E. de, Moreno, J.E. *et al.* (1996). A long pollen record from lowland Amazonia: forest and cooling in glacial times. *Science*, **274**: 85–88.

Cook, P.J. (1992). Phosphogenesis around the Proterozoic–Phanerozoic transition. *Journal of the Geological Society of London*, **149**: 615–620.

Courtillot, V., Besse, J., Vandamme, D. *et al.* (1986). Deccan flood basalts at the Cretaceous/Tertiary boundary? *Earth and Planetary Science Letters*, **80**: 361–374.

Crane P.R., Friis E.M. and Pedersen K.R. (1995). The origin and early diversification of angiosperms. *Nature*, **374**: 27–33.

Crowley, T.J. (1991a). Past CO_2 changes and tropical sea surface temperatures. *Paleoceanography*, **6**: 387–394.

Crowley, T.J. (1991b). Modeling Pliocene warmth. *Quaternary Science Reviews*, **10**: 275–282.

Crowley, T.J. and North, J.R. (1991). *Paleoclimatology*. New York: Oxford University Press.

Crowley, T.J., Mengel, J.G. and Short, D.A. (1987). Gondwanaland's seasonal cycle. *Nature*, **329**: 803–807.

DeAngelis, D.L. (1992). *Dynamics of Nutrient Cycling and Food Webs*. London: Chapman & Hall.

DeLucia, E.H., Hamilton, J.G., Naidu, S.L. *et al.* (1999). Net primary production of a forest ecosystem with experimental CO_2 enrichment. *Science*, **284**: 1177–1179.

Denny, M.W. (1993). *Air and Water: The Biology and Physics of Life's Media*. Princeton: Princeton University Press.

Des Marais, D.J. (1994). Tectonic control of the crustal organic carbon reservoir during the Precambrian. *Chemical Geology*, **114**: 303–314.

Des Marais, D.J., Strauss, H., Summons, R.E. and Hayes, J.M. (1992). Carbon isotope evidence for the stepwise oxidation of the Proterozoic environment. *Nature*, **359**: 605–609.

D'Hondt, S. and Arthur, M.A. (1996). Late Cretaceous oceans and the cool tropic paradox. *Science*, **271**: 1838–1841.

D'Hondt, S., Donaghay, P., Zachos, J.C. *et al.* (1998). Organic carbon fluxes and ecological recovery from the Cretaceous–Tertiary mass extinction. *Science*, **282**: 276–279.

Dowsett, H.J. and Poore, R.Z. (1991). Pliocene sea surface temperatures of the North Atlantic Ocean 3.0 Ma. *Quaternary Science Reviews*, **10**: 189–204.

Dowsett, H.J., Cronin, T.M., Poore, R.Z. *et al.* (1992). Micropaleontological evidence for increased meridional heat transport in the North Atlantic Ocean during the Pliocene. *Science*, **258**: 1133–1135.

Duarte, C.M. and Agustí, S. (1998). The CO_2 balance of unproductive aquatic ecosystems. *Science*, **281**: 234–236.

Dudley, R. (1998). Atmospheric oxygen, giant

Paleozoic insects and the evolution of aerial locomotor performance. _Journal of Experimental Biology_, **201**: 1043–1050.

Dugdale, R.C. and Wilkerson, F.P. (1998). Silicate regulation of new production in the equatorial Pacific upwelling. _Nature_, **391**: 270–273.

Edmond, J.M. (1992). Himalayan tectonics, weathering processes, and the strontium isotope record in marine limestones. _Science_, **258**: 1594–1597.

Erwin, D.H. and Vogel, T.A. (1992). Testing for causal relationships between large pyroclastic volcanic eruptions and mass extinctions. _Geophysical Research Letters_, **19**: 893–896.

Evans, D.A., Beukes, N.G. and Kirschvink, J.L. (1997). Low-latitude glaciation in the Palaeoproterozoic era. _Nature_, **386**: 262–266.

Farquhar, G.D. (1997). Carbon dioxide and vegetation. _Science_, **278**: 1411.

Filippelli, G.M. (1997). Intensification of the Asian monsoon and a chemical weathering event in the Late Miocene–Early Pliocene: implications for Late Neogene climate change. _Geology_, **25**: 27–30.

Fischer, A.G. and Arthur, M.A. (1977). Secular variations in pelagic realm. In H.E. Cook and P. Enos (eds) _Deep-water Carbonate Environments_. SEPM Special Publication **25**: 19–50.

Flower, B. (1999). Warming without CO_2? _Nature_, **399**: 313–314.

François, L.M., Walker, J.C.G. and Opdyke, B.N. (1993). The history of global weathering and the chemical evolution of the ocean–atmosphere system. _Geophysical Monograph_, **74**: 143–159.

Froelich, P.N., Bender, M.L., Luedtke, N.A. _et al._ (1982). The marine phosphorus cycle. _American Journal of Science_, **282**: 474–511.

Gans, C., Dudley, R., Aguilar, N.M. and Graham, J.B. (1999). Late Paleozoic atmosphere and biotic evolution. _Historical Biology_, **13**: 199–219.

Gao, G. (1993). The temperatures and oxygen-isotope composition of Early Devonian oceans. _Nature_, **361**: 712–714.

Garzanti, E. (1993). Himalayan ironstones, 'superplumes,' and the breakup of Gondwana. _Geology_, **21**: 105–108.

Graham, J.B., Dudley, R., Aguilar, N.L. and Gans, C. (1995). Implications of the Late Palaeozoic oxygen pulse for physiology and evolution. _Nature_, **375**: 117–120.

Graus, R.R. (1974). Latitudinal trends in the shell characteristics of marine gastropods. _Lethaia_, **7**: 303–314.

Hallam, A. (1981). _Facies Interpretation and the Stratigraphic Record_. Oxford: W.H. Freeman.

Hallam, A. (1992). _Phanerozoic Sea-level Changes_. New York: Columbia University Press.

Hallam, A. (1994). _An Outline of Phanerozoic Biogeography_. Oxford: Oxford University Press.

Hallock, P. and Schlager, W. (1986). Nutrient excess and the demise of coral reefs and carbonate platforms. _Palaios_, **1**: 389–398.

Haq, B.U., Hardenbol, J.R. and Vail, P. (1987). Chronology of fluctuating sea levels since the Triassic. _Science_, **235**: 1156–1167.

Hay, W.W., Rosol, M.J., Sloan, J.L. II and Jory, D.E. (1988). Plate tectonic control of global patterns of detrital and carbonate sedimentation. In L.J. Doyle and H.H. Roberts (eds) _Carbonate–clastic Transitions_. Amsterdam: Elsevier, pp. 1–34.

Heinrich, B. (1993). _The Hot-blooded Insects: Strategies and Mechanism of Thermoregulation_. Cambridge: Harvard University Press.

Herman, A.B. and Spicer, R.A. (1996). Palaeobotanical evidence for a warm Cretaceous Arctic Ocean. _Nature_, **380**: 330–333.

Hillenius, W.J. (1994) Turbinates in therapsids – evidence for Late Permian origins of mammalian endothermy. _Evolution_, **48**: 207–229.

Hochachka, P.W. and Somero, G.N. (1973). _Strategies of Biochemical Adaptation_. Philadelphia: Saunders.

Hoffman, A., Gruszczynski, M. and Malkowski, K. (1991). On the interrelationship between temporal trends in $\delta^{13}C$, $\delta^{16}O$, and $\delta^{34}S$ in the world ocean. _Journal of Geology_, **99**: 355–370.

Hoffmann, C., Courtillot, V., Feraud, F. _et al._ (1997). Timing of the Ethiopian flood basalt event and implications for plume birth and global change. _Nature_, **389**: 838–841.

Holeton, G.F. (1974). Metabolic cold adaptation of polar fish: fact or artifact? _Physiological Zoology_, **47**: 117–152.

Holland, H.D. (1984). _The Chemical Evolution of the Atmosphere and Oceans_. Princeton: Princeton University Press.

Hooper, P.R. (1990). The timing of crustal extension and the eruption of continental flood basalts. _Nature_, **345**: 246–249.

Horodyski, R.J. and Knauth, L.P. (1994). Life on land in the Precambrian. _Science_, **263**: 494–498.

Hudson, J.D. and Anderson, T.F. (1989). Ocean temperatures and isotopic compositions through time. _Transactions of the Royal Society of Edinburgh (Earth Sciences)_, **80**: 183–192.

Ingram, B.L., Coccioni, R., Montanari, A. and Richter, F.M. (1994). Strontium isotopic composition of mid-Cretaceous seawater. _Science_, 264: 546–550.

Jablonski, D. (1993). The tropics as a source of evolutionary novelty through geological time. *Nature*, **364**: 142–144.

Jackson, T.A. and Keller, W.D. (1970). A comparative study of the role of lichens and 'inorganic' processes in the chemical weathering of Recent Hawaiian lava flows. *American Journal of Science*, **269**: 446–466.

Jeanne, R.L. (1979). A latitudinal gradient in rates of ant predation. *Ecology*, **60**: 1211–1222.

Johnson, K.S., Chavez, F.P. and Friedrich, G.E. (1999). Continental-shelf sediment as a source of iron for coastal phytoplankton. *Nature*, **398**: 697–700.

Kadko, D., Baross, J. and Alt, J. (1995). The magnitude and global implications of hydrothermal flux. *Geophysical Monograph*, **91**: 446–466.

Kafanov, A.I. and Volvenko, I.V. (1997). Bivalve molluscs and Cenozoic paleoclimatic events in the northwestern Pacific Ocean. *Palaeogeography, Palaeoclimatology, Palaeoecology*, **129**: 119–153.

Kaiho, K. and Saito, S. (1994). Oceanic crust production and climate during the last 100 Myr. *Terra Nova*, **6**: 376–384.

Kaufman, A.J., Knoll, A.H. and Narbonne, G.M. (1997). Isotopes, ice ages, and terminal Proterozoic Earth history. *Proceedings of the National Academy of Sciences of the United States of America*, **94**: 6600–6605.

Kemper, E. (1982). Apt und Alb – Beginn einer neuen Zeit. *Geologisches Jahrbuch (A)*, **65**: 681–693.

Knoll, A.H. (1996). Breathing room for early animals. *Nature*, **382**: 111–112.

Knoll, A.H. and Carroll, S.B. (1999). Early animal evolution: emerging views from comparative biology and geology. *Science*, **284**: 2129–2137.

Koch, P.L., Zachos, J.C. and Gingerich, P.D. (1992). Correlation between isotope records in marine and continental carbon reservoirs near the Paleocene/Eocene boundary. *Nature*, **358**: 319–322.

Krantz, D.E. (1991). A chronology of Pliocene sea-level fluctuations: the U.S. Middle Atlantic Coastal Plain record. *Quaternary Science Reviews*, **10**: 163–174.

Kuypers, M.M.M., Pancost, R.D. and Damste, J.S.S. (1999). A large and abrupt fall in atmospheric CO_2 concentration during Cretaceous times. *Nature*, **399**: 342–345.

Labandeira, C.C. (1998). Early history of arthropod and vascular plant associations. *Annual Reviews of Earth and Planetary Science*, **26**: 329–377.

Landis, G.P., Rigby, J.K., Sloan, R.E. *et al.* (1996). Pele hypothesis: ancient atmospheres and geologic-geochemical controls on evolution, survival, and

extinction. In: N. McLeod and G. Keller (eds) *Cretaceous–Tertiary Mass Extinctions: Biotic and Environmental Changes*. London: Norton, pp. 519–556.

Larson, R.L. (1991a). Latest pulse of Earth: evidence for a mid-Cretaceous superplume. *Geology*, **19**: 547–550.

Larson, R.L. (1991b). Geological consequences of superplumes. *Geology*, **19**: 963–966.

Lee K.E. and Slowey N.C. (1999). Cool surface waters of the subtropical North Pacific during the last glacial. *Nature*, **397**: 512–514.

Lenton, T.M. (1998). Gaia and natural selection. *Nature*, **394**: 439–447.

Li, L. and Keller, G. (1998). Abrupt deep-sea warming at the end of the Cretaceous. *Geology*, **26**: 995–998.

Liu, Y.G. and Schmitt, R.A. (1996). Cretaceous Tertiary phenomena in the context of seafloor rearrangements and $p(CO_2)$ fluctuations over the past 200 m.y. *Geochimica et Cosmochimica Acta*, **60**: 973–994.

Lovelock, J.E. and Whitfield, M. (1982). Life span of the biosphere. *Nature*, **296**: 561–563.

Lynch, M. (1999). The age and relationships of the major animal phyla. *Evolution*, **53**: 319–325.

Martin, J.H. and Fitzwater, S.E. (1988). Iron deficiency limits phytoplankton growth in the north-east Pacific subarctic. *Nature*, **331**: 341–343.

Martin, J.H., Coale, K.H., Johnson, K.S. *et al.* (1994). Testing the iron hypothesis in ecosystems of the Pacific Ocean. *Nature*, **371**: 123–129.

Marzoli, A., Renne, P.R., Piccirello, E.M. *et al.* (1999). Extensive 200-million-year-old continental flood basalts of the Central Atlantic Magmatic Province. *Science*, **284**: 616–618.

McCulloch, M.T., Tudhope, A.W., Esat, T.M. *et al.* (1999). Coral record of equatorial sea-surface temperatures during the penultimate deglaciation at Huon Peninsula. *Science*, **283**: 202–204.

McElwain, J.C. (1998). Do fossil plants signal palaeoatmospheric CO_2 concentration in the geological past? *Philosophical Transactions of the Royal Society of London B*, **353**: 83–96.

McElwain, J.C., Beerling, D.J. and Woodward, F.I. (1999). Fossil plants and global warming at the Triassic–Jurassic boundary. *Science*, **285**: 876–879.

McLean, D.M. (1985). Mantle degassing induced dead ocean in the Cretaceous–Tertiary transition. *Geophysical Monograph*, **32**: 493–503.

Melillo, J.M., McGuire, M.D., Kicklighter, D.W. *et al.* (1993). Global climate change and terrestrial net primary production. *Nature*, **363**: 234–240.

Miller, A.I. and Foote M. (1996). Calibrating the

Ordovician radiation of marine life: implications for Phanerozoic diversity trends. *Paleobiology*, **22**: 304–309.

Miller, A.I. and Mao, S. (1995). Association of orogenic activity with the Ordovician radiation of marine life. *Geology*, **23**: 305–308.

Montañez, I.P., Banner, J.P., Osleger, D.A. *et al.* (1996). Integrated Sr isotope variations and sea-level history of Middle to Upper Cambrian platform carbonates: implications for Cambrian seawater $^{87}Sr/^{86}Sr$. *Geology*, **24**: 917–920.

Mora, C.I., Driese, S.G. and Colarusso, L.A. (1996). Middle to Late Paleozoic atmospheric CO_2 levels from soil carbonate and organic matter. *Science*, **271**: 1105–1107.

Morgan, M.E., Kingston, J.D. and Marino, B.D. (1994). Carbon isotopic evidence for the emergence of C_4 plants in the Neogene from Pakistan and Kenya. *Nature*, **367**: 162–165.

Moulton, K.L. and Berner, R.A. (1998). Quantification of the effect of plants on weathering: studies in Iceland. *Geology*, **26**: 895–898.

Nance, R.D., Worsley, T.R. and Moody, J.B. (1986). Post-Archean biogeochemical cycles and long-term episodicity in tectonic processes. *Geology*, **14**: 514–518.

Nisbet, E.G. and Fowler, C.M.R. (1996). Some liked it hot. *Nature*, **382**: 404–405.

Norris, R.D. and Wilson, P.A. (1998). Low-latitude sea-surface temperatures for the mid-Cretaceous and the evolution of planktonic Foraminifera. *Geology*, **26**: 823–826.

Officer, C.B. and Drake, C.L. (1985). Terminal Cretaceous environmental events. *Science*, **227**: 1161–1167.

Olsen, P.E. (1999). Giant lava flows, mass extinctions, and mantle fluxes. *Science*, **284**: 604–605.

Otto-Bliesner, B.L. and Upchurch, G.R. Jr. (1997). Vegetation-induced warming of high-latitude regions during the Late Cretaceous period. *Nature*, **385**: 804–807.

Pagani, M., Arthur, M.A. and Freeman, K.H. (1999a). Miocene evolution of atmospheric carbon dioxide. *Paleoceanography*, **14**: 273–292.

Pagani, M., Freeman, K.H. and Arthur, M.A. (1999b). Late Miocene atmospheric CO_2 concentrations and the expansion of C_4 grasses. *Science*, **285**: 876–879.

Paytan, A., Kastner, M. and Chavez, G.P. (1996). Glacial to interglacial fluctuations in productivity in the equatorial Pacific as indicated by marine barite. *Science*, **274**: 1355–1357.

Paytan, A., Kastner, M., Campbell, D. and Thiemens, M.H. (1998). Sulfur isotopic composition of

Cenozoic seawater sulfate. *Science*, **282**: 1459–1462.

Pearson, P.N. and Palmer, M.R. (1999). Middle Eocene seawater pH and atmospheric carbon dioxide concentrations. *Science*, **284**: 1824–1826.

Pelechaty, S.M. (1996). Stratigraphic evidence for the Siberia–Laurentia connection and Early Cambrian rifting. *Geology*, **24**: 719–722.

Petit, J.R., Jouzel, J., Raynaud, D. *et al.* (1999). Climate and atmospheric history of the past 420,000 years from the Vostok Ice Core, Antarctica. *Nature*, **399**: 429–436.

Podolsky, R.D. (1994). Temperature and water viscosity: physiological versus mechanical effects on suspension feeding. *Science*, **265**: 100–103.

Racki, G. (1998). Frasnian–Famennian biotic crisis: undervalued tectonic control. *Palaeogeography, Palaeoclimatology, Palaeoecology*, **141**: 177–198.

Railsback, L.B., Anderson, T.F., Ackerly, S.C. and Cisne, J.L. (1989). Paleoceanographic modeling of temperature–salinity profiles from stable isotopic data. *Paleoceanography*, **4**: 585–591.

Rampino, M.R. and Caldeira, K. (1993). Major episodes of geologic change: correlations, time structure and possible causes. *Earth and Planetary Science Letters*, **110**: 215–227.

Rampino, M.R. and Self, S. (1992). Volcanic winter and accelerated glaciation following the Toba super-eruption. *Nature*, **359**: 50–52.

Rampino, M.R. and Stothers, R.B. (1988). Flood basalt volcanism during the past 250 million years. *Science*, **241**: 663–668.

Rampino, M.R., Self, S. and Fairbridge, R.W. (1979). Can rapid climatic change cause volcanic eruptions? *Science*, **206**: 826–828.

Raven, J.A. (1998). Extrapolating feedback processes from the present to the past. *Philosophical Transactions of the Royal Society of London B*, **353**: 19–28.

Raymo, M.E. and Ruddiman, W.F. (1992). Tectonic forcing of Late Cenozoic climate. *Nature*, **359**: 117–122.

Raymo, M.E., Grant, B., Horowitz, M. and Rau, G.H. (1996). Mid-Pliocene warmth: stronger greenhouse and stronger conveyor. *Marine Micropaleontology*, **27**: 313–326.

Raymond, A., Kelley, P.H. and Lutken, C.B. (1990). Dead by degrees: articulate brachiopods, paleoclimate and the mid-Carboniferous extinction event. *Palaios*, **5**: 111–123.

Rea, D.K. (1994). The paleoclimatic record provided by eolian deposits in the deep sea: the geologic history of wind. *Reviews of Geophysics*, **32**: 159–195.

Rea, D.K., Zachos, J.C., Owen, R.M. and Gingerich, P.D. (1990). Global change at the Paleocene–Eocene boundary: climatic and evolutionary consequences of tectonic events. *Palaeogeography, Palaeoclimatology, Palaeoecology*, **79**: 117–128.

Renne, P.R. and Basu, A.R. (1991). Rapid eruption of the Siberian Traps flood basalts at the Permo-Triassic boundary. *Science*, **253**: 176–179.

Renne, P.R., Ernesto, M., Pacca, I.G. *et al.* (1992). The age of Parana flood volcanism, rifting of Gondwanaland, and the Jurassic–Cretaceous boundary. *Science*, **259**: 975–979.

Renne, P.R., Zichao, Z., Richards, M.A. *et al.* (1995). Synchrony and causal relation between Permian–Triassic boundary crises and Siberian flood volcanism. *Science*, **269**: 1413–1416.

Retallack, G.J. (1997). Early forest soils and their role in Devonian global change. *Science*, **276**: 583–585.

Retallack, G.J, Veevers, J.J. and Morante, R. (1996). Global coal gap between Permian–Triassic extinction and Middle Triassic recovery of peat-forming plants. *Geological Society of America Bulletin*, **108**: 195–207.

Richards, M.A., Jones, D.L., Duncan, R.A. and DePaolo, D.J. (1991). A mantle-plume initiation model for the Wrangellia flood basalt and other oceanic plateaus. *Science*, **254**: 263–267.

Richter, F.M., Rowley, D.B. and DePaolo, D.J. (1992). Sr isotope evolution of seawater: the role of tectonics. *Earth and Planetary Science Letters*, **109**: 11–23.

Riebessel, U., Wolf-Gladrow, D.A. and Smetacek, V. (1993). Carbon dioxide limitation of marine phytoplankton growth rates. *Nature*, **361**: 249–251.

Runnegar, B. (1982). Oxygen requirements, biology and phylogenetic significance of the Late Precambrian worm *Dickinsonia*, and the evolution of the burrowing habit. *Alcheringa*, **6**: 223–239.

Sager, W.W. and Han, H.C. (1993). Rapid formation of the Shatsky Rise Oceanic Plateau inferred from its magnetic anomaly. *Nature*, **364**: 610–613.

Sancetta, C. (1992). Primary production in the glacial North Atlantic and North Pacific Oceans. *Nature*, **360**: 249–251.

Schopf, T.J.M. (1980). *Paleoceanography*. Cambridge, Mass.: Harvard University.

Schwartzman, D. (1993). Comment on 'Weathering, plants, and the long-term carbon cycle'. *Geochimica et Cosmochimica Acta*, **57**: 2145–2146.

Schwartzman, D. (1998). Life was thermophilic for the first two-thirds of Earth history. In J. Wiegel and M.W.W. Adams (eds) *Thermophiles: The Keys to Molecular Evolution and the Origin of Life?* London: Taylor and Francis, pp. 33–43

Schwartzman, D., McMenamin, M. and Volk, T. (1993). Did surface temperatures constrain microbial evolution? *BioScience*, **43**: 390–393.

Sellwood, B.W., Price, G.D. and Valdes, P.J. (1994). Cooler estimates of Cretaceous temperatures. *Nature*, **370**: 453–455.

Shear, W.A. (1991). The early development of terrestrial ecosystems. *Nature*, **351**: 283–289.

Sheridan, R.E. (1997). Pulsation tectonics as a control on the dispersal and assembly of supercontinents. *Journal of Geodynamics*, **23**: 173–196.

Sigurdsson, H. (1990). Assessment of the atmospheric impact of volcanic eruptions. *Geological Society of America Special Paper*, **247**: 99–110.

Smetacek, V. (1998). Diatoms and the silicate factor. *Nature*, **391**: 224–225.

Stallard, R.F. (1988). Weathering and erosion in the humid tropics. In A. Lerman and M. Meybeck (eds) *Physical and Chemical Weathering in Geochemical Cycles*. Dordrecht: Kluwer, pp. 225–246.

Stallard, R.F. (1992). Tectonic processes, continental freeboard, and the rate-controlling step for continental denudation. In S. Butcher, R.J. Charlson, G.H. Orians and G.V. Wolfe (eds) *Global Biogeochemical Cycles*. London: Academic Press, pp. 93–121.

Street-Perrott, F.A., Huan, Y., Perrott, R.A. *et al.* (1997). Impact of lower atmospheric carbon dioxide on tropical mountain ecosystems. *Science*, **278**: 1422–1426.

Sullivan, C.W., Arrigo, K.R., McClain, C.R. *et al.* (1993). Distributions of phytoplankton blooms in the southern ocean. *Science*, **262**: 1832–1837.

Takeda, S. (1998). Influence of iron availability on nutrient consumption ratio of diatoms in oceanic waters. *Nature*, **393**: 774–777.

Tarduno, J.A., Sliter, W.V., Kroenke, L. *et al.* (1991). Rapid formation of Ontong Java Plateau Aptian mantle plume volcanism. *Science*, **254**: 399–403.

Tarduno, J.A., Brinkman, D.B., Renne, P.R. *et al.* (1998). Evidence for extreme climatic warmth from Late Cretaceous Arctic vertebrates. *Science*, **282**: 241–244.

Thierstein, H.R. (1989). Inventory of paleoproductivity records: the mid-Cretaceous enigma. In: W.H. Berger, V.S. Smetacek and G. Wever (eds) *Productivity of the Ocean: Past and Present*. Chichester: Wiley, pp. 355–375.

Towe, K.M. (1970). Oxygen–collagen priority and the early metazoan fossil record. *Proceedings of the National Academy of Science of the United States of America*, **65**: 781–788.

Treguer, P., Nelson, D.M., van Bennekom, A.J. *et al.* (1995). The silica balance in the world ocean: a reestimate. *Science*, **268**: 375–379.

Tyrrell, T. (1999). The relative influences of nitrogen and phosphorus on oceanic primary production. *Nature*, **400**: 525–531.

Upchurch, G.R. Jr and Wolfe, J.A. (1987). Mid Cretaceous to Early Tertiary vegetation and climate: evidence from fossil leaves and woods. In E.M. Friis, W.G. Chaloner and P.R. Crane (eds) *The Origins of Angiosperms and their Biological Consequences*. Cambridge: Cambridge University Press, pp. 75–105.

Valentine, J.W., Jablonski, D. and Erwin, D.H. (1999). Fossils, molecules, and embryos: new perspective on the Cambrian explosion. *Development*, **126**: 851–859.

Van Valen, L. (1976). Energy and evolution. *Evolutionary Theory*, **1**: 179–129.

Veevers, J.J. (1989). Middle/Late Triassic (230 ± 5 Ma) singularity in the stratigraphic and magmatic history of the Pangean heat anomaly. *Geology*, **17**: 784–787.

Veevers, J.J. (1990). Tectonic-climatic supercycle in the billion-year plate-tectonic eon: Permian Pangean icehouse alternates with Cretaceous dispersed-continent greenhouse. *Sedimentary Geology*, **68**: 1–16.

Vermeij, G.J. (1977). The Mesozoic marine revolution: evidence from snails, predators and grazers. *Paleobiology*, **3**: 245–258.

Vermeij, G.J. (1978). *Biogeography and Adaptation: Patterns of Marine Life*. Cambridge, Mass.: Harvard University Press.

Vermeij, G.J. (1987). *Evolution and Escalation: An Ecological History*. Princeton: Princeton University Press.

Vermeij, G.J. (1993). *A Natural History of Shells*. Princeton: Princeton University Press.

Vermeij, G.J. (1995). Economics, volcanoes, and Phanerozoic revolutions. *Paleobiology*, **21**: 125–152.

Vermeij, G.J. (1996). Animal origins. *Science*, **274**: 525–526.

Vermeij, G.J. (1999). Inequality and the directionality of history. *American Naturalist*, **153**: 243–253.

Vogelmann, A.M., Ackerman, T.P. and Turco, R.P. (1992). Enhancements in biologically effective ultraviolet radiation following volcanic eruptions. *Nature*, **359**: 47–49.

Walsh, J.J. (1988). *On the Nature of Continental Shelves*. San Diego: Academic Press.

Watson, A.J. (1997). Volcanic iron, CO_2, ocean productivity and climate. *Nature*, **385**: 587–588.

Wells, M.L., Vallis, G.K. and Silva, E.A. (1999). Tectonic processes in Papua New Guinea and past productivity in the eastern equatorial Pacific Ocean. *Nature*, **398**: 601–604.

Yapp, C.J. and Poths, H. (1994). Productivity of prevascular continental biota inferred from the $Fe(CO_3)OH$ content of goethite. *Nature*, **368**: 49–51.

Zachos, J.C., Lohmann, K.C., Walker, J.C.G. and Wise, S.W. (1993). Abrupt climate change and transient climate during the Paleogene: a marine perspective. *Journal of Geology*, **101**: 191–213.

Zachos, J.C., Stott, L.D. and Lohmann, K.C. (1994). Evolution of Early Cenozoic marine temperatures. *Paleoceanography*, **9**: 353–387.

Zielinski, G.A., Mayewski, P.A., Meeker, L.D. *et al.* (1997). Record of volcanism since 7000 B.C. from the GISP 2 Greenland ice core and implications for the volcano-climate system. *Science*, **264**: 948–952.

PART 6

The dynamic Earth

13

The interplay of physical and biotic factors in macroevolution

David Jablonski

ABSTRACT

Large-scale evolutionary patterns are shaped by the interplay of physical and biotic processes. We have a new appreciation of the role of physical constraints and perturbations in evolution, and the challenge is to evaluate the roles of physical, intrinsic biotic and extrinsic biotic factors in specific situations. Intrinsic biotic factors such as dispersal ability and environmental tolerance, or at the species or lineage levels geographic range or species richness, clearly influence the origination and extinction rates that underlie the dynamics of evolution above the species level (macroevolution). Such biotic factors determine the differential response of taxa to a physical perturbation, but can be overwhelmed if the perturbation is sufficiently severe or extensive (which helps to explain why mass extinction events can play an important evolutionary role while accounting for only a small fraction of the total extinction in the history of life).

Extrinsic biotic factors such as predation and competition are more difficult to quantify paleontologically, but some long-term changes in the morphology and composition of the biota appear to be driven by such interactions, perhaps mediated locally by physical perturbations. Pervasive incumbency effects at all scales, where established taxa exclude potential rivals, demonstrate both the importance of extrinsic biotic factors and the role of physical factors in opening opportunities for new or marginalized taxa to diversify: the long Mesozoic history of the mammals and their exuberant diversification after the demise of the dinosaurs is only the most famous example, and similar dynamics may occur on local and regional scales as well.

Physical, intrinsic biotic and extrinsic factors have each been afforded a role in many of the large-scale patterns of the fossil record, including: (i) the Cambrian explosion of complex metazoan life, (ii) the 'reef gap', in which the reassembly of diverse benthic communities in clear, shallow tropical waters lags 5–10 million years behind major extinction events, and (iii) the biogeographic pattern of recoveries from mass extinctions, in which regional biotas do not exhibit simultaneous or coordinated re-diversifications following global extinction events.

13.1 Introduction

The study of macroevolution, here defined simply as evolution above the species level, has seen a growing appreciation of the role of physical factors in shaping large-scale evolutionary patterns. Many of the major biotic turnovers, extinctions, and radiations that had once been attributed to direct competitive replacements or adaptive breakthroughs are now seen as physically mediated. Asteroid impacts, abrupt changes in ocean circulation and atmospheric chemistry, and a host of other physical perturbations and secular trends have been emerging as plausible hypotheses for major biotic patterns in the history of life.

Now that the range of potential hypotheses has expanded more fully into the physical

Evolution on Planet Earth
ISBN 0-12-598655-6

realm, the next challenge lies in critically evaluating the relative roles of biotic and physical factors, and their interaction, in a given macroevolutionary situation. (This challenge has arisen repeatedly in evolutionary paleontology, see discussions in Gould, 1977; Allmon and Ross, 1990; Skelton, 1991; Van Valen, 1994.) In this chapter, I review some of the paleontological evidence for the effects of both intrinsic and extrinsic biotic factors, and discuss how the interplay and feedbacks among these biotic and physical factors shape large-scale evolutionary patterns. At the very least, biotic factors can determine the contrasting responses of different taxa to the same physical factor, or the contrasting responses of the same taxon to different physical factors. I conclude by discussing some major evolutionary events where the relative roles of the different factors are controversial, with at least some evidence not only for physical drivers but for either (or both) intrinsic and extrinsic biotic factors. I cannot provide final answers for these problems, but I can lay out some alternatives that underscore the need for interdisciplinary approaches, and comment on some strengths and weaknesses of alternatives.

This chapter will necessarily address both origination and extinction: these are the fundamental terms of the macroevolutionary equation. Large-scale evolutionary patterns are molded by differential birth and death – and their higher-level analogs origination and extinction – and so extinction data are as necessary for macroevolutionary analysis as are data on individual survivorship for a microevolutionary analysis. As discussed below, sometimes extinction directly shapes the direction and fate of a particular lineage, whereas at other times the extinction of co-occurring taxa plays the overriding role. Theoretically, trends at any level might be driven entirely by differential birth/origination, but this is a hypothesis to be tested and not a starting assumption. In any event, the relation between extinction and origination needs to be explored when attempting to understand the interplay of the physical and biotic environment in driving evolution.

13.2 Intrinsic and extrinsic biotic factors and their interaction with abiotic factors

As mentioned above, biotic factors can be partitioned into intrinsic and extrinsic categories. Physical factors can also be divided this way, and intrinsic physical factors are certainly important in evolution, for example as biomechanics affect embryonic development (e.g. Gilbert, 1997; Wolpert *et al.*, 1998) and organismal function (e.g. Rayner and Wootton, 1991; Schmidt-Kittler and Vogel, 1991; Alexander, 1998; and references therein), but here I will focus on the physical factors external to the organism. Extrinsic physical factors can also be subdivided, of course, for example according to rate or spatial scale. These can range from the slow drifting of continents as driven by seafloor spreading (which averages about 5 cm a year, approximately the pace of fingernail growth), to the sudden violence of an asteroid impact; they can be as localized as the shift of a riverbed or the suppression of an oceanic upwelling cell, or as pervasive as a change in the CO_2 content of the atmosphere.

For this discussion, intrinsic biotic factors are aspects of an organism, or of a unit at any other focal level within the biological hierarchy, that affect its probability of becoming extinct or producing descendants. (Besides these aspects of evolutionary tempo, intrinsic factors can also affect evolutionary mode, for example branching vs phyletic evolution, e.g. Jablonski, 1986a.) Extrinsic biotic factors involve the impact of one biological entity upon another, for example via competition or predation. Because life is organized hierarchically, intrinsic and extrinsic status can shift according to focal level; the other members of a breeding population are extrinsic factors for an individual organism, but together may help to determine the intrinsic response of a species to climate change.

13.2.1 Intrinsic biotic factors

A huge array of intrinsic biotic factors have been implicated in differential extinction and speciation probabilities (e.g. Jablonski 1986b, 1995; Pimm *et al.*, 1988; McKinney 1997). Traits that clearly reside at the organismal level include body size, mobility, metabolic rate, physiological tolerance limits, and feeding type (e.g. herbivore, carnivore, parasite). The variation of these factors even among closely related species helps explain why higher taxa such as families, orders and classes are generally poor analytical units for studying evolutionary rate differences, even for groups noted for their high average rates. For example, genus-level origination and extinction rates can vary by an order of magnitude among ammonite superfamilies, and by a factor of four among trilobite orders (Gilinsky, 1994).

Dispersal ability is an interesting intrinsic factor because it often plays a role in determining rates and patterns of gene flow, and geographic ranges, which are aspects of species that are not simply aggregate characteristics of individual bodies. These intrinsic biotic features at the species level can in turn affect speciation and extinction probabilities, just as species richness is a clade-level trait that can influence extinction (and perhaps origination) probabilities (e.g. Jablonski, 1986a, 1987, 1995, 2000; Williams 1992; Grantham, 1995). Views differ on how larval dispersal operates within a hierarchical framework. This is unimportant for the present discussion, so I will simply note that marine species achieve broad geographic ranges and high rates of gene flow by several mechanisms besides larval dispersal (e.g. rafting of bryozoans, Watts *et al*, 1998); those species-level properties, however achieved, are subject to higher-level selection processes, which can oppose or reinforce processes operating on individual-level traits.

Hierarchy arguments aside, fossil larval shells can be used to infer modes of development and are good predictors of evolutionary rates in marine gastropods, where the data are most plentiful. Although there are of course exceptions, species with high-dispersal, planktotrophic larvae tend to be widespread and geologically long-lived, and to speciate infrequently, whereas those with low-dispersal, non-planktotrophic larvae tend to be more narrowly distributed and more extinction-prone,

and to have higher speciation rates, as tested in Cretaceous, Paleogene, and Neogene settings (Scheltema, 1977, 1989; Hansen, 1978, 1980; Jablonski, 1986a, 1995; Gili and Martinell, 1994). For each of these analyses, the species are drawn from a single regional pool, within a single biogeographic province at continental shelf depths within a discrete time frame. Therefore the paleontological data show, at least in part, how taxa with different intrinsic biotic features respond to the same environmental parameters at a particular time.

This differential response, along with a number of others, evidently did not operate during the end-Cretaceous extinction (Jablonski, 1986a, b; Valentine and Jablonski, 1986), although as noted above, it returned in the early Cenozoic. Smith and Jeffery (1998) also found no significant difference in survivorship of echinoids across the Cretaceous–Tertiary boundary according to larval type, although with a p-value of 0.11 they considered this a 'weak correlation'. McGhee (1996: 129) argued for a similar lack of selectivity according to developmental type in the late Devonian mass extinction, although this requires inference of brachiopod developmental types from present-day relatives, which may not be reliable (see Valentine and Jablonski, 1983).

Kammer *et al.* (1998) found in Paleozoic crinoids that habitat generalists had significantly greater species longevities than habitat specialists, at least during times of 'normal' extinction intensities. This is by far the most rigorous test of this intuitively appealing relationship (for earlier observations on mollusks, see Jablonski, 1980; Erwin, 1989; Stanley, 1990; and references therein), and as with molluscan larval modes we are seeing the differential response of taxa within a single region to the challenges of temporal and spatial environmental variation. However, this intrinsic biotic attribute did not confer extinction-resistance upon molluscan genera during the end-Cretaceous extinction (Jablonski and Raup, 1995).

Kammer *et al.* (1998) note that greater species longevity does not necessarily scale up to greater clade longevity: for their crinoids the specialist clades attained higher species richness and became dominant in the late

Paleozoic. Although species-rich clades often tend to be extinction-resistant (an intrinsic biotic feature at the clade level), this effect is not seen for molluscan or echinoid genera at the end-Cretaceous mass extinction (Jablonski, 1986b; Smith and Jeffery, 1998). The benefit of species richness is lost for a number of major clades across mass-extinction boundaries, from end-Ordovician bryozoans to end-Permian brachiopods (Table 2.2 of Jablonski, 1995; Erwin's (1989) gastropods appear to be an exception).

These analyses exemplify two general points: first, that co-occuring taxa can differ in extinction and origination rates in ways that can be related to intrinsic biotic factors, and second, that such differences can be overwhelmed by physical factors when perturbations become sufficiently severe or extensive (see also Jablonski, 1995, 1996; McKinney, 1997; Smith and Jeffery, 1998). This does not mean that all intrinsic factors are ineffective during mass extinction events, however. For example, the presence of a resting cyst in certain phytoplankton life cycles may have promoted diatom and dinoflagellate survival across the K–T boundary (Kitchell *et al.*, 1986; Brinkhuis *et al.*, 1998); this is a nice example of an individual feature molded by natural selection under 'normal' extinction that is also effective under the mass extinction regime.

At least one clade-level feature generally appears to enhance survivorship during mass extinction: geographic range. In end-Cretaceous bivalves, for example, clades (in this case genera) restricted to only one or two biogeographic provinces suffered about 65% extinction, whereas the few clades spread over six or seven provinces showed about 35–45% extinction (Figure 13.1). This large-scale biogeographic effect can be seen for many groups at many extinctions, from end-Ordovician brachiopods, trilobites, bivalves and bryozoans to Permian gastropods to end-Triassic bivalves (Table 2.3 of Jablonski, 1995); Smith and Jeffery's (1998) analysis of K–T echinoids is one of the very few studies that have failed to detect such an effect. McGhee (1996) finds no overall geographic effect in the late Devonian mass extinction, although clades with broad latitudinal ranges survive preferentially to those with broad longitudinal ranges, a pattern perhaps related to tropical perturbations and needing fuller quantitative analysis. Overall, intrinsic biotic factors of different kinds, and at different levels, help to determine survivorship at all spatial scales, from within-province processes to global patterns during the five major mass extinctions that punctuated the Phanerozoic history of life.

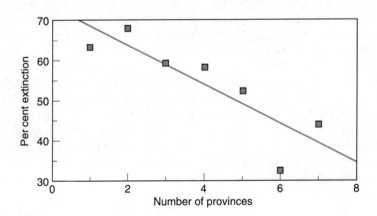

Figure 13.1 Broad geographic range – measured here in terms of biogeographic provinces – significantly enhanced survival of bivalve genera during the end-Cretaceous mass extinction (Kendall's rank test, $p < 0.05$). The plot shows the structure of the data but is inappropriate for regression analysis because error terms increase with the number of provinces (more genera are restricted than widespread); from Jablonski and Raup (1995).

13.2.2 *Extrinsic biotic factors*

Extrinsic biotic factors are more difficult to quantify in the fossil record, because they involve biotic interactions like predation, competition, and even more elusive indirect effects. We can find fossil evidence for momentary, individual interactions, such as drillholes or bitemarks on shells, insect damage on leaves, or overgrowths among competing colonies of corals or bryozoans, but the question is how to scale up such short-term, small-scale encounters to the macroevolutionary arena?

In a series of important contributions, Vermeij (1977, 1987, 1994) argued that evolutionary changes in the effectiveness of shell-penetrating predators drove a change in the structure of marine communities, and in the range of morphologies present in bivalves, gastropods and other marine prey. The biotic interactions engendered by this 'Mesozoic Marine Revolution' are evident, for example, in an increase in the proportion of taxa bearing defensive adaptations such as narrow and reinforced apertures. Some smooth, small, open-aperture forms have persisted (although species with open coiling, in which successive shell whorls are not in contact, have almost vanished), but few observers would confuse an assemblage of shallow-water Carboniferous shells (age *c.* 300 Ma) with one from equivalent, more 'escalated' habitats of Eocene age (age *c.* 50 Ma).

The kinds of morphological changes seen during the Mesozoic Marine Revolution, and their timing relative to the diversification of predators, from bony fish to crabs to carnivorous snails, build a powerful case for an escalation process driven by extrinsic biotic factors. The interactions may have been quite diffuse – large, heterogeneous sets of species impinging on other large, heterogeneous sets of species, rather than the tight, reciprocal predator/prey interactions usually studied by ecologists – but their net effect is quite apparent. Although more work would be welcome, a variety of experimental data corroborate the view that molluscan shells of modern aspect are more resistant to predation than certain shell types more commonly found in Paleozoic assem-blages (e.g. Palmer, 1979; Vermeij, 1982; Harper, 1991; Harper and Skelton, 1993; Miller and LaBarbera, 1995). Still open for testing is exactly how these extrinsic factors interacted with intrinsic ones. The waxing and waning of individual clades, partly a function of intrinsic factors such as those discussed above, might have helped to determine how many species of, say, muricine gastropods, and thus how many species with those extravagant spines, were present at any one time and place.

Not all hypothesized evolutionary responses to predators have been verified, and sometimes further testing may implicate physical factors instead. In Cenozoic terrestrial faunas, for example, Janis and Wilhelm (1993) found that fast-running long-legged ungulates evolved at least 20 million years before the pursuit carnivores that were supposed to drive a coevolutionary process of escalation (e.g. Bakker, 1983). Instead, high-mobility herbivores more likely evolved in response to climatic changes that led to increasingly open grassland habitats.

Even for the Mesozoic Marine Revolution, where the basic faunal changes do appear to be driven by extrinsic biotic factors, physical factors may also have played an important role. First, there is the question of the initiation and expansion of the escalation process. A strictly biotic mechanism might rest entirely on a coevolutionary arms race launched by one or a few shell-penetrating innovations among early Mesozoic predators. An alternative (which of course need not be completely exclusive of the first) is that pulses or long-term trends in nutrient input might have promoted these intensified biotic interactions by fueling more active predators and prey, and perhaps a larger (or more rapidly turning over) prey resource base (e.g. Vermeij, 1995, and this volume, Chapter 12; Bambach, 1993, 1999). Testing these alternatives, or perhaps attempting to quantify their relative contributions, will require multidisciplinary approaches in which independent proxies for nutrient levels are tested against realistic models of biotically driven escalation, in an explicit geographic and temporal framework.

Physical factors might have played a role at the local and regional scale as well. As Miller (1998) notes, few instances of gradual escalation have been documented in a single basin or biogeographic region (Kelley's work (1989, 1991) is an exception, but her studies were in the Miocene, well after the major faunal changes of the Mesozoic and early Cenozoic). Instead, species are largely morphologically static, so that escalation appears to proceed mainly by episodic, local or regional replacements. The argument is largely inferential, based on what we *don't* see, but Roy (1996) provides a concrete example in the geographic variation in the timing and pattern of the replacement of aporrhaid gastropods by their arguably more escalated relatives, the strombids. Their contrasting regional patterns suggest that environmental changes in factors such as climate or productivity can mediate the regional expression of a biotically driven, essentially global macroevolutionary change (Roy, 1996). Thus, even the large-scale biotic changes of the Mesozoic Revolution may require local perturbations to open opportunities for the establishment of more derived taxa. The derived taxa may be better equipped to cope with more escalated predators, but at the local scale incumbents must be removed or reduced before the more effective taxa can become established in a given region. The fossil record is rich in such incumbency effects.

13.2.3 *Incumbency*

Incumbency effects beautifully exemplify the interplay between biotic and physical factors. On the one hand, it is often physical perturbations that break incumbencies, removing dominant forms and opening opportunities for previously minor groups. The spectacular diversification of mammals after the extinction of the dinosaurs is just the most familiar example. The establishment of dinosaurs in the early Mesozoic, and of advanced mammalian carnivores in the mid-Cenozoic also represent biotic replacements and diversifications once thought to be competitively driven but that now appear to have been mediated by extinction of incumbents (see Benton, 1987, 1996; Jablonski 1986b;

Rosenzweig and McCord, 1991; Jablonski and Sepkoski, 1996; McKinney, 1998; Van Valkenburgh, 1999; Sereno, 1999). This is one reason for the evolutionary importance of mass extinctions (cf. MacLeod, this volume, Chapter 14): they account for just a small fraction of the total species extinction over the past 600 Ma (e.g. Raup, 1991), but by removing incumbents – perhaps owing to changes in the effectiveness of intrinsic biotic factors, as discussed above – they open up evolutionary opportunities for other groups that had been minor players.

On the other hand, the limited ecological role and morphological diversity of mammals for the first two thirds of their history, and more generally the very importance of incumbency-breaking events, attest to the importance of extrinsic biotic factors, such as competition, in damping or channeling evolutionary change. Even groups that appear to have enjoyed unimpeded diversification, and thus might seem to have been free of extrinsic biotic effects, show a different story on close examination. For example, marine bivalves (Figure 13.2) diversify exponentially throughout most of their history, suggesting a process driven mostly by intrinsic biotic factors, interrupted by a few sudden downdrops that show the intrusion of physical factors, mass extinction events, into the system. But what about the dynamics immediately after the mass extinctions? The rate of recovery is significantly greater – for a short time – than the general pace of bivalve diversification, and Miller and Sepkoski (1988) argue that these 'hyperexponential' episodes represent the true rate of unimpeded diversification for the clade. The general exponential rate must be a damped one, presumably by biotic interactions so diffuse and pervasive that we can only study their effects via perturbations to the system. Such a pattern might also be generated as an artifact of taxonomy, sampling or preservation around extinction boundaries, but the true termination of many bivalve taxa at or around major extinction horizons (e.g. MacLeod *et al.*, 1997: 278), along with a vacating and rapid refilling of a morphospace defined quantitatively without reference to individual taxa (Lockwood, 1998), suggest that the pattern is real in the bivalves

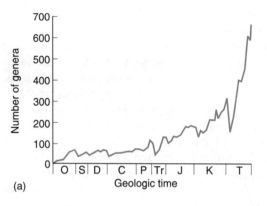
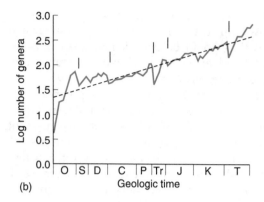

Figure 13.2 Diversity of bivalve genera from the Ordovician to the Recent. Both (a) arithmetic and (b) logarithmic plots illustrate the approximately exponential nature of bivalve diversification after their initial Ordovician radiation. However, recoveries from end-Permian and end-Cretaceous mass extinctions occur at substantially greater rates than the group's basic exponential rate, suggesting that the basic rate is damped by biotic interactions that are reduced during the post-extinction rebounds. Vertical tick marks represent the Big Five mass extinction events. After Miller and Sepkoski (1988).

(see also Patzkowsky's 1995 reanalysis). Most intriguing is that the interactions are not imposing a ceiling on a clade's diversity, but instead are evidently slowing the pace of ongoing diversification in a successful clade. Sepkoski (1996) modeled these more complex dynamics in some detail, providing a more realistic set of expectations at this scale.

13.3 Some large-scale patterns

The preceding sections discuss how intrinsic biotic factors can affect rates and patterns of extinction and origination; how extrinsic biotic factors can both promote and inhibit evolutionary change; and how physical factors can play a pivotal role in when and how the biotic factors operate. For a number of the most important episodes in the history of life, however, the relative contributions of these factors remain controversial. Although I cannot resolve these controversies, it may be useful to discuss them in terms of competing and complementary hypotheses on causal mechanisms.

13.3.1 The Cambrian explosion

Regardless of exactly when the major metazoan lineages actually diverged, the Cambrian explosion represents a uniquely rich and temporally discrete episode of morphological evolution (for recent reviews see Conway Morris, 1998a; Valentine *et al.*, 1999; Knoll and Carroll, 1999). Almost all of the skeletonized metazoan phyla appear within an interval of perhaps 10 My, and the accompanying diversification of microplankton, of forms having agglutinated skeletons (i.e. pieced-together sand grains or skeletal debris), and of behavioral traces all suggest that this is a real evolutionary event and not simply a change in the preservation potential of an already diverse biota.

Decisive tests have been elusive for the many hypotheses put forward for the onset and, equally important, the termination, of the Cambrian explosion. A physical trigger is often invoked. For example, Kirschvink *et al.* (1997) used paleomagnetic data to argue for a 10–15 My episode of rapid plate-tectonic reorganization of early Cambrian continents and suggested that these geographic changes, including attendant shifts in climate and ocean circulation, would have greatly accelerated biological evolution; as the geographic changes slowed, so would evolution. Hoffman *et al.* (1998) revived an old idea relating metazoan diversification to Proterozoic glacial ages, this time presenting evidence for a 'snowball Earth' (that is, a truly global glaciation), and, in a breathtaking extrapolation from Hawaiian *Drosophila* populations, suggested that this would bottleneck metazoan lineages and yield an episode of explosive evolution once conditions relaxed.

The principal candidate for a physical trigger is the crossing of a threshold in atmospheric oxygen, usually set at 10% present atmospheric level (PAL) by analogy to the drop in metazoan diversity when oxygen tension falls below this level in modern seas. However, as discussed by Lenton (this volume, Chapter 3), some evidence suggests that the threshold was crossed too early for oxygen to have been the primary trigger for the explosion itself. Canfield and Teske (1996), for example, estimate 10% PAL at 650 Ma, nearly 100 My too early (see also Knoll, 1996; and Rye and Holland, 1998, who suggest that 10% PAL was reached by 2 billion years ago). Oxygen availability may have been a critical prerequisite for the evolution of large, complex metazoans (Lenton, this volume, Chapter 3; but see Conway Morris, 1998a: 898), but it may not have been the proximate trigger.

In theory, the major biotic hypotheses contrast starkly. Hypotheses based on intrinsic biotic factors, which might be termed the 'loose genes' or genomic hypotheses, postulate the crossing of a threshold in the complexity of developmental systems that allowed organisms to generate more elaborate forms, and in an especially prolific fashion, because those developmental systems were relatively unconstrained (see Valentine and Erwin, 1987; Erwin, 1994, 1999; Arthur 1997). The later damping of the explosion, then, would have occurred as genomes became increasingly burdened by interdependencies among genes and signaling pathways.

The extrinsic biotic hypotheses, which might be called the 'loose niches' or ecological hypotheses, postulate ecological feedbacks both to trigger and to damp the explosion. The evolution of predation and its consequences for driving an evolutionary arms race and for mediating coexistence among competitors represent a potential trigger mechanism (e.g. Stanley, 1976; Bengtson, 1994). The explosion would slow down as the ecological barrel

became full and opportunities became more limited for new biological designs. Thus, the loose-genes hypothesis involves an intrinsic slowing of the generation of evolutionary novelty, independent of ecological changes, whereas the loose-niches hypothesis involves a steady production of evolutionary novelty, but declining success of novel forms owing to extrinsic factors. Despite this clear-cut conceptual distinction, the two hypotheses make very similar predictions in terms of paleontological patterns (Jablonski and Bottjer, 1990; Valentine, 1995).

At the moment, the loose-niches view probably has the edge. Although far more work is needed, present knowledge of molecular developmental biology, combined with phylogenetic data on the relationships of living phyla, suggests that a number of the important developmental pathways that built the organisms of the Cambrian explosion were in place well before the event itself (like oxygen, perhaps, necessary but not sufficient) (see Valentine, 1995; Erwin *et al.*, 1997). However, this view may be premature. We still understand only a fraction of the developmental machinery required to build metazoans, and one inescapable message of the recent work on developmental pathways is that the molecular pathways that generate morphology have evolved in as complex and quirky a fashion as the morphology itself. Early establishment of basic signaling pathways does not necessarily exclude later accumulation of multiple roles for individual genes, for example. One approach to testing genomic vs ecologic hypotheses might be to take a closer look at the kinds of novelties established at different points in a clade's history, eventually with reference to the developmental basis for those novelties (see Jacobs, 1990; Wagner, 1995; Foote, 1999, all three of whom conclude that their evidence supports a genomic hypothesis). For example, Foote (1999) found that post-Paleozoic crinoid diversification was functionally and ecologically prolific but morphologically much more stereotyped than the initial Paleozoic radiation of the group, and concluded that this pattern was more consistent with a genomic hypothesis. Clearly, without a time machine to perform reciprocal transplants between Cambrian and modern seas (to test whether early Paleozoic organisms would be as prolific in their novel morphologies in a more crowded world, or whether modern organisms would spill forth a host of novel forms in the ostensibly more permissive Cambrian seas), a new round of multidisciplinary work will be needed to tackle this fascinating problem.

The 'reef gap'

Following mass-extinction events, the reassembly of complex communities found in clear, shallow, tropical waters often involves a distinct lag time, loosely termed the reef gap (e.g. Hallam and Wignall, 1997) (Figure 13.3). 'Reef' is not being applied in the strict sense of a demonstrably wave-resistant framework, but simply as a general term for the diverse benthic

communities characteristic of tropical onshore habitats over most of geologic time, except for the 10–5 My after major extinction events (e.g. Copper 1988, 1994a, b; Stanley, 1992; Jablonski, 1995; see Webb, 1996, for a somewhat different, microbial perspective; and Wood, 1998, for difficulties of applying strict definitions to ancient associations). Thus, the bizarre rudist bivalves that dominated late Cretaceous tropical onshore settings did not make reefs in the strict sense, but they generated massive, extensive skeletal accumulations in clear-water tropical settings (e.g. Gili *et al.*, 1995). With the demise of the rudists at or near the K–T boundary, diverse and extensive metazoan carbonates were scarce or absent until assemblages dominated by colonial corals began to appear with increasing frequency in the late Paleocene, 10–8 My into the Cenozoic. Coral build-ups are clearly not close analogs of rudist commu-

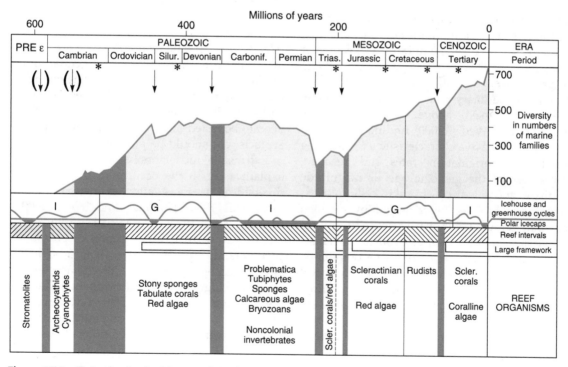

Figure 13.3 Episodes in the history of 'reefs' and other tropical build-ups through geologic time. Arrows represent the major extinctions (those in parentheses are less well documented than the Big Five); asterisks mark small events. Superimposed on the global diversity curve for marine families are vertical black bars representing intervals that lack major reef development. The icehouse/greenhouse curve gives a rough assessment of the global climatic state. As noted in the text, the ability of Cretaceous rudists to build large frameworks is hotly debated. Modified from Stanley (1992).

nities, but the temporal gap is still striking, and similar gaps appear after the other mass extinctions as well.

The reef gap might be the result of physical factors, so that the development of diverse tropical carbonate communities was impeded by the persistence of the physical perturbations that had triggered the extinction itself (e.g. Stanley, 1988; Stanley, 1992). However, isotopic and other time series, along with modeling of, for example, the effects of asteroid impacts and massive volcanic eruptions, suggest that the reef gap persists far longer than the climatic and other disruptions (e.g. for the K–T event, D'Hondt *et al.*, 1996, 1998; Conway Morris 1998a, b; Erwin, 1998: 345). Clearly more work is needed to track the return of physical environmental parameters to their old, or a new, steady state. With improved geochronological methods the necessary data are sure to become available.

Alternatively, the reef gap might be imposed by intrinsic biotic factors. Perhaps the assembly of these tropical communities is determined by the intrinsic diversification rates of their major constituents. This 'waiting time' hypothesis seems weakened, first by the general acceleration of evolutionary rates after mass extinction events discussed above, and second by Kirchner and Weil's (2000) finding that recoveries tend to have a characteristic (*c*. 10 My) lag until peak origination rates are reached, regardless of the geologic age or magnitude of the event (see also Sepkoski, 1998); but the hypothesis has yet to be tested by detailed comparisons of evolutionary rates in the appropriate taxa before and after extinction events.

Extrinsic biotic effects are also a possibility. Perhaps the reef gap is a problem in ecological assembly, with lags set by the time required for complex ecosystems to attain a coherent structure (Talent, 1988; Conway Morris, 1998b). Here, the limiting step is not taxonomic origination, but the reassembly of a complex, stable ecosystem (see also D'Hondt *et al.*, 1998, who suggest that plankton production recovered shortly after the K–T event, while ecosystem structure lagged by almost 3 My). Paleontological testing of this hypothesis will be difficult, but comparative analysis

of the details of reassembly among recovery intervals should help. Analyses might include: detailed time-series for temperature and other environmental parameters (to test the role of physical factors as controls), data on the occurrence and phylogenetic relationships of important taxa in the interval before the community takes on its distinctive flavor (to test for evolutionary lags), and testing whether the duration of the recovery lag is positively related to the complexity of the new system (for which there is some evidence, at least regarding 'reef' vs level-bottom assemblages, Erwin, 1998).

More subtle lags in biotic recovery may occur in level-bottom habitats. For example, after the K–T mass extinction, extratropical, within-habitat molluscan diversity appears to recover within a few million years (see Hansen *et al.*, 1993). But regional diversity, the number of taxa in an entire biogeographic province, evidently does not reach Cretaceous levels until the late Paleocene or early Eocene, on the order of 10 My after the extinction (Hansen, 1988; Jablonski, 1998). This might mean that beta diversity, the differentiation of local faunas among habitats and along environmental gradients, takes longer to recover than alpha, or local diversity. These possibilities need to be tested more rigorously for sampling artifacts, but should be pursued.

Extrinsic biotic factors have been invoked to explain a lag in the occurrence of maximum diversification rates after extinction events (Sepkoski, 1998, Kirchner and Weil, 2000). Diversification itself creates niches, the argument goes, so that the evolution of parasites, predators, and other clades taking advantage of enlarged biotic opportunities will accelerate, leading to higher per-taxon origination rates as the recovery proceeds. Alternatively, however, the lag could result from strong intrinsic differences in diversification rates among clades (see Sepkoski, 1998): given unimpeded post-extinction diversification, the global per-taxon rate will be increasingly dominated by the most rapidly evolving groups, leading to an increase in the overall rate until another extinction, or diversity-dependent feedback, damps the rise of the high-rate taxa. It would be interesting to determine whether this intrinsic biotic factor also plays a role in the recovery of beta and regional diversity, which occurs on a similar timescale, for example by testing whether increases in beta diversity are set by the high-origination taxa.

If either type of biotic factor can be verified for the 'reef gap', or for other lags in the recovery of ecological structure (such as the 'coal gap' after the end-Permian extinction, Faure *et al.*, 1996; Retallack *et al.*, 1996), this would have sobering implications for the restoration ecology of modern ecosystems. It would provide concrete evidence that the recovery of a system that has been degraded below some threshold will lag far behind the removal of the immediate extinction driver. The fossil record presents real opportunities for exploring the magnitudes and types of losses that might impose such recovery lags.

13.3.3 The biogeographic fabric of recoveries

Because the major mass extinctions are global events, presumably driven by physical perturbations at that scale, the general assumption has been that the subsequent recoveries would be global as well. But the molluscan recovery from the end-Cretaceous mass extinction is geographically heterogeneous. Despite indistinguishable extinction intensities at the K–T boundary, faunal dynamics for two tropical regions (North Africa, and the northern margin of the Indian plate including Pakistan) and two temperate regions (the North American Gulf and Atlantic Coastal Plain, and northern Europe) show significant differences, with North America being the odd man out. The evolutionary burst and decline of 'bloom taxa' in the North American early Paleocene documented by Hansen (1988), with other taxa in the region showing slower, steadier diversification, is not seen in the other three regions (Figure 13.4). This pattern holds whether the bloom taxa are treated as a proportion of the biota, or, once the phylogenetic bottleneck is taken into account by estimating minimum surviving lineages at the boundary, as raw species numbers (Jablonski, 1998). It is also unlikely to be an artifact of regional differ-

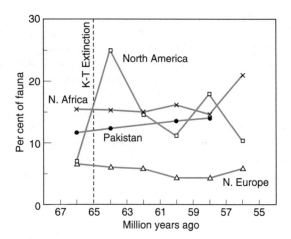

Figure 13.4 Geographic variation in the molluscan recovery from the end-Cretaceous mass extinction. The same 'bloom taxa' that burst and decline immediately after the extinction in North America show significantly less volatile evolutionary behavior in three other regions, as a proportion of species in each fauna as shown here, in raw numbers (as shown by Jablonski, 1998). The reasons for this difference are debated. For details and statistical confidence limits, see Jablonski (1998).

ences in sampling resolution, because in northern Europe at least, the region most likely to conform to North America by reason of proximity and climatic similarity, an earliest Paleocene fauna (the Cerithium Limestone, nannofossil zone NP1) has been well studied by Claus Heinberg (1999), and clearly lacks the expected pulse of bloom taxa. A further contrast among faunas is the source of new taxa: the North American fauna is significantly more subject to invasion from outside the region, whereas the other three faunas accumulate more of their post-extinction diversity by diversification within the region (Jablonski, 1998).

The most obvious physical explanation for this surprising geographic variation in the Paleocene recovery is that North America was nearest to the K–T impact near Chicxulub on the Yucatan Peninsula (which might even have been oblique, with its greatest effects to the northwest, Schultz and D'Hondt, 1996). It might therefore have suffered the greatest perturbation and consequently would have been subject to a most complicated recovery. However, this explanation is undercut by the similarities in extinction intensities among the four regions (Raup and Jablonski, 1993). An alternative might lie in potential geographic differences in the Paleocene recovery environment. By some models, North American forests would have ignited at the time of impact (Toon *et al.*, 1997). Weathering, erosion and transport of the incinerated organic material to the continental shelf might have created a very different nutrient regime around southeastern North America relative to regions whose neighboring lands had not been subject to conflagration. The million-year timescale seems long for this effect, but geochemical analyses aimed at isotopes and biomarkers should permit testing of this hypothesis.

Intrinsic biotic factors might have played a role in the interregional differences in recovery dynamics. For example, the differences might have arisen if the North American set of survivors had features such as low-dispersal larvae that would have promoted rapid speciation and high extinction rates, relative to related taxa in other regions. Both phylogenetic analysis, to confirm the monophyly of the North American bursts of bloom taxa, and a quantitative assessment of the interregional distribution of appropriate character states, are needed here. So far I have seen no evidence suggesting that North America's bloom taxa (turritellid gastropods, and carditid, ostreid, and cucullaeid bivalves) were richer in speciation-promoting intrinsic features such as low-dispersal larvae compared to relatives elsewhere, but comparisons among subclades would certainly be worth pursuing.

Finally, extrinsic biotic factors might have been involved. Similar extinction intensities might yield different long-term consequences among regions according to the ecological roles of the victims or the invaders. For the K–T mollusks, for example, perhaps the North American victims were drawn from the more abundant taxa, or included more keystone species (that is, species whose activities help to structure communities), so that their disappearance was especially disruptive of competitive or predator/prey relationships

and thus enhanced the volatility and invasibility of that region relative to the others. Lockwood (2003) found no correlation between abundance and extinction in North American mollusks, which undermines the simplest version of the extrinsic hypothesis, but more rigorous comparative analyses are needed to rule it out (perhaps, for example, the victims in the other regions are drawn from rare taxa rather than at random – unlikely, given my qualitative observations, but in need of testing). Keystone species are understood in only a few present-day situations, and their loss will be exceedingly difficult to detect in the fossil record, particularly in light of our growing appreciation of the role of indirect effects in structuring communities (e.g. Wootton, 1994). Nevertheless, a more detailed and standardized analysis of the ecological roles of victims, survivors, and invaders in different regions would be extremely valuable. The finding that invaders were not randomly drawn from the pool of survivors, but were significantly more widespread prior to the K–T extinction (Jablonski, 1998), suggests that biotic factors will prove to be important in determining the intensity and asymmetry of post-extinction interchanges. More generally, it suggests a line of research that would contribute to a more complete, general theory of donor/recipient dynamics for biotic invasions in the geologic past and in the accelerating human-mediated invasions of the present day.

The realization that recoveries do not necessarily unfold simultaneously and in a coordinated fashion among regions is mirrored by recent analyses of the geographic fabric of the Ordovician radiations of marine life – a profuse set of variations on the themes initiated during the Cambrian explosion. After correcting for sampling, Miller (1997a, b) found that some regions follow a nearly monotonic diversity increase through the Ordovician, while others peak early and then decline. Thus, like the replacement of aporrhaid gastropods by the strombids, and the region-specific K–T recovery patterns, the global history of the Ordovician diversification is an aggregate of rather disparate patterns on the different continents. Miller (1997a, b; Miller and Mao, 1995) found that the regional diversity trends appear to correlate with the extent or intensity of mountain-building activity, an intriguing physical factor that has been invoked before, but never with such a rich and detailed database (e.g. Lull, 1918; Grabau, 1940; Umbgrove, 1947; Henbest, 1952). Clearly, decomposing global diversity patterns into regional components will be a major area of paleontological research.

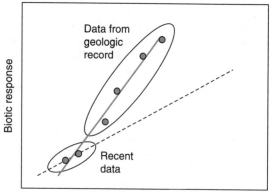

Figure 13.5 The importance of data from the geologic record. The recent and immediate geologic past encompasses a relatively narrow range of situations, leading to potentially faulty extrapolations of biotic responses to physical variables (broken line). With the enlarged range of conditions and biotic responses provided by the geologic record, we can better understand – and perhaps predict – the behavior of the Earth/life system.

13.4 Conclusion

Macroevolutionary patterns are shaped by both physical and biotic factors, and the rich feedbacks between them. The very healthy surge in our interest in, and our ability to quantify rate and magnitude of, physical environmental change has helped to redress a conceptual imbalance that had tended to overemphasize biotic interactions. The growing geologic database on physical factors will provide important opportunities to calibrate magnitudes of perturbations with magnitude and type of biotic response – a long-sought goal anticipated by Raup's (1991, 1992) pioneering 'kill curve' for asteroid impact events. The geologic database also provides a much wider range of parameter values, from temperatures to atmospheric compositions to asteroid impacts, than are available today or in the immediate geologic past, and this expanded range is crucial for a more rigorous understanding of the operation of the Earth/life system (Figure 13.5).

It would be a mistake, however, to allow the pendulum to swing too far in the direction of physical effects or to allow false dichotomies, e.g. that the existence of physical forcing mechanisms or geographic variation in biotic transitions somehow negates the importance of intrinsic biotic factors. Instead, we need to develop protocols for exploring the tension between physical and biotic factors in shaping large-scale evolutionary patterns. These are complementary components of macroevolutionary processes, operating in tandem or in sequence (see also Lister and Rawson, this volume, Chapter 16), and as discussed above their relative strengths will vary with the situa-

tion. Sometimes physical factors will be the ultimate causes of disruptions mediated by proximate biotic causes, as probably seen at the K–T boundary. However, for several reasons it would be an oversimplification to assume that life goes about its business until a rock falls out of the sky, metaphorically or literally, and that physical factors completely reset or even reverse the accumulated biotically driven changes.

First, intrinsic biotic factors apparently determine how co-occurring clades respond to a perturbation, an effect especially strong away from the Big Five mass extinctions, but detectable even at the times of the most extreme global events.

Second, perturbations come at all scales, and they mediate biotic transitions at all scales. A stream bisecting a field means very different things to a mouse and an elephant. But the importance of those perturbations is attributable to their disruption of the incumbent advantage. Incumbency, an extrinsic biotic factor that slows or blocks invasion or diversification, may be such an important component of the macroevolutionary process that local perturbation may be crucial to the establishment of global trends (Miller, 1998). Positive biotic feedbacks may also play a role in diversification rates, as suggested by Sepkoski (1998) and Kirchner and Weil (2000), among others.

Third, major biotic trends can transcend perturbations, even the Big Five mass extinctions. Those trends may suffer momentary setbacks, or at least pauses, but the basic trajectory is recovered. Consider, just for the K–T extinction, the persistence of: the Mesozoic Marine Revolution, discussed above; the onshore/offshore shift or expansion of most post-Paleozoic orders (Jablonski and Bottjer, 1990, 1991); the increasing taxonomic and ecological dominance of cheilostome bryozoans relative to cyclostomes (Lidgard *et al.*, 1993; McKinney *et al.*, 1998); and the rise of flowering plants (Niklas, 1997; Boulter *et al.*, 1998; Lupia, 1999; Magallon *et al.*, 1999). One of the challenges ahead is to understand why such trends persist in the face of major and minor extinction pulses, while others do not. Whatever the ultimate answer, it is clear that physical and biotic factors cannot be reduced to simple, antagonistic alternatives in the explanation of macroevolutionary patterns.

Acknowledgements

I thank the editors for inviting me to participate in this stimulating symposium, and am grateful to Susan M. Kidwell, Michael Foote, Adrian Lister, and Michael J. Benton for valuable reviews. Supported by the National Science Foundation, grants EAR93-17114 and EAR99-03030.

Author's note: The revised version of this chapter was submitted in March 2000. The field has continued to move rapidly, and providing even a partial update on the issues raised here is impossible. See the following references, among many others, for more information and new perspectives on: intrinsic factors in evolution (Jablonski, 2001; Gould, 2002; Jablonski et al., 2003), controls on post-extinction dynamics (Erwin, 2001; Jablonski, 2002; Foote, 2003), the Cambrian explosion (Knoll and Carroll, 1999; Valentine et al., 1999; Thomas et al., 2000; Valentine, 2002), and the history of reefs (Stanley, 2001: Flügel and Kiessling 2002).

References

Alexander, R.McN. (1998). All-time giants: the largest animals and their problems. *Palaeontology*, **41**: 1231–1245.

Allmon, W.A. and Ross, R.M. (1990). Specifying causal factors in evolution: the paleontological contribution. In: R.M. Ross and W.A. Allmon (eds) *Causes of Evolution: A Paleontological Perspective*. Chicago: University of Chicago Press, pp. 1–17.

Arthur, W. (1997). *The Origin of Animal Body Plans*. Cambridge: Cambridge University Press.

Bakker, R.T. (1983). The deer flees, the wolf pursues: incongruencies in predator–prey coevolution. In: D.J. Futuyma and M. Slatkin (eds) *Coevolution*. Sunderland, Mass.: Sinauer, pp. 350–382.

Bambach, R.K. (1993). Seafood through time: changes in biomass, energetics, and productivity in the marine ecosystem. *Paleobiology*, **19**: 372–397.

Bambach, R.K. (1999). Energetics in the global marine fauna: a connection between terrestrial diversification and change in the marine biosphere. *Geobios*, **32**: 131–144.

Bengtson, S. (1994). The advent of animal skeletons. In: S. Bengtson (ed.) *Early Life on Earth*. New York: Columbia University Press, pp. 412–425.

Benton, M.J. (1987). Progress and competition in macroevolution. *Biological Reviews*, **62**: 305–338.

Benton, M.J. (1996). On the nonprevalence of competitive replacement in the evolution of tetrapods. In: D. Jablonski, D.H. Erwin and J.H. Lipps (eds) *Evolutionary Paleobiology*. Chicago: University of Chicago Press, pp. 185–210.

Boulter, M.C., Gee, D. and Fisher, H.C. (1998). Angiosperm radiations at the Cenomanian/Turonian and Cretaceous/Tertiary boundaries. *Cret. Res.*, **19**: 107–112.

Brinkhuis, H., Bujak, J.P., Smit, J. *et al.* (1998). Dinoflagellate-based sea surface temperature reconstructions across the Cretaceous–Tertiary boundary. *Palaeogeogr., Palaeoclimatol., Palaeoecol.*, **141**: 67–83.

Canfield, D.E. and Teske, A. (1996). Late Proterozoic rise in atmospheric oxygen concentration inferred from phylogenetic and sulphur-isotope studies. *Nature*, **382**: 127–132.

Conway Morris, S. (1998a). Palaeontology: grasping the opportunities in the science of the twenty-first century. *Geobios*, **30**: 895–904.

Conway Morris, S. (1998b). The evolution of diversity in ancient ecosystems: a review. *Phil. Trans. Roy. Soc. London B*, **353**: 327–345.

Copper, P. (1988). Ecological succession in Phanerozoic reef ecosystems: is it real? *Palaios*, **3**: 136–152.

Copper, P. (1994a). Ancient reef ecosystem expansion and collapse. *Coral Reefs*, **13**: 3–11.

Copper, P. (1994b). Reefs under stress: the fossil record. *Cour. Forsch.-Inst. Senckenberg*, **192**: 87–94.

D'Hondt, S., King, J. and Gibson, C. (1996). Oscillatory marine response to the Cretaceous–Tertiary impact. *Geology*, **24**: 611–614.

D'Hondt, S., Donaghay, P., Zachos, J.C. *et al.* (1998). Organic carbon fluxes and ecological recovery from the Cretaceous–Tertiary mass extinction. *Science*, **282**: 276–279.

Erwin, D.H. (1989). Regional paleoecology of Permian gastropod genera, southwestern United States and the end-Permian mass extinction. *Palaios*, **4**: 424–438.

Erwin, D.H. (1994). Early introduction of major morphological innovations. *Acta Palaeont. Polonica*, **38**: 281–294.

Erwin, D.H. (1998). The end and the beginning: recoveries from mass extinctions. *Trends Ecol. Evol.*, **13**: 344–349.

Erwin, D.H. (1999). The origin of bodyplans. *Am. Zool.*, **39**: 617–629.

Erwin, D.H. (2001). Lessons from the past: Evolutionary impacts of mass extinctions. Proc. Natl. Acad. Sci. USA, **98**: 5399–5403.

Erwin, D.H., Valentine, J.W. and Jablonski, D. (1997). The origin of animal body plans. *Am. Sci.*, **85**: 126–137.

Faure, K., de Wit, M.J. and Willis, J.P. (1996). Late Permian global coal discontinuity and Permian–Triassic boundary 'events'. In: *Nine Gondwana*. Rotterdam: Balkema, pp. 1075–1089.

Flügel, E. and Kiessling, W. (2002). Patterns of Phanerozoic reef crises. SEPM Spec. Publ. 72691–733.

Foote, M. (1999). Morphological diversity in the evolutionary radiation of Paleozoic and post-Paleozoic crinoids. *Paleobiology*, **25** (Supplement to No. 2): 115pp.

Foote, M. (2003). Origination and extinction through the Phanerozoic: A new approach. J. Geol., **111**: 125–148.

Gilbert, S.F. (1997). *Developmental Biology*. 5th edition. Sunderland, Mass.: Sinauer, 918pp.

Gili, C. and Martinell, J. (1994). Relationship between species longevity and larval ecology of nassariid gastropods. *Lethaia*, **27**: 291–299.

Gili, E., Masse, J.-P. and Skelton, P.W. (1995). Rudists as gregarious sediment-dwellers, not reef-builders, on Cretaceous carbonate platforms. *Palaeogeogr., Palaeoclimatol., Palaeoecol.*, **118**: 245–267.

Gilinsky, N.L. (1994). Volatility and the Phanerozoic decline of background extinction intensity. *Paleobiology*, **20**: 445–458.

Gould, S.J. (1977). Eternal metaphors of paleontology. In: A. Hallam (ed.) *Patterns of Evolution*. Amsterdam: Elsevier, pp. 1–26.

Gould, S.J. (2002). The structure of evolutionary theory. Cambridge, MA: Harvard University Press, 1433pp.

Grabau, A.W. (1940). *The Rhythm of the Ages*. Peking: Henri Vetch, 561pp.

Grantham, T.A. (1995). Hierarchical approaches to macroevolution: recent work on species selection and the 'effect hypothesis'. *Ann. Rev. Ecol. Syst.*, **26**: 301–322.

Hallam, A. and Wignall, P.B. (1997). Mass Extinctions and their Aftermath. Oxford: Oxford University Press, 320pp.

Hansen, T.A. (1978). Larval dispersal and species longevity in Lower Tertiary gastropods. *Science*, **199**: 885–887.

Hansen, T.A. (1980). Influence of larval dispersal and geographic distribution on species longevity in neogastropods. *Paleobiology*, **6**: 193–207.

Hansen, T.A. (1988). Early Tertiary radiation of molluscs and the long-term effects of the Cretaceous–Tertiary extinction. *Paleobiology*, **14**: 37–51.

Hansen, T.A., Farrell, B.R. and Upshaw, B. III. (1993). The first 2 million years after the Cretaceous–Tertiary boundary in east Texas: rate and paleoecology of the molluscan recovery. *Paleobiology*, **19**: 251–265.

Harper, E.M. (1991). The role of predation in the evolution of cementation in bivalves. *Palaeontology*, **34**: 455–460.

Harper, E.M. and Skelton, P.W. (1993). The Mesozoic Marine Revolution and epifaunal bivalves. *Scripta Geol. Special Issue*, **2**: 127–153.

Heinberg, C. (1999). Lower Danian bivalves, Stevns Klint, Denmark: continuity across the K/T boundary. *Palaeogeogr., Palaeoclimatol., Palaeoecol.*, **154**: 87–106.

Henbest, L.G. (1952). Significance of evolutionary explosions for diastrophic division of Earth history – introduction to the symposium. *J. Paleontol.*, **26**: 299–318.

Hoffman, P.F., Kaufman, A.J., Halverson, G.P. and Schrag, D.P. (1998). A Neoproterozoic snowball Earth. *Science*, **281**: 1342–1346.

Jablonski, D. (1980). Apparent versus real biotic effects of transgression and regression. *Paleobiology*, **6**: 397–407.

Jablonski, D. (1986a). Larval ecology and macroevolution of marine invertebrates. *Bull. Mar. Sci.*, **39**: 565–587.

Jablonski, D. (1986b). Background and mass extinctions: the alternation of macroevolutionary regimes. *Science*, 231: 129–133.

Jablonski, D. (1987). Heritability at the species level: analysis of geographic ranges of Cretaceous mollusks. *Science*, 238: 360–363.

Jablonski, D. (1995). Extinction in the fossil record. In: R.M. May and J.H. Lawton (eds) *Extinction Rates*. Oxford: Oxford University Press, pp. 25–44.

Jablonski, D. (1996). Mass extinctions: persistent problems and new directions. *Geol. Soc. Am. Spec. Paper*, 307: 1–11.

Jablonski, D. (1998). Geographic variation in the molluscan recovery from the end-Cretaceous extinction. *Science*, 279: 1327–1330.

Jablonski, D. (2000). Micro- and macroevolution: the infusion of scale and hierarchy into evolutionary biology. *Paleobiology*, 26 (Suppl. to No. 4): 15–52.

Jablonski, D. and Bottjer, D.J. (1990). The ecology of evolutionary innovations: the fossil record. In: M.H. Nitecki (ed.) *Evolutionary Innovations*. Chicago: University of Chicago Press, pp. 253–288.

Jablonski, D. and Bottjer, D.J. (1991). Environmental patterns in the origins of higher taxa: the post-Paleozoic fossil record. *Science*, 252: 1831–1833.

Jablonski, D. and Raup, D.M. (1995). Selectivity of end-Cretaceous marine bivalve extinctions. *Science*, 268: 389–391.

Jablonski, D. and Sepkoski, J.J. Jr. (1996). Paleobiology, community ecology, and scales of ecological pattern. *Ecology*, 77: 1367–1378.

Jablonski, D. (2001). Lessons from the past: Evolutionary impacts of mass extinctions. *Proc. Natl. Acad. Sci. USA*, 98: 5393–5398.

Jablonski, D. (2002). Dead clade walking: survival without recovery after mass extinctions. *Proc. Natl. Acad. Sci. USA*, 99: 8139–8144.

Jablonski, D., Roy, K. and Valentine, J.W. (2003). Evolutionary macroecology and the fossil record. In: T.M. Blackburn and K.J. Gaston (eds) *Macroecology: Concepts and consequences*. Oxford: Blackwell Science, pp. 368–390.

Jacobs, D.K. (1990). Selector genes and the Cambrian radiation of Bilateria. *Proc. Natl. Acad. Sci. USA*, 87: 4406–4410.

Janis, C.M. and Wilhelm, P.B. (1993). Were there mammalian pursuit predators in the Tertiary? Dances with wolf avatars. *J. Mammalian Evol.*, 1: 103–125.

Kammer, T.W., Baumiller, T.K. and Ausich, W.I. (1998). Evolutionary significance of differential species longevity in Osagean–Meramecian (Mississippian) crinoid clades. *Paleobiology*, 24: 155–176.

Kelley, P.H. (1989). Evolutionary trends within bivalve prey of Chesapeake Group naticid gastropods. *Hist. Biol.*, 2: 139–156.

Kelley, P.H. (1991). The effect of predation intensity on rate of evolution of five Miocene bivalves. *Hist. Biol.*, 5: 65–78.

Kirchner, J.W. and Weil, A. (2000). Delayed biological recovery from extinctions throughout the fossil record. *Nature*, 404: 177–180.

Kirschvink, J.L., Ripperdan, R.L. and Evans, D.A. (1997). Evidence for a large-scale reorganization of Early Cambrian continental masses by inertial interchange true polar wander. *Science*, 277: 541–545.

Kitchell, J.A., Clark, D.L. and Gombos, A.M. Jr. (1986). Biological selectivity of extinction: a link between background and mass extinction. *Palaios*, 1: 504–511.

Knoll, A.H. (1996). Breathing room for early animals. *Nature*, 382: 111–112.

Knoll, A.H. and Carroll, S.B. (1999). Early animal evolution: emerging views from comparative biology and geology. *Science*, 284: 2129–2137.

Knoll, A.H. and Carroll, S.B. (1999). Early animal evolution: Emerging views from comparative biology and geology. Science, 284: 2129–2137.

Lidgard, S., McKinney, F.K. and Taylor, P.D. (1993). Competition, clade replacement, and a history of cyclostome and cheilostome bryozoan diversity. *Paleobiology*, 19: 352–371.

Lockwood, R. (2003). Abundance not linked to survival across the end-Cretaceous mass extinction: Patterns in North American bivalves. Proc. Natl. Acad. Sci. USA, 100: 2478–2482.

Lockwood, R. (1998). K–T extinction and recovery: taxonomic versus morphological patterns in veneroid bivalves (Abstract). *Geol. Soc. Am. Abstr. Programs*, 30: A-286.

Lull, R.S. (1918). The pulse of life. In: J. Barrell, C. Schuchert, L.L. Woodruff, R.S. Lull and E. Huntington (eds) *The Evolution of the Earth and its Inhabitants*. New Haven: Yale University Press, pp. 109–146.

Lupia, R. (1999). Discordant morphological disparity and taxonomic diversity during the Cretaceous angiosperm radiation: North American pollen record. *Paleobiology*, 25: 1–28.

MacLeod, N. *et al.* (1997). The Cretaceous–Tertiary biotic transition. *J. Geol. Soc. London*, 154: 265–292.

Magallon, S., Crane, P.R. and Herendeen, P.S. (1999). Phylogenetic pattern, diversity, and diversification of dicots. *Ann. Missouri Bot. Garden*, 86: 297–372.

McGhee, G.R. Jr. (1996). *The Late Devonian Mass Extinction*. New York: Columbia University Press, 303pp.

McKinney, F.K., Lidgard, S., Sepkoski, J.J. Jr. and Taylor, P.D. (1998). Decoupled temporal patterns of evolution and ecology in two post-Paleozoic clades. *Science*, **281**: 807–809.

McKinney, M.L. (1997). Extinction vulnerability and selectivity: combining ecological and paleontological views. *Annu. Rev. Ecol. Syst.*, **28**: 495–516.

McKinney, M.L. (1998). Biodiversity dynamics: niche preemption and saturation in diversity equilibria. In: M.L. McKinney and J.A. Drake (eds) *Biodiversity Dynamics*. New York: Columbia University Press, pp. 1–16.

Miller, A.I. (1997a). Comparative diversification dynamics among palaeocontinents during the Ordovician radiation. *Geobios Mém. Spec.*, **20**: 397–406.

Miller, A.I. (1997b). Dissecting global diversity patterns: examples from the Ordovician Radiation. *Annu. Rev. Ecol. Syst.*, **28**: 85–104.

Miller, A.I. (1998). Biotic transitions in global marine diversity. *Science*, **281**: 1157–1160.

Miller, A.I. and Mao, S. (1995). Association of orogenic activity with the Ordovician radiation of marine life. *Geology*, **23**: 305–308.

Miller, A.I. and Sepkoski, J.J. Jr. (1988). Modeling bivalve diversification: the effect of interaction on a macroevolutionary system. *Paleobiology*, **14**: 364–369.

Miller, D.J., LaBarbera, M. (1995). Effects of foliaceous varices on the mechanical properties of *Chicoreus dilectus* (Gastropoda, Muricidae). *J. Zool.*, **236**: 151–160.

Niklas, K.J. (1997). *The Evolutionary Biology of Plants*. Chicago: University of Chicago Press, 449pp.

Palmer, A.R. (1979). Fish predation and the evolution of gastropod shell sculpture: experimental and geographic evidence. *Evolution*, **33**: 697–713.

Patzkowsky, M.E. (1995). A hierarchical branching model of evolutionary radiations. *Paleobiology*, **21**: 440–460.

Pimm, S.L., Jones, H.L. and Diamond, J. (1988). On the risk of extinction. *Am. Nat.*, **132**: 757–785.

Raup, D.M. (1991). A kill curve for Phanerozoic marine species. *Paleobiology*, **17**: 37–48.

Raup, D.M. (1992). Large-body impact and extinction in the Phanerozoic. *Paleobiology*, **18**: 80–88.

Raup, D.M. and Jablonski, D. (1993). Geography of end-Cretaceous marine bivalve extinctions. *Science*, **260**: 971–973.

Rayner, J.M.V. and Wootton, R.J. (eds) (1991). *Biomechanics in Evolution*. Cambridge: Cambridge University Press.

Retellack, G.J., Veevers, J.J. and Morante, R. (1996). Global coal gap between Permian–Triassic extinction and Middle Triassic recovery of peat-forming plants. *Geol. Soc. Am. Bull.*, **108**: 195–207.

Rosenzweig, M.L. and McCord, R.D. (1991). Incumbent replacement: evidence for long-term evolutionary progress. *Paleobiology*, **17**: 202–213.

Roy, K. (1996). The roles of mass extinction and biotic interaction in large-scale replacements: a reexamination using the fossil record of stromboidean gastropods. *Paleobiology*, **22**: 436–452.

Rye, R., Holland, H.D. (1998). Paleosols and the evolution of atmospheric oxygen: a critical review. *Am. J. Sci.*, **298**: 621–672.

Scheltema, R.S. (1977). Dispersal of marine organisms: paleobiogeographic and biostratigraphic implications. In: E.G. Kauffman and J.E. Hazel (eds) *Concepts and Methods of Biostratigraphy*. Stroudsburg, PA: Dowden, Hutchinson & Ross, pp. 83–108.

Scheltema, R.S. (1989). Planktonic and non-planktonic development among prosobranch gastropods and its relationship to the geographic range of species. In: J.S. Ryland and P.A. Tyler (eds) *Reproduction, Genetics and Distributions of Marine Organisms*. Fredensborg, Denmark: Olsen & Olsen, pp. 183–188.

Schmidt-Kittler, N., Vogel, K. (eds) (1991). *Constructional Morphology and Evolution*. Berlin: Springer, 409pp.

Schultz, P.H. and D'Hondt, S. (1996). Cretaceous–Tertiary (Chicxulub) impact angle and its consequences. *Geology*, **24**: 963–967.

Sepkoski, J.J. Jr. (1996). Competition in macroevolution: the double wedge revisited. In: D. Jablonski, D.H. Erwin and J.H. Lipps (eds) *Evolutionary Paleobiology*. Chicago: University of Chicago Press, pp. 211–255.

Sepkoski, J.J. Jr. (1998). Rates of speciation in the fossil record. *Phil. Trans. Roy. Soc. London B*, **353**: 315–326.

Sereno, P.C. (1999). The evolution of dinosaurs. *Science*, **284**: 2137–2147.

Skelton, P.W. (1991). Morphogenetic versus environmental cues for adaptive radiations. In: N. Schmidt-Kittler and K. Vogel (eds) *Constructional Morphology and Evolution*. Berlin: Springer-Verlag, pp. 375–388.

Smith, A.B. and Jeffery, C.H. (1998). Selectivity of extinction among sea urchins at the end of the Cretaceous period. *Nature*, **392**: 69–71.

Stanley, G.D. Jr. (1992). Tropical reef ecosystems and their evolution. In: W.A. Nierenberg (ed.) *Encyclopedia of Earth System Science, Volume 4*. San Diego: Academic Press, pp. 375–388.

Stanley, S.M. (1976). Fossil data and the Precambrian–Cambrian evolutionary transition. *Am. J. Sci.*, **276**: 56–76.

Stanley, S.M. (1988). Paleozoic mass extinctions: shared patterns suggest global cooling as a common cause. *Am. J. Sci.*, 334–352.

Stanley, S.M. (1990). The general correlation between rate of speciation and rate of extinction: fortuitous causal linkages. In: R.M. Ross and W.A. Allmon (eds) *Causes of Evolution: A Paleontological Perspective*. Chicago: University of Chicago Press, pp. 103–127.

Stanley, G.D., Jr. (ed) (2001). The history and sedimentology of ancient reef systems. New York: Kluwer Academic, 458pp.

Stone, M.H.I. (1998). On predator deterrence by pronounced shell ornament in epifaunal bivalves. *Palaeontology*, **41**: 1051–1068.

Talent, J.A. (1988). Organic reef-building: episodes of extinction and symbiosis? *Senckenbergiana Lethaea*, **69**: 315–368.

Thomas, R.D.K., Shearman, R.M. and Stewart, C.W. (2000). Evolutionary exploitation of design options by the first animals with hard skeletons. Science, **288**: 1239–1242.

Toon, O.B., Zahnle, K., Morrison, D. *et al.* (1997). Environmental perturbations caused by the impacts of asteroids and comets. *Rev. Geophys.*, **35**: 41–78.

Umbgrove, J.H.F. (1947). *The Pulse of the Earth*. 2nd edition. The Hague, Netherlands: Martin Nijhoff, 358pp.

Valentine, J.W. (1995). Why no new phyla after the Cambrian? Genome and ecospace hypotheses revisited. *Palaios*, **10**: 190–194.

Valentine, J.W. and Erwin, D.H. (1987). Interpreting great developmental experiments: the fossil record. In: R.A. Raff and E.C. Raff (eds) *Development as an Evolutionary Process*. New York: Liss, pp. 71–107.

Valentine, J.W. and Jablonski, D. (1983). Larval adaptations and patterns of brachiopod diversity in space and time. *Evolution*, 37: 1052–1061.

Valentine, J.W. and Jablonski, D. (1986). Mass extinctions: sensitivity of marine larval types. *Proc. Natl. Acad. Sci. USA*, **83**: 6912–6914.

Valentine, J.W., Jablonski, D. and Erwin, D.H. (1999). Fossils, molecules and embryos: new perspectives on the Cambrian explosion. *Development*, **126**: 851–859.

Valentine, J.W. (2002). Prelude to the Cambrian explosion. Annu. Rev. Earth Planet. Sci., **30**: 285–306.

Valentine, J.W., Jablonski, D. and Erwin, D.H. (1999). Fossils, molecules and embryos: new perspectives on the Cambrian explosion. Development, **126**: 851–859.

Van Valen, L.M. (1994). Concepts and the nature of selection by extinction: is generalization possible? In: W. Glen (ed.) *Mass Extinction Debates: How Science Works in a Crisis*. Stanford, CA: Stanford University Press, pp. 200–216.

Van Valkenburgh, B. (1999). Major patterns in the history of carnivorous mammals. *Ann. Rev. Earth Planet. Sci.*, **27**: 463–493.

Vermeij, G.J. (1977). The Mesozoic marine revolution: evidence from snails, predators and grazers. *Paleobiology*, 3: 245–258.

Vermeij, G.J. (1982). Gastropod shell form, breakage, and repair in relation to predation by the crab *Calappa. Malacologia*, **23**: 1–12.

Vermeij, G.J. (1987). *Evolution and Escalation*. Princeton, New Jersey: Princeton University Press, 527pp.

Vermeij, G.J. (1994). The evolutionary interaction among species: selection, interaction, and coevolution. *Annu. Rev. Ecol. Syst.*, **25**: 219–236.

Vermeij, G.J. (1995). Economics, volcanos, and Phanerozoic revolutions. *Paleobiology*, **21**: 125–152.

Wagner, P.J. (1995). Testing evolutionary constraint hypotheses with Early Paleozoic gastropods. *Paleobiology*, **21**: 248–272.

Watts, P.C., Thorpe, J.P. and Taylor, P.D. (1998). Natural and anthropogenic dispersal mechanisms in the marine environment: a study using cheilostome Bryozoa. *Phil. Trans. Roy. Soc. London B*, **353**: 453–464.

Webb, G.E. (1996). Was Phanerozoic reef history controlled by the distribution of non-enzymatically secreted reef carbonates (microbial carbonate and biologically induced cement)? *Sedimentology*, **43**: 947–971.

Williams, G.C. (1992). *Natural Selection*. Oxford: Oxford University Press, 208pp.

Wolpert, L., Beddington, R., Brockes, J. *et al.* (1998). *Principles of Development*. Oxford: Oxford University Press, 484pp.

Wood, R. (1998). The ecological evolution of reefs. *Annu. Rev. Ecol. Syst.*, **29**: 179–206.

Wootton, J.T. (1994). The nature and consequences of indirect effects in ecological communities. *Annu. Rev. Ecol. Syst.*, **25**: 443–466.

14

The causes of Phanerozoic extinctions

Norman MacLeod

ABSTRACT

See Plates 20 and 21

In historical sciences such as palaeontology, cause and effect must be established by documenting multiple associations of proposed causes and observed effects in the correct logical sequence. Employing this approach, the Phanerozoic genus-level extinction record can be compared with a variety of environmental-change indices in an attempt to identify general extinction causes. Results of a series of such analyses suggest that the well-established decline in 'background' extinction intensity has been influenced strongly by generalized tectonic and contingent evolutionary–ecological factors whose operation has resulted in (i) an overall increase in the organization and intensity of marine circulation, (ii) an overall decrease in the amount of continent-derived nutrients reaching marine environments, presumably due to the progressive diversification of terrestrial plants, and (iii) an overall increase in intra-oceanic nutrient recycling rates, due largely to the Mesozoic–Cenozoic evolution and subsequent diversification of modern phytoplankton clades.

There is also evidence for similar first-order trends of increasing global average surface temperatures and expansion of tropical habitats over this same Phanerozoic time interval. The operation of these factors would be expected to impart a pattern of directional variation to the history of biodiversification in both planktonic and benthic marine habitats (with consequent macroevolutionary effects on the organisms living there) that should manifest itself in a wide variety of data. Patterns of variation consistent with this hypothesis do indeed appear to be present in the fossil record, especially in the records of marine invertebrate extinctions and diversifications. In addition, a set of relatively shorter-term, tectonically influenced, environmental perturbations (in particular, rapid sea-level fall, continental flood-basalt volcanism) exhibit consistent associations with temporally localized extinction-intensity peaks or so-called 'mass extinction' events. Taken as a whole, these results suggest that the primary controls on long and intermediate-term taxonomic extinction and diversification patterns at all scales are tectonic and likely mediated through the waxing and waning of lineages that occupy the base of marine and terrestrial ecological hierarchies.

14.1 | Introduction

The term extinction elicits an emotional reaction in most people. While extinct modern species (e.g. the Dodo, Great Auk, Tasmanian Wolf) and ancient groups (e.g. dinosaurs[1], ammonites, trilobites) have come to symbolize the fact of extinction, the unfamiliarity of these animals gives many an impression of extinction as an exceptional process that happens to strange creatures that are far away in time and space. Nothing could be further from the truth. Extinction is one of the most common of all ecological–evolutionary processes. Since less than 1% of species that have ever existed on

Evolution on Planet Earth
ISBN 0-12-598655-6

Earth are alive at present, one must agree with palaeontologist David Raup's tongue-in-cheek assessment that, to a first approximation, all life on Earth is extinct. More seriously though, extinction is the all but inevitable corollary of evolution via natural selection in the sense that 'survival of the fittest' logically implies the extinction (either locally or globally) of those individuals or species who lose the evolutionary competition. It is also now commonly recognized that extinction plays an important creative role in evolutionary processes by circumventing the adaptive inertia of incumbency and freeing up resources that can then be taken advantage of by new lineages. Indeed, if the processes that lead to extinction were not as common as they are, it is difficult to see how life could have achieved its present complexity or diversity. Extinction is all around us. It has taken place throughout the history of life. It is inextricably linked to changes in the Earth's physical environment. It is an integral and necessary part of the natural world.

Extinction, coupled with morphological novelty, also creates the basis for the scientific and popular interest in the discipline of palaeontology. If most ancient organisms still existed and were fundamentally similar to their modern counterparts there would be little reason to study the often fragmentary remains that comprise the fossil record. The obvious uniqueness of that record's constituents naturally leads one to wonder how ancient animals and plants lived, what their curious morphological structures were used for, what relations they had to other animals and plants that lived in their time (as well as those of the modern world), and why such impressive and seemingly well-adapted creatures vanished. Among these questions, all but the last have long been the subjects of productive palaeontological research programmes. The study of extinction *per se*, however, has lagged far behind other aspects of palaeontology.

This is not to say that palaeontologists have lacked an appreciation for the importance of extinction. For over 150 years the fact of extinction – as recorded by the patterns of fossil species' last appearance horizons in local stratigraphical sections and cores – has been one of the most common paleontological observations. The entire field of biostratigraphy is predicated on the accurate identification of species' extinction (and appearance) horizons. Indeed, the result of this activity has been the greatest single contribution to the study of Earth history ever made; the geological time-scale. Rather, the problem raised by attempts to understand extinction causes has lain in the nature of the data that must be employed in its study and in the logic that must be used to infer cause and effect in such data.

Briefly put, the problem is that extinctions in the fossil (as well as in the modern) record are identified by what is not observed; by negative evidence. Dependence on negative evidence severely limits our ability to test competing extinction hypotheses. This is because local disappearances can occur for a large number of reasons. For example, a previously common species may 'appear to disappear' from a local stratigraphical succession because the local environment changed causing the resident population to (i) migrate to another (unsampled) locality, (ii) persist in the area but become so rare as to not be evident under sampling programmes designed for its study within previous (more abundant) times, (iii) persist in the area for a time past the last observed sample but have this section of its stratigraphical occurrence removed due to erosion, (iv) persist in the area for a time past the last observed sample but have this section of its stratigraphical occurrence condensed due to a lowering in the sediment-rock accumulation rate, and so on. The physical–ecological processes and physiological–ecological limits that drive local and, ultimately, global populations extinct also usually take place over timescales well below the normal limits of stratigraphical resolution employed for biostratigraphical analysis often altering the character of the last appearance pattern (see MacLeod, 1991 for an example). While a dependence on negative evidence is also true of appearances (where it causes similar problems), in these cases the evolutionary process of ancestry and descent provides means whereby the accuracy of observed first appearance timings can be tested independently (see Benton, 1996).

Unfortunately, no such over-riding theoretical framework is available for addressing the inherent ambiguities of negative evidence in the extinction record or assessing objectively the quality of local last appearance patterns.

The most common result of this apparent unknowability of species extinction causes has been to open the floodgates of speculation that, in other branches of the Earth sciences, are kept more or less shut by traditional commitments to empiricism. Thus, the peer-reviewed extinction literature is replete with imaginative, but fundamentally untestable, extinction causes (e.g. competition with caterpillars, mutagenic cosmic radiation as 'explanations' for the K–T mass extinction, see Benton, 1990). This sort of 'just-so' storytelling, coupled with the practical difficulties involved in using historical data to test alternative models within a formal hypothetico-deductive framework, effectively kept the study of extinction causes out of the scientific mainstream for several generations.

Many have argued that this situation changed with the 1980 proposal of the bolide-extinction hypotheses for the Cretaceous–Tertiary mass-extinction event (also described as the K–T or Maastrichtian event; see Alvarez *et al.*, 1980). In my view such has not been the case. The Alvarez *et al.* (1980) study succeeded in (i) detailing a previously suggested extinction mechanism (asteroid impact, see de Laubenfels, 1956) and (ii) providing physical evidence that the Earth had experienced a bolide impact at the end of the Cretaceous. Since 1980 an enormous research effort has examined both the physical and biotic predictions of this hypothesis. Nonetheless, whereas the physical and mechanistic sides of the Alvarez *et al.* (1980) model quickly achieved consensus status among Earth scientists of all types, the relation between the observed physical and biotic data remains controversial.

The Alvarez *et al.* (1980) model predicts that the K–T extinction event – and perhaps other mass extinction events – was caused by climatic perturbations induced by the collision of the Earth with a bolide (=meteor or comet) approximately 10 km in diameter. If this mechanism is correct it follows logically that a large number of fossil species' last appear-ance horizons should coincide with the layer of impact debris (e.g. iridium-bearing minerals, nickel-rich minerals, shocked minerals, impact glass) known to be distributed globally in uppermost Cretaceous sediments. This fundamental prediction has not been substantiated. Relatively few organismal extinction horizons are known to coincide with the K–T impact debris layer in continuous stratigraphical successions (MacLeod *et al.*, 1997; Hallam and Wignall, 1997). In those instances where a biotic turnover event appears to be coincident with K–T impact debris the magnitude of the biotic event has often been accentuated either by a stratigraphical hiatus (see MacLeod and Keller, 1996; MacLeod *et al.*, 1997; MacLeod, 1998; and references therein) or by biostratigraphers failing to report 'Cretaceous' fossils in overlying Danian sediments because they assume (for the most part without any empirical justification) that all such occurrences to represent older fossils that have been eroded and redeposited in younger sediments (e.g. compare Smit, 1990 with Canudo *et al.*, 1991)[2].

In addition to the demonstrably false assertion that all 'Cretaceous' species recovered from Danian sediments are redeposited (see Barrera and Keller, 1990; MacLeod and Keller, 1994), proponents of a link between the K–T impact and K–T extinctions have long argued that many of the biotic extinctions that appear to pre-date the impact debris layer do so only because of the patchy distribution of species' occurrences in the fossil record – the so-called Signor–Lipps effect (first described by Shaw, 1964, see Ward *et al.*, 1986, 1991; Marshall and Ward, 1996). This proposition neglects to recall Signor and Lipps' (1982) explicit statements that their sampling arguments refer to fundamental uncertainties in the position of last appearance datums in the fossil record and cannot be used to provide post-hoc justification for particular extinction timing patterns within the interval of uncertainty.

In short, 20 years of wrestling with the inherent palaeontological uncertainties of the best-studied mass extinction have failed to resolve the cause(s) of that event. Opinion polls conducted in 1984 and 1996 (see Galvin, 1998 and references therein) support this conclusion in that both indicated a non-specific, multi-casual scenario for the K–T extinctions as enjoying the widest level of support among professional geoscientists, especially palaeontologists. Even the originators of the impact-extinction hypothesis have recently despaired of achieving a final resolution. Referring to marine invertebrates (on the basis of whose fossil record the K–T event is defined as a mass extinction) Alvarez (1997: 16) writes 'perhaps they were victims of food-chain collapse or perhaps their shells were dissolved in acidified sea-water, but no one knows'. Such statements do not amount to compelling causal explanations.

This situation is little better for other Phanerozoic (an interval of time encompassing the approximately 550 million years since shelly fossils become a common feature of the stratigraphical record) stage-level mass extinction events. Book-length treatises on the Permo-Triassic (Erwin, 1993) and Ordovician (McGhee, 1996) extinctions have failed to confirm or refute the idea that these events were caused by the same processes, different processes, or particular combinations of processes. In a recent survey of mass extinction events Hallam and Wignall (1997) argued that fluctuations in sea level have been a primary factor in driving all mass extinction events. But counter-arguments to those advanced by these works abound in the technical literature and it is very difficult for readers not intimately familiar with the discipline of extinction studies, the global stratigraphy of the extinction interval, or the systematics of the organisms involved, to come to grips with the empirical data necessary to arrive at informed, independent opinions. In addition, most contemporary works on extinction causes focus on individual extinction events rather than commonalities existing among events or do so without reference to any commonly agreed-upon standards of hypothesis testing.

Clearly a new approach is needed if the field of extinction studies is to progress toward resolving any of its long-standing debates. This contribution attempts to outline the data and methods that will be needed to create a new, more systematic, integrative, and analytic approach to extinction studies. In particular, it will focus on formal hypothesis testing and use of repeatability as a method of evaluating cause and effect hypotheses in the paleontological record. As David Raup has pointed out (Raup, 1991: 151): 'There is no way of assessing cause and effect [in historical studies] except to look for patterns of coincidence – and this requires multiple examinations of each cause–effect pair. If all extinction events are different the deciphering of any one of them will be next to impossible.' Accordingly, my approach will focus on the entire known Phanerozoic extinction record (as opposed to the record of one or a few individual events) and make comparisons between it and the stratigraphical records of known or suspected extinction causes. It is hoped that the results of such an analysis will broadly indicate the major controls on intermediate-term and long-term Phanerozoic extinction patterns, clarify the nature of the abiotic–biotic feedback mechanisms driving these large-scale patterns, and suggest additional data that might be collected to further test those conclusions.

14.2 The data

The extinction data that form the primary focus of this analysis were originally presented in a short summary paper on genus-level marine Phanerozoic extinction patterns by J.J. Sepkoski, Jr (1994). These data (Figure 14.1) represent an extension of Sepkoski's better-known Permian–Recent extinction data set, upon which mass extinction magnitudes have been estimated and the mass-extinction periodicity hypotheses tested (see Raup and Sepkoski, 1984, 1986; Sepkoski and Raup, 1986; Sepkoski, 1990).

These data are subdivided at the 'stage' level of geological time resolution and expressed as per cent extinctions occurring within these time intervals. Such data have many widely discussed drawbacks (e.g. failure to preserve information concerning the positions of intra-stage extinction horizons, failure to incorporate phylogenetics information on lineage presence during specified time intervals). Nevertheless, they also possess the compensating advantages of being widely available and widely regarded as accurately reflecting the general features of the Phanerozoic extinction record. The quality of these data may be tested by compiling independent summaries of genus-level stratigraphical ranges from the primary literature.

Data summaries such as these are indispensable for extinction-related (and diversification-related) studies. They permit analysts to examine patterns over the broad scope of Phanerozoic history instead of forcing arbitrary, *a priori* decisions regarding the boundaries of various extinction events. It is also worth mentioning that biostratigraphical data available for most marine fossil groups cannot be temporally resolved below the level of the stratigraphical stage. While this prevents causal hypotheses that require substage levels of time resolution from being evaluated, such a restriction is not as problematical as it might appear. Raup's dictum of repeatability suggests that true causal factors should be associated with their hypothetical outcomes throughout Earth history. So long as the proposed causal mechanisms are independent of one another and exhibit different stage-level occurrence patterns, the prediction of non-random association can be evaluated at least at the stage level.

Two primary patterns are immediately apparent from the Sepkoski extinction data (Figure 14.1): (i) a consistent overall decline in extinction intensity from the Palaeozoic to the Cenozoic, and (ii) the fact that the most prominent local peaks in extinction intensity (e.g. Tatarian, Ashgillian, Norian, Maastrichtian) are separated from one another by intervals of relatively lower extinction intensity in an approximately cyclic manner. Both of these major extinction-record features have been noted previously. However, detailed analysis has, for the most part, been confined to family-level data from the Permian–Recent interval. Even more importantly, neither of these extinction-record features has been quantitatively compared to independent patterns of variation drawn from a broad range of environmental proxies.

14.3 The long-term extinction-intensity gradient

It has become something of a tradition among palaeontologists to subdivide the extinction

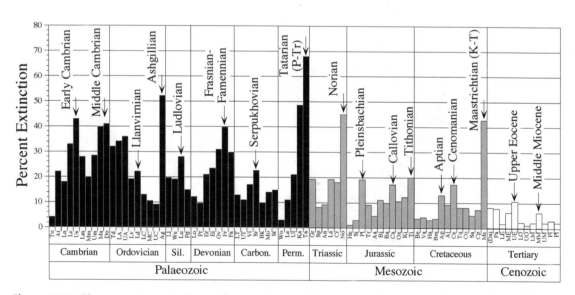

Figure 14.1 Phanerozoic marine invertebrate extinction data (redrawn from Sepkoski, 1994). Arrows mark stages exhibiting local elevations in extinction intensity.

intensity distribution into two subsets, applying the term 'mass extinction' to the multimodal peaks and distinguishing these from the intervening 'background' extinctions. The latter are practically defined as those intervals of relatively low extinction intensity that separate mass extinctions from one another. However, apart from the three, five, or seven largest stage-level extinction peaks, no consensus exists regarding the boundaries of 'mass' and 'background' extinction-intensity classes. This problem is illustrated in Figure 14.2 where the data from Figure 14.1 have been rearranged in rank order to reflect decreasing extinction magnitude.

Since the rank-order extinction magnitude distribution appears continuous, subdivision into separate mass or background classes is necessarily arbitrary. The relative position of an event in the sequence is also affected by the metric used to quantify extinction intensity (see Raup and Sepkoski, 1986). For example, the K–T event ranks third in extinction intensity over the last 250 My when estimated as the per cent family-level extinction (number of

extinction relative to standing diversity), but only sixth if estimated as the per family extinction rate (number of extinctions relative to stage duration relative to standing diversity, see Sepkoski and Raup, 1986). The relative positions of smaller events are even more strongly affected by the type of extinction-intensity index employed. As a result, the extinction-intensity distribution is more reminiscent of a fuzzy spectrum than a precisely defined, discontinuous function. Along this extinction-intensity spectrum mass extinctions and background extinctions occupy opposite poles separated by a more or less smoothly graded transition zone. In terms of causal processes, the continuous nature of the extinction-intensity spectrum suggests that the same types of processes responsible for small events are probably also responsible for the larger events where they likely operate in concert and/or at heightened magnitudes. Irrespective of the continuity of this spectrum, many authors continue to refer misleadingly to many of the smaller events as mass extinctions (e.g. the Pleistocene 'mass extinction'), either for rhetorical effect or

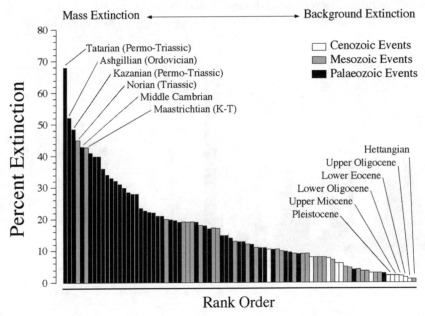

Figure 14.2 Phanerozoic extinction data rearranged by decreasing magnitude (see Figure 14.1). Note continuous nature of the extinction intensity distribution. These data suggest that all Phanerozoic extinctions are the result of the same types of processes varying in intensity over geological time.

to emphasize the severity of extinction effects for particular organismal groups (e.g. the Paleocene–Eocene benthic foraminiferal 'mass extinction').

The decreasing extinction-intensity gradient was first recognized in family-level data within the Permian (Tatarian)–Recent interval. Sepkoski's (1994) generic data shows that this same gradient is also a feature of the genus-level Phanerozoic extinction record (Figure 14.3). In fact, many Palaeozoic background extinction levels are several times larger than most Mesozoic or Cenozoic mass extinction events. Before attempting to account for this trend, though, it must first be determined whether the observed long-term gradient is an intrinsic feature of these data or a statistical artifact.

One obvious factor that might be driving this apparent gradient is a systematic change in the duration of stratigraphical stages over the course of the Phanerozoic. If stage durations have become progressively shorter over this interval – perhaps as a consequence of greater familiarity with Mesozoic and Cenozoic biotas or greater surface outcrop areas of Mesozoic and Cenozoic sediments – the decreasing extinction-intensity gradient may simply reflect the greater proportion of time available in Palaeozoic stages for extinctions to occur. However, a simple graphic inspection of Phanerozoic stage durations (Gradstein and Ogg, 1996) reveals no systematic bias toward longer Palaeozoic stages (Figure 14.4). A more formal regression analysis of stage duration on extinction intensity also returns non-significant results (e.g. for a linear model $r = 0.05$ and $F = 0.220$, DoF $=1$, 80). Thus, systematic biases in stage durations do not account for the Phanerozoic extinction-intensity gradient.

Alternatively, the observed extinction-intensity gradient may simply be the chance result of randomly varying extinction magnitudes. In cases such as this, the null hypothesis is that the apparently directional series actually represents a random walk through time; that is to say, a fortuitous, but by no means exceptional, concatenation pattern. Before we attempt to supply an interpretation for the extinction-intensity gradient, we must first evaluate and be able to reject this random walk hypothesis.

Manley (1997) reviews statistical approaches to the analysis of trends in time-series data using Sepkoski's Tatarian–Recent, family-level data. Application of Manley's (1997) Monte Carlo-based randomization test for time-series trends to determine the significance of the result (Figure 14.5) suggests that the genus-level Phanerozoic data depart even more strongly from the random walk model than the previous Tatarian–Recent family-level data. Given this result there is little doubt that the extinction-intensity gradient is an unusual pattern within these data that is wholly dependent on their ordering.

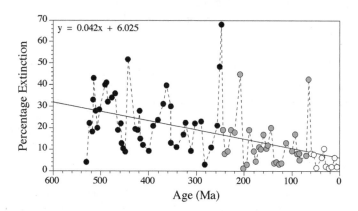

Figure 14.3 Decreasing gradient in Phanerozoic marine invertebrate extinction intensity. This gradient was quantified by linear least-squares regression of extinction intensity on time.

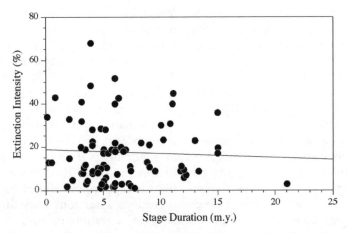

Figure 14.4 Scatterplot of stage duration (Gradstein and Ogg, 1996) versus per cent extinction intensity (Sepkoski, 1994) illustrating overall lack of correlation between these two data sets. The least-squares regression line (intensity on duration) exhibits a slope of 0.19% per My and is not significantly different from 0.0 ($\alpha = 0.95$). See text for discussion.

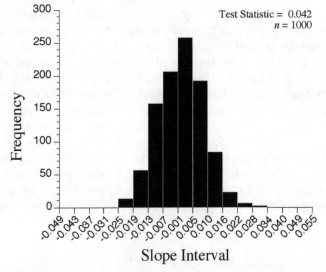

Figure 14.5 Results of 1000 Monte Carlo-simulated linear least-squares regression analyses of randomly ordered extinction intensity on time. This analysis constructed 1000 simulated Phanerozoic extinction histories via randomly sampling (with replacement) from the set of Sepkoski (1994) extinction intensities and then calculated the first-order linear regression of the simulated data on time. Slope values were then tabulated and compared to the observed extinction-intensity gradient in order to evaluate the overall degree of randomized structure that could be expected from the data set. Since the simulated regression slopes are all lower than the observed slope (see Figure 14.3) the significance of the observed pattern is judged to be high ($p < 0.001$).

Since the extinction-intensity gradient appears to be a statistically significant pattern within Sepkoski's genus-level extinction-intensity data set, we are now in a position to search for its cause. There have been various prior attempts to explain the causes of the Tatarian–Recent gradient pattern. Some authors have attributed it to improvements in species fitness

Plate 16 Pterosaurs were an abundant aerial form throughout the Mesozoic. Contrary to popular supposition not all pterosaurs were large, although some later Cretaceous forms became much bigger than any extant bird. The Jurassic pterodactyloid *Pterodactylus kochii* from the Solnhofen limestone of Bavaria was comparable in size to modern thrushes, although like most pterosaurs it had relatively long and pointed wings. (JMVR Bayerisches Staatssammlung, München). **See Chapter 10**

Plate 17 The largest modern flying birds have a maximum mass in the range 12–20 kg. At this size their aerial locomotion is restricted by the problems of taking off, and many large birds use gliding and soaring flight extensively. The lappet-faced vulture *Aegypius tracheliotus* is the heaviest old-World accipitrid vulture, and can take-off only by running into the wind. It will rely on thermal soaring for cross-country transport. (JMVR, Masai Mara, Kenya 1981). **See Chapter 10**

Plate 18 Emperor penguins (*Aptenodytes forsteri*) on sea-ice. Like all polar endotherms, penguins rely on metabolic processes to generate heat internally, and very effective insulation to minimize heat loss. (Copyright British Antarctic Survey). (P. Cooper). **See Chapter 11**

Plate 19 A fish (a juvenile *Pagothenia borchgrevinki*) and a variety of amphipod species resting on ice underwater in Antarctica. The amphipods have body fluids which are broadly isosmotic with seawater and are therefore not in danger of freezing. The fish, however, has dilute blood and would freeze in contact with ice were it not for the presence of a glycoprotein antifreeze in its body fluids. (Copyright British Antarctic Survey). (G. Wilkinson). **See Chapter 11**

Plate 20 Night-time image of a strombolian explosion near the summit of Mount Etna, Italy. Such events are caused by a gas bubble emerging beneath cooling lava. Volcanism brings minerals and gases to the Earth's surface, where they are available to organisms and may influence the composition of the atmosphere. (John Guest). **See Chapters 12 and 14**

Plates 20 and 21   Volcanic activity and massive lava flows have had major impacts through geological time, in creating land masses and islands, making minerals available at the Earth's surface, and affecting climate.

Plate 21   An active lava flow in Hawaii, from an eruption which lasted many years (Professor John Guest, UCL). Lava flows on a massive scale have occurred periodically during Earth's history, helping shape continents. (John Guest).
See Chapters 12 and 14 and 19

Plate 22   The Himalayan mountains, with their foothills, the Churia, in the foreground. The Himalayas and Tibetan Plateau formed some 45 million years ago when the Indian continental plate collided with that of Eurasia. Tectonic processes of mountain-building, and subsequent erosion, have had major impacts on global climate and mineral availability (Adrian Lister).
See Chapters 12, 17 and 19

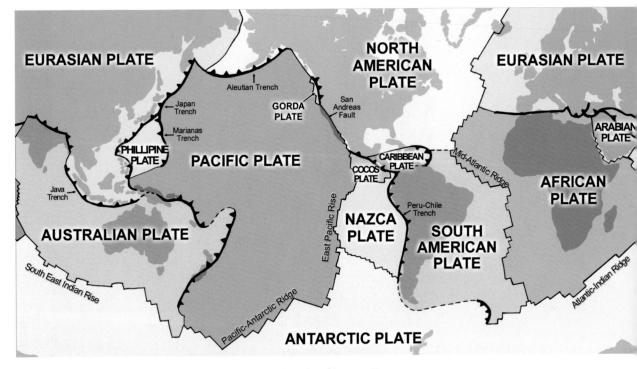

Plate 23 The major tectonic plates and their boundaries. **See Chapter 15**

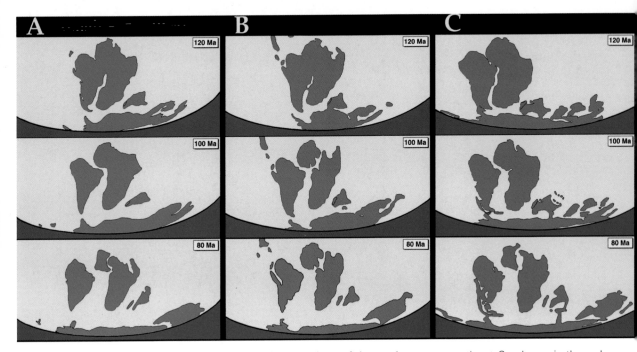

Plate 24 Three paleogeographic reconstructions of the break-up of the southern supercontinent Gondwana in the early, middle, and late Cretaceous. (A) after Smith *et al.*, 1994; (B) after Scotese, 1998; (C) after Hay *et al.*, 1999. Although these three reconstructions are very similar in most respects, they differ in detail and the timing of the final separation of the continents. Paleogeographic reconstructions are fine-tuned and adjusted as additional geologic, geophysical, and even paleontological data are collected. (After Krause *et al.*, 1999). **See Chapter 15**

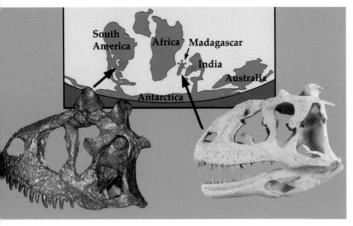

Plate 25 The Maastrichtain (Late Cretaceous) Malagasy theropod *Majungatholus atopus* (right) is most closely related to *Carnotaurus sastrei* (left) from Argentina, and *Indosuchus* and *Indosaurus* (not pictured) from India. Other fossil taxa from Madagascar, such as gondwanatherian mammals and araripesuchid crocodilians, also have closely related species in South America and India. None of these taxa show a close relationship to African taxa. Although these distributions seem incongruous given present day geography, earth history provides a logical explanation. Paleogeographic reconstructions for the late Cretaceous, such as the one shown here for the Campanain, reveal that when *Majungatholus, Carnotaurus, Indosuchus,* and *Indosaurus* were alive, Madagascar had just split from India, which still retained a connection to South America via Antarctica. This allowed Malagasy, Indian, and South American faunas to interact with one another well into the Late Cretaceous exclusive of Africa. **See Chapter 15**

Plate 26 Bones and teeth of mammoth, reindeer and other ice-age animals, dredged by fishing-boats from the floor of the North Sea between England and the Netherlands, form a graphic illustration of sea-level change. During the last glaciation, between about 100,000 and 10,000 years ago, sea levels dropped by 100 metres or more, and the North Sea was a grassy plain inhabited by herds of grazing mammals (Adrian Lister). **See Chapter 16**

Plate 27 Massive changes of sea-level have occurred through geological time. The picture shows a raised beach in South Devon, England. This two-metre thick deposit of sand, pebbles and shells represents the sea-level during a Late Pleistocene interglacial, probably around 200,000 years ago. The beach indicates a sea-level about 8 metres above present (Peter Rawson). **See Chapter 16**

Plate 28 The large mammal fauna from the Thompson Quarry locality, Sheep Creek Formation, Nebraska, late early Miocene (around 17 Ma). The plants at this locality indicate a type of woodland savanna, and the figure shows a transect from woodland (left) to more open grasslands (right), with the mammals of different dietary types (browsers vs mixed feeders or grazers) in their appropriate habitat. The fauna is notable for the much greater numbers of large browsing species than seen in any comparable habitat today, and it is suggested that higher levels of atmospheric CO_2 may have led to greater productivity of dicotyledonous vegetation, thus allowing the habitat to support this variety of mammals. Reproduced with permission from *The Cambridge Encyclopedia of Life Science*, edited by Adrian Friday and David Ingram, Cambridge University Press (1985), pp. 378-379. **See Chapter 17**

Plate 29 During the Pleistocene ice age, massive glaciers extended across huge areas of lowland Europe and North America, obliterating or dividing areas of habitat and, by locking away water, lowering global sea level. The picture shows the Mer de Glace near Chamonix, France, once part of a much larger ice sheet over the Alps and surrounding lowlands (Mike Hambrey). **See Chapter 18**

Plate 30 The sediments of Olorgesailie, southern Kenya rift valley, are one of the best dated terrestrial records containing abundant stone tools and evidence of environmental fluctuation over the past million years. The white sediments are composed of lake diatoms, a form of algae indicative of a fresh water lake. Many changes in lake level are recorded. The excavation, which unearthed stone handaxes and many fossils of zebra, antelopes, and other land mammals, is in a soil that developed approximately 990,000 years ago when the lake temporarily dried up. (Photo credit: Rick Potts). **See Chapter 19**

Plate 31 Village agriculture in the arid environment of northern Syria: two men winnowing dry-farmed wheat using wooden forks, and, beyond, a woman sieving the grain prior to its storage. (Courtesy of Amr N.M. al-Azm). **See Chapter 20**

Plate 32 Terrace cultivation of rice in monsoonal Indonesia: rice seedlings recently transplanted by hand from a seed-bed to leveled and flooded pond fields between the terraces. (Courtesy of Peter Bellwood). **See Chapter 20**

Plate 33 Traditional Mesoamerican agriculture: the three staple crops, maize, common bean and pepo squash, complement each other ecologically and nutritionally and are planted together, with the beans climbing up the maize stalks and the squashes growing in spaces between the maize plants. (David Harris). **See Chapter 20**

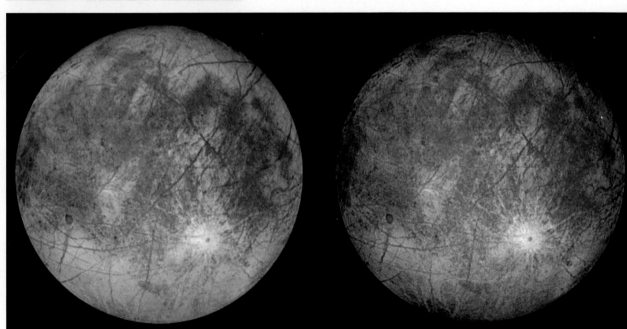

Plate 34 The left image shows the approximate natural colour appearance of Europa. The image on the right is a false-colour composite version combining violet, green and infrared images to enhance colour differences in the predominantly water-ice crust of Europa. Dark brown areas represent rocky material derived from the interior, implanted by impact, or from a combination of interior and exterior sources. Bright plains in the polar areas (top and bottom) are shown in tones of blue to distinguish possibly coarse-grained ice (dark blue) from fine-grained ice (light blue). Long, dark lines are fractures in the crust, some of which are more than 3000 kilometres (1850 miles) long. The bright feature containing a central dark spot in the lower third of the image is 50 kilometres (31 miles) in diameter. This crater has been provisionally named 'Pwyll' for the Celtic god of the underworld. This image was taken on 7 September 1996, at a range of 677 000 kilometres by the solid state imaging television camera onboard the Galileo spacecraft during its second orbit around Jupiter. (Image processed by Deutsche Forschungsanstalt für Luft- und Raumfahrt e.V., Berlin, Germany). **See Chapter 21**

over time (Raup and Sepkoski, 1982; Van Valen, 1984), others to a progressive increase in extinction resistance conferred as a result of (i) an increase in the number of species per clade with consequent extension of the clade's geographic distribution (Sepkoski, 1984; Flessa and Jablonski, 1985), (ii) the progressive radiation of clades into marginal environments where extinction resistance is conferred as a result of improved tolerance to environmental fluctuations (Vermeij, 1987), (iii) greater systematic and taxonomic familiarity with the Cenozoic fauna (Vermeij, 1987), or (iv) more intensive search of Cenozoic sediments as a result of a greater proportion of outcrop exposure (Vermeij, 1987). Note that all of these hypotheses focus on either biotic or systematic/stratigraphical (artifactual) explanations for the gradient as opposed to hypothesizing a link between the gradient and directional patterns of variation in Earth's physical processes.

Martin (1996) has recently challenged these biotic–systematic/stratigraphical artifact interpretations by arguing that patterns of variation in proxies for several abiotic environmental factors exhibit first-order patterns of variation similar to that exhibited by the extinction-intensity gradient. In particular, Martin (1996) argued that a net increase in nutrients delivered to marine habitats over the course of the Phanerozoic was a primary cause for the gradient. This factor would be expected to affect primarily marine phytoplankton, which occupy the basal layers of marine trophic levels and control the rate at which inorganic nutrients are transferred into marine trophic systems.

The marine portion of the organic carbon cycle is particularly interesting in the context of extinctions because it contains a 'weak link' of sorts. Whereas modern marine phytoplankton account for only 0.6% of the global carbon biomass, overall marine productivity is approximately equal to terrestrial productivity (Tappan, 1986). This equality arises as a result of the extremely short cycle time (=high productivity rate) of phytoplankton populations, coupled with the large areal extent of phytoplankton habitats (see Ricklefs, 1979). Consequently, relatively small variations in

phytoplankton abundance and diversity can have dramatic effects on the global carbon cycle and, through that cycle, the entire biosphere. Also, by serving as critical intermediaries in a variety of global geochemical cycles, phytoplankton are capable of changing many fundamental physio-climatic and environmental factors over relatively short geological timescales.

Modern marine phytoplankton communities are nutrient-limited, with phosphorus (P) and iron (Fe) being the principal limiting components (DeAngelis, 1992; Melillo et al., 1993; Takeda, 1998). While P and Fe can be injected directly into the oceans by submarine volcanic activity, most nutrients available to phytoplankton ultimately originate from either the erosion of igneous rocks on the continents, the erosion of igneous rocks forming the submarine basement complex, the injection of dissolved material into deep-water zones at deep-sea vents, and the recycling of unreduced organic materials existing at depth. Thus, processes that disrupt or enhance the delivery of these nutrients to the phytoplankton communities that inhabit relatively shallow depth zones, or that aid or suppress the diversity of such communities, can have a large and geologically immediate influence on a number of important physio-biotic Earth systems. In support of his model Martin (1996) cited first-order trends in three marine environmental proxies: $\delta^{34}S$ (Figure 14.6A, a proxy for marine circulation intensity), $^{87}Sr/^{86}Sr$ (Figure 14.6B a proxy for rates of continental chemical weathering and nutrient runoff), and $\delta^{13}C$ (Figure 14.6C, a proxy for rates of photosynthesis and nutrient recycling efficiency in organic systems).

14.3.1 Sulphur ($\delta^{34}S$) values and ocean circulation

A decreasing trend in sulphur ($\delta^{34}S$) values in marine sediments (Figure 14.6A) is present in data presented by Hallam (1992), suggesting that Early Palaeozoic oceans were much more prone to sluggish circulation and anoxia than Mesozoic or Cenozoic oceans. This interpretation is based on the predicted response of the

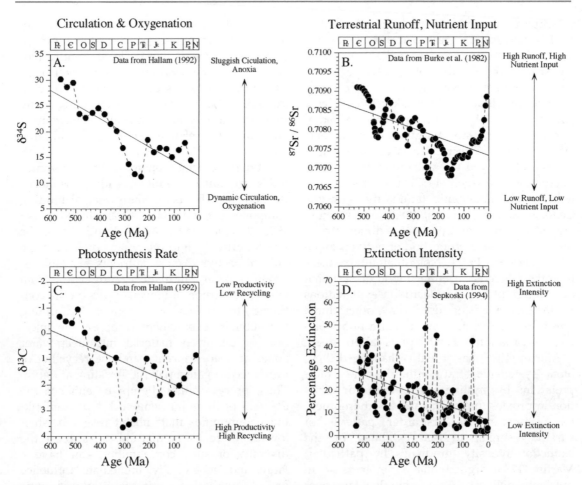

Figure 14.6 Linear least-squares regression analysis of proxies for marine circulation (=nutrient distribution) intensity (A), continent-derived chemical weathering and nutrient delivery (B) and organic productivity/recycling (C) compared to productivity of the Phanerozoic extinction intensity gradient (D). Timescale abbreviations as follows: P_ε – Precambrian; ϵ – Cambrian; O – Ordovician; S – Silurian; D – Devonian; P – Permian; T_R – Triassic; J_R – Jurassic; K – Cretaceous; P_g – Palaeogene; N – Neogene. See text for discussion.

sulphur reservoir isotopic composition to the proliferation of marine regions dominated by SO_4 reduction and is supported additionally by the greater number and larger extent of 'black shale' events recorded in Palaeozoic strata among other data (see Martin, 1996; Hallam and Wignall, 1997, and references therein). In terms of extinction, sluggish Early Palaeozoic marine circulation and a highly stratified water column may have confined high-nutrient bottom waters to anoxic basins, effectively starving the photic zones and deeper shelf habitats. Rising sea level in the Cambro-Ordovician would have increased

rates of circulation, but this might have had an initially detrimental effect as large volumes of anoxic water flooded over the middle and inner continental shelves (Hallam and Wignall, 1997). Interestingly, a period of marine environmental instability appears to be reflected in the extraordinarily high-amplitude fluctuations that characterize the Early Palaeozoic extinction data (see Figure 14.3, 540–350 million years). This Early Palaeozoic subinterval does not appear to exhibit an extinction-intensity trend, lending further support to the interpretation of Early Palaeozoic marine environmental instability.

Relatively low Mesozoic and Cenozoic δ^{34}S values suggest that marine circulation intensities increased in these time intervals relative to the Palaeozoic. This interpretation is consistent with reduced marine extinction intensities owing to a reduced incidence of basin stagnation, improved nutrient distribution, and intensified upwelling. Indeed, if one removes the four most prominent extinction-intensity peaks from the Carboniferous–Recent interval (Figures 14.1, 14.6D), the striking decrease in the amplitude of variations in the extinction signal is as apparent as the extinction-intensity gradient itself. This stabilization of the background extinction-intensity signal, in addition to its progressive decline, may reflect the damping of extinction probabilities by improved intra-ocean nutrient delivery to marine biotas dwelling within the photic zone resulting from a long-term and progressive improvement in marine circulation, especially via upwelling.

14.3.2 *Strontium (^{87}Sr/^{86}Sr) isotopic ratios and nutrient runoff*

The declining first-order trend in marine strontium (^{87}Sr/^{86}Sr) values over the Phanerozoic (Figure 14.6B; Tardy *et al.*, 1989) also may have important implications for understanding the extinction-intensity gradient. Continental igneous rocks are typified by relatively high proportions of ^{87}Sr whereas the hydrothermal input from oceanic crust to seawater injects relatively high proportions of ^{86}Sr. Consequently, the declining first-order trend in the ^{87}Sr/^{86}Sr values of marine sediments suggest a progressive reduction in the amount of continental nutrients delivered to the oceans over the course of the Phanerozoic. Deviations from this first-order trend characterize intervals of relatively increased or decreased runoff, usually due to intervals of active mountain-building (e.g. Raymo *et al.*, 1988; Raymo, 1994).

In the Precambrian to Ordovician, when terrestrial biotas were absent to rudimentary, terrestrial nutrient runoff rates must have been relatively high (Figure 14.6B). Subsequent development of soils, and leaf litter, in addition to the increase in terrestrial biodiversity and biomass, may have had the effect of decreasing the transfer rate of chemically weathered nutrients to marine habitats via runoff. Working against this trend, though, would be an increased rate of terrestrial chemical weathering as a result of plant activity, and the flushing of organic detritus from the continents into the oceans by riverine systems. To the extent that Sr represents a reliable proxy for the history of inorganic nutrient runoff, however, the trend of the Phanerozoic ^{87}Sr/^{86}Sr curve does suggest that some mechanism progressively decreased terrestrial nutrient flux rates to marine environments over time.

Broad deviations from the Phanerozoic ^{87}Sr/^{86}Sr trend line do exist. For example, the Neogene increase in ^{87}Sr/^{86}Sr values represents the effect of the Alpine and Himalayan orogenies that were so massive and so situated as to overwhelm the nutrient sequestration abilities of surrounding terrestrial habitats (Raymo, 1994). Additionally, Vermeij (this volume, Chapter 12, see also Clarke this volume, Chapter 11) argues that Phanerozoic temperature rise drove changes in atmospheric composition that increased rates of chemical weathering and, in turn, increased the delivery of continent-derived nutrients to the oceans. While these inferences may be correct (and much remains to be learned about Earth's geochemical systems), their implications are not consistent with the ^{87}Sr/^{86}Sr data, whose long-term declining trend is quite clear and statistically significant (see below). In addition, one cannot simply select those portions of a data set that conform to the predictions of a hypothesis (e.g. that the Ashgill, Frasnian, and Tatarian mass extinctions occur during local ^{87}Sr/^{86}Sr minima (=inferred terrestrial nutrient input lows)) and fail to discuss the other time intervals in which the predicted relations are not evident (e.g. the Norian and Maastrichtian mass extinctions occur during local ^{87}Sr/^{86}Sr maxima (=inferred terrestrial nutrient input highs)).

The opposing long-term Phanerozoic trends of intra-ocean nutrient delivery to shallow marine habitats (based on the δ^{34}S data that imply a progressive increase in deep-water nutrient flux rates due to upwelling), and terrestrially derived nutrient delivery to those same habitats (based on the $^{87}Sr/^{86}Sr$ data that imply a progressive decrease in the input of continent-derived flux rates, presumably due to the diversification of terrestrial biotas), presents a picture of nutrient availability controls to marine environments that are perhaps best described as existing in a state of dynamic balance between the forces of tectonics and organic evolution. Tectonic forces are responsible for the configuration of the ocean basins (which affects climate patterns of marine circulation), continental positions (which affect global climate), and mountain building (which affects climate and rates of physical weathering), while the processes of organic evolution are responsible for the development of adaptive strategies whose purpose is to store nutrients in particular places (e.g. in organic bodies, on continents via soils, in the marine photic zone via ecological systems). And, of course, biotic systems that become large enough can also exert a direct effect on climate and atmospheric composition.

14.3.3 Carbon ($\delta^{13}C$) values, productivity, and nutrient recycling

Martin's third major line of geochemical evidence for systematic changes in the marine environment lies in the first-order increase in marine carbon ($\delta^{13}C$) isotopic values (Figure 14.6C). A preponderance of 'light' (=negative) $\delta^{13}C$ values in Early Palaeozoic sediments suggests decreased productivity or oxidation of organic carbon reservoirs. In contrast, the 'heavy' (=positive) $\delta^{13}C$ values characteristic of Mesozoic and Cenozoic sediments indicate increased oxygenation of marine carbon reservoirs. This progressive increase in $\delta^{13}C$ values suggests that Early Palaeozoic marine ecosystems were relatively inefficient with low nutrient recycling, low productivity, and short food chains. Any substantial change in this condition requires that the system migrate toward increased rates of conversion of nutrients to a more usable form (e.g. biomass) and increases in the ability of ecosystems to retain nutrients by increasing the efficiency at which organic materials are retained (e.g. via lengthening of

food chains) and recycled (e.g. via ecological specialization of phytoplankton). Such changes would have the effect of making marine ecological systems less dependent on continent-derived nutrient input (which, over time, was being, at least partially, diverted to support terrestrial ecosystems) while at the same time increasing the overall supply of nutrients that underpin increased organic biodiversity and abundance.

14.3.4 Biotic evidence

At present, marine paleontological evidence favours an interpretation of Early Palaeozoic oligotrophic (low productivity, low biodiversity, low abundance) biotas changing to eutrophic biotas (high productivity, high biodiversity, high abundance) in the Mesozoic–Cenozoic. For example, whereas high Early Palaeozoic acritarch diversities (a category of phytoplankton) have been interpreted as responses to nutrient-rich conditions (Tappan, 1986; Bambach, 1993), the dominance of low metabolic-rate animal clades (e.g. brachiopods, bryozoans) in the Early Palaeozoic suggests oligotrophy. There is no question that animal communities of the Mesozoic and Cenozoic were dominated by groups whose modern representatives exhibit higher metabolic rates (e.g. more diversified and specialized zooplankton faunas, higher metabolic-rate benthos, increases in ecological tiering). Therefore, the reversal of marine environmental–ecological conditions between the Palaeozoic and Mesozoic–Cenozoic seems suggestive of a progressive 'eutrophization' of the marine water column. This, in turn, implies increased overall nutrient levels to account for these increases in the biodiversity (Sepkoski, 1981) and ecological stability (e.g. Hutchinson, 1959) of marine invertebrate communities along with a predicted decline in extinction-intensities (Figure 14.3). The fact that this biotic trend appears to run opposite to the decline in $^{87}Sr/^{86}Sr$ values suggests that the Phanerozoic eutrophization of marine biotas was not (or at least not solely) driven by an increase in the amount of terrestrially derived nutrients conserved by marine systems. Instead, what appears to have occurred is an increase in the efficiency with which nutrients were being retained and recycled within shallow marine habitats, perhaps augmented by the enhanced distribution of both terrestrially derived nutrients and marine nutrients stored in deep waters through more intense and efficient circulation.

14.3.5 *Temperature*

Temperature may also have played a decisive role in this system-conversion process. Clarke (this volume, Chapter 11) and Vermeij (this volume, Chapter 12) review the evidence for a progressive increase in tropical sea-surface temperatures and a widening of the equatorial climatic zone from the Palaeozoic to the Recent. In addition to its obvious environmental correlates, increased temperatures permit a greater range of morphological specializations to become physiologically and ecologically – and thus competitively – viable. As a result, the range of successful aptations becomes relatively greater under warm conditions and the balance of extinction/origination shifted positively.

If Early Palaeozoic sea-surface temperatures were cooler than Mesozoic–Cenozoic values, the evident increase in the rate and efficiency of photosynthesis over time might be explained by the increased range of physiological–biochemical specializations that would become available for marine primary producer lineages within the warmer Mesozoic–Cenozoic habitats. In other words, the warmer shallow marine environmental conditions that predominated in the Mesozoic–Cenozoic may have 'set the stage' for contingent evolutionary developments (e.g. the sudden evolution of highly successful marine phytoplankton clades) that were able to take advantage of – indeed, to reinforce – the environemntal feedback mechanisms that tipped the overall marine ecosystem into a new, eutrophic metastable state with consequent macroevolutionary implications for the entire marine biota. These data are also consistent with a wide variety of studies that identify the tropics as centres of nutrient recycling and

evolutionary specialization through diversification (e.g. Jablonski, 1993; Vermeij, 1978, 1993).

14.3.6 *Are these first-order trends deterministic features of the fossil record?*

Though Martin (1996) noted the qualitative similarity of the Phanerozoic extinction-intensity gradient (Figure 14.6D) to long-term changes in these proxies (Figure 14.6A–C), he did not attempt to evaluate their statistical significance or test them against a random time-series model. This is important in that several of Martin's environmental proxy data sets contain relatively small numbers of observations and because these proxy data are generally more variable than the extinction data (see Figure 14.6). Employment of the same regression-Monte Carlo-based analytic strategy used to evaluate the extinction data (discussed above) shows that these first-order geochemical data are statistically distinct from expectations of a random time-series model by considerable margins (Figure 14.7). Consequently, the idea that the decline in Phanerozoic background extinction rates was deterministically driven (at least in part) by progressive abiotic factors – specifically increases in marine circulation rates and (perhaps) sea-surface temperature, along with (i) a decrease in the availability of continent-derived nutrients and (ii) an increase in the intra-oceanic nutrient flux to marine surface waters – must be entertained.

14.3.7 *Evolutionary–ecological implications*

Once the importance of nutrient flux and its controls in marine ecosystems is appreciated, it can be used to develop a variety of hypotheses that can be tested using paleontological data. For example, as the primary producers in marine ecosystems, phytoplankton represent the gatekeepers at the interface of the organic and inorganic realms. If long-term trends in abiotic processes such as nutrient availability and distribution have played a role in influencing macroevolutionary change (e.g. progressively decreased extinction prob-

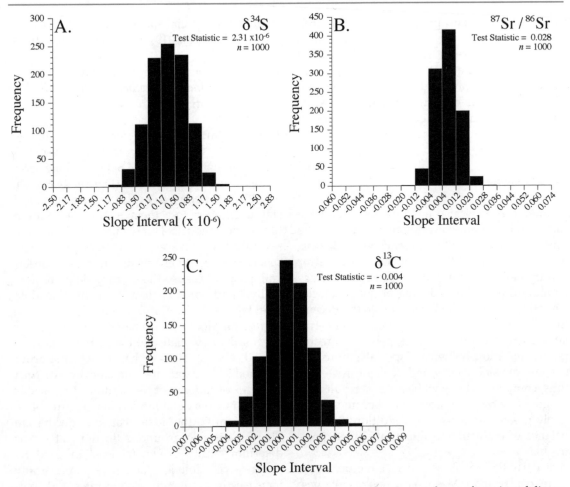

Figure 14.7 Results of Monte Carlo-simulated linear least-squares regression analyses of nutrient delivery proxies (A and B) and productivity proxy data (C). Since the overwhelming majority of all simulations exhibit regression slopes that are greater than (A and B) or less than (C) the observed slope the observed first-order gradients are all judged to be highly significant. The correspondence of these proxy patterns to the extinction intensity gradient suggests that long-term changes in the physical state of marine environments may have played a role in creating or maintaining the extinction-intensity gradient.

abilities), one place to look for evidence of such a relation would be in strong links between the evolutionary–ecological histories of marine phytoplanktonic and benthic invertebrate clades.

Figure 14.8 compares the family-richness histories of Sepkoski's (1981) major evolutionary faunas to the Phanerozoic history of the major marine phytoplankton groups. Fauna I exhibits a striking parallelism to the richness history of Palaeozoic acritarchs. Of course, all trilobites did not consume acritarchs directly. Indeed, the diets of trilobites are presently unknown

(Fortey and Owens, 1999). Nevertheless, as the (apparently) dominant Early Palaeozoic primary producer, increases and decreases in acritarch biodiversity and abundance would be expected to exert a strong influence on the trophic structure of Early Palaeozoic marine biotas, and on the amount of organic carbon residing in marine sediments. It is by these indirect pathways that phytoplankton are linked to marine benthos in the modern biota (see Lambshead and Gooday, 1990; Levinton, 1996), presumably as they were in the past.

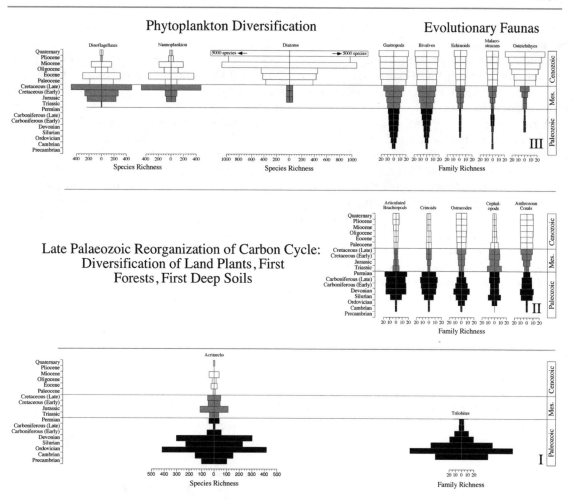

Figure 14.8 Major evolutionary faunas plotted against the diversity history of major phytoplankton clades. Note broad correspondence between diversification patterns of Sepkoski's (1981) evolutionary Faunas I (bottom) and III (top, only dominant constituents shown here) and the diversification histories of the important marine phytoplankton groups. Evolutionary Fauna II does not have an obvious phytoplanktonic analogue, but the upper Palaeozoic (diversification within which forms the identifying attribute of this fauna) was a time of widespread reorganization of the global carbon cycle. These associations do not imply there is a direct ecological relationship between these particular benthic and phytoplankton groups (though this might be the case in certain instances). Rather they show that evolutionary developments in the phytoplankton flora – of which these clade histories are representative – exhibit strong associations with evolutionary developments in the benthic marine invertebrate fauna. See text for discussion.

Comparison of the association between Early Palaeozoic acritarch and trilobite taxic richness patterns is consistent with this prediction. While it can be argued that both acritarch and trilobite clades might have been responding to other environmental changes in the late Early Palaeozoic (e.g. increased glaciation, increased volcanism, increased continentality), this association represents a hypothesis that can be further tested and refined by examining other instances of associated changes in marine phytoplankton and benthic invertebrate evolutionary histories. In this context it is also interesting to note that Mesozoic nannoplankton and dinoflagellates, and Cenozoic diatoms, also undergo explosive diversifications coincident with the equally explosive rise of Sepkoski's Fauna III. Once again, while the argument does not assume a direct trophic relationship between these particular primary producer and consumer groups, it is suggestive of an indirect link.

A more specific exploration of this relation can be found in the systematic study of echinoid evolution during this Mesozoic–Cenozoic interval. The Palaeozoic echinoid fauna was restricted to regular (symmetrical) forms that all exhibited a grazing life habit. But, following the Tatarian extinction event (248 Ma), and occurring at approximately the same time as the initial diversifications of nannoplankton and dinoflagellates (as well as the correlative diversifications of planktonic grazers such as planktonic foraminifera) in the Late Triassic–Early Jurassic, the echinoid clade underwent an unprecedented diversification (Figure 14.9). Significantly, this diversification included this clade's first successful invasion of the deposit-feeding adaptive zone. The association between the echinoid clade's expansion into a previously unoccupied adaptive zone occurring at a time when the diversification of phytoplankton was fundamentally altering the phys-

ical nature of marine benthic environments, and at a time when marine circulation rates and productivity levels were intensifying (Figure 14.6B and D), but continent-derived nutrient input to marine systems – as inferred from the $^{87}Sr/^{86}Sr$ values – was falling (Figure 14.6C) also seems highly suggestive. Valentine and Jablonski (1986) have presented additional evidence for a similar relation between the appearance-diversification of Mesozoic phytoplankton groups and the evolutionary–ecological histories of contemporaneous marine bivalve clades. A similarly striking association of the Cenozoic Osteichtyes (bony fishes) diversification occurring at a time of exponential diatom diversification may be another example of this phytoplankton–nutrient resource link.

Although Sepkoski's (1981) Fauna II is not associated with a diversifying phytoplankton clade, the Late Palaeozoic acme of this fauna does coincide with an interval of fundamental change in the global environment. Evidence for this change includes a switch from the Early Palaeozoic greenhouse to icehouse conditions, decreasing rates of global volcanism, increasing continentality, decreasing rates continental

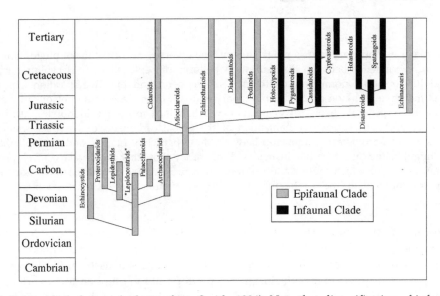

Figure 14.9 Echinoid phylogeny (redrawn from Smith, 1984). Note that diversification of infaunal clades (dark bars) takes place at a time that corresponds to the diversification of Mesozoic phytoplankton groups (e.g. nannoplankton, dinoflagellates), suggesting a macroevolutionary link between environmentally mediated evolution in these two ecologically distinct realms.

runoff, and decreasing concentrations of atmospheric CO_2. In addition, this Late Palaeozoic time interval represents a time of fundamental reorganization of the global carbon cycle driven by the evolutionary diversification of land plants, the spread of the first forests, the development of the first deep soils, and the proliferation of marine, carbonate-mound biomes through the diversification of benthic algae. Note also that this Early–Late Palaeozoic transition interval also appears to mark the onset of the progressively declining extinction intensities, further suggesting that these evolutionary developments in effect created an environmental perturbation buffer that became more effective through time.

In short, there is abundant evidence that (i) all three of Sepkoski's (1981) evolutionary faunas were contained within physio-environmental episodes and characterized by major abiotic and tectonic, as well as biotic, factors, (ii) these episodes were embedded within a progressive sequence of changes in the nature of marine physio-ecological conditions, and (iii) this generalized mechanism of biosphere change implies a progressive reorganization of marine ecologies to adapt to reduced continent-derived nutrient inputs and intensified oceanic circulation patterns, most likely mediated through the offices of primary-producer clades. The fact that the progressive change in per cent extinction intensity also shows a similar progressive patterns of change over the Devonian–Recent interval provides evidence that it too may be a reflection of the biotic response to these physical changes in marine evolutionary–ecological systems.

14.4 'Mass extinction' trends

Sepkoski's extinction (1994) data show 17 different localized increases in extinction intensity. Collectively, these stage-level 'events' represent the mass extinction limb of the Phanerozoic extinction spectrum. Using family- and genus-richness data from the Permian–Recent interval, and a Monte Carlo randomization procedure for the analysis of irregularly spaced time-series, Raup and Sepkoski (1984, 1986) proposed that extinction peaks in this subinterval exhibited a regular 26 million year periodicity. Raup and Sepkoski's interpretation was subsequently challenged on a wide variety of paleontological and analytic grounds (e.g. see Prothero, 1994; Manley, 1997 for reviews). In particular, Manley (1997) has reviewed the statistical problems associated with applying bootstrap and randomization methods to unequally spaced time series (especially those that contain a non-random trend) for the purpose of periodicity testing.

In the context of the present analysis it is important to note that, thus far, the entire periodicity controversy has been confined to this Permian–Recent subinterval. Raup and Sepkoski speculated (but did not demonstrate) that periodicity was not a general feature of the Phanerozoic extinction record. However, application of Fourier analysis to an interpolated evenly spaced series based on the entire Sepkoski (1994) data exhibits relatively strong peaks at periodicities of 35.33 My, 62.25 My, and 132.50 My (Figure 14.10). The former is similar (though not identical to) the 26 My peak predicted on the basis of Raup and Sepkoski's previous analyses.

These peaks in the Fourier spectrum identify harmonic patterns in the extinction data that account for relatively high proportions of the observed variation. They are purely descriptive. It is a different matter to claim that these observations have a deterministic significance, i.e. that they are sufficiently prominent to rule out their being regarded as due to the operation of random factors (see Raup, 1977; Raup and Schopf, 1978). As with the previous extinction-intensity gradient analyses, the problem is that since there is only one fossil record, the traditional replicate tests used to control for the influence of extraneous and systemic variations cannot be carried out. Nevertheless, we can devise a simulation test analogous to that used for the extinction-intensity gradient analysis described above (see Manley, 1997 for details). This test represents the analytic equivalent of 'running the tape of life again and again' to determine how often a particular result turns up. If the observed periodic pattern

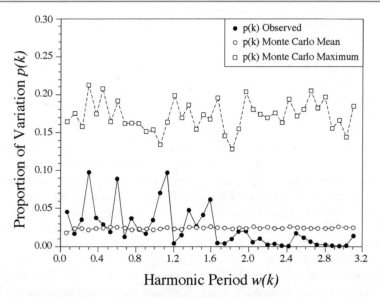

Figure 14.10 Periodogram for Phanerozoic extinction data (see Figure 14.1) showing peaks at $w(k)$ = 0.303 (132.50 My), 0.606 (62.25 My), and 1.136 (35.33 My). (Note: $w(k)$ represents the harmonic period expressed in terms a frequency spectrum 2π units in length and referenced to the number of datapoints [83].) However, 1000 Monte Carlo simulations of these data (see Manley, 1997 for a description of the simulation structure) indicates that random reorderings of these data routinely exhibit periodicities greater than those observed in the Sepkoski data. Comparison of these simulation results to the observed Fourier decomposition of an interpolated extinction-intensity time series using the Boneferroni inequality indicates that none of the observed harmonic frequency peaks is statistically significant. These results, which are consistent with the majority of Late Permian–Recent extinction periodicity analysis (see Prothero, 1994; Manley, 1997), demonstrate that there is no deterministic periodic component to Phanerozoic extinction data.

is sufficiently unlikely (e.g. turns up in less than 5% of the simulations) the result can be judged unambiguously to be a non-intrinsic aspect of the data and so to warrant a process-level interpretation of its own.

Analysis of 1000 Monte Carlo simulations of these data (Figure 14.10) indicates that none of the three major extinction-spectrum harmonic peaks is sufficiently different from those obtained through random simulation to warrant such an interpretation. Since a periodic cause cannot be invoked to constrain the cause of mass extinction events we must look for commonalities within the structure of these events and among their associations with other environmental phenomena in order to understand their cause.

The three best studied mass extinction events are the K–T event, the Tatarian (Permo-Triassic) event and the Frasnian (Late Devonian) event (see Figure 14.1). While each

event involved different groups of organisms, different initial environmental conditions, and had different effects on the subsequent history of life, they share several important similarities.

First, none of these major extinction events was geologically instantaneous. The overwhelming majority of so-called 'victim' clades associated with each event exhibit uncontroversial evidence for prolonged declines lasting for millions of years prior to the various boundary horizons (see MacLeod *et al.*, 1997; Erwin, 1993; and McGhee, 1996 for summaries of the Maastrichtian, Tatarian, and Frasnian–Fammenian events respectively). As mentioned above, some advocates of extraterrestrially forced mass extinctions cite the so-called Signor–Lipps effect in an attempt to justify what is, at best, an ambiguous interpretation of synchrony among observed last appearance datums for a minority of victim clades.

However, this argument (i) ignores the obvious implications of the well-established pre-event declines in these victim clades (see above), (ii) ignores long-standing problems in accurately assigning biotic extinction ages at the levels of resolution demanded by the instantaneous, co-extinction model, and (iii) treats the concept of co-extinction as an assumption made at the outset of analysis instead of a conclusion to be tested and supported (if possible) by analytic results (see MacLeod, 1998).

These major stage-level extinction events (as well as most other mass extinction events) also involved both terrestrial and marine organismal groups. This suggests some type of global environmental change as being responsible for the biotic turnover. Although many smaller (or background) extinction events appear to affect either localized regions or particular biotic groups, this does not necessarily imply a discontinuity of extinction causal processes. Localized events can result from either different causes, or variations in intensities among a series of causes.

Some have suggested that the wholesale collapse of primary productivity (as evidenced by $\delta^{13}C$ fluctuations) may be the causal agent in many, if not all, mass extinction events (e.g. Paul and Mitchell, 1994). But even if this model were true ($\delta^{13}C$ values are not a direct measure of primary productivity and durations of low $\delta^{13}C$ excursions are typically too long to be explained by a simplistic phytoplankton die-off), this represents only a partial explanation for Phanerozoic extinctions since it only addresses itself to their proximal cause. What ultimate causal factors (if any) are associated with the extinction peaks?

Of the many causal mechanisms associated with Phanerozoic mass extinctions, three are by far the most recurrent in the paleontological extinction literature: sea-level fluctuation, continental flood-basalt (CF-B) volcanism, and bolide impact. All three mechanisms imply a variety of environmental changes that might result in extinction. Sea-level fall decreases the area available for marine biotas to colonize. For most of Earth history sea level has stood much higher than it does today, often flooding over the continental platforms and turning vast areas into shallow marine habitat. When continental platforms are flooded, relatively small falls in sea level can dramatically alter marine environmental conditions over large areas (Hallam, 1989). Sea-level fall disrupts marine circulation patterns, tending to isolate marine basins by destroying the relatively shallow connections between them. If marine basins are isolated they can become stagnant and fill with unoxygenated bottom waters that, upon subsequent sea-level rise, could flood over the continental shelves decimating marine populations (Hallam and Wignall, 1997). Sea-level fluctuations also have an effect on climate by altering the concentration of CO_2 in the atmosphere (due to the exposure or isolation of unreduced organic matter) and altering latitudinal temperature gradients (by increasing or decreasing continentality. Additionally, sea-level change can alter the amount of continent-derived nutrients that are delivered to marine habitats by a combination of processes (e.g. exposing rock bodies to erosion, increasing rates of chemical weathering by changing the CO_2 concentration of the atmosphere).

Continental-scale volcanism has many of the same environmental effects as sea-level change, but differs in the processes that bring those effects about. Primarily volcanic eruptions disrupt the environment by injecting large volumes of gas and particulate matter into the atmosphere (McLean, 1985). These materials can increase global temperatures (via injection of greenhouse gases, McLean, 1985), decrease global temperatures (via increasing the dust content of the atmosphere and increasing cloud cover, McLean, 1985), diminish rates of photosynthesis (Deirmendjian, 1973), poison organismal populations (via the release of heavy metals, Hansen, 1991), and increase the incidence of surface radiation (via destruction of the Earth's ozone layer, Stolarski and Butler, 1979; Keith, 1980, 1982).

Bolide impacts produce all the effects of large volcanic eruptions (see MacLeod, 1998 for a review), but concentrate these into a smaller time interval. In addition, bolide impacts may kill organismal populations via (i) the initial heat flash that accompanies the object's entry into the Earth's atmosphere, (ii) increases in the acidity of rainwater (via release of CO_2, NO_2, SO_2; Prinn and Fegley, 1987), and (iii) (perhaps) through the ignition of global wildfires (Wolbach *et al.*, 1985).

Evidence for the operation of all three mechanisms exists throughout Phanerozoic deposits. This allows the stratigraphical records of all three to be quantitatively compared to the Phanerozoic extinction record in order to test for associations. In this context, it is not sufficient merely to demonstrate that a particular mechanism was operating during a particular extinction event or a set of mass extinction events (e.g. Alvarez, 1997). Rather, significance must be judged in the light of correspondences between all available data at a practically realizable level of temporal resolution. Similarly, robust statistical tests of the associations must be carried out to determine whether the observed patterns (many of which are represented by relatively small data samples) could be explained by chance.

Figure 14.11 illustrates the Phanerozoic extinction, bolide impact, continental flood-basalt volcanism, and sea-level records. The bolide impact record exhibits a strikingly low correspondence with the Phanerozoic stage-level extinction record. Although some type of bolide impact occurs in many stages that contain a mass extinction, neither impact size nor impact number exhibits a statistically significant association with extinction magnitude (MacLeod, 1998). To be sure, the Mesozoic–Cenozoic bolide impact record is much more detailed than that for the Palaeozoic. However, the association between bolide impacts and extinction intensities fails to improve if Palaeozoic data are excluded.

Unlike the bolide impact-extinction intensity comparison, large CF-B volcanic events exhibit a near-perfect stage-level association with marked increases in Mesozoic and Cenozoic extinction intensity (Courtillot *et al.*, 1996; MacLeod, 1998). (Note: no comprehensive data are available on the timing of Palaeozoic CF-B volcanic events.) While Wignall (2001) found no consistent scaling relationship between eruption size (as measured by the volume of extruded material) and extinction

magnitude, the strength of the association at the larger end of the extinction scale – where the environmental effects would be expected to be most pronounced – suggests a strong causal relation that can be ultimately traced to Earth's tectonic history (see also Hallam and Wignall, 1984).

Sea-level changes exhibit a mixed association with extinction history. Of the 14 such sea-level falls documented by the Hallam (1992) sea-level curve, seven occur in stages containing an elevated extinction intensity peak (see Figure 14.11). While this is not a statistically significant result, it is interesting to note that the Palaeozoic portion of this distribution exhibits a considerably higher association (five out of seven). Moreover, sea-level falls are associated with each of the three largest stage-level extinction events of the last 250 My. These observations suggest that sea level may play a substantial contributory role in accentuating stage-level extinction rates.

14.5 Conclusion

Examination of the Phanerozoic extinction record shows it to be continuously distributed between large (mass extinction) and small (background extinction) stage-level events. The continuity of this pattern suggests that instead of discrete, causal process classes, Phanerozoic extinctions reflect the operation of a multitude of varying process-level inputs and (most likely) the operation of contingent mediating factors.

A statistically significant linear pattern of declining extinction intensity is present within these data. Since it was first recognized this gradient has been interpreted as a reflection of either macroevolutionary or sampling-taxonomic factors. Nevertheless, comparison of these extinction data with geochemical proxies for marine mixing intensity ($\delta^{34}S$), terrestrial nutrient input rate ($^{87}Sr/^{86}Sr$), and productivity-recycling rate ($\delta^{13}C$) reveals a consistent association between first-order Phanerozoic trends in these proxies that, together, support a model of increased nutrient delivery to shal-

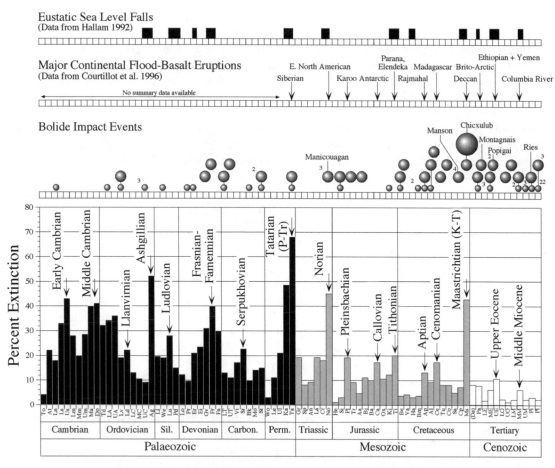

Figure 14.11 Comparison of the stage-level stratigraphical distribution of short-term sea-level falls, continental flood-basalt volcanic events, and bolide impact events with Phanerozoic extinction intensity. Small numbers alongside bolide-impact icons represents multiple impacts of this size in the associated stage. Bolide icon sizes represent crater diameters: small icons = 1–10 km; intermediate icons = 10–100 km; large icon = >100 km. A statistically significant relation exists between local extinction-intensity peaks and Palaeozoic sea-level falls and Permian–Recent continental flood-basalt volcanic events (see MacLeod, 1998). Since these terrestrially based processes represent aspects of the tectonic cycle, and since they would be expected to cause long-term perturbations in marine nutrient concentrations (especially in surface waters), they support the hypothesis of tectonic – as opposed to extraterrestrial – factors as being the primary determinants of life's extinction history.

low marine habitats over the course of this time interval. This long-term trend appears to be driven primarily by (i) increases in rates of physical mixing and (ii) increases in rates of nutrient recycling during the Mesozoic and Cenozoic, relative to Palaeozoic conditions. Other lines of evidence suggest that the declining background extinction gradient appears to have originated in the Late Devonian and that over this same time interval global mean tem-

peratures may have also undergone a long-term, first-order increase. In addition to a reduction in mean extinction rates, these trends would be expected to result in the lengthening of marine food chains, increases in the number of recycler (e.g. deposit feeder) niches, and adaptive improvements in phytoplankton design. All of these predictions are borne out by the fossil record. Together, these associations suggest that tectonic factors (e.g. continental drift) and contingent evolutionary factors (e.g. evolution of trees and forests, evolution of phytoplankton clades) represent the primary classes of causal processes responsible for the background extinction-intensity gradient.

In addition, analysis of the Phanerozoic mass extinction record (=time series of local extinction-intensity peaks) suggests that there is no statistically significant periodicity associated with the Phanerozoic extinction record or any subinterval thereof. This result casts suspicion on the idea of the extraterrestrial forcing of mass extinctions (though non-periodic impact-induced effects remain a possibility for individual mass extinction events, particularly the late Maastrichtian event). Moreover, there appears to be no significant association between extinction intensity and either bolide impact numbers or sizes. On the other hand, the near-perfect correspondence between mass extinctions and occurrences of continental flood-basalt eruptions, along with a highly suggestive correspondence between these larger extinction events and rapid falls in sea level are suggestive of a consistent, cause–effect association. Once again, tectonic processes are implicated as the primary controlling process class most consistently associated with extinctions.

Given the rather clear implication of tectonic factors as the primary causes of both long-term and temporally localized Phanerozoic extinction trends, one may be tempted to conclude that the history of life has been driven primarily by abiotic factors. This would be an oversimplification. By controlling the composition of the atmosphere life has exerted a direct and important influence on tectonic processes (Nisbet, 1987). This effect comes about through the atmospheric maintenance of a surface temperature regime that allows for the presence of large quantities of liquid water at and near the Earth's surface. In the absence of life Earth's atmosphere would be dominated by CO_2 and H_2O (vapour), resulting in a Venus-like runaway greenhouse and consequent rapid dehydration of the crust and upper mantle. These effects would severely curtail many tectonic processes as they have been curtailed on Venus. For the last 3.8 billion years life has set the level of atmospheric CO_2 and main-

tained the Earth's surface temperature and pressure within the limits necessary to maintain water in its liquid state. This has both prolonged and diversified the planet's suite of tectonic phenomena well beyond what would have been expected otherwise (Nisbet, 1987). Those tectonic processes, in turn, have played a predominant role in controlling long-term and short-term trends in biotic processes such as extinction and evolution.

Endnotes

[1] Strictly speaking dinosaurs are not extinct because a direct descendant of a dinosaur lineage still survives, indeed flourishes, in the modern world – birds.

[2] While several recent studies have demonstrated that some reworked Cretaceous specimens are indeed present in lower Tertiary sediments (e.g. Huber, 1996), this observation was never a source of dispute. Rather the reworking controversy revolves about the question of whether *all* of the 'Cretaceous' specimens that numerically dominate sediments located in the first few centimetres of the Tertiary, which often exhibit exquisite degrees of preservation fully comparable to 'non-reworked' Tertiary species, and which exhibit isotopic and biogeographic patterns of variation that are difficult to accommodate under a reworking model (see Barera and Keller, 1990; MacLeod and Keller, 1994; Smit and Zachariasse, 1996; Smith and Jeffrey, 1998) should be regarded as being reworked.

Acknowledgements

Conversations with Andrew Smith, Ronald Martin, Euan Nisbet, David Polly, and Adrian Lister, in addition to comments by an anonymous reviewer, contributed to the ideas expressed herein. However, none of these individuals necessarily accepts the views expressed above and any errors of fact or interpretation rest with the author. Computer programs and data sets used in the analysis of the Phanerozoic extinction data are available from the PaleoNet Pages WWW site (http://www.ucmp.berkeley.edu/Paleonet, PaleoNet West; http://www.nhm.ac.uk/hosted_sites/

paleonet, PaleoNet East). This is a contribution from the Palaeontology & Stratigraphy Programme, Earth Materials, History, and Resources Research Theme of The Natural History Museum, London.

References

Alvarez, L.W., Alvarez, F., Asaro, F. and Miche, H.V. (1980). Extraterrestrial cause for the Cretaceous–Tertiary extinction. *Science*, **208**: 1095–1108.

Alvarez, W. (1997). *T. Rex and the Crater of Doom*. Princeton: Princeton University Press.

Bambach, R.K. (1993). Seafood through time: changes in biomass, energetics, and productivity within the marine ecosystem. *Paleobiology*, **19**: 372–397.

Barrera, E. and Keller, G. (1990). Foraminiferal stable isotope evidence for gradual decrease of marine productivity and Cretaceous species survivorship in the earliest Danian. *Paleoceanography*, **5**: 867–870.

Benton, M.J. (1990). Scientific methodologies in collision: the history of the study of the extinction of dinosaurs. *Evolutionary Biology*, **24**: 371–400.

Benton, M.J. (1996). Diversity in the past: comparing cladistic phylogenies and stratigraphy. In J. Colbert and R. Barbault (eds) *Aspects of the Genesis and Maintenance of Biological Diversity*. Oxford: Oxford University Press, pp. 19–40.

Burke, W.H., Denison, R.E., Hetherington, E.A. *et al.* (1982). Variation in $^{87}Sr/^{86}Sr$ throughout Phanerozoic time. *Geology*, **10**: 516–519.

Canudo, J.I., Keller, G. and Molina, E. (1991). Cretaceous/Tertiary boundary extinction pattern and faunal turnover at Agost and Caravaca, S.E. Spain. *Marine Micropaleontology*, **17**: 319–341.

Courtillot, V., Jaeger, J.-J., Yang, Z. *et al.* (1996). The influence of continental flood basalts on mass extinctions: where do we stand? In G. Ryder, D. Fastovsky and S. Gartner (eds) *The Cretaceous–Tertiary Event and Other Catastrophes in Earth History*. Geological Society of America Special Paper, **307**: 513–525.

de Laubenfels, M.W. (1956). Dinosaur extinction – one more hypothesis. *Journal of Paleontology*, **36**: 207–218.

DeAngelis, D.L. (1992). *Dynamics of Nutrient Cycling and Food Webs*. London; Chapman Hall.

Deirmendjian, D. (1973). On volcanic and other particulate turbidity anomalies. In: H.E. Landsberg and J.Y. Meigham (eds) *Advances in Geophysics*. New York and London: Academic Press, pp. 267–269.

Erwin, D.H. (1993). *The Great Paleozoic Crisis: Life and Death in the Permian*. New York: Columbia University Press.

Flessa, K.W. and Jablonski, D. (1985). Declining Phanerozoic background extinction rates: effect of taxonomic structure? *Nature*, **313**: 216–218.

Fortey, R.A. and Owens, R.M. (1999). Feeding habits in trilobites. *Palaeontology*, **42**: 429–465.

Galvin, C. (1998). The great dinosaur extinction controversy and the K–T research program in the late 20th century. *Earth Sciences History*, **17**: 41–55.

Gradstein, F.M. and Ogg, J. (1996). A Phanerozoic time scale. *Episodes*, **19**: 3–5.

Grieve, R., Rupert, J., Smith, J. and Therriault, A. (1996). The record of terrestrial impact cratering. *GSA Today*, **5**: 193–195.

Hallam, A. (1989). The case for sea-level change as a dominant causal factor in mass extinctions of marine invertebrates. *Philosophical Transactions of the Royal Society of London, Series B*, **325**: 437–455.

Hallam, A. (1992). *Phanerozoic Sea-level Changes*. New York: Columbia University Press.

Hallam, A. and Wignall, P.B. (1997). *Mass Extinctions and their Aftermath*. Oxford: Oxford Science Publications.

Hansen, H.J. (1991). Diachronous disappearance of marine and terrestrial biota at the Cretaceous–Tertiary boundary. In: Z. Kielan-Jaworowska, N. Heintz and H.A. Nakrem (eds) *Fifth Symposium on Mesozoic Terrestrial Ecosystems and Biota. Extended Abstracts*. Oslo: Contributions from the Paleontological Museum, University of Oslo, No. 364, 31–32.

Huber, B.T. (1996). Evidence for planktonic foraminifer reworking versus survivorship across the Cretaceous–Tertiary boundary at high latitudes. In: G. Ryder, D. Fastovsky and S. Gartner (eds) *The Cretaceous–Tertiary Event and Other Catastrophes in Earth History*. Boulder: Geological Society of America Special Paper, **307**: 319–334.

Hutchinson, G.E. (1959). Homage to Santa Rosalita, or why there are so many kinds of animals. *American Naturalist*, **93**: 145–159.

Jablonski, D. (1993). The tropics as a source of evolutionary novelty: The post-Palaeozoic fossil record of marine invertebrates. *Nature*, **364**: 142–144.

Keith, M.L. (1982). Violent volcanism, stagnant oceans and some inferences regarding petroleum, strata-bound ores and mass extinction. *Geochimica Cosmochimica Acta*, **46**: 2631–2637.

Lambshead, P.J. and Gooday, A.J. (1990). The impact

of seasonally deposited phytodetritus on epifaunal and shallow infaunal benthic foraminiferal populations in the bathyal northeast Atlantic: the assemblage response. *Deep-Sea Research*, **37**: 1263–1283.

Levinton, J.S. (1996). Trophic group and the end-Cretaceous extinction: did deposit feeders have it made in the shade? *Paleobiology*, **22**: 104–112.

MacLeod, N. (1991). Punctuated anagenesis and the importance of stratigraphy to paleobiology. *Paleobiology*, **17**: 167–188.

MacLeod, N. (1998). Impacts and marine invertebrate extinctions. In: M.M. Grady, R. Hutchinson, G.J.H. McCall and D.A. Rotherby (eds) *Meteorites: Flux with Time and Impact Effects*. London: Geological Society of London, 217–246.

MacLeod, N. and Keller, G. (1994). Comparative biogeographic analysis of planktic for a-miniferal survivorship across the Cretaceous/Tertiary (K/T) boundary. *Paleobiology*, **20**: 143–177.

MacLeod, N. and Keller, G. (eds) (1996). *Cretaceous-Tertiary Mass Extinctions*. W.W. Norton & Company, Inc., New York, 575pp.

MacLeod, N., Rawson, P.F., Forey, P.L. *et al.* (1997). The Cretaceous–Tertiary biotic transition. *The Journal of the Geological Society of London*, **154**: 265–292.

Manley, B.F.J. (1997). *Randomization, Bootstrap and Monte Carlo Methods in Biology*. London; Chapman Hall.

Martin, R.E. (1996). Secular increase in nutrient levels through the Phanerozoic: implications for productivity, biomass, and diversity of the marine biosphere. *Palaios*, **11**: 209–219.

Marshall, C.R. and Ward, P.D. (1996). Sudden and gradual molluscan extinctions in the latest Cretaceous of western European Tethys. *Science*, **274**: 1360–1363.

McGhee, G.R., Jr. (1996). *The Late Devonian Mass Extinction: The Frasnian/Famennian Crisis*. New York: Columbia University Press.

McLean, D.M. (1985). Mantle degassing unification of the trans-K–T geobiological record. In M.K. Hecht, B. Wallace and G.T. Prance (eds) *Evolutionary Biology, Volume 9*. New York: Plenum Press, pp. 287–313.

Melillo, J.M., McGuire, M.D., Kicklighter, D.W. *et al.* (1993). Global climate change and terrestrial net primary production. *Nature*, **363**: 234–240.

Nisbet, E.G. (1987). *The Young Earth: An Introduction to Archaean Geology*. London: Allen & Unwin.

Paul, C.R.C. and Mitchell, S.F. (1994). Is famine a common factor in marine mass extinctions? *Geology*, **22**: 679–682.

Prinn, R.G. and Fegley, B., Jr. (1987). Bolide impacts, acid rain, and biospheric traumas at the Cretaceous–Tertiary boundary. *Earth and Planetary Science Letters*, **83**: 1–15.

Prothero, D.R. (1994). *The Eocene–Oligocene Transition: Paradise Lost*. New York: Columbia University Press.

Raup, D.M. (1977). Stochastic models in evolutionary paleontology. In: A. Hallam (eds) *Patters of Evolution as Illustrated by the Fossil Record*. Amsterdam: Elsevier, pp. 59–78.

Raup, D.M. (1991). *Extinction: Bad Genes or Bad Luck*. New York: W.W. Norton and Co.

Raup, D.M. and Schopf, T.J.M. (1978). Stochastic models in paleontology: a primer. (Course notes for a workshop on 'Species as Particles in Time and Space' held at the US National Museum, Smithsonian Institution, 5–16 June 1978.)

Raup, D.M. and Sepkoski, J.J., Jr. (1982). Mass extinctions in the marine fossil record. *Science*, **215**: 1501–1503.

Raup, D.M. and Sepkoski, J.J., Jr. (1984). Periodicity of extinctions in the geologic past. *Proceedings of the Natural Academy of Sciences*, **81**: 801–805.

Raup, D.M. and Sepkoski, J.J., Jr. (1986). Periodic extinction of families and genera. *Science*, **231**: 833–836.

Raymo, M.E. (1994). The initiation of Northern Hemisphere glaciation. *Ann. Rev. Earth and Planet Sci.*, **22**: 353–383.

Raymo, M.E., Ruddiman, W.F. and Froelich, P.N. (1988). Influence of late Cenozoic mountain building on ocean geochemical cycles. *Geology*, **16**: 649–653.

Rhyther, J.H. (1956). Photosynthesis and fish production in the sea. *Science*, **166**: 72–76.

Ricklefs, R.E. (1979). *Ecology*, Second Edition. Concord, Mass.: Chiron Press.

Sepkoski, J.J., Jr. (1981). A factor analytic description of the Phanerozoic marine fossil record. *Paleobiology*, **7**: 36–53.

Sepkoski, J.J., Jr. (1982). A compendium of fossil marine families. *Milwaukee Public Museum Contributions in Biology and Geology*, **51**: 1–125.

Sepkoski, J.J., Jr. (1984). A kinetic model of Phanerozoic taxonomic diversity III. Post-Paleozoic families and mass extinctions. *Paleobiology*, **10**: 246–267.

Sepkoski, J.J., Jr. (1990). The taxonomic structure of periodic extinction. In: V.L. Sharpton and P.D. Ward (eds) *Global Catastrophes in Earth History: An Interdisciplinary Conference on Impacts,*

Volcanism, and Mass Mortality. Boulder: Geological Society of America Special Paper, **247**: 33–44.

Sepkoski, J.J., Jr. (1994). Extinction and the fossil record. *Geotimes*, **March**: 15–17.

Sepkoski, J.J., Jr. and Raup D.M. (1986). Periodicity in marine extinction events. In: D.K. Elliott (ed.) *Dynamics of Extinction*. New York: Wiley-Interscience, pp. 3–36.

Shaw, A. (1964). *Time in Stratigraphy*. New York: McGraw-Hill.

Signor, P.W. III and Lipps, J.H. (1982). Sampling bias, gradual extinction patterns and catastrophes in the fossil record. In: L.T. Silver and P.H. Schultz (eds) *Geological Implications of Impacts of Large Asteroids and Comets on the Earth*. Boulder: Geological Society of America Special Paper, **190**: 291–296.

Smit, J. (1990). Meteorite impact, extinctions and the Cretaceous–Tertiary boundary. *Geologie en Mijnbouw*, **69**: 187–204.

Smit, J. and Zachariasse, W.J. (1996). Planktic foraminifera in the Cretaceous/Tertiary boundary clays of the Geulhemmerberg (Netherlands). *Geologie et Mijnbouw*, **75**: 187–191.

Smith, A. (1984). *Echinoid Paleobiology*. London: George Allen & Unwin.

Smith, A.B. and Jeffrey, C.H. (1998). Selectivity of extinction among sea-urchins at the end Cretaceous period. *Nature*, **392**: 69–71.

Stolarski, R.S. and Butler, D.M. (1979). Possible effects of volcanic eruptions on stratospheric minor constituent chemistry. *Pure and Applied Geophysics*, **117**: 486–497.

Takeda, S. (1998). Influence of iron availability on nutrient composition ratio of diatoms in oceanic waters. *Nature*, **393**: 774–777.

Tappan, H. (1986). Phytoplankton: below the salt at the global table. *Journal of Paleontology*, **60**: 545–554.

Tappan, H. and Loeblich, A.R. (1973). Evolution of the oceanic plankton. *Earth Science Reviews*, **9**: 207–240.

Tardy, Y., N'kounkou, R. and Probst, J.-L. (1989). The global water cycle and continental erosion during Phanerozoic time (570 My). *American Journal of Science*, **289**: 455–483.

Valentine, J.W. and Jablonski, D. (1986). Mass extinctions: selectivity of marine larval types. *Proceedings of the National Academy of Sciences, USA*, **83**: 6912–6914.

Van Valen, L. (1984). A resetting of Phanerozoic community evolution. *Nature*, **307**: 50–52.

Vermeij, G.J. (1978) *Biogeography and Adaptation: Patterns of Marine Life*. Harvard University Press, Cambridge, Mass.

Vermeij, G.J. (1987). *Evolution and Escalation: An Ecological History of Life*. Princeton, New Jersey: Princeton University Press.

Vermeij, G.J. (1993) *A Natural History of Shells*. Princeton: Princeton University Press, 207 pp.

Ward, P., Wiedmann, J. and Mount, J.F. (1986). Maastrichtian molluscan biostratigraphy and extinction patterns in a Cretaceous/Tertiary boundary section exposed at Zumaya, Spain. *Geology*, **14**: 899–903.

Ward, P.D., Kennedy, W.J. MacLeod, K.G. and Mount, J.F. (1991). Ammonite and inoceramid bivalve extinction patterns in Cretaceous/Tertiary boundary sections of the Biscay region (southwestern France, northern Spain). *Geology*, **19**: 1181–1184.

Wignall, P.B. (2001). Large igneous provinces and mass extinctions. *Earth Science Reviews*, **53**: 1–33.

Wolbach, W.S., Lewis, R.S. and Anders, E. (1985). Cretaceous extinctions: evidence for wildfires and the search for meteoritic material. *Science*, **230**: 167–170.

15

Drifting continents and life on Earth

Catherine A. Forster

ABSTRACT

See Plates 23–25

Earth and biotic history are irrevocably entwined. Plate tectonic processes have shaped the surface of the Earth and its environments since before life began, and have profoundly influenced the lives of the organisms that have existed upon it. The physical rift of a once contiguous landmass can divide populations of species, allowing each to evolve along a separate pathway, while simultaneously joining once divided marine biotas. The reconfiguring of landmasses can affect the ability of animals to migrate, disperse to new areas, or shift their home range.

Separate landmasses may unite via plate collision or the emergence of land bridges, allowing once disparate faunas to interact. The resultant competition can lead to extinction for some taxa, or rapid diversification and spread for others. Importantly, tectonic shifts often drive ecological change through modification of climates, sea level, ocean circulation patterns, temperature, glaciations, and even atmospheric CO_2 levels, influencing even those taxa not directly touched by tectonic events. Thus, the tectonic history of the Earth is a primary catalyst effecting the evolution and distribution of life through time.

15.1 Introduction

The Earth moves under our feet. In some places the ground shifts suddenly and cataclysmically as shallow- to deep-focus earthquakes. Elsewhere, volcanoes may rumble to life, blowing apart mountains and covering the landscape with their debris. Although most parts of the world appear placid and stable, they too are on an achingly slow journey to somewhere else. The skin of the Earth is a mobile and dynamic carapace driven by the continuous processes of plate tectonics.

Plate tectonic processes have been reinventing the surface of the Earth almost since its formation. Throughout time, tectonic plates have slipped slowly across the surface of our planet, carrying on their backs chunks of both continent and ocean floor as they collide, separate, and slip past each other. Colliding plates may build mountain ranges, destroy crust, give birth to strings of volcanoes, and accrete landmasses into enormous supercontinents. Diverging plates may split or fragment continental areas into numerous landmasses and islands and create new sea floor. Ocean basins are likewise affected. They may be riven by continental collision or amalgamated into enormous basins as continents pull apart and diverge. Wandering continents and ocean basins may, over time, radically shift their latitudes and longitudes, passing through numerous climatic regimes as they creep across the

Evolution on Planet Earth
ISBN 0-12-598655-6

globe. This giant, spherical jigsaw puzzle of ocean floor and continents has changed continually through time.

Since life began in the Precambrian it has responded to this continual morphing of the Earth's surface. Organisms are well designed for living in their natural environments and respond to changes by altering their design, either morphologically or behaviorally, to better suit the new circumstances. Tectonically driven environmental change can alter the distributions and movements of organisms, push organisms to shift home ranges, force morphological change leading to speciation, or lead to extinction when organisms either cannot adapt or adapt quickly enough to keep pace with the changes. Thus the effects of plate tectonic processes – geographic, geologic, and climatic – have acted as powerful catalysts for the evolution and movement of organisms throughout time.

Tectonic processes have the power to change local and global climate, ocean current and circulation patterns, sea level, glaciation, and even atmospheric gas content. Each of these factors can directly impact both the evolution and distribution of organisms. For example, increased rates of tectonic plate spreading and periods of extensive continental break-up have been correlated with global sea-level rise. Rising sea levels drown lowland regions, divide continental areas, encourage more equable climates, and alter the extent of continental shelves which affects life in the sea and on land. Even the precise positions of the continents around the globe can affect climate (e.g. Schneider and Londer, 1984). The positioning of continents over the poles encourages the development of polar ice caps, leading to a drop in sea level and increased pole-to-pole temperature gradients. Both bursts of speciation and diversification (e.g. the 'Cambrian Explosion' of early life in the seas) and mass extinctions (e.g. the Cretaceous–Tertiary boundary extinctions that decimated the dinosaurs) have been linked by some researchers to swift environmental changes driven by plate tectonic processes.

The surficial geography of the Earth, controlled in large part by plate tectonic processes,

can facilitate or limit the movements of organisms. On a broad scale, the break-up of a landmass or ocean basin may geographically restrict organisms on each fragment, resulting in increased endemism (distribution of a species over a restricted area). Conversely, the convergence of continents or ocean basins allows dispersal between once disparate areas, increased cosmopolitanism (distribution of a species over a large area), and new interactions between organisms.

By altering geography and environments, plate tectonic processes may provoke changes in the distributions of organisms. The creation of barriers (e.g. mountains, ocean basins) can disrupt or divide a taxon's range through a process called vicariance (e.g. Croizat, 1964; Rosen, 1978). Once divided, each part of the previously cosmopolitan population is theoretically free to evolve along its separate pathway through allopatric speciation (speciation in geographic isolation). But organisms don't always stay in one place – they migrate and alter their ranges through the process of dispersal. A changing local environment may provoke an organism to shift its home range to track its preferred ecological space. Alternatively, satellite populations may move away from their home range, sometimes by dispersing across a barrier. Once a barrier is breached, this population, now isolated from its mother population, may diverge and speciate.

The complex relationships between life and Earth histories are the subject of historical biogeography. Biogeography combines information on the phylogenetic (genealogical) relationships of organisms, the patterns of their distributions, and the evolution of the Earth to examine cause and effect between these patterns and processes. Not only do biogeographers examine factors that affected the evolution and distribution of recent organisms, but paleobiogeographers search for connections between life and Earth histories in deep time. Biogeographers are slowly compiling snapshots of this interplay between life and Earth history throughout time by asking: just how has the Earth moved? How has this movement changed the landscape, environments, and climate? How have these changes affected the distribution and evolution of the organisms that inhabit the Earth?

15.2 The process of plate tectonics

The outer layer of the Earth is formed by a rigid layer of rock called the lithosphere. It is about 70 km thick where it forms ocean floor, and about 100 to 150 km thick where it forms the lighter, bulkier continental crust. This entire outer shell of lithosphere is broken into more than two dozen jagged plates of varying sizes (Condie, 1997) (Figure 15.1). These rigid lithospheric plates ride on the partially molten, underlying asthenosphere. Each plate moves independently of the others, and is capable of changing direction, speed, and its type of boundary. The slow, deliberate movement of the plates varies from place to place, ranging from about 1 to 18 cm per year (McKenzie and Richter, 1976).

Most of the action occurs at the boundaries between plates. Three general types of boundaries are defined according to the relative motion between adjacent plates:

1. Transform faults, or fracture zones: where plates slide past one another. The San Andreas Fault, which runs like a rusty zipper through southern California, is a transform fault produced by the Pacific plate sliding northwest along the western margin of the North American plate. The fitful stop-and-go movement along the San Andreas Fault creates sudden releases of pressure manifest as shallow-focus earthquakes, some of ferocious intensity.

2. Convergent boundaries, or subduction zones: where adjacent plates move towards one another. The leading edge of one plate may override another, driving the overridden plate down into the hot mantle where it is slowly melted and resorbed. Convergent boundaries produce deep-sea trenches, volcanic arcs, folded mountain ranges, and shallow- and deep-focus earthquakes. For example, Krakatau, Tambora, and other volcanoes strung out from Malaysia to New Guinea, are produced by the subduction of the Australian Plate beneath the Eurasian Plate.

3. Divergent zones, or spreading centers: where adjacent plates move away from one

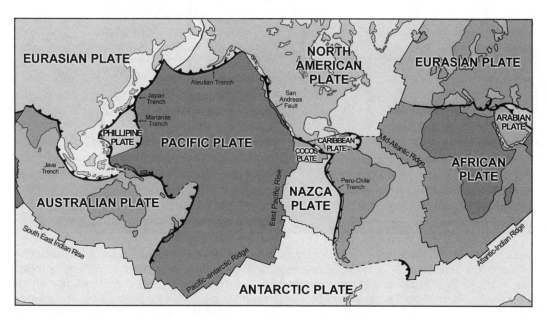

Figure 15.1 The major tectonic plates and their boundaries.

another and new lithosphere is created. As two plates diverge, partially molten mantle material wells up to create new lithosphere that is accreted to the trailing edge of each plate. This upwelling occurs along linear mid-oceanic ridges, or seafloor spreading centers. Related to this process are hot spots, localized plumes of magma that well up from the mantle to punch up within a plate. These intra-plate volcanic plumes are thought to be fixed relative to the mantle, and continue to punch through the crust as the lithospheric plate rides over it. Long-lived hot spots can leave a trail of extinct, progressively older volcanoes trailing away in the direction of plate movement. The Emperor Seamounts (to the west) and Hawaiian Islands (to the east) are such a chain, with the easternmost island of Hawaii still expanding from lava flows as it continues over the hot spot (Carson and Clague, 1995).

The history of plate movement and major topographic features is charted by paleogeographers using geophysical, geological, and sometimes paleontological data. Maps of continental positions and plates through time are continually refined as new data become available (e.g. Scotese, 1991, 1998; Smith *et al.*, 1994; Hay *et al.*, 1999) (Figure 15.2). This paleogeographic information is used by biogeographers to interpret the distributional and evolutionary histories of organisms.

15.3 Life on a changing Earth

15.3.1 *Volcanism: evolutionary and biogeographic effects*

The internal heat of the Earth vents itself through spreading centers, hot spots, volcanoes, and hydrothermal springs. Oceanic

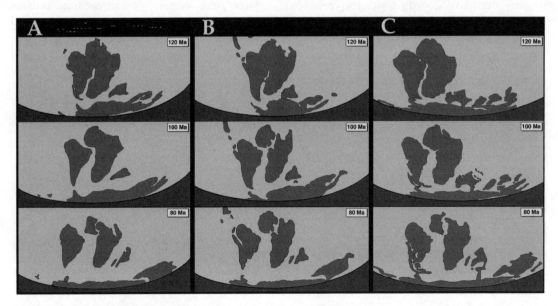

Figure 15.2 Three paleogeographic reconstructions of the break-up of the southern supercontinent Gondwana in the Early, Middle, and Late Cretaceous. A after Smith *et al.*, 1994; B after Scotese, 1998; C after Hay *et al.*, 1999. Although these three reconstructions are very similar in most respects, they differ in detail and the timing of the final separation of the continents. Paleogeographic reconstructions are fine-tuned and adjusted as additional geologic, geophysical, and even paleontological data are collected. After Krause *et al.*, 1999.

spreading centers and intra-plate hot spots sometimes generate only heated waters as underwater hot springs. These intense regions are rich in heavy metals (metal sulfides) and high temperatures (about 350°C), producing an environment lethal to most marine life. But in 1977, biotas specifically adapted to these extreme conditions were discovered 2500 m down in the Galapagos Rift off the coast of Ecuador. These are the deep-sea hydrothermal vent faunas, found exclusively at mid-ocean spreading ridges and undersea hot spots.

Vent communities evolved to adapt to the unique environment produced directly by plate tectonic volcanism. Over 90% of vent species are restricted to vent habitats, and are notably distinct from surrounding marine life (Tunnicliffe and Fowler, 1996). Although rich in over 500 described species, including unicellular types, hydrothermal vent faunas are overwhelmingly composed of arthropod, mollusk and polychaete taxa. The vents are an oasis for giant clams, crabs, tube worms, barnacles, and various scavengers and grazers. The novelty of these highly endemic vent communities is not surprising given their adaptations to a deep-sea world of chemical and thermal extremes.

And what about the distributions of the vent faunas? The historical pattern of spreading ridges and hot spots also appears to have directed the distribution and evolution of vent faunas. Tunnicliffe and Fowler (1996) examined the taxic composition of known vent faunas and used clustering methods to group 'most similar' faunas. By comparing these clusters to plate tectonic reconstructions through time, they discovered that vent faunas are most similar to those closest along their spreading ridge, rather than closest geographically. Since nearly all vent taxa rely primarily on larval dispersal, vent faunas appear to travel exclusively along mid-oceanic ridges, with larval forms migrating across the friendly, hot, metal-rich environment to colonize new areas. Taxa do not easily cross open ocean to another spreading center, even though the direct distance may be shorter. For example, the vent faunas of the 21 degree East Pacific Rise are most similar to those from the Galapagos rather than the closer Juan de Fuca vent communities. This is because the 21 degree East Pacific Rise fauna is connected to the Galapagos fauna via a mid-ocean ridge, whereas it is separated from the closer Juan de Fuca fauna by the San Andreas Fault, the land-locked portion of the plate boundary. The evolution and biogeographic pattern of vent faunas are thus best explained by the tectonically driven development and configuration of the Mesozoic and Tertiary mid-ocean ridges.

Vent communities show an obvious, visible relationship between faunal evolution, distribution, and plate tectonic volcanism. But the effects of volcanism on life, both physically and environmentally, are legion and sometimes subtle. On land, volcanoes can produce rapid, cataclysmic events with disastrous local effects including pyroclastic and lava flows, gas clouds, airborne ash, earthquakes, and tsunamis. Volcanic devastation has affected the course of human civilization through time. For example, the 1815 eruption of Tambora (Indonesia) killed 12 000 people outright, and another 44 000 are estimated to have perished from famine due to local crop destruction. The enormous volume of ash and aerosol Tambora put into the atmosphere is thought to have contributed to unusual global weather patterns the following year. Reduced temperatures were recorded in North America, Europe, and China throughout 1816, where it became known as the 'year without a summer'. Killing frosts and an extremely hard winter resulted in widespread crop and livestock loss. Famine and deaths from starvation were recorded in Europe, Iceland, and North America, and soup kitchens were opened in New York City to feed the hungry. In France, the high food prices engendered by the crop failures of 1816, compounded by Napoleon's loss at Waterloo the previous year, led to widespread inflation, rioting, food-related crimes, and political unrest (Fisher *et al.*, 1997).

As volcanic activity has changed human history, so it has been an agent for evolutionary change, effecting the dispersal of organisms as well as acting as agents for vicariance and speciation. Volcanism can destroy then orchestrate subsequent colonization, evolution, and distributions of organisms. For example, the North

Island of New Zealand has undergone numerous major and minor episodes of volcanic activity. A major eruption of Taupo in AD 186 deposited a thick blanket of ash, killing and altering habitat over a large portion of the island. Ash-laden river systems carried the devastation downstream, choking off and sterilizing some rivers all the way to the sea. McDowall (1996) discovered that in areas where an entire river system was covered by Taupo ash, freshwater fish faunas are composed entirely of diadromous species (fish having part of their life cycle in the ocean). In river systems where at least part of the river catchment escaped the ash, a mix of diadromous and non-diadromous fishes (fish completely confined to fresh water) are found. McDowall concluded that recolonization of river systems completely wiped out by Taupo ash was solely by fishes that could re-enter via the sea to disperse up the barren rivers.

Volcanic activity can erect barriers that act as vicariant agents for speciation. The island of Dominica is part of the Lesser Antilles, a chain of small, volcanic islands in the eastern Caribbean Ocean. Dominica hosts a single, endemic species of the lizard *Anolis oculatus* which varies across the island in morphology (color, size, scalation). Based on morphology and mitochondrial DNA (cytochrome *b*) analyses, *Anolis* is divisible into four morphologically and spatially distinct morphotypes (Malhotra and Thorpe, 2000). When Malhotra and Thorpe (2000) compared the distribution of these four morphotypes to recent geographic features, no barriers were evident. However, they discovered that the dividing lines between these four morphotypes were closely correlated to the location of major recent pyroclastic flows (28 000 years old) that once dissected the island. Although other environmental and selective factors have since molded the evolution of *Anolis* on Dominica, the vicariance produced by pyroclastic flows, dividing and separating populations of *Anolis,* helps explain the evolutionary divergence of this species into four distinct morphotypes via allopatric speciation.

Volcanic activity can also provide a blank canvas for evolution as new land is accreted onto existing landmasses, or completely new islands are formed. The Hawaiian Islands provide a wonderful laboratory for studying speciation and biogeographic processes on recent volcanic islands. Many of the extremely diverse native plants and animals on the Hawaiian Islands, including species of insects, birds, spiders, ferns, and various angiosperms, are closely related but endemic to each island. The westernmost island of Kauai is the oldest in the chain at about 5 million years old. The easternmost island of Hawaii, the largest and youngest in the archipelago, began forming only about half a million years ago (Carson and Clague, 1995).

The original colonization of the Hawaiian Islands was by long-distance dispersal from as far away as Southeast Asia and North America. As new islands were created by hot spot activity on the eastern margin of the archipelago and older ones eroded and subsided, taxa continued to spread and diversify. New colonizations through long-distance dispersal occurred, but newer islands were primarily colonized by immigrants from the older islands through short distance (waif) dispersal. Many biogeographic studies have shown a distinct trend of dispersal followed by speciation from older-to-younger islands (west to east) in the archipelago (e.g. flowering plants, Funk and Wagner, 1995; *Drosophila,* DeSalle, 1995; spiders, Gillespie and Croom, 1995). That is, the most basal taxa in a lineage inhabit one of the older islands, while progressively more derived species are found on younger and younger islands. Thus species appear to have dispersed in conveyor-belt fashion to the newly formed islands to the east as they became inhabitable, with speciation events happening as new islands were colonized.

15.3.2 Isolation and islands

Islands, whether formed by volcanism or landmass fragmentation, are geographically isolated, insular areas that can affect the evolution of their biotas in a number of ways. Allopatric speciation, adaptive radiation to fill vacant ecological niches, changes in morphology (e.g. size, defense mechanisms), and persistence of 'relic' taxa have all been tied to evolution in isolation.

One well-studied phenomenon involves 'island dwarfing', the tendency for large-bodied animals to become miniatures of their mainland compatriots (Lister and Rawson, this volume, Chapter 16). Why this should happen is not fully understood although many explanations have been proposed to varying success. These include adaptation to high caloric foods (Prothero and Sereno, 1982), habitat perturbation (Pregill, 1986), adaptation to rocky terrane (Sondaar, 1977), maintenance of large breeding populations in limited space (Hooijer, 1967), and character displacement due to interspecific competition (Heaney, 1978). Probably no single hypothesis can account for all insular dwarfism.

Within the past few hundred thousands of years, a number of islands in the Mediterannean, including Mallorca, Crete, Corsica, Sicily, Sardinia, Malta, Tilos, and Cyprus, were home to a host of Tom Thumb-sized animals. The sub-fossil bones of meter-high elephants, sheep-sized hippos, dwarf pigs, and tiny deer have been recovered from numerous sites on these islands (e.g. Simmons, 1983; Theodorou, 1988; Reese, 1989; Diamond, 1992). Other islands have their dwarves as well: Madagascar was home to pigmy hippos (Dewar, 1984), Sulawesi had small elephants (Hooijer, 1982), and Papua New Guinea supported a pigmy cassowary (Rich *et al.*, 1988).

Even dinosaur island dwarves have been identified. At the close of the Mesozoic, much of southern Europe was inundated by the sea and divided into a string of large and small islands. The Hateg region of Romania was an island in the trans-European archipelago 70 million years ago (latest Cretaceous), and boasted a diverse fauna including dinosaurs, crocodilians, turtles, birds, and pterosaurs (Weishampel *et al.*, 1993; Benton *et al.*, 1997). The dinosaurs include two ornithopods, *Telmatosaurus* and *Rhabdodon*, the predatory theropod *Megalosaurus*, the sauropod *Magyarosaurus*, and the armored ankylosaur '*Struthiosaurus*'. All the Hateg dinosaurs are far smaller than closely related mainland taxa, prompting numerous paleontologists to sug-

gest they were insular dwarves (e.g. Nopcsa, 1934; Weishampel *et al.*, 1993; Benton *et al.*, 1997).

Not all 'islands' need to be surrounded by water. Climatic changes can lead to the constriction or dissection of terrestrial environments producing isolated, spatially restricted refuges. Refuges can produce the same geographic isolation of organisms as islands. For example, Ford (1980) suggested that callitrichid monkeys, which include the tiny New World marmosets and tamarinds, are 'phyletic dwarves'. To explain their tiny size, Ford (1980) suggested that arid conditions in the Late Tertiary and Quaternary may have reduced and isolated forest regions into small refugia, limiting resources and pushing callitrichids towards dwarfism to survive.

A similar refugia hypothesis has been offered to explain the remarkable diversity of birds in the tropical rainforests of Amazonia. Nores (1999) discovered that zones of high bird diversity were concentrated in areas over 100 m in elevation. Citing evidence of tectonically driven, global sea-level rise of approximately 100 m in the Late Tertiary and Early Quaternary, Nores concluded that these tectonically driven marine transgressions would have divided Amazonia into a number of islands and archipelagos. He hypothesizes that the formation of these islands segregated the biota, encouraging widespread allopatric speciation and independent diversification of bird faunas in each refuge.

Both ocean basins and lakes can become restricted in size and isolated as well, affecting biotic evolution as if they too were islands. For example, Verheyen and colleagues (1996) studied eretmodine cichlid fishes in Lake Tanganyika, employing both phylogeny (mitochondrial DNA sequences) and Earth history to examine the evolutionary processes responsible for their diversification. They discovered that at least two cichlid lineages have diversified into three genetically distinct groups. Geological data show an extreme drop in lake level during the Pleistocene, creating a vicariant event which divided Lake Tanganyika into three separate basins. These three paleobasins correspond to the distributions of the three genetically distinct groups in each cichlid lineage. Verheyen and colleagues concluded that isolation into three paleolakes, subsequent to Pleistocene lake level drop, was an important factor in the evolution and diversification of these cichlid fishes.

15.3.3 *Evolutionary change on large scales*

The biogeographic (dispersal, vicariance) and evolutionary (speciation, extinction, natural selection) processes that operate at the level of islands and refugia also occur at far grander spacial scales. Large-scale tectonic processes, such as whole-scale continental rifting or collision, have the ability to effect an enormous expanse of geography and a large number of taxa.

Large-scale continental fragmentation can have near global consequences for organisms. As mentioned previously, sea-level rise may have helped fuel the explosive radiations of faunas in the Early Cambrian. However, Lieberman (1997) has proposed continental fragmentation itself as a causal factor of greater importance. Using data from abundant, rapidly diversifying lineages of olenellid trilobites, Lieberman concluded that their distribution, and by inference their biotic evolution, was most influenced by rifting patterns themselves. Rapid continental fragmentation in the Early Cambrian, which further fragmented the supercontinent Rodinia, created additional coastline and continental shelf areas, facilitating the spread and radiation of olenellid trilobites into these new areas. Lieberman further hypothesizes that sea-level fluctuation, while influential, may have played only a secondary role in the Cambrian Explosion.

In another study invoking continental fragmentation as a causal effect on evolution, Heads (1999) examined *Abrotanella*, a small, cushion-forming composite plant restricted to mountain terrains in southern South America (seven species) and Australasia (three species). He found the distribution of *Abrotanella* species exhibit clear patterns of vicariance, suggesting that speciation followed the fragmentation of its trans-Pacific ancestral range. This vicariant event may have been the final phase of the break-up of the southern supercontinent of Gondwana in the Late Cretaceous. Moreover, distributions of *Abrotanella* species today often appear correlated with tectonically produced features such as plate margins, transform faults, and orogenic terranes. For example, *A. rostrata*, from the South Island of New Zealand, is endemic to an almost linear strip lying between the Alpine Fault and the Moonlight Tectonic Zone. This vicariant interpretation is strengthened by repeated, identical patterns of distribution in other lineages of organisms, including many other land plants (e.g. DeVore and Stuessy, 1995).

The effects of the fragmentation of Gondwana can also be seen on the Cretaceous faunas themselves. Krause *et al.* (1999) summarized the affinities and biogeographic patterns revealed by the Late Cretaceous (Maastrichtian) Maevarano Formation fauna of northwestern Madagascar. The closest relatives of the Maevarano theropod and sauropod dinosaurs, as well as some of its mammals and crocodilians, have been found in India and South America, rather than on the much closer continent of Africa (see plate section, Plate 24). While the distributions of these closely related species seem incongruous given current geography, they must be placed in a historical perspective. In the Late Cretaceous, just prior to the final break-up of Gondwana, Madagascar was attached to the western margin of the Indian subcontinent. Indo-Madagascar retained a subaerial connection to South America via Antarctica. However, Africa had been isolated from these continents since it rifted off South America in the Early Cretaceous (Figures 15.2 and 15.3). Despite present day geography, the persistent link between Madagascar, India, and South America well into the Late Cretaceous allowed intimate contact between their biotas exclusive of Africa.

When two landmasses merge, portions of their biotas begin to interact as taxa emigrate into the new territory. This may throw biotas out of equilibrium as immigrants and native taxa begin to interact, resulting in competition for resources between similarly adapted organisms, altered predator/prey interactions, and additional impact on the ecosystem through modification by some taxa. These interactions

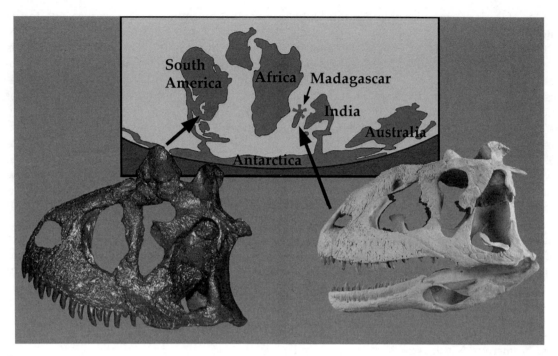

Figure 15.3 The Maastrichtian (Late Cretaceous) Malagasy theropod *Majungatholus atopus* (right) is most closely related to *Carnotaurus sastrei* (left) from Argentina, and *Indosuchus* and *Indosaurus* (not pictured) from India. Other fossil taxa from Madagascar, such as gondwanatherian mammals and araripesuchid crocodilians, also have closely related species in South America and India. None of these taxa show a close relationship to African taxa. Although these distributions seem incongruous given present-day geography, Earth history provides a logical explanation. Paleogeographic reconstructions for the Late Cretaceous, such as the one shown here for the Campanain (after Hay *et al.*, 1999), reveal that when *Majungatholus*, *Carnotaurus*, *Indosuchus*, and *Indosaurus* were alive, Madagascar had just split from India, which still retained a connection to South America via Antarctica. This allowed Malagasy, Indian, and South American faunas to interact with one another well into the Late Cretaceous exclusive of Africa.

may spell success for some organisms, but decimation or extinction for others.

One success story following plate collisions involves reef-forming zooxanthellate corals, which reach their zenith of diversity today in Southeast Asia. The fossil record shows Southeast Asia was notably depauperate in these corals prior to the Miocene. Wilson and Rosen (1998), who examined the origins of the coral fauna of Southeast Asia, suggested that this rarity of zooxanthellates resulted from Southeast Asia's geographic isolation from coral-rich areas. However, as the Australian Plate moved northward it carried continental fragments that collided with and were accreted onto Southeast Asia. These fragments carried with them zooxanthellate corals, ending the isolation of Southeast Asia and spreading these taxa throughout that region. The rather haphazard geometry of the colliding continental fragments created a jagged, geographically complex region with extensive shallow-water areas. The dissected nature of the new Southeast Asia resulted in localized geographic isolation which, in turn, encouraged speciation and greatly increased the endemic diversity of the shallow-water zooxanthellate corals. Wilson and Rosen (1998) thus concluded that the accretion of plate fragments onto Southeast Asia played a crucial role in instigating and controlling its current high diversity of zooxanthellate corals.

One of the best examples of large-scale biotic interchange due to continental amalgamation is the formation of the Isthmus of Panama in the Late Pliocene (2.5 million years ago) which sutured North America to South America. This resulted in wholesale dispersal of terrestrial and freshwater organisms in both directions as the Great American Biotic Interchange (e.g. Simpson, 1947, 1980; Patterson and Pascual, 1972; Webb, 1985, 1991). Although numerous taxa have been studied with regards to this interchange, the most thoroughly investigated group are the mammals.

Prior to the Late Pliocene, both North and South America had distinct mammalian faunas. After the suturing of these continents, widespread reciprocal dispersal forever altered their faunal compositions. South American immigrants to the north included capybaras, porcupines, glyptodonts, armadillos, ground sloths, possums, manatees, vampire bats, and anteaters. Numerous North American taxa crossed the isthmus and spread into South America, including camels, horses, deer, tapirs, gomphotheres, cricetid rodents, procyonid and mustelid carnivores, and tayassuid artiodactyls (Webb, 1985, 1991).

Initially, the interchange between North and South America appears reciprocal, with approximately the same number of taxa moving in each direction. But after about 1 million years, a decided shift in this balance occurred. North America mammals continued to diversify and spread in South America, whereas South American mammals either went extinct or remained small components of northern faunas. Today, only 10% of mammal genera in North America originated in South America, whereas nearly half the genera of extant South American mammals are derived from northern immigrants (Webb, 1991). Some of the most distinctive mammals in South America today, the camelids (guanaco, vicuña, alpaca), are descended from northern immigrants now extinct in their homeland.

To account for the seeming imbalance in immigration and success, as well as explain why some taxa never emigrated at all, Webb (1991) proposed an intriguing ecogeographic model. Webb suggested that the growing imbalance in faunal exchange was due to the expansion of arid conditions as glaciation progressed further north. This aridity dried parts of Central and South America, decreasing the amount of rainforest while increasing temperate savannah conditions. Since far more of North America was previously covered by savannah, it provided a ready-made biota already adapted to the expanding temperate regions of Central and South America. Northern savannah dwellers had a huge, new territory opened to them, while the more tropical faunas of South American had only a small part of North America available to conquer. Additionally, immigrants to both continents consisted of taxa existing adjacent to the isthmus, emphasizing the importance of proximity as a control on emigration. Therefore, both geographic and environmental changes converged to allow the biotic interchange and control immigration.

15.3.4 Tectonic controls of climate, sea level, glaciation and ocean currents

Previous examples have offered glimpses of how climatic change (e.g. North–South American interchange) have shaped biotic evolution and distribution. It is fairly trivial to point out that most organisms, adapted to their present environment, would be pushed either to adapt or move if crucial changes in climatic parameters such as moisture, temperature, and seasonality occurred. Examples of the power of climate change can be seen today. For example, the Sahara Desert continues to encroach into the semiarid Sahel region in northern Africa, slowly eating away at these borderlands as the desert expands southward. Areas that are now deep, dune covered desert were in the recent past open grassland containing rich faunas including lion, giraffe, antelope, and ostrich. Today all that remains of these once diverse faunas in the Sahara are the animals left cavorting in petroglyphs in the deep desert (Figure 15.4).

Climate is an extremely dynamic system developed from several complex, interconnected components and feedback mechanisms including atmospheric gas content, sea level,

Figure 15.4 Petroglyphs deep in the Sahara desert in Algeria. Early hunters stalk an ostrich with their hunting dog (top), and antilope and bovids grace a rock (middle) near the town of Aflou; long horned bovids (bottom) march across a boulder near Taghit. These savannah animals, and others such as giraffe and lion, were displaced to the south by the creeping desertification of the Sahara region within the past few thousands of years.

ocean current patterns, latitude, and topography (e.g. position of mountain ranges) (e.g. Schneider and Londer, 1984). Plate tectonic processes influence all of these components making them intimately involved in the development, sustenance, and alterations of short- and long-term climatic regimes. Delimiting causal factors from feedback mechanisms and effects in climate change can be extremely difficult. Nonetheless, the effects of plate tectonics on climate change, whether direct or indirect, are well documented.

First, plate tectonic volcanism directly impacts atmospheric gas composition and global temperatures. Today's atmosphere is largely volcanic in origin (Condie, 1997). Enhanced volcanism along spreading centers, convergent plate boundaries, and hot spots can change the composition of the atmosphere by increasing the rates of emission of CO_2, SO_2, H_2S, Cl, F, and other trace gases (e.g. Holland, 1984). The level of CO_2 is particularly important due to the greenhouse effect. Atmospheric CO_2 absorbs radiation, releases heat back into the atmosphere, and reflects radiation towards the Earth's surface. Elevated CO_2 levels due to increased volcanic activity lead to greater heat input to the surface of the Earth and warmer global temperatures. Although CO_2 levels are mediated by other factors as well (photosynthesis, surficial weathering of silicate rocks, respiration, precipitation of carbonates) (e.g. Holland, 1978; Goodess *et al.*, 1992; Condie, 1997), numerous studies and models indicate the primary factor in CO_2 levels remains seafloor spreading rates and continental land area (e.g. Frakes, 1979; Berner *et al.*, 1983).

Volcanism also influences climate by affecting sea level. Increased activity along spreading ridges produces a large volume of buoyant new oceanic crust, displacing water, raising eustatic (global) sea level, and flooding continental areas. Often increased periods of spreading also correlate with continental fragmentation as plates are pushed apart and new ocean basins are created. Continental flooding divides landmasses (vicariance) and leads to endemism on land and increased cosmopolitanism in oceans. The increase in aerial extent of the oceans dampens short-term temperature

fluctuations to produce more equable climates and reduce seasonal temperature variations.

Sea-level changes also have been linked to radiations and extinctions. For example, Prothero (1985) linked the extinction of gigantic mammals in North America to large sea-level falls in the Oligocene. Marine regressions (sea-level falls) have been correlated with mass extinctions in the Phanerozoic by a number of researchers (e.g. Newell, 1967; Jablonski, 1986; Paulay, 1990; Hallam and Wignall, 1999). These studies usually link biotic diversity with area, citing the reduction in continental shelf extent as the prime culprit for extinction. But others have argued against this correlation. For example, Stanley (1986) points out that no mass marine extinction occurred in the Oligocene despite this being a time of the largest sea-level fall. Similarly, Valentine and Jablonski (1991) demonstrate that species extinction rates were low through the Quaternary sea-level change as species appear to continue around islands where the shelves tend to remain extensive. While other factors certainly contribute to marine radiations and extinctions, all mass marine extinctions in the Phanerozoic nevertheless appear to correlate with an episode of sea-level fall (e.g. Jablonski, 1986) (Figure 15.5).

Heat distribution and temperature, particularly in the oceans, also are greatly influenced by circulation patterns. Approximately 70% of the Earth's surface is covered by water today. Seawater has a low and invariable albedo and gives up its heat 2–5 times slower than land (Frakes, 1979). Oceans, operating as huge reservoirs for the Earth's heat, are thus able to dampen short-term fluctuations in climate. Most heat is held in the surface waters, and is input primarily in the equatorial zone and transferred poleward through ocean currents. Thus currents act to distribute the heat held in the ocean waters. For example, currents carrying waters from equatorial zones can transport warm surface waters to areas of low temperatures (e.g. the Gulf Stream along the eastern coast of North America) and enhance high-latitude atmospheric heating via evaporation (Frakes, 1979). Major ocean current patterns are functions primarily of

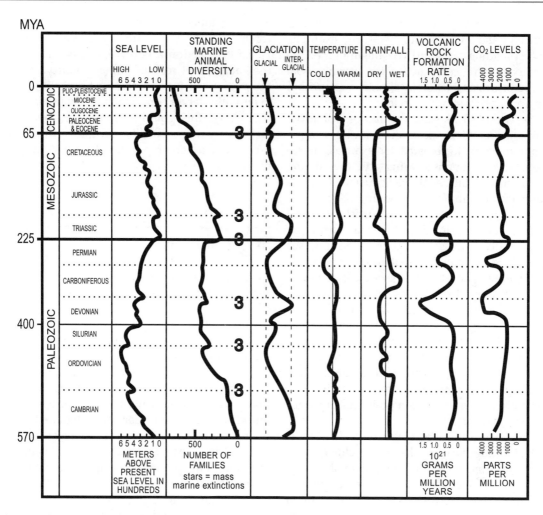

Figure 15.5 Variations in various factors involved in climate and/or driven by plate tectonic processes from the Cambrian through Recent. From left: sea level after Hallam (1984); standing marine diversity by family with mass extinctions denoted by stars, after Raup and Sepkoski (1982); glacial vs interglacial periods with major glacial maxima denoted by circles, after Frakes (1979); mean global temperature, after Frakes (1979); mean global rainfall, after Frakes (1979); CO_2 concentrations, after Schneider and Londer (1984); volcanic activity, after Schneider and Londer (1984).

the positions of the continents. As plate tectonic processes change continental positions, oceanic circulation patterns and heat flow are altered.

Taking all major factors into consideration, increased tectonic plate spreading, high sea level, high atmospheric CO_2 levels, and a tendency to global equability and warm temperatures should be correlated. Conversely, lowered spreading rates should correlate to

drops in sea level, exposure of more continent area, increased uptake of CO_2 via weathering, a drop in atmospheric CO_2 levels, an increase in seasonal temperature variation as climates become more continental, global cooling and glaciation, and marine extinctions (Figure 15.5). Although our current rate of volcanism is probably not sufficient to affect global climate to any major degree (Toon, 1984), ample evidence exists that the amount of global volcanic activity has changed through time (e.g. Wyrwoll and McConchie, 1986). For example, high spreading rates and volcanism in the Late Cretaceous are correlated with warm global temperatures and equable climates even within the arctic and antarctic circles.

As with CO_2 levels, control of sea level, temperature, ocean circulation, and glaciation is not soley by tectonism. For example, tectonoeustacy (sediment loading and the sinking of cooling crustal belts), glacioeustacy (locking water into ice and thus removing it from global circulation), and glacial isostacy (local loading and unloading of the Earth by weighty ice sheets) all help determine sea level. It is estimated that if today's polar ice caps were to melt, sea level would rise nearly 200 m. These other factors can greatly influence short-term changes in climate parameters, and probably longer-term alterations as well. Nevertheless, tectonism and spreading rates are correlated to long-term sea level, glaciation, global temperature, shelf flooding, and marine mass extinctions (Figure 15.5), and remain a powerful influence on the environment.

Finally, one of the best examples of the effect of tectonically driven change in climate, sea level and glaciation on a taxon involves our own species. We *Homo sapiens* may ultimately owe our existence to tectonism. Three million years ago ice caps developed over the North and South Poles, accompanied by global sea-level fall, increased pole-to-pole temperature gradients, and reduced rainfall. This encouraged the spread of arid conditions through much of central Africa, constricting forests and spreading open woodlands and grasslands. Forest dwelling taxa were forced either to contract their ranges to the remaining refuges, or come up with new strategies to adapt to the more open habitat.

Homo appeared approximately 2.5 million years ago at the time of a local glacial maximum. Fossil pollen records and soil carbon isotopes indicate this was a cool, dry period with reduction of the forests through much of Africa (e.g. Bonnefille, 1983; Vrba, 1985; Cerling, 1992). Researchers (e.g. Brain, 1981; Stanley, 1992) have suggested these severe climatic and vegetation changes resulted in 'climatic forcing', with the shrinking forests driving *Homo* to abandon arboreal habitats and activities and adapt to more open country. These adaptations led to increased bipedality, and eventually larger brains and advanced tool use. Potts (this volume, Chapter 19) emphasizes increasing *variability* in these environmental factors as an important driver of human evolution.

15.4 Conclusion

In the past, the evolution and distribution of living organisms has been examined in terms of recent events and geography. However, all organisms have histories, and all regions of the Earth have changed through time. To fully understand the evolution and distributions of organisms it is paramount to take history, both of the organism and the Earth, into account. In doing so, the historical patterns of the co-evolution of life and the Earth are explored, and the causal factors driving evolutionary change and distributions may be identified. This holds true not only for recent organisms, but for extinct, fossil forms as well.

Earth and life have co-evolved through time because organisms adapt to their environments and respond to changes in their surroundings. Alterations in the geography and environments of Earth are driven by the engine of plate tectonics, which has been reshaping the Earth almost since its formation. Endemism, cosmopolitanism, isolation, natural selection and speciation, adaptive radiations, adaptation, dispersal, and extinction can all be triggered by tectonically driven environmental and geographic change. Some of these changes result directly resulting from plate tectonic processes, such as vicariance due to lava flows. Other changes are indirect, such as the spread of arid conditions that alters rainfall and temperature. Whether direct or indirect, short or long term, plate tectonics has always been and remains a powerful catalyst for biotic change.

References

Benton, M.J., Cook, E., Grigorescu, D. *et al.* (1997). Dinosaurs and other tetrapods in an Early Cretaceous bauxite-filled fissure, northwestern Romania. *Paleogeography, Paleoclimatology, Paleoecology*, **130**: 275–292.

Berner, R.A., Lasaga, A.C. and Garrels, R.M. (1983). The carbonate–silicate geochemical cycle and its effects on atmospheric carbon dioxide over the past 100 million years. *American Journal of Science*, **283**: 641–683.

Bonnefille, R. (1983). Evidence for a cooler and drier climate in the Ethiopian uplands towards 2.5 Myr ago. *Nature*, **303**: 487–491.

Brain, C.K. (1981). Hominid evolution and climatic change. *South African Journal of Science*, **77**: 104–105.

Carson, H.L. and Clague, D.A. (1995). Geology and biogeography of the Hawaiian Islands. In: W.L. Wagner and V.A. Funk (eds) *Hawaiian Biogeography*. Washington: Smithsonian Institution Press, pp. 14–29.

Cerling, T.E. (1992). Development of grasslands and savannas in East Africa during the Neogene. *Paleogeography, Paleoclimatology, Paleoecology*, **97**: 241–247.

Condie, K.C. (1997). *Plate Tectonics and Crustal Evolution*. Oxford: Oxford University Press.

Croizat, L. (1964). *Space, Time, Form: The Biological Synthesis*. Published by the author, Caracas, Venezuela, 881 pp.

DeSalle, R. (1995). Molecular aproaches to biogeographic analysis of Hawaiian Drosophilidae. In: W.L. Wagner and V.A. Funk (eds) *Hawaiian Biogeography*. Washington: Smithsonian Institution Press, pp. 72–89.

DeVore, M.L. and Stuessy, T.F. (1995). The place and time of origin of the Asteraceae, with additional comments on the Calyceraceae and Goodeniaceae. In: D.J.N. Hind, C. Jeffrey and G.V. Pope (eds) *Advances in Compositae Systematics*. Kew: Royal Botanical Gardens, pp. 23–40.

Dewar, R.E. (1984). Extinctions in Madagascar. In: P.S. Martin and R.G. Klein (eds) *Quaternary Extinctions*. Tucson: The Universiy of Arizona Press, pp. 574–593.

Diamond, J.M. (1992). Twilight of the pigmy hippos. *Nature*, **359**: 15.

Fisher, R.V., Heiken, G. and Hulen, J.B. (1997). *Volcanoes: Crucibles of Change*. Princeton: Princeton University Press.

Ford, S.M. (1980). Callitrichids a phyletic dwarfs, and the place of the Callitrichidae in Platyrrhini. *Primates*, **21**: 31–43.

Frakes, L.A. (1979). *Climates Through Geologic Time*. Amsterdam: Elsevier.

Funk, V.A. and Wagner, W.L. (1995). Biogeography of seven ancient Hawaiian plant lineages. In: W.L. Wagner and V.A. Funk (eds) *Hawaiian Biogeography*. Washington: Smithsonian Institution Press, pp. 160–194.

Gillespie, R.G. and Croom, H.B. (1995). Comparison of speciation mechanisms in web-building and non-web-building groups within a lineage of spiders. In: W.A. Wagner and V.A. Funk (eds) *Hawaiian Biogeography*. Washington: Smithsonian Institution Press, pp. 121–146.

Goodess, C.M., Palutikof, J.P. and Davies, T.D. (1992). *The Nature and Causes of Climate Change*. London: Belhaven Press.

Hallam, A. (1992). *Phanerozoic Sea-Level Changes*. New York: Columbia University Press.

Hallam, A. (1994). *An Outline of Phanerozoic Biogeography*. Oxford: Oxford University Press.

Hallam, A. and Wignall, P.B. (1999). Mass extinction and sea-level change. *Earth-Science Review*, **48**: 217–250.

Hay, W.W., de Conto, R.M., Wold, C.N. *et al.* (1999). An alternative global Cretaceous paleogeography. In: E. Barrera and C. Johnson (eds) *Evolution of the Cretaceous Ocean-Climate System*. Geological Society of America Special Paper, **332P**: 1–48.

Heads, M. (1999). Vicariance biogeography and terrane tectonics in the South Pacific: analysis of the genus *Abrotanella* (Compositae). *Biological Journal of the Linnean Society*, **67**: 391–432.

Heaney, L.R. (1978). Island area and body size of insular mammals: evidence from the tricolored squirrel (*Callosciurus prevosti*) of Southeast Asia. *Evolution*, **32**: 29–44.

Holland, H.D. (1978). *The Chemistry of the Atmosphere and Oceans*. New York: John Wiley and Sons.

Holland, H.D. (1984). *The Chemical Evolution of the Atmosphere and Oceans*. Princeton: Princeton University Press.

Hooijer, D.A. (1967). Indo-Australian insular elephants. *Genetica*, **38**: 143–162.

Hooijer, D.A. (1982). The extinct giant land tortoises and the pygmy stegodont of Indonesia. *Modern Quaternary Research in Southest Asia*, **7**: 171–176.

Jablonski, D. (1986). Causes and consequences of mass extinctions. In: D.K. Elliott (ed.) *Dynamics of Extinction*. New York: John Wiley and Sons, pp. 183–229.

Krause, D.W., Rogers, R.R., Forster, C.A. *et al.* (1999). The Late Cretaceous vertebrate fauna of Madagascar: implications for Gondwanan biogeography. *GSA Today*, **9**: 1–7.

Lieberman, B.S. (1997). Early Cambrian paleogeography and tectonic history: a biogeographic approach. *Geology*, **25**: 1039–1042.

Malhotra, A. and Thorpe, R.S. (2000). The dynamics of natural selection and vicariance in the Dominican anole: patterns of within-island molecular and morphological deivergence. *Evolution*, **54**: 245–258.

McDowall, R.M. (1996). Volcanism and freshwater fish biogeography in the northeastern North Island of New Zealand. *Journal of Biogeography*, **23**: 139–148.

McKenzie, D.P. and Richter, R. (1976). Convection currents in the Earth's mantle. *Scientific American*, **89**: 72–89.

Newell, N.D. (1967). Revolutions in the history of life. Geological Society of America Special Paper, **89**: 63–91.

Nopsca, F. (1934). The influence of geological and climatological factors on the distribution of non-marine fossil reptiles and Stegocephalia. *Quarterly Journal of the Geological Society of London*, **90**: 76–140.

Nores, M. (1999). An alternative hypothesis for the origin of Amazonian bird diversity. *Journal of Biogeography*, **26**: 475–485.

Patterson, B. and Pascual, R. (1972). The fossil mammal fauna of South America. In: A. Keast, F.C. Erk and B. Glass (eds) *Evolution, Mammals, and Southern Continents*. Albany: State University of New York Press, pp. 247–309.

Paulay, G. (1990). Effect of late Cenozoic sea-level fluctuations on the bivalve faunas of tropical oceanic islands. *Paleobiology*, **16**: 415–434.

Pregill, G. (1986). Body size of insular lizards: a pattern of holocene dwarfism. *Evolution*, **40**: 997–1008.

Prothero, D.R. (1985). North American mammalian diversity and Eocene–Oligocene extinctions. *Paleobiology*, **11**: 389–405.

Prothero, D.R. and Sereno, P.C. (1982). Allometry and paleoecology of medial Miocene dwarf rhinoceroses from the Texan Gulf coastal plain. *Paleobiology*, **8**: 16–30.

Raup, D.M. and Sepkoski, J.J. (1982). Mass extinctions in the marine fossil record. *Science*, **215**: 1501–1503.

Reese, D.S. (1989). Tracking the extinct pygmy hippopotamus of Cyprus. *Field Museum of Natural History Bulletin*, **60**: 22–29.

Rich, P.V., Plane, M. and Schroeder, N. (1988). A pigmy cassowary (*Casuarius lydekkeri*) from the late Pleistocene bog deposits at Pureni, Papua New Guinea. *BMR Journal of Australian Geology and Geophysics*, **10**: 377–389.

Rosen, D.E. (1978). Vicariant patterns and historical explanation in biogeography. *Systematic Zoology*, **27**: 159–188.

Schneider, S.H. and Londer, R. (1984). *The Coevolution of Climate and Life*. San Francisco: Sierra Club Books.

Scotese, C. (1991). Jurassic and Cretaceous plate tectonic reconstructions. *Paleogeography, Paleoclimatology, Paleoecology*, **87**: 493–501.

Scotese, C. (1998). Continental drift (0–750 million years), a quicktime computer animation. University of Texas at Arlington PALEOMAP Project.

Simmons, A.H. (1983). Extinct pygmy hippopotamus and early man in Cyprus. *Nature*, **333**: 554–557.

Simpson, G.G. (1947). Holarctic mammalian faunas and continental relationships during the Cenozoic. *Bulletin of the Geological Society of America*, **58**: 613–688.

Simpson, G.G. (1980). *Splendid Isolation: The Curious History of South American Mammals*. New Haven: Yale University Press.

Smith, A.G., Smith, D.G. and Funnell, B.M. (1994). *Atlas of Mesozoic and Cenozoic Coastlines*. Cambridge: Cambridge University Press

Sondaar, P.Y. (1977). Insularity and its effect on mammal evolution. In: M.K. Hecht, P.C. Goody and B.M. Hecht (eds) *Major Patterns of Vertebrate Evolution*. New York: Plenum Press, pp. 671–707.

Stanley, S.M. (1986). *Extinction*. New York: Scientific American Library.

Stanley, S.M. (1992). An ecological theory for the origin of *Homo*. *Paleobiology*, **18**: 237–257.

Theodorou, G.E. (1988). Environmental factors affecting the evolution of island endemics: the Tilos example from Greece. *Modern Geology*, **13**: 183–188.

Toon, O.B. (1984). Sudden changes in atmospheric composition and climate. In: H.D. Holland and A.F. Trendall (eds) *Patterns of Change in Earth Evolution*. Berlin: Springer-Verlag, pp. 41–61.

Tunnicliffe, V. and Fowler, C.M.R. (1996). Influence of sea-floor spreading on the global hydrothermal vent fauna. *Nature*, **379**: 531–533.

Valentine, J.W., Jablonski, D. (1991). Biotic effects of sea level change: the Pleistocene test. *Journal of Geophysical Research, B, Solid Earth and Planets*, **96**: 6873–6878.

Verheyen, E., Rüber, L., Snoeks, J. and Meyer, A. (1996). Mitochondrial phylogeography of rock-dwelling cichlid fishes reveals evolutionary influence of historical lake level fluctuations of Lake Tanganyika. *Philosophical Transactions of the Royal Society of London B*, **351**: 797–805.

Vrba, E. (1985). Ecological and adaptive changes associated with early hominid evolution. In: E. Delson (ed.) *Ancestors: The Hard Evidence*. New York, American Museum of Natural History, pp. 63–71

Webb, S.D. (1985). Late Cenozoic animal dispersals between the Americas. In: F.G. Stehli and S.D. Webb (eds) *The Great American Biotic Interchange*. New York: Plenum Press, pp. 356–386.

Webb, S.D. (1991). Ecogeography and the Great American Interchange. *Paleobiology*, **17**: 266–280.

Weishampel, D.B., Norman, D.B. and Grigorescu, D. (1993). *Telmatosaurus transsylvanicus* from the Late Cretaceous of Romania: the most basal hadrosaurid dinosaur. *Paleontology*, **36**: 361–385.

Wilson, M.E.J. and Rosen, B.R. (1998). Implications of paucity of corals in the Paleogene of SE Asia: plate tectonics or Centre of Origin? In: R. Hall and J.D. Holloway (eds) *Biogeography and Geological Evolution of SE Asia*. Leiden: Backhuys Publishers, pp. 165–195.

Wyrwoll, K.-H. and McConchie, D. (1986). Accelerated plate motions and rates of volcanicity as contols on Archean climates. *Climate Change*, **8**: 257–265.

16

Land/sea relations and speciation in the marine and terrestrial realms

Adrian Lister and Peter Rawson

ABSTRACT

See Plates 26 and 27

The relative level of land and sea has changed frequently through geological time, triggered by a complex of factors including tectonic movement and periodic glaciation. These changes had many effects of potential evolutionary significance: shifting the extent and global distribution of different habitat types, both terrestrial and marine; opening up dispersal routes or isolating populations; altering patterns of ocean circulation and hence climate.

The Bering Strait provides an example of alternation between land connection and open seaway in the Late Cenozoic: some marine forms dispersed and speciated, but land-bridges were generally too short-lived to produce speciation among mammals. Late Jurassic to Early Cretaceous ammonites show repeated patterns of dispersal (at times of sea-level rise), often causing extinction among endemic faunas, and diversification of isolated or semi-isolated forms (at times of sea-level fall).

Islands of the continental shelf, repeatedly isolated and conjoined during Pleistocene sea-level oscillations, provide a fine-scale study, and resulted in speciation or subspeciation, leading, for example, to the diversity of mammals of Southeast Asian archipelagos today.

Collation of these and other case histories suggests that isolation of the order of 10^4–10^5 years has been required for speciation in mammals and ammonites, probably longer in other groups such as bivalves.

16.1 Introduction

Changes in sea level, both global and regional, have been a recurrent feature of Earth history. On a long-term geological timescale, present-day global sea level is relatively low, hence the continental crust is mainly emergent and the submerged continental shelves are generally narrow. In contrast, there were intervals in the Phanerozoic when high sea levels led to the flooding of extensive areas of continental crust to form broader shelves and major epi-continental shelf seas. At such times, once continuous landmasses shrank and often became fragmented. Conversely, on a lower-order, Quaternary timescale, current sea level is perceived as relatively high, in comparison with recent glacial episodes when it has been even more deeply depressed, opening up further land areas and connections. In concert with major tectonic movements, sea-level change has thus had a significant impact on the extent and distribution of biotic environments, especially in shallow marine and low-lying terrestrial areas.

Evolution on Planet Earth
ISBN 0-12-598655-6

The effects of a change in sea level are, in terms of isolation and connection, inverse for marine and terrestrial organisms. Sea-level rise may allow previously separated marine biota to mix, while splitting formerly continuous terrestrial populations. Fall has the reverse effect. An important feature of these changes is that they are often cyclic. In this chapter we review some well-documented examples of the effects of sea-level change on the distribution and evolution of terrestrial vertebrates (mammals), and on marine invertebrates (molluscs). We emphasize the rate and amplitude of environmental change, in relation to the degree and duration of the evolutionary response.

16.2 Sea-level change over geological time

Changes in global or regional sea level in relation to land have occurred throughout the geological record, at varying frequencies, amplitudes and rates. They are best documented for the Phanerozoic era, the last 540 million years of Earth's history, for which interval a global sea-level curve has been compiled (Figure 16.1). The two Phanerozoic first-order cycles and the numerous second-order (*c.* 10–20 million year) cycles are of eustatic, i.e. global, nature. Some third-order (1–10 million year) cycles also appear to be eustatic, but other third-order and also the smaller-scale cycles (less than 1 million years) recognized by sequence stratigraphers and sedimentologists are harder to interpret. Some are global, but others appear to be developed only regionally. Interpretation is complicated by uncertainties of correlation in the smaller-scale cycles.

The amplitude and rate of sea-level changes are highly variable, depending largely on the underlying causal factors. First- and second-order cycles generally represent long-term changes in amplitude measured in scores to hundreds of metres. Those on a smaller temporal scale normally represent only a few metres to a few tens of metres, but at times (e.g. in the Quaternary, when glacioeustasy

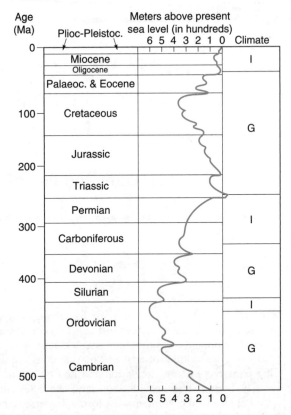

Figure 16.1 Global eustatic sea-level curve for the Phanerozoic, after Hallam (1992). G = 'Greenhouse World', I = 'Icehouse World'.

was the driving force), changes of 100 metres or more were achieved in a few thousand years.

Interpreting the causes of sea-level change is difficult, since the relative heights of land and sea are affected by many factors. A variety of possible causes have been proposed; they are discussed in detail by Hallam (1992) and only the more widely accepted possible mechanisms are outlined here.

The first mechanism is change in the volume of oceanic ridges as a result of variation in the rate of sea-floor spreading. This would cause changes in the volume of the 'container' – the ocean basins – rather than of the seawater. It represents a slow, long-term effect, changing sea level at a rate of about 1 cm per thousand years, over several million years.

During Earth's 'icehouse' phases, the expansion of ice sheets led to rapid and significant change in sea level. During this process the global volume of seawater is reduced, much water becoming locked into ice during cold phases. The amplitude of such glacioeustatic changes is up to a few hundred metres.

The more frequent, often lower amplitude sea-level changes represented by the higher-order cycles of sequence stratigraphy are more difficult to explain. Many appear linked with the Milankovitch orbital forcing cycles (Hewitt, this volume, Chapter 18), and are thus climate-driven. In the Quaternary (and presumably other icehouse phases), waxing and waning ice caps caused geologically rapid cycles of sea-level change (Figure 16.2). At other times, the mechanism by which climate change affects sea level is not well understood.

On a regional scale the vertical (isostatic) movement of landmasses also affects relative sea level. The present-day uplift of much of northern Europe is an isostatic response to the melting of ice caps covering much of the area until some 15 000 years ago. Other tectonic movements may produce dramatic shifts in land/sea relations quite locally. Additional processes affecting regional (as opposed to global) sea level include changing rates of sedimentation and sea-floor subsidence.

It is clear that an array of physical factors can act in a cascade in more complex feedback processes, one precipitating the next. Sea-level change, viewed from this perspective, is a secondary- or tertiary-level effect, taking its cue ultimately from crustal movement and astro-

nomical climate forcing. These factors set up a framework in which more proximal, largely biotic factors will operate, but constraining and influencing in important ways the evolutionary products of those biotic factors.

Some of the major ways in which sea-level change may influence evolution (and extinction) are:

● change in area or distribution of terrestrial and marine habitats
● creation of terrestrial migration routes and isolation of marine basins during sea-level fall
● fragmentation of terrestrial habitats and opening of marine migration routes during sea-level rise
● indirect effects on regional or global climate via changes in ocean circulation patterns, oceanic/continental influence on landmasses, etc.

In the following selected case histories, we contrast the role of sea-level change on terrestrial and marine organisms, and examine the relationship between the timing and magnitude of sea-level cause and evolutionary effect.

16.3 The example of the Bering Strait: contrasting effects on marine and terrestrial biota

Two of the most celebrated land connections in geological history are those that joined great continents. The first, the Panama isthmus, connected North and South America in the Pliocene, in an event of essentially plate tectonic origin (Forster, this volume, Chapter 15). The second, at the Bering Strait (Figure 16.3), has alternately connected and separated Eurasia and North America, with global sea-level change as the primary controlling factor. Both of these regions clearly illustrate the inverse effect on the marine and terrestrial biota. As Sher (1999) has graphically described it, the opening of these seaways put the traffic lights at green for marine life and at red for the terrestrial biota, while their closure had the

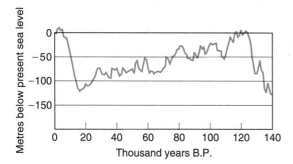

Figure 16.2 Global eustatic sea-level curve for the last interglacial-glacial cycle of the Quaternary, based on planktonic and benthonic ^{18}O data (after Shackleton, 1987).

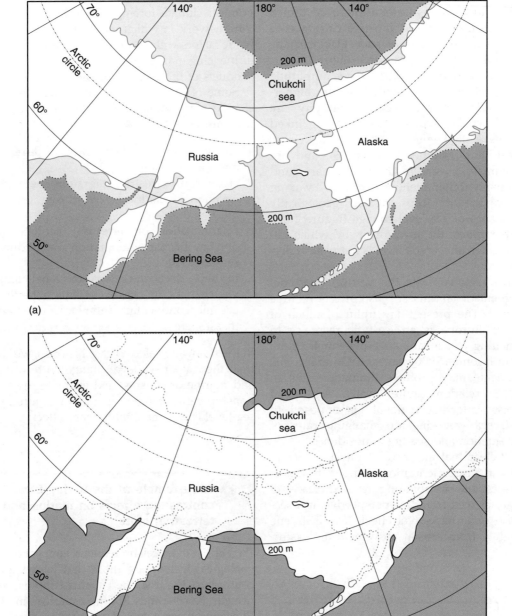

Figure 16.3 Alternating history of the Bering land connection. (a) Present-day: seaway open, land connection closed; (b) at the Last Glacial Maximum *c*. 18 ka: seaway closed, land connection open, based on the current 200 m submarine contour.

opposite effect. But whereas the Panama isthmus has been permanently closed, the alternating traffic lights at the Beringian crossroads has effectively imposed a cyclical pattern on both marine and terrestrial migrants.

16.3.1　*Marine taxa*

In the Tertiary, a permanent land connection joined Eurasia and North America into one supercontinent until the Miocene, and allowed exchange of terrestrial animals and plants while sealing off the marine connection between the North Pacific and North Atlantic–Arctic Oceans (Marincovich and Gladenkov, 1999). These authors date the marine opening to some time between 4.8 and 7.4 million years ago, based on the arrival in the Pacific of the Arctic–North Atlantic bivalve mollusc *Astarte*. Before this event, molluscan faunas in these two oceanic realms evolved separately. The opening of the Strait produced a dramatic change in the composition of Arctic, North Atlantic and North Pacific shallow-water marine faunas. As a result, much of the modern Arctic Ocean molluscan fauna is dominated by taxa of Pacific origin. The anatomy of these changes has been examined for various groups. Examples of originally Pacific taxa which have diversified in the Atlantic include the eel-grass *Zostera* and the sand-dollar *Echinarachinus*. In a detailed study of molluscs, Vermeij (1991) found that many more Pacific taxa migrated to the Arctic—Atlantic than the other way round. In well over half of the cases, in both directions, no morphological species-level divergence has taken place between the two oceans, including species present in both areas since the Pliocene. Where species-level evolution has occurred, it appears in most cases to be anagenetic in the sense that a single ancestor gave rise to a single descendent. In some cases, however, cladogenesis has occurred, for example in *Littorina* (*Neritrema*), where a single Pacific invader has diversified into five North Atlantic species. Vermeij (1991) found that the regions with the lowest representation of trans-Beringian invaders are those in which the invaders have undergone the most species-level evolution, perhaps because of niche availability.

16.3.2 *Terrestrial taxa*

The degree to which sea-level change will have important biological effects will clearly depend greatly on regional topography. The areal expansion or contraction of land or continental shelf will depend on local sea-floor and terrestrial contours. In the case of the Bering Strait, the topography is fairly level. Minimal sea depth at the present day and in the recent past is only 50 metres over a large area of shelf, so that the 100-metre-plus sea-level drop of the Quaternary Ice Ages repeatedly exposed a vast area of low-lying land (Figure 16.3). In other words, the topography here is set up to be extremely sensitive, so that only minor changes in the physical 'driver' can have potentially major effects. Sher (1999) has pointed out that it is oversimplistic to regard Beringia as simply a land-bridge between two continents: when sea level was low, it was a huge territory in its own right, in which Arctic-adapted species evolved, and periodically spread east and/or west into the New and Old Worlds respectively. An example is provided by the origin of the woolly mammoth *Mammuthus primigenius* (Lister and Sher, 2001).

As a result, the established series of North American Land-Mammal Ages is essentially based on the successive appearances of an array of immigrant taxa from Eurasia (Woodburne, 1987). For the later Cenozoic (Pliocene and Pleistocene), after the initial breaching of the Bering Strait, important mammalian migrations have been provisionally pigeonholed into periods of low global sea level (Sher, 1999). The Irvingtonian Land-Mammal Age, for example, beginning close to the base of the Pleistocene at *c.* 1.6–2.0, has been characterized by the entry of true horses (*Equus*), mammoths (*Mammuthus*) and field voles (*Microtus*), while the succeeding Rancholabrean has been characterized by the appearance of *Bison* and many mammalian species closely similar to modern forms. However, dating is poor, and it is possible that these taxa entered over a range of sea-level lows through the Middle Pleistocene.

An interesting paradox in the case of Beringia is that many of the Plio-Pleistocene mammalian species which migrated between Eurasia and North America are essentially temperate in climatic preference, whereas the land-bridge was present during the colder episodes when sea levels fell, not during the warmer intervals when the ranges of such species might have extended northwards towards Beringia. For cold-adapted migrants like the woolly mammoth, this does not pose a problem. For temperate to boreal ones, like red deer (which migrated from Eurasia to North

America some time in the Middle to Late Pleistocene but whose northward limit on both continents today is far south of Beringia), we must invoke time-lags between climate warming, ice-cap melting, and sea-level rise. In other words, there was a window of opportunity after the climate had warmed to allow northward migration, but before the land link had been severed. Perhaps the most famous example of this is the Clovis people, whose migration into North America from Siberia is dated to about 12 000 radiocarbon years BP – when climate was as warm as today, following a very rapid increase in temperature about a thousand years before, but while the Bering bridge was still in operation.

Once into the 'new' continent, many of the mammalian species evolved into endemic species. These evolutionary radiations clearly required the existence of the land-bridge as a permitting factor. To what extent the evolutionary diversification required or was facilitated by the subsequent rise in sea level, flooding the Strait, is more debatable. An example is the lineage of moose-like deer *Cervalces* (or *Alces*) (Lister, 1993). This genus evolved in Eurasia in the Early to Middle Pleistocene, and subsequently, at a time of low sea level, the species *C. latifrons* migrated into North America, where it produced an endemic species in the temperate east, *C. scotti*. In this case, the drop in sea level allowed the migration of the ancestral species, without which the new species could never have evolved, but it is doubtful that the subsequent flooding of the Strait had anything to do with the origin of *C. scotti* thousands of miles away: other, regional factors must have brought about its isolation and evolution.

To determine if any species have originated as a direct result of flooding of the Strait, we consider living mammalian species whose current distribution is limited to the Beringian, or more broadly the Arctic, zone. In this way, we exclude species which, like *Cervalces latifrons* discussed above, are endemic to one continent but probably evolved far away from the Beringian region. When this exercise is undertaken for the entire living North American and northern Eurasian mammal fauna (Corbet, 1978; Burt and Grossenheider, 1987), hardly any cases of likely sibling species pairs across the Bering Strait can be found. Many taxa are conspecific across the Strait, such as moose *Alces alces*, musk-ox *Ovibos moschatus*, caribou/reindeer *Rangifer tarandus*, arctic fox *Alopex lagopus*, wolf *Canis lupus*, northern vole *Microtus oeconomus* and ground squirrel *Spermophilus parryi*. Other North American Beringian endemics, such as the hare *Lepus othus* or lemming *Dicrostonyx groenlandicus*, have a clear candidate ancestor or sister-group within North America itself. There remain only three candidates for sibling species pairs across the Bering Strait. For two of these, the North American form is so similar to the Eurasian one that their separate status has been questioned (Kurtén and Anderson, 1980). They are the lemming *Lemmus trimucronatus* (possibly conspecific with the Siberian *L. sibiricus*), and the wolverine *Gulo luscus* (possibly conspecific with the Eurasian *G. gulo*). At most, they are probably best regarded as subspecies. In both of these cases also, the genus has been present in North America since at least the Middle Pleistocene; in the case of *Gulo*, the progenitor *G. schlosseri* extended across Eurasia and North America. The remaining genus is the lynx, whose Eurasian and North American species (*L. lynx* and *L. canadensis* respectively) are generally regarded as distinct (Kurtén and Anderson, 1980), but whose progenitor *L. issiodorensis* was spread across both continents in the Middle Pleistocene.

In conclusion, there is no example of a Beringian mammal species that has evolved since the last flooding of the Strait. This is perhaps unsurprising, since this event occurred only about 8000 years ago. Indeed, throughout the Quaternary, intervals of isolation, though frequent, have been short in comparison with intervals of connection and potential gene flow. The case of the moose (*Alces alces*) is instructive, because this species is thought to have entered North America for the first time only 10 000 years ago, shortly before the last severance of the land connection. Moose from northeast Siberia and Alaska are virtually indistinguishable morphologically (Heptner *et* al., 1988), and mitochondrial DNA indicates very recent separation (Hundertmark *et al.*, 2002), but in microsatellites, fast-evolving nuclear genes, divergence is evident (Lister and van Pijlen, in prep.). But when it comes to subspeciation or sibling speciation, as in *Lemmus, Gulo* and *Lynx*, this seems to have taken at least since the Middle Pleistocene, not less than 200 000 years ago and possibly longer. Many cycles of sea-level rise and fall occurred during this interval, but it is impossible to say to what extent these may have contributed to the process of taxonomic divergence. More likely, they quenched any incipient divergence by allowing introgression; permanently isolated populations might well have diverged more rapidly.

16.3.3 Climatic and coastline effects

As well as its effect in isolating landmasses, there are important climatic implications of sea-level change, again exemplified by the breaching of the Bering Strait (Sher, 1997, 1999). The establishment of water flow between the Pacific and Arctic Oceans would have strongly influenced general ocean circulation and also regional climate, impacting the terrestrial as well as marine realms. The open connection between two huge water masses with different temperatures would, in all likelihood, have resulted in more turbulent atmospheric conditions, increased precipitation and lower summer temperatures. Combined with the generally low temperature at these high latitudes, the resulting increased humidity could eventually have led to the emergence of the characteristic treeless steppe- and tundra-like communities of the Arctic, and their wide dispersal over adjacent areas. This set the stage for the role of Beringia in the Pliocene and Pleistocene as an area where many arctic-adapted taxa evolved from more temperate stocks (Sher, 1986; Lister and Sher, 2001).

Beringia also exemplifies the way in which the precise shape of the coastline – determined by regional topography – can have important effects. Even when a land connection existed, the shape of its northern and southern coastlines formed a 'bottleneck' half-way along

(Sher, 1984; Yurtsev, 1974) (Figure 16.3). The closer approach of the sea at this point produced a locally more mesic climate and resultant wet tundra (Elias *et al.*, 1996), which acted as a filter to faunal migration between the drier, more steppic habitats on either side (Guthrie, 2001). Although this may have been less of a barrier for large mammals, it may have led to evolutionary endemism of steppic insects between eastern and western Beringia – this is a subject of current research (Alfimov and Berman, 2001; Alfimov, pers. comm., 2002).

16.3.4 Discussion

At first sight, the effect of sea-level change at an intersection like the Bering Strait would appear to be inverse for the marine and terrestrial biota. However, the repeated, cyclic nature of the sea-level change means that both marine and terrestrial realms were alternately connected, then separated. Over time, therefore, the effects on the two realms were similar. The cycles of dispersal and isolation effectively provided an ongoing engine for biotic diversification in both the marine and terrestrial realms, except that they are out of phase by half a cycle. However, the shape of the Quaternary climate curve disturbs this symmetrical picture. The periods warm enough to flood the Strait (the highest peaks in Figure 16.2) were much shorter than the intervening cold episodes when dry land emerged. This appears to have allowed greater opportunities for speciation in the marine than in the terrestrial realm, even allowing for the fact that the marine organisms considered here are invertebrates, the terrestrial ones mammals, traditionally regarded as having a much higher evolutionary rate.

16.4 The example of Early Cretaceous ammonites: sea-level effects in the marine realm

Ammonites are an extinct, widely distributed cephalopod group that show great and often rapid evolutionary diversification. They have formed the basis of many studies of evolution in the fossil record, and played a significant role in the formulation or testing of many of the nineteenth century evolutionary 'laws' such as Dollo's law of evolutionary irreversibility, and Cope's rule of increasing body size through evolution. They also had one of the most rapid rates of species turnover among marine invertebrates (Hallam, 1987), so providing promising examples for the detailed investigation of physical influences on evolution.

Distinct ammonite 'bioevents' have long been recognized, when sudden major turnovers in ammonite faunas occurred. The turnover is usually marked by extinction of some lineages, followed by rapid multiplication of others, often in the wake of an evolutionary innovation but with a greater or lesser timelag. The turnover is often linked to a major dispersal event, leading to the almost instantaneous spread of particular genera far beyond their previous geographic range. But the extent of many such generic 'spreads' has yet to be determined, and the exact nature and cause of the turnovers remains controversial. In a major review of Jurassic ammonite bioevents, Westermann (1993) noted how palaeontologists have long recognized that many bioevents coincide with major lithological changes that indicate widespread marine transgression. Both Westermann (1993) and Rawson (1993) regarded sea-level rise as a major factor controlling the dispersal and evolution of ammonites but, as Westermann (1993: 188) noted, 'the causal inter-relations between sea-level events and the bio-events of cladogenesis and extinction remain poorly understood'.

Study of the biogeographic evolution of ammonites has shown that for mid-Jurassic to Barremian (Early Cretaceous) times (c. 175 Ma to 120 Ma) two ammonite realms can be recognized. The Boreal Realm occurred at high northerly latitudes, covering present-day Arctic areas and extending southward over northern Europe, much of Siberia, and the Pacific coast of northern North America. The Tethyan Realm extended over the rest of the world (Figure 16.4). The boundaries between the two marine realms were partially defined

by landmasses. But even where there was marine connection the boundaries were often quite sharp though they oscillated in position through time. In Europe, the most extensive marine connection between the two realms occurred through the Early and Late Jurassic. During the Early Jurassic in particular, there were several intervals of ammonite migration from the Mediterranean Province to northwest Europe, the migrant (frequently just a single genus or even a single species) often appearing with a sea-level rise and then evolving *in situ* for a few million years before being replaced abruptly by the next immigrants (Hallam, 1987). By mid-Jurassic times there was a well-developed Boreal fauna, and northwest Europe became an area of overlap where Boreal faunas were dominant but at times were replaced by Tethyan-derived migrants before those in turn were followed by another immigration from the north.

A global fall in sea level across the Jurassic–Cretaceous boundary (Figure 16.1) led to greater geographical isolation of the Boreal Realm. Nevertheless there were some marine connections with the Tethyan Realm, mainly narrow seaways that sometimes closed. Sea levels gradually rose again through the Early Cretaceous. The pattern of migration of Early Cretaceous ammonites from one realm to another across Europe and their importance for correlation has been documented in detail by Rawson (1993, 1994), who showed that migration occurred mainly at times of rapid sea-level rise, even when seaways remained open between the rises. On a global scale, Rawson (1993) noted a relationship between global sea-level rise and the rapid geographical spread of some genera. Such 'global spreads' often became extinct quickly outside their normal area of distribution, but sometimes gave rise to thriving new local assemblages by allopatric speciation.

The effects on ammonite evolution of two Early Cretaceous sea-level rises are reviewed here; the review is an updated version of summaries published by Rawson (1993, 2000), expanded here to place greater emphasis on the evolutionary consequences of sea-level rises.

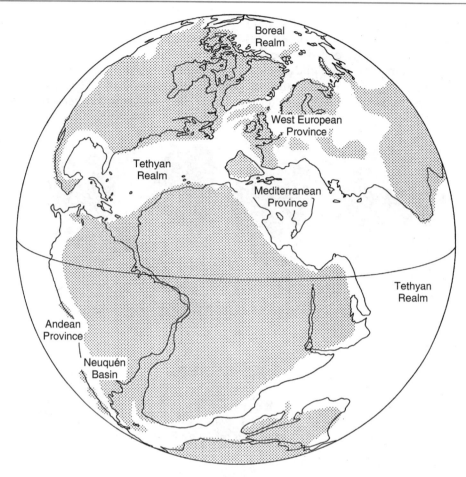

Figure 16.4   Early Cretaceous (Hauterivian) biogeography: the Tethyan and Boreal realms, with provinces mentioned in the text. Map modified from Rawson (1993, Fig. 10.1).

16.4.1   *The Early Hauterivian sea-level rise (c. 132 Ma)*

This event is only well documented across Europe, and may represent a regional rather than global rise in the relative height of sea to land. For a short time immediately before the sea-level rise, northwest Europe (the West European Province) was populated almost exclusively by *Olcostephanus*, which had migrated there from the Mediterranean Province of the Tethyan Realm. In the latter area *Olcostephanus* co-occurred with a much more varied fauna, including Neocomitidae. As sea-level rose, some Neocomitidae spread northwards to totally replace the West European *Olcostephanus*. There is little or no overlap between the successive faunas. It is difficult to tell in such a situation if there was extinction prior to replacement, or if there was an ecologically significant but geologically very short period of overlap, and perhaps competitive displacement, which has not been preserved. To what extent the newcomers occupied the same ecological niche as the outgoing forms is also difficult to ascertain, but the broad environmental context – muddy sea floors – did not alter greatly. The earliest immigrants appear close to Mediterranean forms of *Neocomites* (*Teschenites*), but they quickly gave rise to *Endemoceras*, which apparently arose *in situ* by allopatric speciation during the high sea-level event. *Endemoceras* then evolved in the West European Province for a period of about 2 million years. Its main lineage changed, through three nominal chronospecies, from an evolute (loosely coiled), coarsely ribbed form to a more involute (tightly coiled), finely ribbed form (Figure 16.5). This line also gave rise to loosely coiled, tuberculate forms ('*Acanthodiscus*') and to uncoiled, strongly tuberculate ones (*Distoloceras*) – perhaps ten species in all, over a period of around 3 million years. Individual species in the main lineage lasted no more than 0.75 million years, while those derived from it were of even shorter

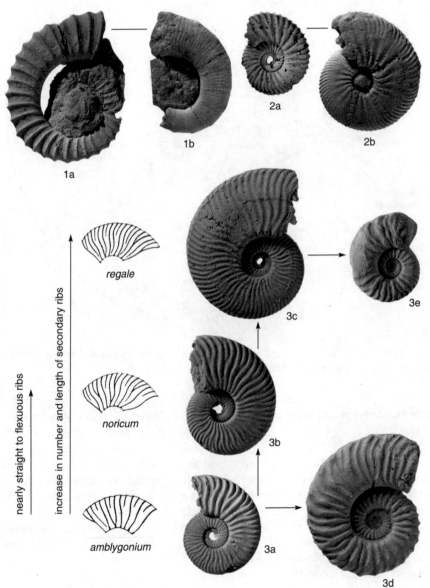

Figure 16.5   Hauterivian (Early Cretaceous) ammonites from the West European Province. 1. The uncoiling forms *Aegocrioceras spathi* (1a) and *Crioceratites duvali* (1b). 2. Early and late members of the *Simbirskites* clade: *Simbirskites* (*Speetoniceras*) *inversum* (2a) and *Simbirskites* (*Craspedodiscus*) *gottschei* (2b). 3. The *Endemoceras* lineage and selected offshoots: *E. amblygonium* (3a), *E. noricum* (3b) and *E. regale* (3c). An early speciation event is marked by the evolution of *Distoloceras* (3d) from *E. amblygonium*, and a later event by the evolution of '*Acanthodiscus*' (3d) from *E. regale*. All photos approx ×0.75.

duration. This diversification occurred during a period of falling sea level, but apparently without significant subdivision of the species' ranges into distinct basins.

Shortly after the initial migration northwards, the Neocomitidae became extinct in their ancestral area, the Mediterranean Province. This is evidenced by a hiatus in their occurrence through the *loryi* zone, except for some relict *Saynella* in the lower part of the zone (Bulot *et al.*, 1993). They also appear to have died out in most other Tethyan areas, though exact dating of some faunas is uncertain. Northwest Europe, Argentina, and California to Oregon, were the only areas where they continued to evolve during the Early Hauterivian, until the next sea-level rise led to renewed faunal turnover.

16.4.2 *The mid-Hauterivian sea-level rise (c. 130 Ma)*

This appears to have been a global event and at least within Europe took place in two closely spaced pulses (Rawson, 1995). Neocomitids made a brief reappearance in southern France in mid-Hauterivian times (*nodosoplicatum* zone), believed to mark a southerly migration from the West European Province (Thieuloy, 1977). At the same time, the neocomitids were becoming extinct in the latter area. Again, it is hard to say if the extinction preceded, or was caused by, the immigration of new forms.

The mid-Hauterivian event led to the spread of an uncoiling genus, *Crioceratites* (Figure 16.5), over much of the world. *Crioceratites* had first appeared in Late Valanginian times (*c.* 134 Ma), probably originating in the Mediterranean Province of Tethys. But its main evolutionary development occurred in response to its widespread dispersal during the initial mid-Hauterivian sea-level rise. Then different forms evolved in several disparate parts of the world. The best known faunas are in the Mediterranean and West European Provinces, and in the Neuquén Basin of Argentina (Andean Province). In the Mediterranean Province the lineage continued until Late Barremian times (*c.* 124 Ma) and a wide variety of forms occur,

assigned to several different genera. In *Crioceratites* alone, around 20 species arose in an interval of *c.* 9 Ma, and there were around eight other genera besides. In the Neuquén Basin the first species to appear (*Crioceratites apricus*) is virtually indistinguishable from Mediterranean forms, but later species became increasingly morphologically distinct. In the West European Province, a distinctive, short-lived (*c.* 1 million years) endemic crioceratitid genus, *Aegocrioceras* (Figure 16.5), appeared (Rawson, 1975; Kemper, 1992). At the time the English part of the fauna was monographed (Rawson, 1975) the origin of the genus was problematic; three species appeared in the sequence suddenly, only a few centimetres above the last neocomitid ammonites at a level of total ammonite turnover. Subsequently, rare *Crioceratites nolani*, a species known from the Mediterranean Province, have been found at an intervening level (Doyle, 1989, and Rawson collection) and provide the likely ancestral form from which *Aegocrioceras* apparently evolved through allopatric speciation. The English sequence is condensed but is unlikely to represent more than a million years. *Aegocrioceras* diversified in the West European Province into a total of about seven species, although no more than three existed at any one time.

The first phase of the mid-Hauterivian sea-level rise saw the immigration to the West European Province not only of forms from the south, but from the north as well, with a brief appearance of the boreal genus *Simbirskites* (*Speetoniceras*) (Figure 16.5), although it failed to diversify at that time. Subsequently, however, throughout the province, the *Aegocrioceras* fauna was replaced by a renewed immigration of boreal *Simbirskites*, marking the second phase of the mid-Hauterivian sea-level rise. *Simbirskites* also occupied the Russian Platform (East European Province), but while the initial species appear to be the same in the two provinces, later forms, covering some 3–4 million years, came to differ. In particular, the umbilical tubercle, which rests below the umbilical seam in many of the West European forms (the likely primitive condition), is placed against the umbilical seam in the East European forms (Rawson, 1971). This feature appeared early in the East European radiation and is seen in all subsequent species. In other features, however, such as size increase and development of discoidal forms (Figure 16.5) at the end of each lineage, there appears to be a degree of parallel change between the two regions. These changes occurred while sea level was gradually falling.

16.4.3 Discussion

Taxa migrating into a new area in response to a sea-level rise either quickly became extinct, or completely replaced the pre-existing local faunas – often to evolve by allopatric speciation into endemic taxa. From the data available (the almost instant appearance of an endemic form), the initial speciation was probably stimulated by dispersal to a new area rather than by partial isolation during the subsequent sea-level fall. During the latter periods, however, the newcomers continued to diversify, even though in the Early Cretaceous examples discussed above, connecting seaways often remained open and limited faunal exchange still occurred between adjacent basins. However, in the case of northwest Europe, its position as a semi-isolated basin at that time may have contributed to the diversification of successive immigrant taxa there. Typically, the initial radiation after an immigration event was extremely rapid (tens of thousands of years). Species then endured for varying lengths of time, some up to 750 000 years, others shorter.

Why high sea levels should effectively control dispersal remains a matter for speculation. In none of the cited examples did sea level rise actually open up new seaways, although it must have widened and deepened them. It may be that the rapid rise of sea level led to short-lived changes in current direction and/or strength, facilitating dispersal in certain directions; or to the spread of warmer or cooler water masses, allowing the spread of stenothermic groups. Subsequent regression may have led to a critical shallowing, producing ecological barriers to ammonite spread even though the seaways were still open.

Ammonites were undoubtedly more environmentally sensitive than many other invertebrate groups (Hallam, 1987). Unfortunately it is difficult to speculate what ecological factors operated to limit their movement through the seaways, since the mode of life of ammonites is still open to debate. Some were certainly nektobenthonic, but many may have been pelagic, 'divided almost equally among swimmers, drifters and vertical migrants' (Westermann, 1996). The Early Cretaceous migrants reviewed above embrace forms of very differing morphologies, including an uncoiled genus (*Crioceratites*) that may have been one of the 'drifters'. It is therefore difficult to perceive any pattern which may explain which forms were more prone to migration or subsequent radiation.

Similarly, the proximate causes of speciation and diversification following dispersal to a new area require further research. Possible causal factors include both biotic and physical factors, such as vacant niches, competition with imcumbent species, or the change in physical conditions.

16.5 The example of offshore and oceanic islands: isolation of terrestrial biota

Islands provide exciting opportunities for studying the evolution of terrestrial organisms. The isolation of island populations, which may lead to the origin of endemic forms, can clearly be strongly influenced by sea level, as well as by tectonic and other factors. Moreover, islands vary in their size, distance from the mainland, and surrounding bathymetry (and hence the probability of links to other land masses with sea-level change), allowing us to examine the influence of these factors, and the duration of isolation, on evolution.

This discussion is limited to islands that are sufficiently close to the mainland (or to an archipelago providing stepping-stones to the mainland) that occasional colonization by non-volant vertebrates has been possible. Colonization itself can be critically dependent on sea-level history, and may occur by two main routes. In the first, the island is permanently isolated by sea from the mainland, at such a distance that colonization by rafting or swimming is a possible but very rare event, providing sufficient isolation for endemism to develop (Sondaar, 1977). In this situation, sea-level change may influence the possibility of colonization or the degree of isolation by narrowing or widening the seaway. Sondaar cites the example of Cyprus, where no terrestrial mammals are known before the Pleistocene, when lowered sea level presumably first allowed sweepstakes dispersal. Endemic dwarf hippo and elephant then evolved. A second mode of origin for endemic populations occurs when an island was at first connected to the mainland, sharing its fauna, but by relative sea-level rise, the island became isolated, carrying its cargo of fauna which, with time, developed endemic features. Examples are Sardinia, connected to the mainland until the Early Pliocene, and populated by the ancestors of a later endemic fauna (Sondaar, 1977); and Jersey, isolated from France only for short periods during Pleistocene high sea-level stands (Lister, 1996; see below).

16.5.1 Europe

In the Pliocene and Pleistocene, many islands around the world developed endemic vertebrate faunas. The areas that have attracted the most attention are the Mediterranean and Southeast Asia. An outstanding example is provided by the area of the Gargano, now part of the mainland of eastern Italy, but for much of the Pliocene and Early Pleistocene (c. 5–1 Ma) an offshore archipelago (de Guili *et al.*, 1990). Several lineages of rodents and insectivores produced strongly endemic forms, such as the giant hedgehog *Deinogalerix*, which is so bizarre that its taxonomic position in relation to mainland groups is quite unclear (Butler, 1980). Among other endemics, *Deinogalerix* is represented in the Plio-Pleistocene of the Gargano by no fewer than five species in two co-existing lineages; the murid rodent *Microtia* by four phyletic lines. The time span in which this fauna evolved is uncertain, but it is presumed

to be in the c. 5 Ma interval between an Early Pliocene regression, and the Early Pleistocene fossiliferous deposits (de Guili *et al.*, 1990). De Guili *et al.* (1990) speculate that these radiations occurred allopatrically as populations were isolated on nearby islands of the archipelago, the resulting species becoming sympatric when minor regressions linked the islands.

Moving to a shorter time span, the deer fauna of Crete was studied by de Vos (1984), who identified seven or eight endemic species. Dating is poor, but these species apparently all evolved during the Middle and Late Pleistocene – a matter of a few hundred thousand years. They have distinctive antlers, postcranial proportions and other features. De Guili *et al.* (1990) speculate that one or two colonizing species (arriving by sweepstakes dispersal as there was no land link) may have spread out over the island's coastal plain, but that subsequent sea-level rise inundated the plain and isolated the deer in separate valleys, where they differentiated into separate species.

An even more rapid event was documented by Lister (1989, 1996), where a dwarfed form of red deer (*Cervus elaphus*) evolved on the small island of Jersey (English Channel) in less than 6000 years of the Last Interglacial, about 120 000 years ago (Figure 16.6). Here, the island's deer descended directly from full-size animals present on Jersey during the low sea level of the preceding cold phase, when Jersey was part of the mainland. The animals became strongly dwarfed, with associated allometric effects, but there were no adaptive or other novel changes and the form can be regarded as a subspecies (Lister, 1989). This can be contrasted with the Cretan deer, which show distinctive anatomical features including unique display organs (antlers), indicative of full speciation. The difference between the two situations may be partly due to the much larger size of Crete, but mainly to the longer duration available – probably several hundred thousand years (Lister, 1996).

16.5.2 Southeast Asia

The islands of Southeast Asia have provided a rich source of data. Van den Bergh *et al.* (1995)

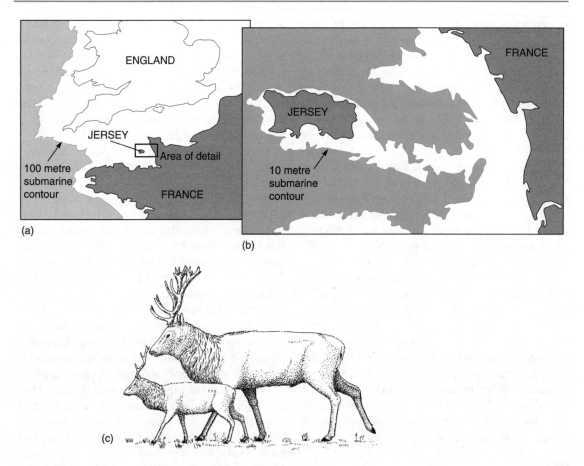

Figure 16.6 (a) Jersey as part of the north European landmass during the last glaciation; (b) the current – 10 m contour around Jersey, showing the isthmus that would form with only a small drop in sea level; (c) reconstruction of dwarf Jersey deer in comparison with its mainland progenitor (*Cervus elaphus*). Note reduced body size and simplified antlers. From Yalden (1999).

show that Java, isolated by sea during the early part of its history, developed a typically limited and 'unbalanced' island mammal fauna with endemic elements during the period *c.* 1.5–0.8 Ma, but this was replaced after that date by a more extensive fauna more typical of mainland Southeast Asia. The 0.8 Ma date corresponds to a marked lowering of eustatic sea level. Based largely on modern distributions, Heaney (1986) has gone some way toward quantifying patterns of endemic mammalian speciation in Southeast Asia in relation to changing sea level. Thousands of islands in the Philippine archipelago, for example, were joined into fewer, larger islands during the low sea-level stands of the Late Pleistocene: these

can be approximately modelled using the current 120 metre submarine contour. The limits of these Late Pleistocene islands are found to correspond exactly with modern faunal regions, each of which is characterized by 70% or more endemic species of mammals, but is internally homogeneous (Heaney, 1986: Figure 16.7). Other groups (amphibians, reptiles and birds) follow the same pattern. The Philippine islands as a whole support at least 17 endemic genera of rodents as well as many endemic species of more widespread genera. By contrast, islands of the nearby continental shelf (e.g. Borneo, Java and Sumatra), which were connected to mainland Southeast Asia during each low stand (most recently *c.* 18 000

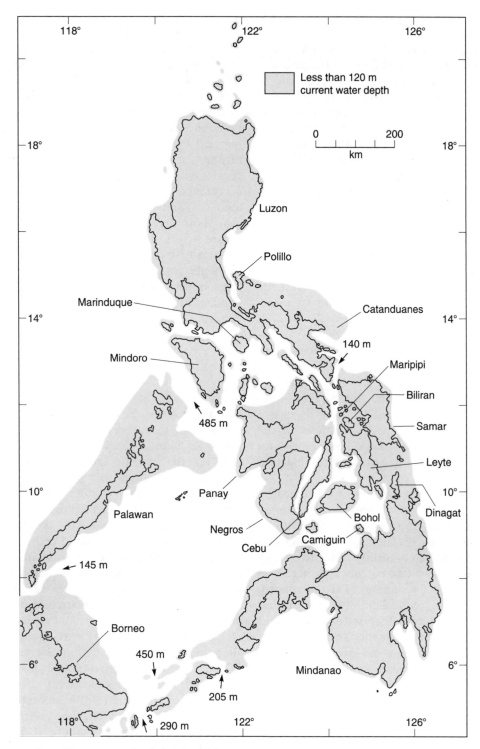

Figure 16.7 Late Pleistocene islands of the Philippines (shaded), based on the −120 m submarine contour. Modern faunal provinces are highly concordant with the limits of these 'palaeo-islands'. After Heaney (1986).

years ago) have a far lesser proportion of endemics. An interesting intermediate condition is provided by the island of Palawan, between Borneo and the other Philippine islands. The channel between Borneo and Palawan is about 145 m deep, greater than the 120 m depth reached during the Late Pleistocene, but less than the 160 m sea-level drop of the Middle Pleistocene. Heaney (1986) suggests that the large number of endemic *species* but low number of endemic *genera* (compared to the more isolated Philippine islands) is consistent with Palawan's separation from Borneo for about 160 000 years.

The Philippine archipelago also illustrates the effect of local geography and sea levels in determining the sources and rates of mammalian colonization. Highly distinctive endemic mammalian faunas can evolve and be maintained for long periods between islands only a few kilometres apart. Heaney (1986) calculates that successful colonization across channels only 5–25 km wide during the Pleistocene has occurred at rates of only one species per 250 000–500 000 years. The diversity of endemics is largely accounted for by subsequent radiation. Thus, phylogenetic analysis of the 40 murid rodent species of the Philippines suggests they arose from five to seven colonization events.

Depending on the location of other nearby islands, and on ocean currents, an island may receive its colonizers from one or more directions, determining its faunal composition and hence the biotic factors further influencing evolution. Heaney also finds a generally high correlation between the number of island species, and of endemics, with island area, conforming to island biogeography theory (MacArthur and Wilson, 1967). The fluctuations in island size caused by Pleistocene sea-level oscillations will have caused concomitant shifts in community diversity and composition.

16.5.3 *Patterns of change*

Mammals endemic to smaller islands show some consistent patterns in their morphological evolution. The most obvious is size change, with smaller mammals in general tending to increase in size, larger ones to decrease. This 'island rule' has generally been explained in terms of resource limitation (especially in the case of large mammals), and release from predator pressure and interspecific competition (especially in the case of small mammals) (Foster, 1964; Heaney, 1978; Lomolino, 1985). Sondaar (1977) also points out that the sweepstakes route of island colonization selects for large mammals that are good swimmers, explaining the repeated occurrence of hippos, elephants and deer among island faunas. The endemic ungulates also show predictable adaptive changes, consistent across taxa and islands. These characteristics include shortening of distal limb elements, fusion of ankle bones and increased hypsodonty (tooth crown height), which Sondaar (1977) associates with the commonly rugged terrain and tougher, more restricted vegetation on islands compared to the mainland. Hippos become more terrestrially adapted, as the islands are usually poorly supplied with lakes or rivers. In other words, islands have produced both radiation and a strong degree of parallel evolution.

To what extent does the study of island endemics illuminate evolutionary diversification on the continental landmasses? First, despite the plethora of endemic forms on islands around the world, there does not appear to be any documented case of an island endemic successfully colonizing the mainland, let alone radiating there. Island endemics tend to go extinct either at times of environmental change, or when invaded by mainland competitors or predators (Sondaar, 1977; Lister, 1996). They are not, therefore, a source of mainland diversity. Second, we may ask to what extent offshore or oceanic islands provide models for allopatric speciation on the continents, where 'islands' of habitat (terrestrial or freshwater) occur and are often implicated in discussions of allopatric speciation? The answer to this question is at present elusive. Compared to the controlled island situation, it is much more difficult to give a detailed account of the history of isolation of a particular mainland habitat area, or of the genetic isolation of the target species, through sufficient palaeontological time. Modern patterns of vicariance, or the geographical distribution of fossils, may hint at allopatric speciation, but a detailed historical account of the process is at present generally beyond our grasp.

16.6 Conclusion

Drawing together the varied case histories presented above, we can list in more detail the various ways in which sea-level change may influence evolution:

- Changing the area or distribution of terrestrial or marine habitats, such as coastal plain, neritic zone, continental shelf and epicontinental shelf seas, or islands. This will influence niche availability and community composition and hence biotic interactions which in turn affect evolution.
- Breaking a land barrier to marine organisms or sea barrier to land organisms, thus creating migration routes.
- Altering the shape of the coastline, promoting evolution by, for example, creating isolated populations in pockets of coastal plain.
- Changing the dimensions of the barrier which has to be crossed by dispersing populations, such as seaways to islands, width of an isthmus, or depth of a seaway. This influences which species can cross, how often they will cross, and the degree of genetic isolation once there.
- Creating isolated or semi-isolated areas such as marine basins or islands, where allopatric speciation may occur.
- Repeatedly joining and separating land or sea areas. Depending on the frequency of change, this may inhibit or aid species-level evolution according to the pattern of genetic mixing.
- Altering patterns of ocean circulation or the distribution of cold and warm waters, thus affecting the dispersal of marine organisms. In turn, such changes may affect regional or even global climate, and hence the distribution of terrestrial vegetation and faunas.

How the various potential causal factors impact on evolution is a complex, multifactorial problem. It is clear that in many ways sea-level change occupies an intermediate position in the chain of cause and effect. It is caused by more distal forcing factors, such as glaciation or tectonic movement; and it sets a framework within which more proximal physical or biotic factors may operate, such as substratum type or community composition. Repeating themes of the case histories examined in this survey are the two couplets of dispersal-speciation and speciation-extinction. The latter is the basic engine of all macroevolution (Jablonski, this volume, Chapter 13), but the former is particularly linked to physical drivers like sea-level change.

The case histories discussed above demonstrate that sea-level change appears to trigger evolution in some examples but not in others. Timescale, taxonomic group, regional biota, habitat structure, and other factors may all influence the outcome.

Our ammonite examples show rapid dispersal followed by evolutionary radiations in response to rises in sea level. Conversely, Valentine and Jablonski (1991) found that in another molluscan group, the Pleistocene to Recent marine bivalves of California, similar communities re-formed with each sea-level high stand over the past million years: the fauna appeared to have absorbed the area and habitat effects of sea-level change by latitudinal migration alone, not by evolution or extinction. This could be an effect of the slightly shorter timescale of the Pleistocene example compared to the Cretaceous ones. But it more probably reflects the more rapid rate of evolution of ammonites. Individual ammonite 'populations' normally show much more variation than do bivalves and hence may be genetically predisposed to more rapid evolution. The variation often reflects plasticity in the timing of development of individual morphological characteristics, which 'can be a source of morphological and taxonomic diversity' (Yacobucci, 1999: 69). It is instructive also to compare the mammalian speciation on Southeast Asian islands since the Middle Pleistocene (200 000 years or so), with the lack of apparent speciation across the Bering Strait in an equivalent period. Whereas the major island groups were permanently isolated from each other during that period, Beringia was repeatedly rejoined after each short interval of isolation, so that any incipient divergence was lost by interbreeding. The episodes of Beringian separation, of the order of 10 000 years each, were evidently not long enough for species-level divergence. Thus we can begin to obtain an estimate of a typical duration of isolation for mammalian speciation: of the order 10^4–10^5 years. It is more difficult to make a comparable estimate for the ammonoid examples but it is likely to be in the region of 10^5 years at most.

References

Alfimov, A.V. and Berman, D.I. (2001). Beringian climate during the Late Pleistocene and Holocene. *Quaternary Science Reviews*, **20**: 127–134.

Bulot, L.G., Thieuloy, J.-P., Blanc, E. and Klein, J. (1993). Le cadre stratigraphique du Valanginien supérieur et de l'Hauterivien du Sud-Est de la France: définition des biochronozones et caractérisation de nouveaux biohorizons. *Géologie Alpine*, **68**: 13–56.

Burt, W.H. and Grossenheider, R.P. (1987). *Peterson Field Guides: Mammals*. Boston: Houghton Mifflin.

Butler, P.M. (1980). The giant erinaceid insectivore *Deinogalerix* Freudenthal, from the Upper Miocene of Gargano, Italy. *Scripta Geologica*, **57**: 1–72.

Corbet, G.B. (1978). *Mammals of the Palearctic Region*. London: British Museum (Natural History).

De Guili, C., Masini, F. and Torre, D. (1990). Island endemism in the eastern Mediterranean mammalian paleofaunas: radiation patterns in the Gargano Paleo-Archipelago. *Academia Nazionale dei Lincei, Atti dei Convegni Lincei*, **85**: 247–262.

De Vos, J. (1984). The endemic Pleistocene deer of Crete. *Verh. Kon. Ned. Akad. Wet., Afd. Natuurk. Erste Reeks*, **31**: 1–100.

Doyle, J. C. (1989). The stratigraphy of a late Lower Hauterivian horizon in the Speeton Clay formation (Lower Cretaceous) of East Yorkshire. *Proceedings of the Geologists Association*, **100**: 175–182.

Elias, S., Short, S.K., Nelson, R. and Birks, H.H.

(1996). Life and times of the Beringian Land Bridge. *Nature*, **382**: 60–63.

Foster, J.B. (1964). The evolution of mammals on islands. *Nature*, **202**: 234–235.

Guthrie, R.D. (2001). Origin and causes of the mammoth steppe: a story of cloud cover, woolly mammoth tooth pits, buckles, and inside-out Beringia. *Quaternary Science Reviews*, **20**: 549–574.

Hallam, A. (1987). Radiations and extinctions in relation to environmental change in the marine Lower Jurassic of north west Europe. *Paleobiology*, **13**: 152–168.

Hallam, A. (1992). *Phanerozoic Sea-Level Changes*. New York: Columbia University Press, x + 266pp.

Heaney, L.R. (1978). Island area and body size of insular mammals: evidence from the tri-colored squirrel (*Callosciurus prevosti*) of Southeast Asia. *Evolution*, **32**: 29–44.

Heaney, L.R. (1986). Biogeography of mammals in SE Asia: estimates of rates of colonization, extinction and speciation. *Biol. J. Linn. Soc.*, **28**: 127–165.

Heptner, V.G., Nasimovich, A.A., Bannikov, A.G. (1988). *Mammals of the Soviet Union, vol. 1: Artiodactyla and Perissodactyla*. Washington, DC: Smithsonian Institution.

Hoedemaeker, Ph.J. (1995). Ammonite evidence for long-term sea-level fluctuations between the 2nd and 3rd order in the lowest Cretaceous. *Cretaceous Research*, **16**: 231–241.

Hundertmark, K.J., Shields, G.F., Udina, I.G. *et al.* (2002). Mitochondrial phylogeography of moose (*Alces alces*): Late Pleistocene divergence and population expansion. *Mol. Phylogenet. Evol.*, **22**: 375–387.

Kemper, E. (1992). *Die Tiefe Unterkreide im Vechte-Dinkel-Gebiet*. Het Staringmonument te Losser, 95pp., 66 pls.

Kurten, B. and Anderson, E. (1980). *Pleistocene Mammals of North America*. New York: Columbia University Press.

Lister, A.M. (1989). Rapid dwarfing of red deer on Jersey in the Last Interglacial. *Nature*, **342**: 539–542.

Lister, A.M. (1993). Evolution of mammoths and moose: the Holarctic perspective. In: R.A. Martin and A.D. Barnosky (eds) *Morphological Change in Quaternary Mammals of North America*. New York: Cambridge University Press, pp. 178–204.

Lister, A.M. (1996a). Dwarfing in island elephants and deer: processes in relation to time of isolation. *Symposia of the Zoological Society of London*, **69**: 277–292.

Lister, A.M. (1996b). Sea-levels and the evolution of island endemics: the dwarf red deer of Jersey. *Geol. Soc. Spec. Publ.*, **96**: 151–172.

Lister, A.M. and Sher, A.V. (2001). The origin and evolution of the woolly mammoth. *Science*, **294**: 1094–1097.

Lomolino, M.V. (1985). Body size of mammals on islands: the island rule re-examined. *Amer. Naturalist* **125**: 310–316.

MacArthur, R.H. and Wilson, E.O. (1967). *The Theory of Island Biogeography*. Princeton: Princeton University Press.

Marincovich, L. and Gladenkov, A.Y. (1999). Evidence for an early opening of the Bering Strait. *Nature*, **397**: 149–151.

Rawson, P.F. (1971). Lower Cretaceous ammonites from north-east England: the Hauterivian genus *Simbirskites*. *Bulletin of the British Museum (Natural History) Geology*, **20**: 25–86.

Rawson, P.F. (1975). Lower Cretaceous ammonites from north-east England: the Hauterivian heteromorph *Aegocrioceras*. *Bulletin of the British Museum (Natural History) Geology*, **26**: 129–159.

Rawson, P.F. (1993). The influence of sea level changes on the migration and evolution of Lower Cretaceous (pre-Aptian) ammonites. In: M.R. House (ed.) *The Ammonoidea: Environment, Ecology and Evolutionary Change*. Systematics Association Special Volume, **47**: 227–242.

Rawson, P.F. (1994). Sea level changes and their influence on ammonite biogeography in the European Early Cretaceous. *Palaeopelagos Special Publication No. 1*: 317–326.

Rawson, P.F. (2000). The response of Cretaceous cephalopods to global change. In: S.J. Culver and P.F. Rawson (eds) *Biotic Response to Global Change. The last 145 Million Years*. Cambridge: Cambridge University Press, pp. 97–106.

Shackleton, N.J. (1987). Oxygen isotopes, ice volume and sea level. *Quaternary Science Reviews*, **6**: 183–190.

Sher, A.V. (1984). The role of Beringian land in the development of Holarctic mammalian fauna in the Late Cenozoic. In: V.L. Kontrimachivus (ed.) *Beringia in the Cenozoic Era*. New Delhi: Amerind Publishing Co., pp. 296–316.

Sher, A.V. (1986). On the history of the mammal fauna of Beringida. *Quartärpaläontologie*, **6**: 185–193.

Sher, A.V. (1997). Late Quaternary extinction of large mammals in northern Eurasia: a new look at the Siberian evidence. In: B. Huntley *et al.* (eds) *Past and Future Rapid Environmental Changes: The Spatial and Evolutionary Responses of Terrestrial Biota*. Nato ASI Series 1, **47**: 319–339. Heidelberg: Springer.

Sher, A.V. (1999). Traffic lights at the Beringian crossroads. *Nature*, **397**: 103–104.

Sondaar, P.Y. (1977). Insularity and its effect on mammalian evolution. In: M.K. Hecht, P.C. Goody and B.M. Hecht (eds) *Major Patterns in Vertebrate Evolution*. New York: Plenum, pp. 671–707.

Thieuloy, J.-P. (1977). Les ammonites boréales des formations Néocomiennes du sud-est Français (Province Subméditerranéenne). *Géobios*, **10**: 395–461.

Valentine, J.W. and Jablonski, D. (1991). Biotic effects of sea level change: the Pleistocene test. *J. Geophys. Res.* **96**: 6873–6878.

Van den Bergh, G.D., Sondaar, P.Y., de Vos, J. and Aziz, F. (1995). The proboscideans of the south-east Asian islands. In: J. Shoshani and P. Tassy (eds) *The Proboscidea: Evolution and Palaeoecology of Elephants and their Relatives*. Oxford: Oxford University Press, pp. 240–248.

Vermeij, G.J. (1991). Anatomy of an invasion: the trans-Arctic interchange. *Paleobiology*, **17**: 281–307.

Westermann, G.E.G. (1993). Global bio-events in mid-Jurassic ammonites controlled by seaways. In: M.R. House (ed.) *The Ammonoidea: Environment, Ecology and Evolutionary Change*. Systematics Association Special Volume, **47**: 187–226.

Westermann, G.E.G. (1996). Ammonoid life and habitat. In: N.H. Landman, K. Tanabe and R.A. Davis (eds) *Ammonoid Paleobiology*. New York and London: Plenum Press, pp. 608–707.

Woodburne, M.O. (ed.) 1987). *Cenozoic Mammals of North America: Geochronology and Biostratigraphy*. Berkeley: University of California Press.

Yacobucci, M.M. (1999). Plasticity of developmental timing as the underlying cause of high speciation rate in ammonoids: an example from the Cenomanian Western Interior Seaway of North America. In: F. Olóriz and F.J. Rodríguez-Tovar (eds) *Advancing Research on Living and Fossil Cephalopods*. New York: Kluwer Academic/Plenum Publishers, pp. 59–76.

Yalden, D. (1999). *The History of British Mammals*. London: Poyser.

Yurtsev, B. (1974). Steppe communities in the Chukotka tundra and Pleistocene 'tundra-steppe'. *Botanicheskii Zhurnal*, **59**: 484–501.

PART 7

Climate

17

Tectonics, climate change, and the evolution of mammalian ecosystems

Christine Janis

ABSTRACT

See Plates 22 and 28

The diversity and radiation of Cenozoic mammals has clearly been influenced by climatic changes, which in turn have been driven by tectonic events related to continental movements. The initial diversification of larger, more ecologically diverse mammals than those seen in the Mesozoic, following the extinction of the dinosaurs at the end of that era, was in the 'hot house' global conditions of the early Cenozoic. The inception of the cooler 'ice house' world of the later Cenozoic at around 51 Ma, with temperatures in higher latitudes plummeting through the late Eocene, resulted in increased zonation of vegetation and types of mammals. For example, primates were now restricted to the tropics. However, the late Eocene marked the time of great diversification of ungulates (hoofed mammals) with teeth indicative of a leaf-eating diet, with vegetation being more available for consumption by mammals in a more seasonal environment.

Recurrence of a warming trend in the Miocene, possibly combined with drying, resulted in the spread of grasslands and a rebound in mammalian diversity from an Oligocene low. However, mammalian diversity declined in the later Miocene, in combination with declining temperatures and possibly also with changing atmospheric conditions.

Studies of the diets of ungulates reveal a diversity of browsers in the mid-Miocene that is unlike that in any present-day environment, and may be reflective of non-analogous vegetational habitats under conditions of greater atmospheric carbon dioxide. The present-day mammal diversity is reduced from that of the mid-Miocene, and today's mammals inhabit a world that is cold, arid, and highly vegetationally zoned in comparison with most of the Cenozoic.

17.1 Introduction

Changes in the movements of the continents during the Cenozoic (the past 65 million years) brought profound changes to the global climate, influencing the evolutionary patterns of all organisms. The radiation of modern mammals in the Cenozoic represents only a third of the total history of mammals (with the earliest mammals known from the Late Triassic), but it was only following the demise of the dinosaurs at the end of the Cretaceous that mammals radiated into the variety of body sizes and trophic types that are familiar today. The evolutionary history of Cenozoic mammals is linked to the tectonic changes of the Cenozoic for two main reasons.

The first reason relates to biogeography, which will only be considered briefly here (see Janis, 1993, for further detail, and also Forster, this volume, Chapter 15, for a more general discussion of the biogeographic effects of continental drift). First, the break-up of the continents led to the isolation of faunal communities. Today only the mammal faunas of Australia and Madagascar are really isolated from the rest of the world. Each continental region has its own distinct mammalian fauna today, but earlier in the Cenozoic the faunas of South America, North America, Eurasia, and Africa were more distinctly different from each other than they are today. Continental movements also influenced these patterns of geographic isolation, such as the collision of Africa and Eurasia with faunal mixing apparent by the early Miocene, or the Pliocene Great American Interchange with the formation of the Isthmus of Panama.

The second reason relates to climatic change, and will be the focus of discussion here. The progressive break-up of the continents, and the northward movement of the major continental masses, resulted in a change in global climatic regimes with greatly increased latitudinal temperature gradients and increased aridity. There was major shift in the early Tertiary, commencing in the mid-Eocene, from the 'hot house world' of the Mesozoic and earliest Cenozoic to the 'ice house world' of the later Cenozoic that characterizes the world today (Prothero and Berggren, 1992). Mammals have been affected by these climatic changes in part by the simple fact of temperature tolerance, but more profoundly by the effect of these climatic changes on the global vegetation.

17.2 Tectonic and climatic changes during the Cenozoic

17.2.1 The initial conditions and a summary of Cenozoic tectonics

At the start of the Cenozoic the position of the continental masses was quite different from that seen today (see Figure 17.1). Although the supercontinent Pangaea had been breaking up through most of the Mesozoic, many remnants of this agglomeration of continents remained. The northern continents of North America and Eurasia had broken away from the southern Gondwanaland, but remained conjoined (as Laurasia) across Greenland and Beringia. Gondwana had broken up to the extent that Africa and India were no longer linked to the other continents, but they had not yet made contact with Eurasia, and the Tethys Sea stretched across the entire southern border of Eurasia. South America, Antarctica, and Australia were still connected to each other, and there was no connection between South America and North America. There was no ice at the poles, and the global climate was fairly warm and equable, probably due to the mixing of polar and equatorial waters with low latitudinal temperature gradients.

In contrast, today's world is a very cold and dry place in comparison with the earliest Cenozoic, with much steeper latitudinal zonation of climatic regimes. The continued break-up of the southern continents led to the isolation of Antarctica over the South Pole, and this, in combination with a dramatic fall in levels of atmospheric CO_2, led to the formation of an Antarctic ice cap by the Oligocene (De Conto and Pollard, 2003). Similarly the continued northern movement of Laurasia, and the rifting between North America and Eurasia leading to the isolation of Greenland, led to the formation of an Arctic ice cap during the Pliocene. Changes in oceanic circulation resulted from the isolation of circumpolar currents, with higher latitude cooling. Tectonic events also disrupted patterns of equatorial currents: the joining of Africa to Eurasia during the Oligocene truncated the Tethys seaway into the Mediterranean; the northern movement of Australia resulting in the closing-off of the Indonesian seaway in the Miocene; and the formation of the Isthmus of Panama in the Pliocene resulted in the establishment of our present-day Gulf Stream, with resultant warming of western Eurasia.

Further climatic effects can be related to uplift of mountain ranges, especially during the Miocene with the uplift of the Rocky Mountain and Andean Cordillerian region in North and South America, and the uplift of the Himalayas in Asia, the latter event related to the collision of India with Eurasia. This may

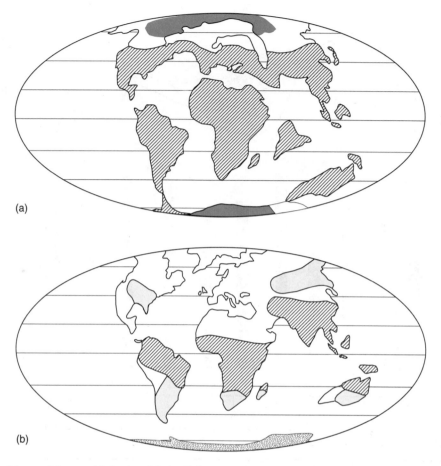

Figure 17.1 Maps of the world during (a) the Paleocene (around 60 million years ago) and (b) the middle Miocene (around 15 million years ago). (Modified from Osborne and Tarling, 1996, by Brian Regal). Key: single hatching = tropical or tropical-like forests; light shading = grasslands (savannas); dark shading (Paleocene) = polar broad-leaved deciduous forests; dotted (Miocene) = polar ice.

have affected the distribution of global rainfall, and would have resulted in a rain-shadow cast on the leeward side of the mountain ranges.

Figure 17.2 shows that the general global diversity of mammals has been strongly influenced by Cenozoic climate change. (Note: the Buchardt (1978) paleotemperature curve, derived from oxygen isotope ratios for the North Sea area, is used here to illustrate general patterns of response of mammalian diversity to climate change. Although there are more recent paleotemperture curves available, and for the global climate – e.g. Miller *et al.*, 1987 – our knowledge of the basic pattern of paleotemperature change over the Cenozoic remains more or less the same.) The following types of

mammals were excluded from the analysis. (i) Mammals with Australian origins: there is almost no Cenozoic fossil record in Australia prior to the late Oligocene–early Miocene. Thus their inclusion would bias the data towards increased Neogene values. (ii) Volant or aquatic mammals: bats have a very poor fossil record in comparison with terrestrial groups, and the record of the presence of a family might mean the existence of a single tooth; aquatic mammals (cetaceans, pinnipeds, sirenians, and desmostylians) may have been subject to different evolutionary pressures from terrestrial mammals due to the nature of their habitat. (iii) Mammals with their first origin in the Pleistocene or Holocene–Recent:

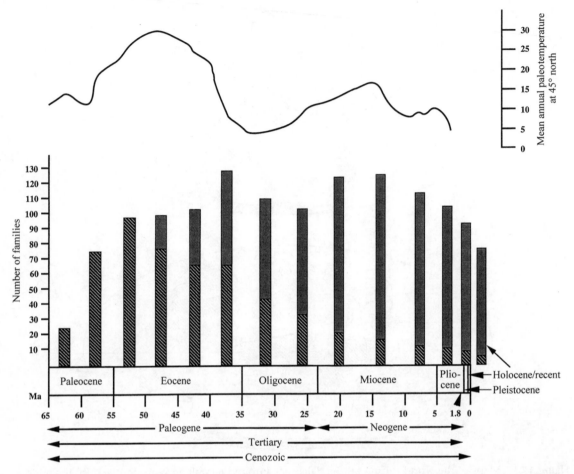

Figure 17.2 Familial diversity of mammals (data from Stucky and McKenna, 1993) shown along with the paleotemperature curve (modified from Buchardt, 1978). Ma = millions of years ago. Hatched areas = families of 'archaic' mammals (defined here as those originating in the Paleocene or early Eocene); shaded areas = families of 'modern' mammals (defined here as those with origins post-dating the early Eocene). For the types of mammals excluded from the analysis, see text.

such late first appearances most likely reflect inadequacies in the fossil record, such as the first appearance of the four families of lemurs in the Holocene because of the lack of a prior Cenozoic record in Madagascar. Inclusion of such families will bias the data towards modern richness. This exclusion of families represents 66 out of a total of 337, or 19.5%.

Here I use the number of mammalian families (from Stucky and McKenna, 1993) as a proxy for mammalian diversity. Although there are various problems with using higher taxa (rather than genus or species data) in this fashion to illustrate diversity, even paraphyletic

groupings can be shown to recapture underlying actual patterns (Sepkoski and Kendrick, 1993). The diversity patterns illustrated in Figure 17.2 are meant only to show in a very general fashion how overall, worldwide mammalian diversity has varied over the Cenozoic. The correlation of mammalian evolution with climatic change has been questioned recently, most notably by John Alroy and co-workers (e.g. Alroy *et al.*, 2000), and also by Donal Prothero (e.g. Prothero, 2001). Both authors address the issue of species-level changes in North American mammals. Prothero's major concern is that the most dramatic climatic event of the Cenozoic, the profound temperature drop at the very start of the Oligocene, is not accompanied by any notable change in species diversity or composition. While I admit that this discovery is a little surprising, there is clearly a major effect on European species at this time (Hooker, 2000). Moreover, the North American fauna underwent profound changes during the slower decline in temperature over the preceding 10 million years, and it may be the case that by the end of the Eocene the fauna was sufficiently cold-adapted that further temperature decline produced little response.

Alroy's criticism of the correlation of mammal evolution with climatic change is of a different nature, and relies primarily on his statistical correlation of the species-level diversity curve with the paleotemperature curve as derived from the oxygen isotope record of benthic foraminifera. Even if one were to have complete faith in Alroy's sampling-standardized diversity curve (e.g. Alroy 2000), the issue of whether the absolute value of mammalian species diversity can be seen to follow in lock-step with the variations in paleotemperature is a very narrow definition of the correlation between mammalian evolution and climate. This paper addresses broader, large-scale evolutionary changes (e.g. loss of temperate latitude primates, or the rise of savanna types of ecosystems) that are clearly related to changes in global climate.

Mammalian familial diversity shows an increase with the increasing temperatures of the early Eocene, a decrease with Oligocene cooling, a further increase with renewed early Miocene warming, and a subsequent decrease with late Cenozoic cooling. The only anomalous relationships of the mammalian diversity to the paleotemperature curve is in the late Eocene, where decreasing temperatures to an Eocene–Oligocene boundary minimum coincide with a maximum diversity of mammalian families in the latest Eocene. However, this paradoxical correlation can be understood if the mammals are split into two groups: 'archaic' families, defined here as having their origin in the Paleocene or earliest Eocene, and 'modern' families, defined here as having their origin subsequent to the earliest Eocene. With

this division it is clear that the late Eocene maximum is the result of two competing trends: a leveling off in the decline of the archaic forms, which commenced in the middle Eocene and continued more sharply in the Oligocene, and the increase in diversification of the modern forms. The late Eocene thus represents an evolutionary 'crossover' of diversification events: the final significant holdover of the archaic mammals, whose inception was in the 'hot house world' of the earlier Cenozoic (e.g. Northern Hemisphere marsupials and higher-latitude primates), and the radiation of the modern families better adapted for the 'ice house world' developed during the mid-Eocene (e.g. rodents and folivorous ungulates – the rise of the latter forms will be explored more fully in a later section).

17.2.2 A synopsis of Cenozoic changes in climatic and vegetational regimes

This section is designed to provide a brief overview of how changes in climate and vegetation have resulted from tectonic change, and how these in turn have affected mammalian evolutionary patterns. Figure 17.3 summarizes some of these trends, illustrating changes in vegetation and mammal diversification in correlation with changing paleotemperatures. A more comprehensive bibliography of original literature sources can be found in Prothero and Berggren (1992), Janis (1993) and Prothero (1998).

17.2.2.1 Paleocene and early Eocene

As previously discussed, global climatic conditions in the Paleocene were fairly equable, with low pole-to-equator temperature gradients and the absence of polar ice or winter frosts (see Figure 17.1a). Forests resembling those found today in the tropical areas extended into much higher latitudes. At higher latitudes still, in the areas covered today by tundra and taiga, were woodlands resembling those now found in subtropical latitudes. At the poles was a type of vegetation unknown today, termed 'polar broad-leaved deciduous forests': this vegetation

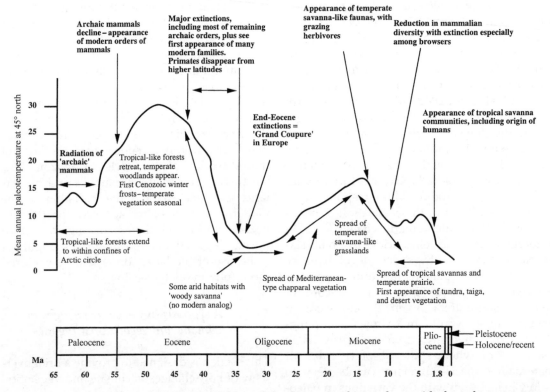

Figure 17.3 Faunal (bold face) and floral events of the Cenozoic, shown along with the paleotemperature curve (modified from Buchardt, 1978). Ma = millions of years ago.

would have been adapted to regimes of seasonal darkness. Although the general climate would have been similar to that of the Late Cretaceous, the vegetation differed in that it appeared to be mainly closed-canopied forest with little open areas. This may have been due to an absence of cropping by dinosaurian herbivores, and/or to a difference in the seasonal pattern of precipitation. The radiation of Paleocene mammals was largely of those with no close living relatives (although a notable exception was primitive, weasel-like members of the order Carnivora). Many of the mammals were partially or fully arboreal, and most were fairly small (under 10 kilograms). Diets were largely generalized omnivorous/herbivorous or omnivorous/carnivorous: there were few specialized predators, and few true folivores (leaf-eaters) until the end of the epoch (this phenomenon is examined more fully in a later section). The largest terrestrial mammals were

bear-sized, herbivorous pantodonts, ungulate-like animals that have been broadly considered as 'browsers'.

At the end of the Paleocene there was a rise in temperature, reaching a Cenozoic maximum near the end of the early Eocene. Two main tectonic reasons have been advanced for this event. First, rifting between Greenland and Norway, resulting in increased seafloor hydrothermal activity and an increase in atmospheric CO_2, with a declining pole-to-equator temperature gradient. Second, a shift in deep ocean circulation from a temperature-driven regime to a salinity-driven one, with the intrusion of warm, salty water into higher latitudes. A more recent explanation is the degassing of biogenic methane from hydrates buried on continental shelves (Bains *et al.*, 1999). The effect of this warming on the vegetation was for the even higher-latitude extension of the tropical-like forests, into the confines of the Arctic circle.

Many of the modern orders of mammals appeared around this time, such as the rodents in the late Paleocene, and the primates and the artiodactyl and perissodactyl ungulates (the modern types of hoofed mammals) in the earliest Eocene. In contrast, the more archaic types of mammals of the Paleocene declined, although they were still relatively abundant in the early part of the Eocene, especially in North America. The mammalian communities suggest a rather different vegetational structure than in the Paleocene, with a greater number of terrestrial forms suggesting a more open-canopied structure to the forests, perhaps the result of more seasonal patterns of rainfall.

17.2.2.2 Mid–late Eocene and Oligocene

From a maximum around the early–middle Eocene boundary global temperatures fell through the later part of the Eocene to reach a minimum shortly after the Eocene–Oligocene boundary. This dramatic temperature change, which resulted in the formation of our modern world which is exceedingly cool and dry in comparison with the Mesozoic and early Cenozoic, was originally explained by events such as shifts in the Earth's axis or a meteorite impact. However, it is now considered that tectonic events are a sufficient explanation. Profound changes in oceanic circulation resulted from the separation of Greenland from Norway, and of Australia from Antarctica, leading to the formation of cold bottom water and sea ice at the poles, cooling the higher latitudes. A circum-Antarctic current was fully established by the Oligocene, and an Antarctic ice cap was definitely in place by the start of that epoch, with evidence of glaciation on Antarctica in the middle Eocene.

The higher-latitude vegetation would have experienced winter frosts in the later Eocene; the tropical-like forests retreated to the equatorial latitudes, broad-leaved temperate deciduous woodland spread across the middle latitudes, while a new type of vegetation was apparent in the higher latitudes (e.g. across northern North America and Eurasia), mixed coniferous and deciduous temperate woodland. Although the first grass fossils date from the Eocene, and there appear to have been areas of more arid, open habitat during the Eocene and Oligocene, these were not true grasslands; instead, they appear to be a type of habitat unknown today, termed 'woody savanna', lacking a grassy cover (Leopold *et al.*, 1992).

More modern types of mammals radiated in the later Eocene, including folivorous ungulates and rodents, and some members of the order Carnivora evolved into large-bodied predators. The radiation of folivorous mammals suggests seasonality in the vegetation, as will be discussed later. Larger body sizes became common among many types of mammals, with the late Eocene brontotheres attaining the size of modern rhinos.

The Oligocene world was cooler and drier than the world of the earlier Cenozoic, but still fairly equable by today's standards: there were no deserts, and no tundra or taiga at high latitudes. A mid-Oligocene peak in Antarctic glaciation appeared to have little direct effect on the already cold-adapted fauna and flora, but may have had an indirect one via the results of tying up much of the world's water in the ice cap: a worldwide drop in sea level and continental aridity. A number of Oligocene mammals show adaptations for open, arid climates, such as longer legs and high-crowned teeth more resistant to abrasion, but there were not as yet true grasslands nor grazing mammals.

17.2.2.3 Miocene

Temperatures started to rise again in the early Miocene, although there was still a steep latitudinal thermal gradient. This warming may have been related to the separation of Antarctica and South America, resulting in the isolation of the cold circumpolar current from the rest of the world. Antarctic glaciation was also at minimum levels during the early Miocene. This warming trend resulted in the re-expansion of the tropical-like forests into the subtropical zones, and the development of a new type of arid-adapted vegetation along the western sides of continents, Mediterranean-type woodland (also known as thorn scrub or chaparral). With the increased warming and continuing aridity woodland savanna, a habitat with a mix of trees, shrubs and grasses, started to develop in the middle latitudes; associations of plants and/or animals of this type of ecosystem were seen in the Great Plains area of North America, in Argentina, and in central Asia. Evidence from paleosols suggests that sod-forming short grasslands were present in the Great Plains from around 20 Ma (Retallack, 1997). However, there was still no evidence of true grazing among the herbivores; some herbivores developed high-crowned (hypsodont) cheek teeth at this point, but other craniodental indicators of grazing were not present, and these animals were probably mixed grazer/browsers, at best.

The later Miocene cooling, which has continued with a brief respite in the early Pliocene until the present day, was probably caused by a number of tectonic factors. First, there were disruptions to the equatorial oceanic circulation caused by the continued movement of continents. During the early Miocene the Tethys Sea became closed off to form the Mediterranean with the final joining of Africa to Eurasia, and the northern movement of Australia resulted in the closure of the Indonesian seaway. Colder polar conditions were marked by the development of a circum-North Atlantic at the North Pole with the continuing isolation of Greenland, and during the late Miocene the Antarctic ice sheet expanded to beyond its present extent. Second, there were episodes of uplift in the Rocky Mountain and Andean Cordillerian regions, and in the Himalayas when India collided with Asia, which would have resulted in drying via a rainshadow effect.

The response of the vegetation to the climatic changes of the later Miocene was the spread of the first true grasslands in the middle latitudes, with ecosystems resembling those of present-day East Africa in Argentina and the Great Plains area of North America. A pronounced shift occurred in the vegetation at around 7 Ma; isotopic evidence from soil carbonates and ungulate dental enamel indicates that grasses had shifted from using the C_3 carbon cycle for photosynthesis to using the C_4

cycle (see Ehleringer *et al.*, 1991; MacFadden and Cerling, 1994 for a review of the biochemistry). Low CO_2 levels and high temperatures favor the use of the C_4 cycle by plants today. C_4 plants (mainly grasses) predominate in tropical habitats at lower altitudes. The C_3/C_4 shift corresponds with a major vegetational community change in North America, with the introduction of more northern types of grasses (interpreted as a shift in climatic regime from summer-wet to summer-dry conditions, Leopold and Denton, 1987), and paleosol evidence for a shift from short to tall grasslands (Retallack, 1997).

There is controversy about the reasons for the C_3/C_4 transition. During the early 1990s it was considered that the early Miocene levels of atmospheric CO_2 levels were around three times their present (preindustrial) level (Berner, 1991, 1998; see also Chaloner, this volume, Chapter 5). A later Miocene decline was related to the weathering effects of tectonic uplift (Ruddiman, 1997), and a fall below some critical threshold value was considered to be the trigger for the biochemical changes in the plants (Ehleringer *et al.*, 1991). However, more recent geochemical work has led to the conclusion that atmospheric CO_2 levels were at present day values by the early Miocene (Pagani *et al.*, 1999; Pearson and Palmer, 2000). This issue is discussed further in a later section.

The spread of true grasslands resulted in the evolution of true savanna-adapted faunas among the mammals, with the radiation of forms with unequivocal adaptations for grazing, primarily among the North American horses and some extinct South American endemic ungulates. These types of ecosystems were not yet present in Africa, which retained more forested and tropical woodland types of habitats, and the Eurasian ecosystems of the earlier late Miocene were not true savannas (Solounias and Dawson-Saunders, 1988); Eurasian savanna-adapted association of mammals were not apparent until after the C_3/C_4 transition at 7 Ma (Barry *et al.*, 1985). However, the late Miocene was also a time of declining mammal diversity, with extinctions especially among the browsing mammals with the transformation of the more woodland savanna habitats into open, grassland savannas. The significance of these faunal changes is discussed further in a later section.

17.2.2.4 *Plio-Pleistocene*

There was a brief warming period in the early Pliocene, but temperatures plunged again in the later part of the epoch, probably related to disruptions in equatorial oceanic circulation following the formation of the Isthmus of Panama linking North and South America. An Arctic ice sheet was established during the Pliocene at around 2.75 million years ago, its formation possibly triggered by short-term climatic fluctuations that have been observed at this time (Willis *et al.*, 1999). These cycles were apparently of shorter duration than the Milankovitch climatic short-term cycles (tens of thousand of years in duration) that relate to various irregularities of the Earth's orbit. Milankovitch cycles have probably affected global climate since the Earth's creation (Hewitt, this volume, Chapter 18). However, with the presence of an ice cap at the North Pole, and the concentration of continental masses in the higher latitudes of the Northern Hemisphere, these cycles would now act to throw the higher latitudes into periodic episodes of glaciation, the so-called 'ice ages' of the Pleistocene. A major effect of these ice ages was not only glaciation in the higher latitudes but aridity in the lower latitudes, because of the water tied up in the polar ice.

Many of the world's major types of vegetation, which are adapted to the cold, arid world of the present day, made their first appearance near the end of the Cenozoic. Northern latitude tundra and taiga first appeared in the latest Miocene; deserts, and temperate grasslands such as prairie, steppe, and pampas are of Plio-Pleistocene origin. With the expansion of these less-productive types of habitats at the expense of savanna, subtropical woodland, and tropical forest, the diversity of mammals has decreased. Some productive types of Plio-Pleistocene ecosystems, such as the high-latitude 'mammoth steppe' of eastern Asia and Alaska of the last glacial period (e.g. Guthrie, 1982), are missing from today's world. The

increase in body size in many Pleistocene mammals was probably a response to coping with nutrient-poor, seasonal habitats. Debate continues as to whether the extinction of many of these large mammals (the 'megafauna') at the end of the Pleistocene 10 000 years ago was the result of climatic changes or due to the influence of human hunting (e.g. Alroy, 2001).

One effect of the continued cooling and drying of the Plio-Pleistocene, probably aided by the rise of the East African Rift system that cast a rainshadow onto eastern Africa, was the establishment of tropical savanna habitats in Africa. The evolution of these habitats has been tied into the origin and radiation of humans (e.g. Vrba *et al.*, 1989; see Potts, this volume, Chapter 19).

17.3 Patterns of mammal evolution reveal details about climatic changes

The previous section detailed how understanding of tectonic events can aid our understanding of patterns of climatic change that influenced Cenozoic mammal evolution. In this section I provide two examples from my own work (the first from Janis, 1997 and Janis, 2000; the second from Janis *et al.*, 2000, 2002) that show that reciprocal illumination is possible in the understanding of mammal evolution in the context of climatic changes. That is, not only can climatic changes help us to understand events in the mammalian fossil record, but the patterns in the record themselves can help further our detailed understanding about climatic events.

Both examples are taken from evolutionary patterns in hoofed mammals in North America, using excellent and newly available data on generic ranges (Janis *et al.*, 1998). Hoofed mammals are mostly herbivorous, and thus their patterns of abundance and diversity provide data on the type of vegetation present. The craniodental morphology of ungulates is tightly correlated to the feeding behavior in living ungulates, and can be used to make determinations about the feeding

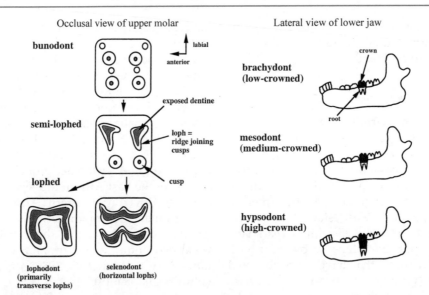

Figure 17.4 Schematic diagrams of ungulate teeth to explain morphological terminology. On the left-hand side: occlusal views of an upper molar showing differences in crown morphology in bunodont (omnivorous and/or frugivorous (fruit-eating)), semi-lophed (frugivorous/folivorous), and lophed (folivorous or leaf-eating) mammals. Lophodont molars (found in horses, rhinos, tapirs, hyraxes, elephants) and selenodont molars (found in camels, cattle, antelope, deer, giraffe) are different variants on the basic lophed form. On the right-hand side: lateral views of lower jaws (anterior to the left) showing differences in molar crown height (with jawbone below gum line cut away) in brachydont (low-crowned browsers), mesodont (medium-crowned browsers or mixed feeders), and hypsodont (high-crowned mixed feeders or browsers) mammals. Note that ungulates with all types of occlusal patterns can have brachydont teeth, but that only lophed taxa exhibit mesodont or hypsodont teeth.

habits of extinct taxa (Janis, 1995). Figure 17.4 diagrams some aspects of ungulate molars, showing how occlusal morphology can be used to determine broad dietary patterns, and how molar crown height can provide information about the abrasiveness of the diet; ungulates with an abrasive diet, such as grass or gritty vegetation in open habitats, must have higher-crowned teeth so that the dentition can withstand a lifetime of wear.

17.3.1 *Changes in occlusal morphology in North American Paleogene herbivorous mammals, and the development of climatic seasonality*

17.3.1.1 *Materials and methods*

Dental morphologies were classified into three basic molar types (see Figure 17.4). Bunodont teeth have isolated, rounded cusps that act to compress and pulp non-brittle, non-fibrous food such as fruit and roots; such teeth are seen today in pigs and many primates. Lophed teeth have the individual cusps joined by high-crested ridges or lophs, which act to shear and shred flat, fibrous food such as leaves, and grass blades; such teeth are seen today in specialized herbivores such as horses and deer. Semi-lophed teeth intermediate in morphology are seen today in mammals with a diet of fruit and soft browse, such as mouse deer and tree hyraxes. Bunodont teeth represent the primitive type: lophed teeth have evidently been derived from bunodont teeth on many different occasions, and many different detailed occlusal patterns may be identified among lophed molars (Jernvall *et al.*, 1996). The majority of present-day ungulates have lophed molars (see Figure 17.5).

The taxa in this study include both true ungulates and ungulate-like mammals (a

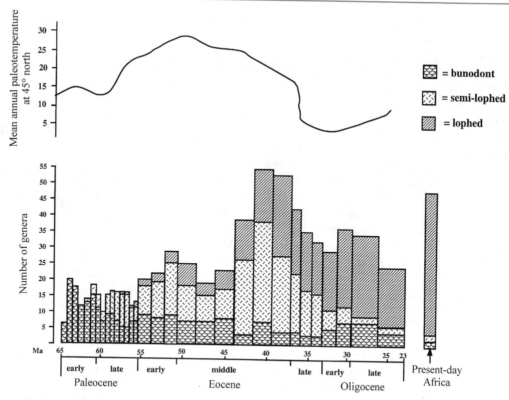

Figure 17.5 Types of occlusal patterns in Paleocene ungulates and ungulate-like mammals (data from Janis, 1997/1998), shown along with the paleotemperature curve (modified from Buchardt, 1978). The different widths of the rectangles reflect the differences in time intervals sampled: differences in length of time intervals and other sources of sampling biases were not found to affect the overall pattern (see Janis, 1997/1998). Ma = millions of years ago.

paraphyletic assemblage of the orders Taeniodonta, Tillodontia, Pantodonta, and Dinocerata; see chapters by Lucas in Janis *et al.*, 1998). These groups all have Paleocene origins and their range extends into the middle Eocene. The Paleocene ungulates comprised the basal 'Order Condylarthra', a paraphyletic assemblage of lineages that mainly had bunodont dentitions, suggestive of an omnivorous diet. Most condylarth families are known from the Eocene, but all were extinct by the Oligocene. True ungulates first appeared in the early Eocene and comprise members of the extant orders Perissodactyla and Artiodactyla. The more modern radiation of extant families, such as rhinos, peccaries, camelids and ruminant artiodactyls dates from the late Eocene.

17.3.1.2 Results and discussion

The earliest Cenozoic ungulates (this term will include 'ungulate-like mammals' throughout) all had a bunodont form of molar occlusal morphology, suggesting that all forms were omnivorous in their diet. These represent condylarths, all of uniformly small body size (i.e. less than 25 kg, most less than 5 kg: patterns in body sizes are examined in more detail elsewhere, see Janis, 1997). Some larger-sized, semi-lophed taxa appeared in the mid-Paleocene, representing the various ungulate-like mammals, which were probably not like any modern ungulate in their dietary behavior, although some pantodonts and dinoceratans might have been large, folivores semi-aquatic (riverine). It was not until the latest

Paleocene, coincident with the start of a major warming trend, that the first fully lophed ungulates were apparent (see Figure 17.5).

As only lophed ungulates have teeth adapted for eating leaves to any great extent, it appears that the first 10 million years of ungulate radiation was largely bereft of true folivores. This, in turn, suggests certain features about the nature of the Paleocene vegetation. Paleobotanical evidence suggests that the forests were dense but of low taxonomic diversity, containing plants with large seeds probably available year-round (Wing, 1998). The decrease in the numbers of bunodont taxa in the later Paleocene is coincident with falling paleotemperatures, suggesting that the availability of seeds may have been decreasing under these climatic conditions. In turn the larger, semi-lophed forms ungulate-like mammals, possibly somewhat folivorous in their habits, increased in diversity during this time. The lack of mammalian specialized folivores in the Paleocene (also true among the arboreal forms, and in European faunas; Stucky, 1990; Collinson and Hooker, 1991) suggests a non-seasonal climate where the plants were not deciduous, where the leaves were heavily protected against herbivory. Most of mammals of the time would have been too small to have been able to utilize such fibrous plant material (see Kay and Covert, 1984). Note, however, that the lophed ungulates of the latest Paleocene were small animals (5–10 kg), suggesting that vegetational conditions were changing.

Ungulate diversity increased in the early Eocene, to reach an initial peak at the end of the early Eocene, coincident with the Cenozoic temperature maximum. This diversity rise was the result of an increase in the numbers of lophed and semi-lophed taxa; bunodont ungulates continued at their original, late Paleocene diversity. During the early Eocene, and into the early middle Eocene the bunodont taxa consisted of surviving condylarths joined by some early artiodactyls, all still of small body size. The semi-lophed taxa were in the main small to medium-sized condylarths and perissodactyl ungulates, such as equids and brontotheres (extinct, rhino-like forms), with a few larger ungulate-like mammals (mostly semi-

aquatic). The fully lophed taxa were a mixture of condylarths and tapiroid perissodactyls, also of small to medium body size. A similar pattern of increase in lophed taxa was also seen in Europe and Asia at this time: in fact, the patterns reported here for the Paleogene in general appear to be true across the Northern Hemisphere, with some minor variations (Collinson and Hooker, 1991; Jernvall *et al.*, 1996).

This increase in partially and fully folivorous ungulates in the early part of the Eocene, as suggested by the increase in the number of forms with semi- and fully lophed teeth, is coincident with rising temperatures, and is suggestive of a change in the vegetational habitat. This time also evidenced increasing seasonality of precipitation (Wing, 1998), which in turn would have resulted in a more deciduous type of vegetation. Seasonal rainfall might also have resulted in the trees being more widely spaced, allowing for more low-level undergrowth. All of these predictions are matched by the evolutionary patterns of Eocene mammals, which reveal a diversity of terrestrial, folivorous forms. Eocene faunas were quite different from those of the Paleocene, where the majority of the mammals present were arboreal or semi-arboreal, even among the ungulates. The mammalian faunas of the tropical-like forests of the Eocene also contained a much larger diversity of terrestrial folivores and frugivores, suggesting a different, more open habitat structure to present-day equatorial forests; there was also a much greater diversity of arboreal insectivores (Andrews, 1992).

Distinct changes in patterns of ungulate trophic diversity were seen in the later part of the middle Eocene. Ungulate diversity greatly increased, with a change in the proportion of trophic types. The numbers of bunodont ungulates fell, reflecting the decline in numbers and extinctions among the primitive artiodactyls and the surviving condylarths. The bunodont taxa of the late Eocene and Oligocene were the larger-sized artiodactyl suines (including peccaries and extinct forms such as anthracotheres and entelodonts). The increase in ungulate numbers reflects increases among the lophed and, especially, the semi-lophed taxa. The semi-lophed taxa of the later Eocene were two very different types of animals that did not survive the epoch; the very large (1 to 3 ton) brontotheres and the small (5–10 kg) primitive artiodactyls. The lophed taxa at this time were primarily perissodactyls, including rhinos and equids.

This increase in diversity was coincident with decreasing paleotemperatures, and this phenomenon is also seen in the general diversity of mammals at the family level, as previously discussed (see Figure 17.2). Figure 17.5 shows that the later Eocene changes in diversity comprise two competing trends, as previously discussed. These are the extinction of more archaic mammals (in this case, the condylarths and the ungulate-like mammals), and the radiation of more modern types (in this case, the artiodactyls and perissodactyls with lophed teeth). The increasing seasonality of the later Eocene, as reflected by changes in the vegetation from more a tropical to more temperate nature (Wolfe, 1985; Wing, 1998), would have favored the ability of folivorous mammals to utilize the herbage. In seasonal environments there is greater differentiation in the fiber content between leaf and stem, and folivores are now able to select less fibrous parts of the plant (see discussion in Janis, 1997).

However, the great increase in diversity in small lophed artiodactyls in the late middle Eocene is not suggestive of changes caused directly by a cooling climate. An increase in the numbers of small rodents with semi- or fully lophed teeth was also apparent at this time (Collinson and Hooker, 1991; Stucky, 1990). Both types of mammals would have been well adapted to a climate characterized by greater aridity rather than by cooling, specializing on seeds and buds (see Janis, 2000, for a more extensive discussion).

With the steeper decline in paleotemperatures in the late Eocene the numbers of ungulates also declined. This reduction in diversity was largely due to extinctions among the semi-lophed forms. This late Eocene pattern persisted, with a few fluctuations, through the Oligocene. The remaining artiodactyls were now predominantly lophed forms, with the exception of the persistently bunodont suoids.

The late Paleogene diversity of trophic types was broadly similar to that of the present day: a few bunodont and semi-lophed, more omnivorous taxa, and those bunodont taxa present being fairly large (pig-sized), with the great majority of taxa being fully lophed, specialized folivores.

By the late Eocene the vegetation in North America was predominantly temperate, with mixed coniferous and deciduous woodlands, with winter frosts probably present (Wolfe, 1985; Wing, 1998). The extinctions among the semi-lophed taxa, with some exceptional losses such as the very large (3–4 ton) rhino-like brontotheres at the end of the Eocene, suggests that a pattern of cold seasonality was established in the vegetational habitat. There was no longer non-fibrous material available year-round to meet the dietary needs of these folivorous/frugivorous taxa. This pattern continued through the Oligocene, with probable increasing aridity during the epoch, as previously discussed.

17.3.1.3 Summary and conclusions

Analysis of the patterns of trophic diversity among the North American Paleogene ungulates, as evidenced by their molar morphologies, enables a more detailed picture to be built up about climatic and vegetational change on this continent. The lack of fully lophed taxa (folivores) in the Paleocene suggests that, whatever the forest structure, the leaves were not generally available for consumption by mammals, especially by small mammals (some of the larger semi-lophed ungulate-like mammals of the later Paleocene may have been partially folivorous). This in turn suggests a non-seasonal habitat where there would be heavy investment by the plants in their leaves and little differentiation in fiber between leaf and stem.

The increase in diversity among semi- and fully lophed forms in the early Eocene suggests a change in the degree of plant seasonality, echoed by paleobotanical evidence of more seasonal precipitation, with a more open nature of the vegetational habitat to allow for a diversity of terrestrial herbivores. However, the retention of a moderate diversity of small bunodont taxa suggests the retention of many succulent elements in the vegetation. The increase ungulate diversity in the later Eocene paralleled declining paleotemperatures, reflecting increased seasonality of the vegetation and thus greater availability of structural plant parts of relatively low fiber content. However, the increased diversity of the semi-lophed ungulates in the late middle Eocene is suggestive of a vegetational habitat where drying, rather than cooling, was the predominant feature. Drying or cooling is also suggested by the reduction in diversity of bunodont taxa at this time, with the replacement of the smaller, archaic forms such as condylarths with larger-sized, suoid artiodactyls. Finally, the extinctions among the semi-lophed taxa in the late Eocene and into the Oligocene indicate that cold, seasonal climates now dominated the vegetational habitat.

17.3.2 Changes in molar crown height in North American Neogene mammals, and declining levels of atmospheric carbon dioxide

17.3.2.1 Materials and methods

The dental morphologies examined in this study are those of molar crown height (see Figure 17.4). Almost all of the North American taxa have fully lophed teeth, although the brachydont forms include a few bunodont suoids, represented only by peccaries after the early late Miocene. Brachydont, mesodont, and hypsodont molars correlate broadly with browsing, mixed feeding, and grazing diets, although note that many present-day hypsodont mammals are actually mixed feeders in open habitats (see Janis, 1995). However, brachydont taxa are almost always browsers.

Among Neogene taxa, the brachydont ungulates were predominantly browsing horses, deer-like artiodactyls, primitive camelids, a variety of rhinos, and mammutid proboscideans. The mesodont taxa were primarily small antilocaprids, oreodonts, early more derived horses (e.g. *Merychippus*), derived camelids, and gomphotheriid proboscideans. Hypsodont taxa were later more derived horses, larger antilocaprids (pronghorns), the rhinoceros *Teleoceras*, and derived gomphotheres. Of these taxa, it is likely that only

the horses contained any exclusive grazers resembling present-day horses. Recent isotopic studies (MacFadden *et al.*, 1999) have revealed that late Miocene horses had a diversity of diets.

17.3.2.2 Results

The generic richness of North American ungulates started to rise in the early Miocene from the overall low diversity in the late Oligocene (see Figures 17.5 and 17.6), reaching a Cenozoic maximum in the early middle Miocene (Figure 17.6) that was also coincident with the paleotemperature maximum for the Neogene (Figure 17.2). Following this peak, ungulate diversity declined at a more or less steady rate, reaching levels at the end of the Pliocene comparable to those of the late Oligocene. However, the rise and fall of the diversities of the ungulates grouped by their molar crown heights – brachydont, mesodont, and hypsodont – varied from one another.

Until the late early Miocene brachydont taxa comprised the majority of the fauna. While the first hypsodont taxa appeared in the late middle Eocene they were not the ancestors of the Neogene hypsodont forms; rather they comprised a specialized group of animals that were probably adapted to arid climates – leptaucheniine oreodonts, stenomyline camelids, and the hypertragulid *Hypisodus*. Almost none of these taxa survived past the earliest Miocene. The late early Miocene is notable for its sharp increase in diversity, with the initial appearance and radiation of the mesodont forms. This diversity rise heralded the start of the North American savanna ecosystem that dominated the Great Plains area for most of the rest of the Miocene (Janis, 1993). Hypsodont taxa first increased in diversity at the start of the middle Miocene.

These various tooth height categories were fairly stable with respect to membership. Some hypsodont taxa, such as the horses and the antilocaprids, were derived from existing mesodont taxa, but this represented relatively discrete events rather than a steady progression from mesodonty to hypsodonty. The hypsodont rhino and proboscideans arrived via immigration. The members of the brachy-

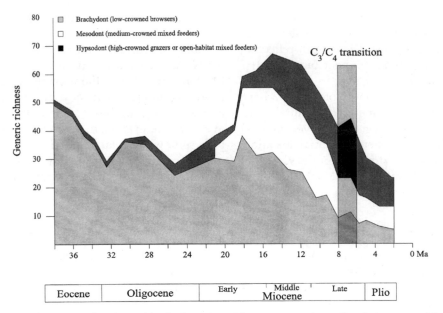

Figure 17.6 Molar crown height in North American Neogene ungulates (hoofed mammals) at the continent-wide generic level. Adapted from Janis *et al.* (2000). The gray rectangle indicates the C_3/C_4 shift. Ma = millions of years ago.

dont fauna did not evolve over time into permanent members of the mesodont fauna; they either became extinct without issue (e.g. browsing horses and deer-like ruminants) or they persisted at low diversities (peccaries, tapirs, and mammutid proboscideans).

The brachydont ungulates declined in diversity from an early Miocene peak, when they represented around 60% of the taxa present, to a late Pliocene low, when they were less than 20% of the faunal diversity. Major extinctions occurred at 17 Ma (in the late early Miocene), 11 Ma (at the end of the early Miocene), and 9 Ma (at the end of the early late), and a minor event occurred at 6 Ma, in the latest Miocene. Mesodont ungulate declined from a peak in diversity in the middle Miocene to a low in the late Pliocene, with major extinctions at 12 Ma and 9 Ma. Hypsodont taxa did not experience any major extinction events during the late Miocene. There was an apparent decline at around 7 Ma, but that reflects a decrease from an earlier brief rise in diversity rather than an increase in extinction rates. Hypsodont taxa did not undergo any significant extinctions

until the early Pliocene at around 4.5 Ma. This pattern of changes in continent-wide, generic diversity in ungulates of different tooth crown height is paralleled by changes in species alpha diversity in individual localities in the Great Plains region (Janis *et al.*, 2000, 2002).

17.3.2.3 Discussion

As previously discussed, the middle and late Miocene of North America was the time of the predominance of savanna-like ecosystems, even if the composition of taxa present (in terms of ecomorphology as well as taxonomic composition) cannot be precisely analogized with any present-day savanna ecosystem (Janis, 1993). The rise of the grazing mammals, especially the horses, has become a textbook story in discussions of climatic and evolutionary change. This savanna-like ecosystem apparently faltered in the late Miocene, with a decline in overall diversity and especially among the brachydont taxa, and by the Pliocene a more modern type of grasslands fauna was established, possibly representing a more prairie-like habitat.

A new twist to this well-established tale has come from the recent literature on the C_3/C_4 transition (see previous discussion), reflecting a change in grass photosynthetic biochemistry, and coincident with a change in the predominant vegetation. These new isotopic data have prompted claims that the C_3/C_4 shift is causally related, via its effects on the vegetational habitat, to the late Neogene mammalian extinctions, especially the extinctions of the variety of species of hypsodont horses (MacFadden and Cerling, 1994; Cerling *et al.*, 1997, 1998). However, the detailed faunal data (Figure 17.6) do not show major extinctions at the generic level at this point. Instead they show that the major extinctions either happened prior to the C_3/C_4 transition (for the brachydont and mesodont taxa) or a couple of million years after it (for the hypsodont taxa). A similar lack of mammalian extinctions at 7 Ma in Eurasian faunas is reported by Köhler *et al.* (1998). The major extinction of ungulates occurred at around 5 Ma, at the end of the Miocene. This episode does not precisely correlate with any vegetational events, although these extinctions may represent some form of delayed response to the C_3/C_4 transition.

A potentially interesting, but hitherto largely ignored pattern in these ungulate diversity data is the paleoenvironmental signal carried by the decline in the diversity of the brachydont taxa (= browsers), which occurred in the 10 million years prior to the C_3/C_4 transition.

The general decline of numbers of mammalian taxa in the late Neogene has usually been interpreted as reflecting a global drying trend (see Janis, 1993), and there has been the tacit assumption that as the numbers of browsing ungulates declined with the encroachment of savannas, the grazers radiated to take their place in the ecosystem. However, this is not the actual pattern observed, either at the continent-wide, generic level (Figure 17.6), nor at the level of species in individual localities (Janis *et al.*, 2000, 2002, in press). Rather, the overall decline in diversity is due to the decline of the brachydont taxa (browsers) alone, with the numbers of taxa of other feeding types (mesodont and hypsodont) remaining rela-

tively steady during the late Neogene. A particularly striking feature is the very high numbers of browsers in the individual communities in the late early and early middle Miocene sites. While the maximum number of browsing ungulates in any present-day local habitat is eight, these communities had an average of nine and a maximum of 19 probable browsers, while retaining a similar number of mixed-feeding and/or grazing taxa (around a dozen) to those seen in present-day habitats (Janis *et al.*, 2000, 2002, in press). Mid Miocene communities elsewhere in the world also appear to be 'over-rich' in the numbers of browsers (e.g. Kay and Madden, 1997, and data reviewed in Janis *et al.*, in press).

The implication of these 'over-rich' paleo-communities is that there were levels of primary productivity of the dicotyledonous vegetation (i.e. 'browse', leaves of trees and shrubs) that do not exist in today's world. (Other possible causes of these overly species-rich ungulate communities, including possible taphonomic and sampling effects, are discsussed, and dismissed, in Janis *et al.*, in press.) An obvious possible explanation for enriched primary productivity would be increased levels of atmospheric CO_2, as C_3 plants grown under conditions of elevated CO_2 show increased photosynthetic rates and growth, and greater resistance to water loss (Bazzaz *et al.*, 1997; DeLucia *et al.*, 1999; Koerner, 2001). The possibly global faunal response to mid Miocene conditions would also be better explained by global factors such as increased atmospheric CO_2 than by local factors. Although this explanation is at variance with current geochemical interpretation of modern day levels of atmospheric CO_2 at this time (e.g., Pagani *et al.*, 1999; Pearson and Palmer, 2000), the notion of higher mid Miocene CO_2 is also supported by some paleobotanical studies (e.g. Kürsher, 1997; Retallack, 2001).

17.3.2.4 Summary and conclusions

The generic diversity of ungulate mammals is highly informative about climatic changes in the late Neogene, especially when the types of different tooth crown height (reflecting dietary differences) are considered separately. The C_3/C_4 transition in the late Miocene may have had an important effect on the vegetation, but does not appear to have had an immediate, direct effect on the ungulate fauna. In contrast, the decline in the abundance and relative faunal proportions of the browsing ungulates in the 10 million years prior to the C_3/C_4 transition may reflect the effect of declining levels of atmospheric carbon dioxide on plant productivity prior to the point when levels fell so low as to promote a change in grass photosynthetic biochemistry.

17.4 Conclusion

The general pattern of Cenozoic mammal diversity and radiation shows a strong correlation with paleoclimate, which in turn can be shown to be related to tectonic events. Changes in global distribution of vegetational types (see Figure 17.1), resulting from paleoclimatic changes, can explain much of the pattern of mammalian morphological diversification and adaptation. The 'archaic' mammals of the Paleocene and Eocene were those adapted for globally tropical conditions, with the absence of open habitats such as grasslands. Their demise in the later Eocene has often been interpreted as due to competition with more modern mammals, which first appeared in the early Eocene. However, the pattern of replacement of more archaic mammals with more modern ones (see Figure 17.2) clearly shows a 'double wedge' pattern, which is not indicative of competitive interaction (Benton, 1987). The changeover between archaic and more modern mammals is better interpreted as reflecting the change from the 'hot house' world of the Mesozoic and early Cenozoic to the 'ice house' world of the later (post-middle Eocene) Cenozoic. Loss of many types of early Cenozoic mammals is most probably related to the loss of habitats that have no later Cenozoic analog, such as the high-latitude tropical-like forests (see Figure 17.1).

The cooling and drying of the later Cenozoic resulted in the more open habitats familiar today, with grasslands appearing in the Miocene (see Figure 17.1), and deserts, tundra and taiga in the Plio-Pleistocene. With these more open habitats came the evolution of mammals adapted to fibrous diets and the need to range over long distances to forage: larger-bodied, longer-legged herbivores such as antelope and modern horses date from the mid-Miocene, as do the modern types of carnivores that prey on them.

Today's world is very cold and dry in comparison with most of the Cenozoic, with much more exaggerated climatic zonation. This zonation itself adds to the diversity of types of mammals present. The equatorial forests retain mammals similar in some ways to Eocene ones, such as mouse deer and prosimian primates; the middle latitudes house mammals of the ecomorphological types that primarily appeared during the opening of habitats during the Miocene, such as sheep, pigs, and foxes; and the newly formed tundra and desert habitats contain novel types of mammals such as polar bears, reindeer and oryx.

Finally, because the teeth of ungulates (hoofed mammals) can reveal so much about their diets, studies of the evolutionary history of these mammals can provide refinement to our understanding of patterns of climatic changes. The diversification of early Cenozoic ungulates indicates that leafy foliage was unavailable for mammalian diets prior to the Eocene. The rise in Eocene global temperatures may have been accompanied by a change in the seasonality of rainfall distribution, resulting in a more deciduous type of vegetation with more palatable leaves. Changes in inferred diets of ungulates in the later Eocene indicate that drying, as well as cooling, was an important factor in the paleoenvironmental changes of higher latitudes. The diversification of later Cenozoic ungulates shows that the early Miocene world was over-rich in browsers in comparison with any known habitat, indicative of greater plant productivity perhaps resulting from higher levels of atmospheric CO_2. The subsequent decline in ungulate diversity (and in mammalian diversity in general) from a middle Miocene peak may be related to declining levels of atmospheric CO_2 as well as to declining temperatures and the formation of polar ice caps.

Acknowledgments

I thank Adrian Lister and Jerry Hooker for comments on the manuscript.

References

Alroy, J. (2000). Successive approximations of diversity curves: ten more years in the library. *Geology*, **28**: 1023–1026.

Alroy, J. (2001). A multispecies overkill simulation of the end-Pleistocene megafaunal mass extinction. *Science*, **292**: 1893–1896.

Alroy, J., Koch, P.L. and Zachos, J.C. (2000). Global climate change and North American mammalian evolution. In: D.H. Erwin and S.L. Wing (eds) *Deep Time*, supplement to *Paleobiology* **26(4)**, 259–288.

Andrews, P. (1992). Community evolution in forest habitats. *Journal of Human Evolution*, **22**: 423–438.

Bains, S., Corfield, R.M. and Norris, R.D. (1999). Mechanisms of climate warming at the end of the Paleocene. *Science*, **285**: 724–727.

Barry, J.C., Johnson, N.M., Raza, S.M. and Jacobs, L.L. (1985). Neogene mammalian faunal change in southern Asia: correlations with climatic, tectonic, and eustatic events. *Geology*, **13**: 637–640.

Bazzaz, F.A., Baslow, S.L., Berntson, G.M. and Thomas, S.C. (1997). Elevated CO_2 and terrestrial vegetation: implications for and beyond the global carbon budget. In: B. Walker and W. Steffan (eds) *Global Change and Terrestrial Ecosystems*. Cambridge: Cambridge University Press, pp. 43–76.

Benton, M.J. (1987). Progress and competition in macroevolution. *Quarterly Review of Biology*, **58**: 29–51.

Berner, R.A. (1991). A model for atmospheric CO_2 over Phanerozoic time. *American Journal of Science*, **291**: 56–91.

Berner, R.A. (1998). The carbon cycle and CO_2 over Phanerozoic time: the role of land plants. *Philosophical Transactions of the Royal Society of London, Series B, Biological Science*, **353**: 75–82.

Buchardt, B. (1978). Oxygen isotope paleotemperatures in the North Sea area. *Nature*, **275**: 121–123.

Cerling, T.E., Harris, J.M., MacFadden, B.J. *et al.* (1997). Global vegetation change through the Miocene/Pliocene boundary. *Nature*, **389**: 153–158.

Cerling, T.E., Ehleringer, J.R. and Harris, J.M. (1998). Carbon dioxide starvation, the development of C_4 ecosystems, and mammalian evolution. *Philosophical Transactions of the Royal Society, London, series B*, **353**: 159–171.

Collinson, M.E. and Hooker, J.J. (1991). Fossil evidence of interactions between plants and plant-eating mammals. *Philosophical Transactions of the Royal Society of London, Series B*, **333**: 197–208.

DeConto, R.M., and Pollard, D. (2003). Rapid Cenozoic glaciation of Antarctica induced by declining atmospheric CO_2. *Nature*, **421**: 245–249.

DeLucia, E.H., Hamilton, J.G., Naidu, S.L. *et al.* (1999). Net primary production of a forest ecosystem with experimental CO_2 enrichment. *Science*, **284**: 1177–1179.

Ehleringer, J.R., Sage, R.F., Flanagan, L.B. and Pearcy, R.W. (1991). Climate change and the evolution of C_4 photosynthesis. *Trends in Ecology and Evolution*, **6**: 95–99.

Guthrie, R.D. (1982). Mammals of the mammoth steppe as paleoenvironmental indicators. In: D.M. Hopkins, J.V. Matthews Jr, C.E. Schweger and S.B. Young (eds) *Paleoecology of Beringia*. New York: Academic Press, pp. 259–298.

Hooker, J.J. (2000). Paleogene mammals: crisis and ecological change. In: S.J. Culver and P.F. Rawson (eds) *Biotic Response to Global Change: the Last 145 Million Years*. Cambridge, Cambridge University Press, pp. 333–349.

Janis, C.M. (1993). Tertiary mammals in the context of changing climates, vegetation, and tectonic events. *Annual Review of Ecology and Systematics*, **24**: 467–500.

Janis, C.M. (1995). Correlations between craniodental morphology and feeding behavior in ungulates: reciprocal illumination between living and fossil taxa. In: J.J. Thomason (ed.) *Functional Morphology in Vertebrate Paleontology*. Cambridge: Cambridge University Press, pp. 78–98.

Janis, C.M. (1997). Ungulate teeth, diets, and climatic changes at the Eocene/Oligocene boundary. *Zoology*, **100**: 203–220.

Janis, C.M. (2000). Patterns in the evolution of herbivory in large terrestrial mammals: the Paleogene of North America. In: H.-D. Sues and C. Labanderia (eds) *Origin and Evolution of*

Herbivory in Terrestrial Vertebrates. Cambridge: Cambridge University Press, pp. 168–221.

Janis, C.M., Damuth, J. and Theodor, J. (2000). Miocene ungulates and terrestrial primary productivity: where have all the browsers gone? *Proceedings of the National Academy of Sciences*, **97**: 7899–7904.

Janis, C.M., Damuth, J. and Theodor, J.M. (2002). The origins and evolution of the North American grassland biome: the story from the hoofed mammals. *Palaeogeography, Palaeoclimatology, Palaeoecology*, **177**: 183–198.

Janis, C.M., Damuth, J. and Theodor, J.M. (in press). The diversity of Miocene browsers, and implications for habitat type and primary productivity in the North American grassland biome. *Palaeogeography, Palaeoclimatology, Palaeoecology*.

Janis, C.M., Scott, K.M. and Jacobs, L.L. (eds) (1998). *Evolution of Tertiary mammals of North America*. Cambridge: Cambridge University Press.

Jernvall, J., Hunter, J.P. and Fortelius, M. (1996). Molar tooth diversity, disparity, and ecology in Cenozoic radiations. *Science*, **274**: 1489–1492.

Kay, R.F. and Covert, H.H. (1984). Anatomy and behavior of extinct primates. In: D.J. Chivers, B.A. Wood and A. Bilsborough (eds) *Food Acquisition and Processing in Primates*: New York: Plenum Press, pp. 467–508.

Kay, R.F. and Madden, R.H. (1997). Mammals and rainfall: paleoecology of the middle Miocene at La Venta (Colombia, South America). *Journal of Human Evolution*, **32**: 161–199.

Koerner, C. (2001). Biosphere responses to CO_2 enrichment. *Ecological Applications*, **10**: 1590–1619.

Köhler, M., Moyà-Solà, S. and Agusti, J. (1998). Miocene/Pliocene shift: one step or several? *Nature*, **393**: 126.

Kürschner, W.M. (1997). The anatomical diversity of recent and fossil leaves of the durmast oak (*Quercus petraea* Lieblein/*Q. pseudocastanea* Goeppert) – implications for their use as biosensors of palaeoatmospheric CO_2 levels. *Palaeogeography, Palaeoclimatology, Palaeoecology*, **96**: 1–30.

Leopold, E. and Denton, M.F. (1987). Comparative age of grassland and steppe east and west of the northern Rocky Mountains. *Annals of the Missouri Botanic Garden*, **74**: 841–867.

Leopold, E., Liu, G. and Clay-Poole, S. (1992). Low biomass vegetation in the Oligocene? In: D.R. Prothero and W.A. Berggren (eds) *Eocene–Oligocene Climatic and Biotic Evolution*. Princeton: Princeton University Press.

MacFadden, B.J. and Cerling, T.E. (1994). Fossil horses, carbon isotopes and global change. *Trends in Ecology and Evolution*, **9**: 481–486.

MacFadden, B.J., Solounias, N. and Cerling, T.E. (1999). Ancient diets, ecology, and extinction of 5 million-year old horses from Florida. *Science*, **283**: 824–827.

Miller, K.G., Fairbanks, R.G. and Mountain, G.S. (1987). Tertiary oxygen isotope synthesis, sea level history, and continental margin erosion. *Paleoceanography*, **2**: 1–19.

Osborne, R. and Tarling, D. (1996). *The Historical Atlas of the Earth*. New York: Henry Holt.

Pagani, M., Freeman, K.H. and Arthur, M.A. (1999). Late Miocene atmospheric CO_2 concentrations and the evolution of C_4 grasses. *Science*, **285**: 876–879.

Pearson, P.N. and Palmer, M.R. (2000). Atmospheric carbon dioxide concentrations over the past 60 million years. *Nature*, **406**: 695–699.

Prothero, D.R. (1998). The chronological, climate, and paleogeographic background to North American mammalian evolution. In: C.M. Janis, K.M. Scott and L.L. Jacobs (eds) *Evolution of Tertiary Mammals of North America*. Cambridge: Cambridge University Press, pp. 9–36.

Prothero, D.R. (1999). Does climatic change drive mammalian evolution? *GSA Today*, **9(9)**: 1–7.

Prothero, D.R. and Berggren, W.A. (eds) 1992). *Eocene–Oligocene Climatic and Biotic Evolution*. Princeton: Princeton University Press.

Retallack, G.J. (1997). Neogene expansion of the North American prairie. *Palaios*, **12**: 380–390.

Retallack, G.J. (2001). A 300-million-year record of atmospheric carbon dioxide from fossil plant cuticles. *Nature*, **411**: 287–290.

Ruddiman, W. (ed.) (1997). *Tectonic Uplift and Climatic Change*. New York: Plenum Press.

Sepkoski, J.J. Jr. and Kendrick, D.C. (1993). Numerical experiments with model monophyletic and paraphyletic taxa. *Paleobiology*, **19**: 168–184.

Solounias, N. and Dawson-Saunders, B. (1988). Dietary adaptation and palaeoecology of the Late Miocene ruminants from Pikermi and Samos in Greece. *Palaeogeography, Palaeoclimatology, Palaeoecology*, **65**: 149–172.

Stucky, R. (1990). Evolution of land mammal diversity in North America during the Cenozoic. In: H.H. Genoways (ed.) *Current Mammalogy, Vol. 2*, pp. 375–432.

Stucky, R.K. and McKenna, M.C. (1993). Mammalia. In: M.J. Benton (ed.) *The Fossil Record, 2*. London: Chapman and Hall, pp. 739–777.

Vrba, E.S., Denton, E.H. and Prentice, M.L. (1989).

Climatic influence on early hominid behavior. *Ossa*, **14**: 127–156.

Willis, K.J., Kleczkowski, A., Briggs, K.M. and Gilligan, C.A. (1999). The role of sub-Milankovitch climatic forcing in the initiation of the Northern Hemisphere glaciation. *Science*, **285**: 568–571.

Wing, S.L. (1998). Tertiary vegetation of North America as a context for mammalian evolution. In: C.M. Janis, K.M. Scott and L.L. Jacobs (eds) *Evolution of Tertiary Mammals of North America*. Cambridge: Cambridge University Press, pp. 37–65.

Wolfe, J.A. (1985). Distribution of major vegetational types during the Tertiary. *Geophysical Monographs*, **32**: 357–375.

18

Ice ages, species distributions, and evolution

Godfrey Hewitt

ABSTRACT

See Plate 29

The Quaternary period from some 2.4 million years ago (Ma) has seen increasing fluctuations in climate producing several major ice ages. These events caused great changes in species distributions as deduced from the fossil record, and these range changes had effects on the distribution of genetic variation within species. Modern DNA techniques allow measurement of the divergence among populations and reconstruction of species biogeography through the recent ice ages.

The picture for terrestrial organisms – plants, vertebrates and invertebrates – is becoming much clearer, and forms the focus of this chapter. The DNA data also suggest that the Quaternary period saw the divergence of many subspecies and some species, and this occurred under the influence of these major climatic oscillations.

18.1 Glaciation

The Earth cooled through the Tertiary (65–2 Ma) leading to the ice ages of the Quaternary period (2 Ma–present). There is also evidence for extensive glaciation in the Permian period (290–245 Ma) and before, but such distant events provide little detail about the processes involved in the formation of new species. This Tertiary cooling would seem to be due to a complex of processes, particularly the great plate tectonic movements, which reorganized land and sea form and position (Forster, this volume, Chapter 15). These in turn affected ocean currents, the distribution of heat and biotic productivity. The previously forested Antarctic saw its first glaciers in the early Oligocene (38 Ma), deep water temperatures fell by 5°C, and the southern ice cap continued to grow. The first evidence of glaciation in the Arctic is found in Greenland (7 Ma) and then the North Atlantic and Pacific (5 Ma): this had grown to produce moderate-sized ice sheets by 2.4 Ma, close to the beginning of the Quaternary (Williams *et al.*, 1998).

Global temperatures have fluctuated many times between cold and warm conditions in the Quaternary. Furthermore, these have been increasing in amplitude in the last 1 My to produce the recent major ice ages with ice reaching East Anglia and Wisconsin! The evidence for these events comes from many sources, both physical and biological. First, there are direct geological effects of ice sheets far beyond their present extent. The evidence also includes carbon and oxygen isotope levels, variation in CO_2 levels, magnetic and mineral signatures, animal and plant remains; these have been analysed from cores of the sea bed, ice sheets, lakes and on land (Williams *et al.*, 1998). A large body of data has been generated in the last 25 years, and this is providing increasingly coherent descriptions and explanations. The conclu-

Evolution on Planet Earth
ISBN 0-12-598655-6

sions from this field are exciting and challenging, particularly when applied in a consideration of processes of divergence and speciation. The period is one during which genetical and evolutionary studies indicate that many current subspecies and some species have formed. Some fossil studies also provide evidence of speciation and extinction (Lister, 1993), but other studies on fossil beetles show very little morphological divergence that would indicate speciation during the Quaternary period (Coope, 1994).

18.2   Milankovitch cycles

These ice sheets grew and receded in regular cycles. From about 2.5–0.9 Ma this was dominated by a roughly 41 ka cycle, but thereafter into the present a 100 ka cycle becomes apparent for the ice ages, and they become more severe. This periodicity suggests some internal and/or external mechanisms are controlling these fluctuations. The most likely candidates as drivers of these ice age cycles are the regular variations in the orbit of the Earth around the Sun (Hays *et al.*, 1976) (Figure 18.1).

    The Earth's orbit describes an ellipse, which regularly grows fatter and thinner. This main orbital eccentricity has a 100 ka cycle which causes variations in the insolation on the Earth. This insolation is also modified by the variation in the Earth's axial tilt (41 ka) and precession (23 ka) due to wobbling of the axis. These variations change which parts of the Earth are closest to the Sun and where in the ellipse this occurs. The Croll–Milankovitch theory attributes climatic changes to these orbital oscillations. These three cycles are not in phase and so their individual changes in insolation when combined produce a complex oscillation in insolation intensity. This predicted pattern is matched fairly closely by the data from deep sea cores. It is still not clear how relatively small variations of a few per cent in insolation trigger such major responses and many possibilities involving complex feedback mechanisms are being suggested and modelled (Williams *et al.*, 1998).

Such orbital forcing of climatic changes should be long term and extend back well before the Pleistocene ice ages, since it is the product of the solar system's dynamics. There is now much evidence for such Pre-Quaternary climatic oscillations with periodicities between 100 and 20 ka (see Bennett, 1997 for a review). Perhaps the most comprehensive demonstration of such persistent cyclic climatic oscillations comes from the composite variance spectrum for seven $\delta^{18}O$ data sets over 130 My, which has peaks at 100, 41 and 23 ky (Shackleton and Imbrie, 1990). Other data come from many sources in different places, indicating that orbitally forced climatic changes occurred in all parts of the world and back into the Paleozoic (300 Ma). Thus frequent major climatic oscillations are a fundamental component of the environment which has affected the evolution of species and moulded their genomes for aeons. One might expect that the survivors have some adaptation to these events. They should be tolerant of climatic change, capable of changing their range, and have genetic mechanisms that allow adaptation.

18.3   The last glacial cycle

Detailed study of the last ice age (*c.* 115–10 ka) (Figure 18.2) is revealing surprising oscillations of climate on a millennial scale within the 100 ka ice age cycle. Recent technical advances have made it possible to extract ice cores of some 2 km in length, and analyse the annually layered snow for entrapped gas composition, isotopes, acidity, dust and pollen content. These long cores can sample ice to over 200 ka and now even 400 ka (see Stauffer, 1999), but usually cover to about 125 ka which is the time of the last (Eemian) interglacial. Such long cores have been drilled in both the Arctic and Antarctic, with the Vostok and Greenland ice cores being particularly long and informative (Jouzel *et al.*, 1993; Dansgaard *et al.*, 1993). In addition to the continuity of the record, they offer particularly fine temporal resolution.

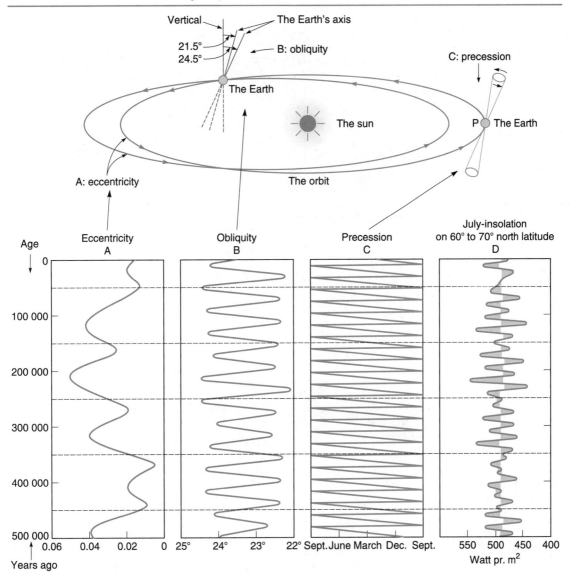

Figure 18.1 Regular variations in the Earth's orbit that on the Croll–Milankovitch theory determine the major climatic oscillations through the Quaternary. A: Eccentricity – the orbit's centre varies on a 100 000 y cycle. B: Obliquity – the tilt of the Earth's axis varies on a 41 000 y cycle. C: Precession – the wobble of the Earth's axis which has a 23 000/19 000 y cycle. P: Perihelion – point in the orbit where the Earth is nearest the Sun. Fluctuations in these factors are shown below over the last 500 ky and the resulting effects on insolation in northern latitudes. From Andersen and Borns, 1994.

The Greenland Ice Core Project (GRIP) provides detailed evidence of frequent rapid changes in temperature as measured by changes in the oxygen isotope ratio ($\delta^{18}O$) (Alley, 2000). Some 24 warmer periods (interstadia) are identified through the ice age (Dansgaard *et al.*, 1993), in which the average temperature rises rapidly by some 7°C over 10–20 years and continued for 500–2000 years. The cooling back to glacial (stadia) conditions some 13°C colder than now was apparently slower in most cases. The causes of such cycles are still debated, but are clearly correlated with the periodic release of enormous volumes of ice

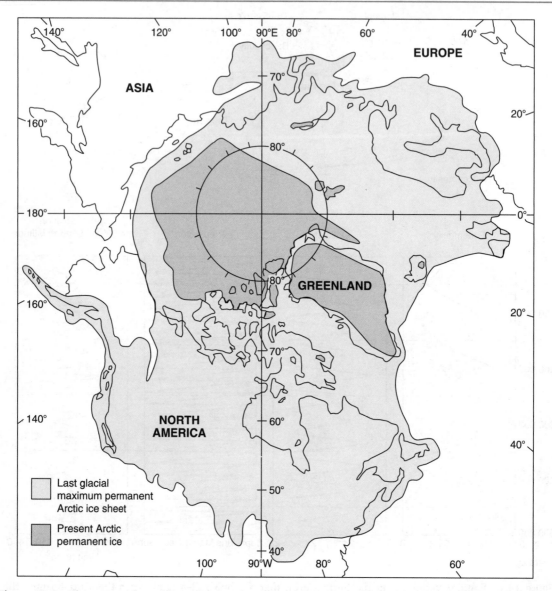

Figure 18.2 The extent of permanent land and sea ice in the Arctic now and at the last glacial maximum. From Williams *et al.* (1998).

into the North Atlantic from Canadian and Scandinavian glaciers. Since these discoveries, ice and sediment cores from around the world have demonstrated the global impact of these events on climate (e.g. Adkins *et al.*, 1997; Behl and Kennett, 1996; Hughen *et al.*, 1996; Schulz *et al.*, 1998; Williams *et al.*, 1998).

It seems likely that major climatic changes, which are so dramatic in the North Atlantic, are connected through the oceanic circulation

system. The thermohaline circulation in the North Atlantic bringing warm water from the subtropics may be switched off in cold periods, and when switched on again provide rapid warming as in the millennial scale switches described above – the Dansgaard–Oeschger (D–O) events (e.g. Bjorck *et al.*, 1996). Models involving all the major oceans support this (Sakai and Peltier, 1996). Furthermore, on a longer, broader scale, since the southern

Atlantic provides heat for the North Atlantic circulation, the amount of heat circulating in the southern oceans could control the growth of ice in the north. This southern heat source could vary with the Milankovitch cycle of precession (McIntyre and Molfino, 1996), and so orbital cycles of insolation may thus drive the major glacial cycles and ice ages.

Such major oscillations in temperature will greatly affect terrestrial life and should leave a similar record. Indeed previous records of pollen and beetle remains have shown rapid changes in species composition at a site, indicating rapid climatic change (Coope, 1977; Guiot *et al.*, 1989). Long pollen cores reaching 140 ka contain a number of switches from cold to temperate species (Beaulieu and Reille, 1992). As more of these longer pollen cores are analysed it will be possible to document the effects of the D–O oscillations on the vegetation in various parts of the world, and hence analyse the species' response to such ancient climatic changes. We may note here that ice core ages are given in absolute years, while most biotic studies (e.g. pollen, beetles) use radiocarbon dating, which over these time periods may be some 10% less. Radiocarbon ages are used below except where indicated.

18.4 Range changes

Such major climatic reversals, particularly between ice age maxima and warm interglacials (Figure 18.2) have caused large changes in the distribution of species, as evidenced in cores and other fossil sequences from land, lakes and the sea. Some pollen records stretch back beyond the Eemian interglacial and our knowledge of recent events since the glacial maximum (18 ka) is very detailed in some places (Coope, 1994; Beaulieu and Reille, 1992; Webb and Bartlein, 1992). The glaciations involved large increases and spreading of the polar ice sheets, so that temperature zones, and marine and terrestrial life, were compressed toward the equator. Mountain blocks such as the Alps, Pyrenees, Andes, Rockies, Sierras and those in Northeast Asia also had significant glaciation and the accumulation of the large

volume of land ice reduced sea levels by up to 150 m (Denton and Hughes, 1981; Tooley, 1993). This joined many lands that are currently separated by shallower seas, such as Britain to Europe, Alaska to Siberia, Australia to Papua-New Guinea and Southeast Asia to Indonesia. Terrestrial organisms might use these bridges to migrate, while the movement of the marine organisms would be restricted during these periods (Lister and Rawson, this volume, Chapter 16). The lower temperatures of ice sheets and changed ocean heat conveyor were also associated with changes in wind and rainfall, with many regions being drier including the tropics of Africa and South America (Williams *et al.*, 1998).

These several climatic effects were felt differently in various parts of the globe, modified by regional differences in land form, ocean currents and latitude. Species responded differently, and their range changes were particular to their regional geography and climate. Recently a new powerful genetic tool has been added to these studies, the use of DNA sequence to produce genealogical lineages in relation to their geographical distribution – that is the phylogeographies of species (Avise, 1994). This approach can tell us much about past evolution and colonization, particularly when combined with fossil evidence (Hewitt, 1996). It therefore seems best to treat these matters regionally, and then look for generalities. Europe and North America have been studied most extensively for both fossil and genetic data. There is now interesting fossil and genetic data from Australia, some from South America and a little from Africa, Southeast Asia and Japan.

18.5 Europe

At the height of the last 'Weischelian' ice age (20 ka) (Figure 18.2), the Scandinavian ice sheet covered much of Britain and northern Europe down to near Warsaw at 52°N. Eastwards of Moscow across Siberia it was less extensive and west of the Taymyr Peninsula only the highlands had ice caps. The previous Saalian ice sheet had come further south in western

Russia. Permafrost extended south of the ice to the south of France, Alps and Black Sea in Europe, and across to Mongolia and Manchuria in the east (Frenzal *et al.*, 1992). As a consequence of these cold conditions in the north, the organisms now living in temperate Europe would have colonized from the south as the climate warmed. From the pollen record, the expansion up eastern Europe was apparently more rapid for some species (Huntley and Birks, 1983). There is currently little permafrost and tundra left in northern Europe west of the Urals, so that most species adapted to these habitats are no longer found here. Permafrost and tundra still occur in northeast Russia.

18.5.1 *In the north*

La Grande Pile in northeast France at 48°N lies in the centre of this ice age permafrost belt. Fossils of beetles found in cores there from toward the end of the last glaciation were nearly all cold-adapted species that now occur in the far north of Scandinavia and Russia (e.g. the carabid *Drachiela arctica*) with some also in Alpine habitats (Ponel, 1997). Many other species in other groups have this classic Arctic–Alpine distribution, and would have similar extensive range changes because of major climatic oscillations.

Pollen core analysis tells us that the herb and tree species that now cover northern Europe mostly had ice age refugia in Iberia, Italy, the Balkans and possibly near the Caucasus. The extensive network of cores across Europe allows the detailed reconstruction of the advance of species as the climate warmed and the ice began to retreat some 15 ka (Huntley and Birks, 1983; Huntley, 1990; Tzedakis and Bennett, 1995). It is significant that each species expanded individually in response to its own environmental conditions, and produced changing mixtures of vegetation with no modern analogue (Huntley, 1990). The speed of the advance was also remarkable, with many species averaging 50–500 m/year over the plains of Europe. Species like pine and hazel advanced at 1500 m/year at times, and alder reached 2000 m/year (Bennett, 1986). Beetles apparently expand very quickly

and can track climatic changes more closely. Fossils of Mediterranean beetle species are found in Britain by 13 ka (Coope, 1990). The individual response of beetle species, rather than as communities, was also noted and discussed by Coope.

The initial advance was particularly rapid up the eastern side of Europe between the Caspian and White seas and to a lesser extent up the western seaboard as compared with central Europe. Alder and spruce reached 60°N in the east by 13 ka, and pine, oak, elm and alder are recorded in Brittany, Ireland and Scotland (Huntley and Birks, 1983).

The pollen and beetle records both clearly show the reversal of this northern advance by the sharp cold spell around 11 ka – the Younger Dryas, which lasted for some 1500 years (Birks and Ammann, 2000; Williams *et al.*, 1998). The ice readvanced, tundra spread south again through France, birches that had reached northern Europe died out and in the south, pine and oak retreated. The Younger Dryas is the most recent of the major temperature oscillations seen in the GRIP core, and is also recorded in other parts of the world including marine sediment cores off Pakistan (Schulz *et al.*, 1998). It would be most useful to have both long fossil pollen and beetle cores, as at La Grand Pile (Beaulieu and Reille, 1992; Ponel, 1997) from locations all over Europe, since this provides an accessible model of species' responses to rapid major climatic changes.

After the Younger Dryas reversal, colonization continued apace and the vegetation broadly resembled today's by 6000 BP, although natural range changes continued and continue to occur (Huntley and Birks, 1983). Man's activities have also modified some species distributions.

18.5.2 *In the south*

The southern peninsulas of Europe that acted as refugia were also warming and populations of some species died out as the tolerance range of the species moved north. Many species have distributions which run in latitudinal bands across Europe and Asia showing north and south range limits, for example birds and butterflies (Harrison, 1982; Higgins and Hargreaves, 1983), and related species frequently have Boreal, Nemoral and Mediterranean ranges. Their adaptation to such conditions means that their ranges would generally move north and south as the climate oscillated.

The particular geography of these southern refugia is significant in a number of ways; they are mountainous and bounded by the Mediterranean sea, Black Sea and big mountain ranges running east–west like those of the Caucasus, Balkans, Alps and Pyrenees. These variously act as major barriers to dispersal from North Africa and Arabia, and also hinder the northward spread from the southern European refugia across the plains of Europe and western Russia. The southern mountains also provided suitable conditions for many cool-adapted species when the climate warmed, and these could climb these as well as migrate northward. Many species currently have Arctic–Alpine or Nemoral–Sierran distributions as evidence of this movement. When the climate turned colder again the northern populations would go extinct, while those in the southern mountains would descend to repopulate the refugia (Hewitt, 1993). There are some long pollen cores from central Italy and northern Greece which show continual presence of species in changing abundance through cold–warm cycles (Tzedakis, 1994). Truly Arctic and Boreal species such as the beetles *Diacheila arctica* or *Hippodamia arctica* would need to migrate south and east in a cold reversal to their stadial ranges beyond the ice. With the series of ice ages and repeated major D–O events, the varied topography and climate of southern Europe provides greater opportunity of finding suitable nearby habitat in which to survive. We might expect this topographical variety to retain genetic diversity and produce divergent lineages by repeated allopatry through the Quaternary (Hewitt, 1996).

18.5.3 Genetic effects of colonization

The rapid expansion northward as the climate warmed would be from populations at the north of a species refugial range, and would involve sequential colonizations by long-distance migrants (Hewitt, 1993). This involves a series of founder events that leads to loss of alleles and increased homozygosity. Modelling and simulations of such range expansion shows that leptokurtic distribution of dispersal distances, with its long-distance component, produces large areas of homozygosity as compared with stepping-stone and normal (Gaussian) modes of dispersal. These areas of homozygosity persist and increase in size and level with time (Ibrahim *et al.*, 1996). Smaller, secondary oscillations increase the effect. This predicts that the rapid long-distance colonization of Northern Europe or any other similar colonization would produce broad genome areas of reduced variability (Figure 18.3).

An increasing number of studies now show lessened genetic diversity in northern colonized regions (Hewitt, 1996). On the other hand, population movements within the mountainous refugial regions of southern Europe would have been shorter and slower and retained more genetic diversity. These two extreme types of population movement have been called 'Pioneer and Phalanx' (Nichols and Hewitt, 1996). Precisely what model between these two extremes applies in

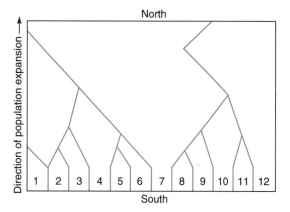

Figure 18.3 As climate warms post-glacially, genotypes 1–12 expand northwards from the northern edge of their refugia in the south. Genotype 7 by chance comes to occupy most of the expanding front through long-range (leptokurtic) dispersal, and thereby hinders the advance of other genotypes arriving later from behind.

any area will depend on the individual species dispersal and reproduction, as well as topography and climate.

18.5.4 Colonization routes

It has become apparent over the last 20 years that many species are geographically subdivided by narrow hybrid zones. These are deduced to be the products of secondary contact between two genomes or taxa that have diverged in separate refugia and met as they expanded their ranges (Hewitt, 1988). While the characters traditionally used to differentiate the forms are morphological, chromosomal and enzymatic, new DNA methods now provide more powerful markers, which can also be used to study the phylogeography of these genomes. Using DNA similarity it is possible to deduce from which ice age refugia populations emanated that now occupy different parts of the species range. A few studies are emerging with a reasonably broad sampling across Europe that use mt DNA sequence differences to detect phylogeographic relationships (Hewitt, 1996; Taberlet *et al.*, 1998). A clear example using a nuclear DNA sequence is the common meadow grasshopper of Europe, *Chorthippus parallelus* (Cooper *et al.*, 1995). Using this DNA sequence, the genome of this species is divided into at least five major geographic regions, Iberia, Italy, Greece, Turkey and the rest of northern Europe including wes-

tern Russia. The DNA haplotypes (sequence variants) across all northern Europe showed little diversity and were similar to those in the Balkans; this clearly indicates that the colonization of Europe after the last ice age was from a Balkan refugium. Spain, Italy, Greece and Turkey each contain a large proportion of unique haplotypes, demonstrating that they contained refugia from which the species did not expand. Narrow hybrid zones between the Spanish and French genomes and Italian and French/Austrian genomes have been described along the Pyrenees and Alps where the expanding genomes met (Hewitt, 1990; Flanagan *et al.*, 1999); and others may well occur where distinct genomes meet in southern Europe. The pattern of postglacial colonization from refugia by this species is thus deduced from both the palynology of plants with which it is associated and its own genetics, and is dominated by the Balkan expansion (Figure 18.4). Notably the pollen record shows a rapid advance of plants from their eastern refugia northward.

The likely colonization routes of some 12 species for which sufficient molecular data are now available across Europe reveal some interesting features (Hewitt, 1999). They confirm Iberia, Italy, and the Balkans as distinct major glacial refugia for most species; while Turkey, Greece and the Caucasus also appear significant. They contributed differently to the colonization of Europe by various species, but

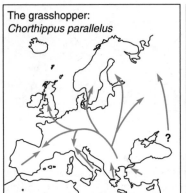  

Figure 18.4 Postglacial colonization routes deduced from DNA similarity and combined fossil evidence, for three paradigm organisms, the grasshopper, the hedgehog and the bear.

some patterns are suggested. Italy contributed little and Spain somewhat more, partly due to the Alps and Pyrenees, while the Balkans sourced northern populations of most species except where blocked by a rapid eastern genome advance. Three paradigm patterns are indicated – the grasshopper, the hedgehog and the bear (Figure 18.4). The hedgehog colonized northern Europe from all three major southern refugia and the bear from only a western and an eastern one. The alder, beech and crested newt are similar to the grasshopper, the oaks and silver fir are like the hedgehog, and shrews and water voles resemble the bear. It is important to note that such molecular information allows a fine subspecific resolution that is not possible with the morphology of fossil pollen, but it will be some time before such a comprehensive geographic coverage is achieved for molecular work.

18.5.5 *Refugial genetics*

It has been noted that the southern regions of Europe that acted as refugia are mountainous, and species survived climatic oscillations by migrating up and down them. Genetic and pollen data indicate that they survived there for several ice ages, during which time they could diversify and diverge genetically. The existence of relatively large populations and the slow migration required would maintain greater allelic diversity, and when two somewhat diverged genomes met as they migrated after a climatic reversal, the hybrid zone formed could act as a barrier to mixing. Different mountain blocks and major valleys may maintain diverging genomes (Figure 18.5). Certainly there is greater genetic and taxonomic diversity in these regions as compared with the north. At the continental scale, the present interglacial hybrid zones between the major subspecific genomes in many species indicate that the series of major Quaternary climatic oscillations could have caused the refugial populations in Spain, Italy, the Balkans and the East to have repeatedly expanded northward across Europe, met, but not mixed to any extent. It is repeated

allopatry protected by hybrid zones (Hewitt, 1989, 1993).

18.5.6 *Adaptation with range change*

In the timescale of the ice ages, characters such as pollen morphology or beetle exoskeleton seem to change little. It is argued that their climatic adaptations also remain very similar, thereby allowing the reconstruction of past climates by the common present-day tolerances of the species (e.g. Atkinson *et al.*, 1987; Huntley and Webb, 1988). It is clear, however, that some inherited attributes do change during range expansion, adapting the species to new conditions. For example, the reproductive biology of many plants is different across their range from south to north (Gray, 1997), with northern populations that migrated from southern refugia having genetic adaptations suited to different light and temperature regimes, for example they can be much slower to first flowering. Similarly, recent plant invasions from Europe to North America have already established such clines of flowering time across their new range. The life history of invading insects has also adapted rapidly, such as the cornborer moth, *Ostrinia nubialis*, which following its introduction in 1910 has spread down the eastern United States and genetically modified its diapause to suit warmer climates (Showers, 1981). Of course, when the climate reverses and the ranges move south these characters must change back, but not necessarily along the same genetic pathway. The genetic structure may diverge with repeated range changes, while the character states appear so similar.

Nonetheless, the morphological stability and species constancy seen in the Quaternary fossil insect record, with apparently little extinction, is remarkable (Coope, 1994). This morphological constancy is maintained while genetic divergence proceeds; an apparent paradox that requires resolution. The fossil work mostly concerns beetles from northern latitudes of Europe and America. These regions were greatly affected by the glaciations, with an Arctic fauna south of the ice sheets in France and the midwest United States. In the postglacial

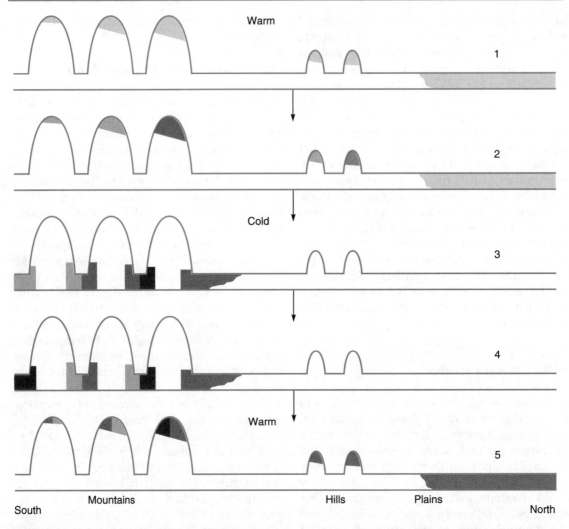

Figure 18.5   A cross-section through Europe with southern mountains and northern plains. An organism's distribution (e.g. a meadow grasshopper) moves up and down the mountains with climatic oscillations (2–3, 4–5). One genome from the northern edge of the refugium colonizes all the northern expansion range (1, 5). Genomic modifications are prevented from mixing by hybrid zones (3, 4, 5). Genomes persist and accumulate in southern mountain regions, but go extinct in the north in an ice age.

warming it would be the populations from the northern edge of the species refugia that colonized northward, or in the case of Beringia, westward. These species tracked their environment and were adapted to it, apparently requiring little morphological change. But separate populations were diverging genetically, as demonstrated in several mammals, birds and fish where DNA sequence has been examined (see section 18.7). Furthermore, within southern refugial areas such as Spain, Italy, the

Balkans and the southern states of the United States species are divided into patchworks of genomes with considerable genetic divergence of Pleistocene age, including beetles such as *Timarcha* in Spain (Gomez-Zurita *et al.*, 2000). Many of these show morphological divergence also, and morphological speciation is found in beetles such as *Calathus* and *Tarphius* on the Canary Islands that can be ascribed to the last 2 My from DNA divergence (Emerson *et al.*, 1999).

In the south of Europe, populations in refugial areas such as Iberia and the Balkans will have experienced different conditions from each other, both in the glacial and interglacial periods. It seems quite possible that different adaptations would be selected in the two refugia, or that similar adaptations could evolve through different genetic controls, particularly over several ice ages. There are a number of cases where geographic races or subspecies with significant adaptive differences have apparently diverged in the Quaternary, as, for example, in grasshoppers, newts, bees and trout in Europe, and butterflies, woodrats, sticklebacks and char elsewhere (Hewitt, 1996).

18.6 North America

18.6.1 *Ice and refugia*

When considering the effects of Quaternary climatic oscillations on range changes and genetic structure on the North American continent, there are many common features with Europe as a result of latitude, size and the periodicity of these global variations. There are, however, significant differences. At the height of the last ice age (Wisconsin) the great Laurentide ice sheet extended down south of the Great Lakes to some 40°N; the Scandinavian sheet reached only 52°N (Figure 18.2). Some previous glaciations reached further south in Illinois, Kansas and Nebraska. In the west the Cordilleran sheet was contiguous with the Laurentide and reached down into Washington State to some 47°N (Frenzel *et al.*, 1992). However, the permafrost south of the ice was not as extensive as in Europe; indeed some broad-leafed and conifer forest reached near Washington DC and St Louis in the east and midwest, with only a relatively narrow belt of tundra before the ice. The tundra was more extensive in the Rockies and west, and the arid salt flats of today were large pluvial lakes. Interestingly, part of Alaska was ice free and linked by land to eastern Siberia, due to lowered sea level, to form Beringia. This acted as a refugium and route of entry

for some cold-adapted species, but most species survived suitable distances south of the great ice sheets.

In the east, most broad-leafed vegetation was reduced to the southern United States with Florida as a significant refugium (Webb, 1988). Unlike Europe, which has the Mediterranean Sea as a southern barrier, the land extends south through Texas into Mexico and beyond, which could provide a route for some species. However, this was very dry during the glacial period; but even so the mountains here and in the western states could have provided the opportunity for altitudinal migrations as in southern Europe. Also unlike Europe, the major mountain ranges, Appalachian and Rockies, run north–south, not east–west, so do not block a northward postglacial advance.

18.6.2 *Pollen and beetles and pack rats*

As with Europe there is a substantial pollen record for much of North America, particularly the northeast, which provides a similarly detailed reconstruction of past vegetation and climate (Davis, 1984; Prentice *et al.*, 1991; Webb, 1988). There is also growing evidence from fossil beetles (Ashworth, 1997; Morgan, 1997), and interestingly from pack rat middens in the arid southwest where plant macrofossils are preserved by urine (Betancourt *et al.*, 1990). These allow reconstruction of past biota and deduced climate, particularly from the last ice age and through the postglacial warming to the present. As the ice was retreating 15 ka BP from its maximum, species advanced from the south rapidly. It had receded to the Canadian Shield by 11 ka, but did not disappear from Labrador until 6–7 ka BP. As in Europe, the rates of advance measured in the pollen record were very rapid for some species. Thus, in the east, beech spread from Tennessee to Canada at an average of 150 m/y, but faster at times (Bennett, 1985), and white spruce may have spread at up to 2000 m/y as Canada became clear of ice (Ritchie and MacDonald, 1986). The pollen, beetle and midden records show that the mixtures of species often had no modern analogues, each species responding separately. Likewise climate models based on a range of paleoclimatic indicators produce unique patterns over the continent that are the products of insolation, precipitation and wind as well as geography (Bartlein, 1997). Nonetheless the composite data demonstrate east–west bands of climate advancing from south to north in the east and central parts of the continent, while in the mountainous west the distribution of climate was quite discontinuous.

As in Europe, the beetle fossil record in the midwest (Ashworth, 1997) shows remarkably rapid changes in species composition and hence in the species ranges. Interestingly, Arctic–Boreal species were south of the ice between 21 ka and 14 ka, but then they disappeared suddenly with the warming. They did not apparently move north, and probably the deglaciated north was colonized from Beringia in the northwest Arctic. Some cold-adapted species south of the ice apparently did survive by moving high into the Rockies and Appalachian mountains (Ashworth, 1997; Morgan, 1997), which provides a similar altitudinal refugial model as for the mountains of southern Europe. During this warming period after 14 ka the beetle data from the Great Lakes region show a great mixture of species presently in tundra, boreal, prairie and Great Lakes environments (Morgan, 1997), and hence community relationships were greatly changing and different from today. Thus both pollen and beetle evidence indicate turbulence in species movements, ranges and mixtures.

As mentioned, the fossil data for the mountainous and arid southwest regions of North America suggest that they had a drier, more discontinuous climatic picture (Bartlein, 1997). The pack rat middens from the southwest deserts provide plant material that indicates pine and juniper retreating from the lowlands to higher elevations, as woods, grasses, scrub and desert invaded with the warming. In such a large topologically diverse region there is considerable spatial and specific variation in temporal response to climate change, with disappearance and movement of species. The available data for the Chihuahua, Sonora, Mohave, Great Basin and Colorado regions

have been reviewed by Bennett (1997). Once again movement of species up and down mountains, as well as over some distance, is produced by rapid major climatic oscillation.

18.6.3 *Genetics of range change*

Advances in obtaining DNA sequence data are promoting the investigation of genetic relationships within and among species across their ranges. Such phylogeographic studies are producing a flood of papers, particularly from North America (Avise, 1998). Some of these, where sampling is adequate, provide data pertinent for the study of major range changes and their genetic consequences. Thus a number of recent studies report lower genetic diversity in northern populations that have expanded from more southerly refugia after the ice age. This was noted a decade ago with allozyme studies in, for example, sheep and pines, and fits with models of rapid colonization (Hewitt, 1989, 1996 for references). To these examples can be added those of the coral *Balanophyllia elegans* on the Pacific coast (Hellberg, 1994), the wireweeds *Polygonella* spp (Lewis and Crawford, 1995), *Pinus albicaulis* in the western mountains (Jorgensen and Hamrick, 1997), *Pinus virginiana* from the southeast Appalachians (Parker *et al.*, 1997), the herb *Asclepias exalta* with a south Appalachian refugium (Broyles, 1998), the snail *Nucella emarginata* expanding up the Californian coast (Marko, 1998), five species of plants from the Pacific northwest (Soltis *et al.*, 1997) and several species of fishes (see Bernatchez and Wilson, 1998, for a recent review). One of the earliest detailed reports of this phenomenon of allele loss in northward colonization was that of the lodgepole pine *Pinus contorta latifalia* which now occurs all up western North America (Cwynar and MacDonald, 1987). This expanded some 2200 km from its ice age refugium to the central Yukon, and, as well as losing allozyme alleles as it did so, it increased the dispersal power of its seed with a modified mass to wing ratio, providing an elegant example of adaptation.

As in Europe, North America contains many hybrid zones, where the ranges of two diverged genomes, races, subspecies, or even species meet and hybridization occurs (Barton and Hewitt, 1985; Harrison, 1993). It is considered that these were produced as diverged refugial populations expanded their ranges after the last ice age and collided. In many cases different genomes produce unfit hybrids and the zones may maintain roughly the same position until the climate and habitat change again (Hewitt, 1988). In a fascinating review of the distribution of animals and trees across North America, Remington (1968) identified 'suture zones of hybrid interaction between recently joined biotas', which are regions where a large number of hybrid zones for different species occur (Figure 18.6). A number are associated with major physical and ecological barriers, such as the Appalachian Mountains, the Rocky Mountains, the Sierra Nevada and Blue Mountains running north–south, and the Florida Borders and the Texan Plateau in the south. While Remington argued that these contacts are possibly recent in origin,

it is most likely that these are interfaces between regional biotas, which met and hybridized soon after the last ice age. The north–south mountain ranges would separate populations of the same species expanding out from different glacial refugia from the south, allowing them to meet only after the climate had warmed considerably; similar in some ways to the Pyrenees or Alps in Europe. The suture zones in the south of North America are possibly analogous to clusters of hybrid and contact zones in the south of Europe, where the more northern taxa in the refugia expanded and, by their presence, blocked any advance of their related genomes from further south in Iberia, Italy, Greece and Turkey (Hewitt, 1993, 1996, 1999). A number of species have been shown to have divergent northern and southern mtDNA lineages in eastern USA that make contact in northern Florida and the southeast states in the region of this suture zone (see Avise, 1994). This type of explanation, where

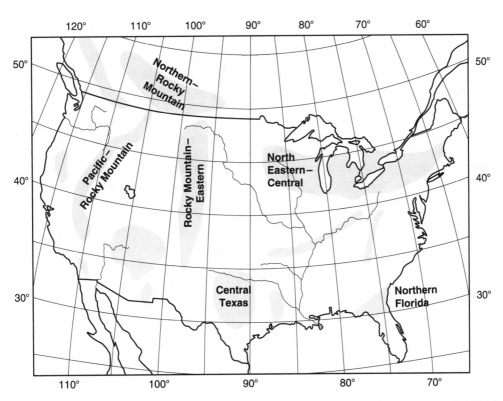

Figure 18.6 The major and minor suture zones in North America described by Remington (1968). These are regions where there is overlap between biotic assemblages where many species and subspecies hybridize.

the northern expansion prevented advance by the southern genome, probably applies to the suture zones running east–west from New England across the Great Lakes to the Dakotas. These broadly coincide with the limits of the ice sheets, and populations in these regions, or just south, would have colonized the deglaciated areas first, thereby blocking subsequent colonization from more southerly genomes and taxa. This range differentiation would also involve adaptation of the genomes to colder and warmer habitats. There may be similar reasons for the suture zone running from Saskachewan along the east side of the Rockies through Alberta, but the presence of the mountains and intricacy of possible refugial sites make for a complex situation.

18.7 The Arctic

The colonization of the deglaciated northern regions and the Arctic is of particular interest, and there are a number of recent studies using DNA markers to investigate the phylogeography of present-day species. There is already a considerable body of data for fish, which reveals several important features (Bernatchez and Wilson, 1998). These authors reviewed the genetic variation in 42 species (all but one from North America) and showed that there was much less nucleotide diversity in species with more northern medians of their distributions. Significantly, there was a change to lower diversity around 46°N, which is the latitude of the maximum extent of the Laurentide ice sheet, and this clearly agrees with the results of rapid colonization discussed for Europe. Furthermore, sister species in the north were much more similar than in the south, i.e. their phylogenetic trees were shallower suggesting recent divergence and recolonization. The area occupied by equivalent genetic clades was also an order of magnitude greater in the deglaciated regions than in the southern non-glaciated ones; this too agrees with the northern colonizations modelled and recorded in Europe (Figure 18.3) (Hewitt, 1996, 1999; Ibrahim, *et al.*, 1996). In the case of many of

these northern fish species, the large and mobile proglacial lakes produced by the melting ice would have greatly assisted their dispersal to produce these extensive genome ranges in North America (Bernatchez and Wilson, 1998). The large meltwater lakes in Eurasia acted similarly (Vainola, 1995). Careful comparative analyses of such growing data sets promises to advance greatly the explanation of the distribution of genetic diversity and phylogeographic patterns.

North of the Cordilleran ice sheet, parts of Alaska were free of ice during the glaciation, and due to lowered sea levels the Bering Strait was a wide land bridge to eastern Siberia. This bridge repeatedly allowed the passage of northern organisms between the two main continents during the Pleistocene, including mankind. This is reflected in the genetic phylogenies of several species groups that have related taxa in both continents (Lister and Rawson, this volume, Chapter 16). Besides acting as a bridge for terrestrial organisms, it would also periodically have blocked the passage of Arctic marine organisms. Beringia itself acted as a refugium and possible origin for cold-adapted species, and the genetic data from freshwater fish provide nice examples of the colonization of deglaciated northern regions from here (Bernatchez and Wilson, 1998). For example, lake whitefish, *Coregonus clupeaformis*, and the lake trout *Salvelinus namaycush*, both have genotype distributions that demonstrate Beringian refugia, as well as Mississippian, Great Lakes and Atlantic ones south of the ice. The postglacial colonization was extensive for genotypes from some refugia; for example, in lake trout the Beringian genotype occurs across to Quebec (Wilson and Hebert, 1996). But the routes of colonization from the refugia were very different for each species, as was also noted for European colonizations.

Phylogeographic information for Arctic organisms other than fish is becoming available, some of the taxa holarctic or circumpolar. The geographic subdivision and extent of genetic divergence at *cyt b* mtDNA (cytochrome b mitochondrial DNA) suggest that various guillemot taxa have diverged and speciated due to refugial separation by Pleistocene glaciations (Friesen *et al.*, 1996; Kidd and Friesen, 1998). The common (*Uria aalge*) and Brunnick's (*U. lomvia*) guillemots probably diverged near the beginning of the Quaternary. Within both species the Atlantic and Pacific populations are distinct molecular clades with divergence indicating separation some 300 ka. Since then they appear to have been separated by the Pleistocene glaciers and Bering land bridge. Likewise the genome structure and distribution of circumpolar migratory dunlins, which comprise several morphological subspecies identified as distinct mtDNA clades, would seem to have been produced by repeated fragmentation in isolated tundra refugia during the Late Pleistocene (Wenink *et al.*, 1996). The circumpolar reindeer or caribou, *Rangifer tarandus*, has some six extant subspecies, which from mtDNA sequence appear to have diverged in the Pleistocene. Most interestingly, the small bodied high Arctic subspecies of reindeer on the Canadian Archipelago and East Greenland (extinct) probably evolved from North American subspecies, while the similar small subspecies on Svalbard is more related to the Eurasian subspecies (Gravlund *et al.*, 1998). These colonizations were probably only some 10 ka, just after the ice retreated. Fossils show that the brown bears used the Bering land bridge and expanded from Beringia after the ice sheet melted, extending right down to Mexico. Several subspecies are recognized, and their mtDNA phylogeny reveals three clades, one Siberian–West Alaskan, one central Alaskan–Canadian and a more divergent (older) one on the southeast Alaskan islands. Interestingly, this latter includes the very recently evolved polar bear with only 1.1% *cyt b* mtDNA divergence, placing its separation from the brown bear some 300 ka (Talbot and Shields, 1996). There is also informative work on lemmings (Eger, 1995; Fedorov, 1998), *Daphnia* (Colbourne *et al.*, 1998), beluga whales (Corry-Crowe *et al.*, 1997) and beetles (Ashworth, 1996) that further demonstrates the striking effects that Quaternary ice ages have had on genetic structure and speciation in these polar regions.

18.8 Tropics

The major climatic oscillations of the Quaternary affected all parts of the world. During the last ice age conditions became colder and drier with the extension of deserts and savannah and the reduction of rain forest in the tropical latitudes. There is unfortunately little pollen data for these forests and it has been argued from areas of overlapping species endemism that the Amazon rain forest survived in a number of localized refugia, with most in mountain areas (Haffer, 1969, 1997). However, recent pollen data throw this into doubt and indicate that temperature was lower by some 6°C in the mountains, causing forest species to descend to the lowland. Also rainfall was variable and higher locally in the lowlands, so that some forest could have survived there (Bush and Colinvaux, 1990; Colinvaux, 1997). This suggests that forest species may have survived more widely in small patches.

Similar arguments apply to the rain forests in Africa, Southeast Asia and Australia, and palynological analysis of the rain forest near Cairns in northeast Queensland exemplifies this (Kershaw, 1986; Hopkins *et al.*, 1993). Here the refugial areas within the present rain forest, which had high rainfall in the ice age and are species rich, were subject to *Eucalyptus* fires as evidenced by charcoal deposits. This clearly suggests that these 'refugial' areas had open *Eucalyptus* woodland and the rain forest species must have survived the fires as small pockets, possibly in gullies and creeks (Williams *et al.*, 1998). The effects of such climatically driven contractions and expansions on genome structure and genetic diversity have been addressed in recent surveys of *cyt b* mtDNA in several birds, reptiles and frogs characteristic of these forests (Joseph *et al.*, 1995; Schneider *et al.*, 1998). In this strip of forest just north of Cairns, a break between the north and south clusters of refugial areas coincides with concordant phylogenetic divergence in these species that is large enough for Pliocene or early Pleistocene separation on the standard molecular clock. Within each of these refugial clusters the distribution of mtDNA variation shows that some species have low genetic diversity as a result of population contractions, while others have managed to maintain high diversity by different haplotypes surviving in various patches. The details of the pattern vary among species, reflecting both chance and their adaptive differences. From these data it is also possible to construct possible routes of colonization and suggest putative locations of hybrid zones, as in Europe and North America. This combination of palynology and genetics is particularly pleasing as it allows an investigation of divergence and speciation in the tropics, similar to that in higher latitudes. The tropics are of course very speciose but have not been studied extensively.

Investigation of the Amazon region is very difficult, on account of its size, poor pollen record of refugia and inadequate taxonomy and collection. Even so, mtDNA data from small rodents and marsupials are providing novel insights into the faunal evolution of the region (Da Silva and Patton, 1998). DNA lineage divergence provides a maximum estimate of age of speciation, since gene lineage splitting may occur before two species form. Furthermore, the processes of lineage sorting and speciation take some time to complete (Avise *et al.*, 1998). Da Silva and Patton (1998) analysed the *cyt b* gene in some 35 species from 15 genera and found high sequence divergence among many taxa, placing their origins in the Pliocene (5–2 Ma), somewhat earlier than was thought. When their phylogeographies are compared, only some rivers appeared as major transition zones. Thus taxa separated by the lower River Negro and middle River Solimoes showed average sequence divergence of 9.72% and 7.73% respectively, while other major rivers were lesser or insignificant divides. The riverine barrier hypothesis for the structuring and speciation of Amazonian taxa clearly requires rethinking. Another striking result of these phylogeographic comparisons is a strong phylogenetic break across the Jura River, some 7.92% average divergence involving 11 taxa, which coincides with the Iquitos Arch, a major geological structure that

formed as a consequence of the last Andes uplift in the Late Tertiary. Such genetic results are clearly very revealing and more detailed surveys and phylogeographies are needed in order to assess the effects of Quaternary climatic changes, along with other factors, on divergence and speciation in this vast region.

It has been noted that related species of birds in lowland tropical forests of South America and Africa show considerable genetic (DNA) divergence, placing their origin more than 6 Ma (Fjeldsa, 1994; Fjeldsa and Lovett, 1997); they are 'old' like many of the small mammal taxa in Amazonia. On the other hand, tropical mountain areas contain clusters of recently diverged bird DNA lineages, many within the Pleistocene on a molecular clock, along with some old and relictual species. This has led to the proposition that such mountain regions provide moist habitat stability through ice age arid periods, allowing the survival of lineages as well as their generation. Further evidence for this pattern comes from recent mtDNA studies on African greenbuls in the East Africa mountain ring (Roy, 1997), and supports such mountain regions as centres of speciation through Pleistocene climatic changes. The plants and insects from these mountains have very different life histories, which may well affect their rate of divergence and speciation; it would be instructive to examine their molecular relationships in the light of Quaternary climatic oscillations. In summary, it seems that as with southern Europe (Figure 18.5), mountains provide the topographic variety and habitat persistence for organisms, lineages and taxa to survive and diverge.

18.9   Speciation

Clearly the climatic oscillations of the Pliocene and Pleistocene have caused many major range changes in all parts of the world. Organisms have necessarily moved, adapted or gone extinct, and so most extant species have survived many climatic shifts. Each major glacial and interglacial period would cause organisms to invade new territory, where they would meet new environments and new neighbours. These challenges would provide the opportunity for selection and drift to cause genomes to diverge and possibly speciate (Hewitt, 1993).

There is growing evidence that natural selection can act very rapidly to produce differentiation and speciation when organisms enter new environments. Convincing experimental cases have been produced for insects on new plant hosts, lizards on islands, and fishes in new habitats – literally occurring within our lifetime (e.g. Orr and Smith, 1998). Particularly pertinent for a consideration of the effects of ice ages on speciation are the sympatric ecomorphs (differentiated forms adapted to different ecologies) in sticklebacks, Arctic charr and lake white fish which live in lakes formed after the ice retreated (e.g. Bernatchez and Wilson, 1998; Schluter, 1998). In most cases these are considered species and phylogenetic analyses show these species pairs to be very closely related and probably postglacial in origin. Furthermore, separate lakes can have similar ecomorph pairs that are derived from different lineages, so the process of postglacial differentiation has occurred several times. Such recent adaptive divergence has been phylogenetically corroborated in other organisms, such as the small bodied tundra reindeer and most spectacularly in the cichlid fish species flocks of the young African rift lakes (Meyer, 1993).

It seems unlikely that these new forms and species will survive subsequent major climatic change, particularly those now in higher latitudes. But occasionally, by chance location and fortunate movement one may live on, to found a distinct lineage and taxon. In lower latitudes, particularly in mountain regions, there may be greater opportunity for such new species to survive a major climatic change (Figure 18.5) (Hewitt, 1996) and this may in part explain the great genetic diversity and species richness towards the tropics.

With repeated ice ages and interglacials, genomes that have diverged in refugia will expand their ranges only to be reduced to refugia again. The fossil and genetic evidence indicates that the varied topography of southern Europe served such continual and repeated refugial functions. The eastern and western

mountains of North America, the Andes, the African mountains and the Australian Divide would appear to have done the same. So while most of the adaptation and speciation of each expansion is obliterated by each major climatic reversal, divergence may accumulate in these continual refugial regions. The expanding postglacial genomes need not mix when they meet, due to protection by hybrid zones, and speciation may occur in refugial areas by repeated alloparapatry (population isolation and contact on expansion) through climatically induced range changes (Hewitt, 1989, 1993, 1996). This is clearly a tenable hypothesis in Europe, and probably also in other regions of the world.

It has been emphasized that species respond individually to climate change and conditions in each ice age were somewhat different. Consequently the selection experienced by different geographic genomes will vary through time, driving them down different evolutionary pathways. A corollary of this is that most species are formed by different combinations of classical modes of speciation, i.e. allopatry, parapatry and sympatry, and this involves drift, selection and hybridization to varying degrees. The organism's biological attributes and geographical location will determine how it responds to climate change, and which paths of divergence its genomes follow to full speciation.

Thus it seems that speciation can occur rapidly within an interglacial, or it may be more extended over several ice ages. Sequence differences can give estimates of when gene flow ceased among populations, races and species and divergence occurred, assuming a molecular clock. A compilation of the molecular data for divergence in subspecies and species complexes showed that many species were formed through the Pliocene and Pleistocene (Hewitt, 1996). This is highlighted by the varying depth of phylogenetic divergence among subspecies forming hybrid zones, indicating that some sister genomes have only recently separated and others have been effectively separate for millions of years. This is clearly seen in European taxa, where it can also indicate when the progenitors of the present taxa colonized the southern European peninsulae. This was at different times for different species, and they have been diverging in these refugia over a various number of ice ages (Hewitt, 1999). Similar deductions may be made for refugial areas in other parts of the world when such detailed comparative phylogenetic data are available.

The depth of phylogenetic divergence between lineages and sister species can also give an estimate of the length of time for speciation to occur, again assuming a molecular clock. The available data on birds, mammals, frogs, reptiles and fishes have recently been collated and analysed for some 431 species (Avise *et al.*, 1998). This shows that genetic divergence between sister species ranges fairly evenly through the Pleistocene, Pliocene and late Miocene. Within species, genetic divergence between major lineages or phylogroups is concentrated in the Pleistocene with some extending back through the Pliocene. Reasonable estimates of mean duration of speciation are around 2 Ma, although cold-blooded animals may be slower. Thus speciation times vary, and the process has been proceeding apparently unabated through the Quaternary climatic changes. The orbitally driven glaciations have caused major changes in the distribution of organisms, thereby providing the opportunity for strong selection and physical isolation to generate genomic divergence, reproductive barriers and new species. The climatic shifts affect regions of the world differently, producing different patterns and the pathways to speciation are many. It seems likely that these major climatic changes have produced many new races and species, but whether this rate is accelerated compared with previous periods is difficult to ascertain and open to debate. While an apparent species constancy in morphology has been noted over the Quaternary in northern realms (cf. Bennett, 1997), clearly populations of species in different refugia have diverged genetically, and in some cases morphologically, physiologically and behaviourally in this period. The morphological constancy observed in some species suggests stabilizing selection, and needs to be interpreted in the light of the genetic evidence.

18.10 Summary

The Earth's orbit around the sun shows three regular variations, precession, tilt and eccentricity, which occur in 23 000, 41 000 and 100 000 y cycles modifying the insolation of the Earth. The Croll–Milankovitch theory attributes climatic changes to these cycles, and in particular the major ice ages of the Late Pleistocene which occur every 100 ky in line with this orbital forcing. Recent analysis of marine sediments and deep ice cores has revealed large very rapid climatic changes of some 1500 y within the 100 ky ice age cycle – the Dansgaard–Oeschger cycles.

Such ice ages caused great changes in distributions of species. These are being studied in detail from the previous interglacial (130 ka) to the present through fossils in cores from land and sea (e.g. pollen, beetles and forams). The glaciations involved large increases in the polar ice sheets, with vegetation zones being compressed toward the equator, and regional changes in rainfall. Warm interglacials produced a recolonization into higher latitudes and a rise in sea level.

Shorter climatic oscillations nested within the major ice ages caused rapid changes in species ranges. The dynamics of the contractions and expansions are expected to have consequences for the genetic content of the populations involved. Thus rapid colonizations may involve genetic drift and lead to a reduction in genetic diversity. There is increasing evidence that this has occurred for many species in higher latitudes as a consequence of postglacial colonization.

The form of the land and water influences the pattern of range changes, where mountains and straits with oscillating ice caps and sea levels provide refugia, passages and barriers. These interact with latitudinal climatic variation to produce different movement patterns in distinct geographic regions.

Modern DNA techniques are allowing the study of species genotypes across their range, and such phylogeography provides information on past range changes and divergence. Detailed studies are beginning to accumulate for Europe and North America. These allow us to deduce ice age refugia and postglacial colonization patterns for population genomes within species.

In Europe they confirm Iberia, Italy and the Balkans as the major sources of colonists for the north, with three broad paradigm patterns of expansion in the grasshopper, hedgehog and bear. The mountainous regions of southern Europe have acted as refugia and accumulated genetic differences over several ice ages. In North America, in addition to the colonization from several refugia south of the ice sheets, several species have distinct genomes coming from Beringia, northwest of the ice sheets. In both these continents species are seen to be subdivided into a patchwork of genomes that was produced by their expansion from distinct glacial refugia moderated by local topography. The few studies from the Tropics are already contributing significantly to the resolution of debate on refugial hypotheses and the role of mountains in divergence and species distribution.

DNA data also measure divergence among populations over several ice ages and suggest how ice age range changes may be involved in speciation. Colonizing adaptation and host shifts can produce species rapidly, while parapatry, repeated allopatry and divergence in refugia accumulate differentiation through several ice ages. DNA divergence and speciation have been proceeding throughout the Quaternary, and orbitally driven climatic changes are fundamental to these processes.

References

Andersen, B.G. and Borns, H.W. (1994). *The Ice Age World*. Oslo: Scandinavian University Press.

Adkins, J.F., Boyle, E.A., Keigwin, L. and Cortijo, E. (1997). Variability of the North Atlantic thermohaline circulation during the last interglacial period. *Nature*, **390**: 154–156.

Alley, R.B. (2000). Ice-core evidence of abrupt climate changes. *Proceedings of the National Academy of Sciences USA*, **97**: 1331–1334.

Ashworth, A.C. (1996). The response of arctic Carabidae (Coleoptera) to climate change based on the fossil record of the Quaternary Period. *Annales Zoologici Fennici*, **33**: 125–131.

Ashworth, A.C. (1997). The response of beetles to Quaternary climatic changes. In: B. Huntley *et al.* (eds) *Past and Future Rapid Environmental Changes*. NATO ASI Series Vol. I **47**. Berlin: Springer-Verlag, pp. 119–128.

Atkinson, T.C., Briffa, K.R. and Coope, G.R. (1987). Seasonal temperatures in Britain during the past 22,000 years, reconstructed using beetle remains. *Nature*, **325**: 587–592.

Avise, J.C. (1994). *Molecular Markers, Natural History and Evolution*. New York: Chapman and Hall.

Avise, J.C. (1998). The history and preview of phylogeography: a personal reflection. *Molecular Ecology*, **7**: 371–379.

Avise, J.C, Walker, D. and Johns, G.C. (1998). Speciation durations and Pleistoene effects on vertebrate phylogeography. *Proceedings of the Royal Society of London B*, **265**: 1707–1712.

Bartlein, P.J. (1997). Past environmental changes: characteristic features of Quaternary climate variations. In: B. Huntley *et al.* (eds) *Past and Future Rapid Environmental Changes*. NATO ASI Series Vol. I **47**. Berlin: Springer-Verlag, pp. 11–30.

Barton, N.H. and Hewitt, G.M. (1985). Analysis of hybrid zones. *Annual Review of Ecology and Systematics*, **16**: 113–148.

Beaulieu, J.L. de and Reille, M. (1992). The last climatic cycle at La Grande Pile (Vosges, France): a new pollen profile. *Quaternary Science Reviews*, **11**: 431–438.

Behl, R.J. and Kennett, J.P. (1996). Brief interstadial events in the Santa Barbara basin, NE Pacific, during the past 60 kyr. *Nature*, **379**: 243–246.

Bennett, K.D. (1985). The spread of *Fagus grandifolia* across eastern North America during the last 18,000 years. *Journal of Biogeography*, **12**: 147–164.

Bennett, K.D. (1986). The rate of spread and population increase of forest trees during the postglacial. *Philosophical Transactions of the Royal Society, London, B*, **314**: 523–531.

Bennett, K.D. (1997). *Evolution and Ecology: The Pace of Life*. Cambridge: Cambridge University Press.

Bernatchez, L. and Wilson, C.C. (1998). Comparative phylogeography of nearctic and palearctic fishes. *Molecular Ecology*, **7**: 431–452.

Betancourt, J.L. van Devender, T.R. and Martin, P.S. (eds) (1990). *Packrat Middens: The Last 40000 Years of Biotic Change*. Tucson: University of Arizona Press.

Birks, H.H. and Ammann, B. (2000). Two terrestrial records of rapid climatic change during the glacial-Holocene transition (14,000–9,000 calendar years B.P.) from Europe. *Proceedings of the National Academy of Sciences USA*, **97**: 1390–1394.

Bjorck, S., Kromer, B., Johnsen, S. *et al.* (1996). Synchronized terrestrial-atmospheric deglacial records around the North Atlantic. *Science*, **274**: 1155–1160.

Broyles, S.B. (1998). Postglacial migration and the loss of allozyme variation in northern populations of *Asclepias exalta* (Asclepiadaceae). *American Journal of Botany*, **85**: 1091–1097.

Bush, M.B. and Colinvaux, P.A. (1990). A pollen record of a complete glacial cycle from Lowland Panama. *Journal of Vegetation Science*, **1**: 105–118.

Colbourne, J.K., Crease, T.J., Weider, L.J. *et al.* (1998). Phylogenetics and evolution of a circumarctic species complex (Cladocera: *Daphnia pulex*). *Biological Journal of the Linnean Society*, **65**: 347–365.

Colinvaux P.A. (1997). An arid Amazon? *Trends in Ecology and Evolution*, **12**: 318–319.

Coope, G.R. (1977). Fossil coleopteran assemblages as sensitive indicators of climatic change during the Devension (last) cold stage. *Philosophical Transactions of the Royal Society, London, B*, **280**: 313–340.

Coope, G.R. (1990). The invasion of Northern Europe during the Pleistocene by Mediterranean species of Coleoptera. In: F. di Castri, A.J. Hansen and M. de Bussche (eds) *Biological Invasions in Europe and the Mediterranean Basin*. Dordrecht; Kluwer, pp. 203–215.

Coope, G.R. (1994). The response of insect faunas to glacial–interglacial climatic fluctuations. *Philosophical Transactions of the Royal Society, London, B*, **344**: 19–26.

Cooper, S.J.B., Ibrahim, K.M. and Hewitt, G.M. (1995). Postglacial expansion and genome subdivision in the European grasshopper Chorthippus parallelus. *Molecular Ecology*, **4**: 49–60.

Corry-Crowe, G.M., Suydam, R.S., Rosenberg, A. *et al.* (1997). Phylogeography, population structure and dispersal patterns of the beluga whale *Delphinapterus leucas* in the western Nearctic revealed by mitochondrial DNA. *Molecular Ecology*, **6**: 955–970.

Cwynar, L.C. and MacDonald, G.M. (1987). Geographical variation of lodgepole pine in relation to poulation history. *The American Naturalist*, **129**: 463–469.

Da Silva, M.N.F. and Patton, J.L. (1998). Molecular phylogeography and the evolution and conservation of Amazonian mammals. *Molecular Ecology*, **7**: 475–486.

Dansgaard, W., Johnsen, S.J., Clausen, H.B. *et al.*

(1993). Evidence for general instability of past climate from a 250-kyr ice-core record. *Nature*. **364**: 218–220.

Davis, M.B. (1984). Holocene vegetational history of the eastern United States. In H.E. Wright (ed.) *Late Quaternary Environments of the United States*. Vol. 2. *The Holocene*. London: Longman, pp. 166–181.

Denton, G.H. and Hughes, T.J. (eds) (1981). *The Last Great Ice Sheets*. New York: Wiley.

Eger, J.L. (1995). Morphometric variation in the Nearctic collared lemming (*Dicrostonyx*). *Journal of Zoology London*, **235**: 143–161.

Emerson, B.C., Oromi, P. and Hewitt, G.M. (1999). MtDNA phylogeography and recent intra-island diversification of Canary Island Calathus beetles (Carabidae). *Molecular Phylogenetics and Evolution*, **13**: 149–158.

Fedorov, V.B. (1998). Contrasting mitochondrial DNA diversity estimates in two sympatric genera of Arctic lemmings (*Dicrostonyx*: *Lemmus*) indicate different responses to Quaternary environmental fluctuation. *Proceedings of the Royal Society of London B*, **266**: 621–626.

Fjeldsa, J. (1994). Geographical patterns for relict and young species of birds in Africa and South America and implications for conservation priorities. *Biodiversity and Conservation*, **3**: 207–226.

Fjeldsa, J. and Lovett, J.C. (1997). Geographical patterns of old and young species in African forest biota: the significance of specific montane areas as evolutionary centres. *Biodiversity and Conservation*, **6**: 323–244.

Flanagan, N.S., Mason, P.L., Gosalvez, J. and Hewitt, G.M. (1999). Chromosomal differentiation through an Alpine hybrid zone in the grasshopper *Chorthippus parallelus*. *Journal of Evolutionary Biology*, **12**: 577–585

Frenzel, B., Pecsi, M. and Velichko, A.A. (1992). *Atlas of Paleoclimates and Paleoenvironments of the Northern Hemisphere. Late Pleistocene–Holocene*. Geographical Research Institute, Hungarian Academy of Sciences, Budapest. Gustav Fischer Verlag: Stuttgard.

Friesen, V.L., Montevecchi, W.A., Baker, A.J. *et al.* (1996). Population differentiation and evolution in the common guillemot *Uria aalge*. *Molecular Ecology*, **5**: 793–805.

Gomez-Zurita, J., Petitpierre, E. and Juan, C. (2000). Nested cladistic analysis, phylogeography and speciation in the *Timarcha goettingensis* complex (Coleoptera, Chrysomelidae). *Molecular Ecology*, **9**: 557–570.

Gravlund, P., Meldgaard, M., Paabo, S. and

Arctander, P. (1998). Polyphyletic origin of the small-bodied, high-arctic subspecies of tundra reindeer (*Rangifer tarandus*). *Molecular Phylogenetics and Evolution*, **10**: 151–159.

Gray, A.J. (1997). Climate change and the reproductive biology of higher plants. In: B. Huntley *et al.* (eds) *Past and Future Rapid Environmental Changes*. NATO ASI Series Vol. I **47**. Berlin: Springer-Verlag, pp. 371–380.

Guiot, J., Pons, A., de Beaulieu, J.L. and Reille, M. (1989). A 140,000 year continental climate reconstruction from two European pollen records. *Nature*, **338**: 309–313.

Haffer, J. (1969). Speciation in Amazonian forest birds. *Science*, **165**: 131–137.

Haffer, J. (1997). Alternative models of vertebrate speciation in Amazonia: an overview. *Biodiversity and Conservation*, **6**: 451–476.

Harrison, C. (1982). *An Atlas of the Birds of the Western Palearctic*. London: Collins.

Harrison, R.G. (1993). *Hybrid Zones and the Evolutionary Process*. New York: Oxford University Press.

Hays J.D., Imbrie J. and Shackleton N.J. (1976). Variations in the Earth's orbit: pacemaker of the ice ages. *Science*, **194**: 1121–1132.

Hellberg, M.E. (1994). Relationships between inferred levels of gene flow and geographic distance in a philopatric coral, *Balanophyllia elegans*. *Evolution*, **48**: 1829–1854.

Hewitt, G.M. (1988). Hybrid zones – natural laboratories for evolutionary studies. *Trends in Ecology and Evolution*, **3**: 158–167.

Hewitt, G.M. (1989). The subdivision of species by hybrid zones. In: D. Otte and J. Endler (eds) *Speciation and its Consequences*. Sunderland, Massachusetts: Sinauer Associates, pp. 85–110.

Hewitt, G.M. (1990). Divergence and speciation as viewed from an insect hybrid zone. *Canadian Journal of Zoology*, **68**: 1701–1715.

Hewitt, G.M. (1993). Postglacial distribution and species substructure: lessons from pollen, insects and hybrid zones. In: D.R. Lees and D. Edwards (eds) *Evolutionary Patterns and Processes*. Linnean Society Symposium Series **14**: 97–123. London: Academic Press.

Hewitt, G.M. (1996). Some genetic consequences of ice ages, and their role in divergence and speciation. *Biological Journal of the Linnean Society*, **58**: 247–276.

Hewitt G.M. (1999). Post-glacial recolonization of European Biota. *Biological Journal of the Linnean Society* **68**: 87–112.

Higgins, L.G. and Hargreaves, B. (1983). *The Butterflies of Britain and Europe*. London: Collins.

Hopkins, M.S., Ash, J., Graham, A.W. *et al.* (1993). Charcoal evidence of the spatial extent of the *Eucalyptus* woodland expansions and rainforest contractions in North Queensland during the late Pleistocene. *Journal of Biogeography*, **20**: 357–372.

Hughen, K.A., Overpeck, J.T., Peterson, L.C. and Trumbore, S. (1996). Rapid climate changes in the tropical Atlantic region during the last deglaciation. *Nature*, **380**: 51–54.

Huntley, B. (1990). European vegetation history: palaeovegetation maps from pollen data – 1300 yr BP to present. *Journal of Quaternary Science*, **5**: 103–122.

Huntley, B. and Birks, H.J.B. (1983). *An Atlas of Past and Present Pollen Maps for Europe*. Cambridge: Cambridge University Press.

Huntley, B. and Webb, T. (1988). *Vegetation History*. Dordrecht: Kluwer.

Ibrahim, K., Nichols, R.A. and Hewitt, G.M. (1996). Spatial patterns of genetic variation generated by different forms of dispersal during range expansion. *Heredity*, **77**: 282–291.

Jorgensen, S.M. and Hamrick, J.L. (1997). Biogeography and population genetics of white bark pine, *Pinus albicaulis*. *Canadian Journal of Forest Research*, **27**: 1574–1585.

Joseph, L., Moritz, C. and Hugall, A. (1995). Molecular data support vicariance as a source of diversity in rainforests. *Proceedings of the Royal Society of London B*, **260**: 177–182.

Jouzel, J., Barkov, N.I., Barnola, J.M. *et al.* (1993). Extending the Vostok ice-core record of palaeoclimate to the penultimate glacial period. *Nature*, **364**: 407–412.

Kershaw, A.P. (1986). The last two glacial—interglacial cycles from north eastern Queensland: implications for climatic change and Aboriginal burning. *Nature*, **322**: 47–49.

Kidd, M.G. and Friesen, V.L. (1998). Analysis of mechanisms of microevolutionary change in *Cepphus* guillemots using patterns of control region variation. *Evolution*, **52**: 1158–1168.

Lewis, P.O. and Crawford, D.J. (1995). Pleistocene refugium endemics exhibit greater allozymic diversity than widespread congeners in the genus *Polygonella* (Polygonaceae). *American Journal of Botany*, **82**: 141–149.

Lister, A. (1993). Patterns of evolution in Quaternary mammal lineages. *Linn. Soc. Symp. Ser.* **14**: 71–93.

Marko, P.B. (1998). Historical allopatry and the biogeography of speciation in the prosobranch snail genus *Nucella*. *Evolution*, **52**: 757–774.

McIntyre, A. and Molfino, B. (1996). Forcing of Atlantic equatorial and subpolar millennial cycles by precession. *Science*, **274**: 1867–1870.

Meyer, A. (1993). Phylogenetic relationships and evolutionary processes in East African cichlid fishes. *Trends in Ecology and Evolution*, **8**: 279–284.

Morgan, A.V. (1997). Fossil Coleoptera assemblages in the Great Lakes region of North America: past changes and future prospects. In: B. Huntley *et al.* (eds) *Past and Future Rapid Environmental Changes*. NATO ASI Series Vol. I **47**. Berlin: Springer-Verlag, pp. 129–142.

Nichols, R.A. and Hewitt, G.M. (1994). The genetic consequences of long distance dispersal during colonization. *Heredity*, **72**: 312–317.

Orr, M.R. and Smith, T.B. (1998). Ecology and speciation. *Trends in Ecology and Evolution*, **13**: 502–506.

Parker K.C., Hamrick J.L., Parker, A.J. and Stacey, E.A. (1997). Allozyme diversity in *Pinus virginiana* (Pinaceae): intraspecific and interspecific comparisons. *American Journal of Botany*, **84**: 1372–1382.

Ponel, P. (1997). The response of Coleoptera to late-Quaternary climate changes: evidence from north-east France. In: B. Huntley *et al.* (eds) *Past and Future Rapid Environmental Changes*. NATO ASI Series Vol. I **47**. Berlin: Springer-Verlag, pp. 143–151.

Prentice, I.C., Bartlein, P.J. and Webb, T. (1991). Vegetation and climate change in eastern North America since the last glacial maximum. *Ecology*, **72**: 2038–2056.

Remington, C.L. (1968). Suture-zones of hybrid interaction between recently joined biotas. *Evolutionary Biology*, **2**: 321–428.

Ritchie, J.C. and MacDonald, G.M. (1986). The patterns of post-glacial spread of white spruce. *Journal of Biogeography*, **13**: 527–540.

Roy, M.S. (1997). Recent diversification in African greenbuls (Pycnonotidae: *Andropadus*) supports a montane speciation model. *Proceedings of the Royal Society of London B*, **264**: 1337–1344.

Sakai, K. and Peltier, W.R. (1996). A multibasin reduced model of the global thermohaline circulation: paleoceanographic analyses of the origins of ice-age climate variability. *Journal of Geophysical Research*, **101**: (C10) 22535–22562.

Schluter, D. (1998). Ecological causes of speciation. In: D.J. Howard and S.H. Berlocher (eds) *Endless Forms: Species and Speciation*. Oxford: Oxford University Press, pp. 114–129.

Schneider, C.J., Cunningham, M. and Moritz, C. (1998). Comparative phylogeography and the history of endemic vertebrates in the Wet Tropics rainforests of Australia. *Molecular Ecology*, **7**: 487–498.

Schulz, A., von Rad, V. and Erlenkeuser, H. (1998). Correlation between Arabian Sea and Greenland climatic oscillations of the past 110,000 years. *Nature*, **393**: 54–57.

Shackleton, N.J. and Imbrie, J. (1990). The $\delta^{18}O$ spectrum of oceanic deepwater over a five-decade band. *Climatic Change*, **16**: 217–230.

Showers, W.B. (1981). Geographic vaiation of the dispause response in the European Corn Borer. In: R.F. Denno and H. Dingle (eds) *Insect Life History Patterns*. New York: Springer-Verlag, pp. 97–111.

Soltis, D.E., Gitzendanner, M.A., Strenge, D.D. and Soltis, P.S. (1997). Chloroplast DNA intraspecific phylogeography of plants from the Pacific Northwest of North America. *Plant Systematics and Evolution*, **206**: 353–373.

Stauffer, B. (1999). Cornucopia of ice core results. *Nature*, **399**: 412–413.

Taberlet, P., Fumagalli, L., Wust-Saucy, A.G. and Cossons, J.-F. (1998). Comparative phylogeography and postglacial colonization routes in Europe. *Molecular Ecology*, **7**: 453–464.

Talbot, S.L. and Shields, G.F. (1996). Phylogeography of Brown Bears (*Ursos arctos*) of Alaska and paraphyly within the Ursidae. *Molecular Phylogenetics and Evolution*, **5**: 477–494.

Tooley, M.J. (1993). Long term changes in eustatic sea level. In: R.A. Warwick, E.M. Barrow and T.M.L. Wigley (eds) *Climate and Sea Level Change: Observations, Projections and Implications*. Cambridge: Cambridge University Press, pp. 81–107.

Tzedakis, P.C. (1994). Vegetation change through glacial–interglacial cycles: a long pollen perspective. *Philosophical Transactions of the Royal Society of London Series B*, **345**: 403–432.

Tzedakis, P.C. and Bennett, K.D. (1995). Interglacial vegetation succession: a view from southern Europe. *Quaternary Science Reviews*, **14**: 967–982.

Vainola, R. (1995). Origins and recent endemic divergence of a Caspian *Mysis* species flock with affinities to the 'glacial relict' crustaceans in Boreal lakes. *Evolution*, **49**: 1215–1223.

Webb, T. (1988). Eastern North America. In B. Huntley and T. Webb (eds) *Vegetation History*. Dordrecht: Kluwer Academic, pp. 388–414.

Webb, T. and Bartlein, P.J. (1992). Global changes

during the last 3 million years: climatic controls and biotic responses. *Annual Reviews of Ecology and Systematics*, **23**: 141–173.

Wenink, P.W., Baker, A.J., Rosner, H.-U. and Tilanus, M.G.J. (1996). Global mitochondrial DNA phylogeography of holarctic breeding dunlins (*Calidris alpina*). *Evolution*, **50**: 318–330.

Williams, D., Dunkerley, D., DeDeckker, P. *et al.* (1998). *Quaternary Environments*. London: Arnold.

Wilson, C.C. and Hebert, P.D.N. (1996). Phylogeographic origins of lake trout (*Salvelinus namaycush*) in eastern North America. *Canadian Journal of Fisheries and Aquatic Sciences*, **53**: 2764–2775.

19

Environmental variability and its impact on adaptive evolution, with special reference to human origins

Richard Potts

ABSTRACT

See Plates 21, 22 and 30

The history of life on Earth has been powerfully influenced by environmental change. Oxygen enrichment of the atmosphere, break-up of continental masses, and development of global ice-house conditions are examples of long-term shifts that have impacted the survival of organisms. Enduring trends shape the habitats and evolutionary opportunities available to living things. The continual molding of adaptations via directional selection in relation to environmental trends and specific habitats is thus one of the fundamental principles of evolutionary biology.

Environmental variability, however, also demands our attention. Short-term fluctuation and habitat perturbations are an ever-present dimension of environmental change, and may have far-reaching evolutionary consequences.

Habitat variability presents an important adaptive question. How have organisms accommodated through genetic evolution to unstable conditions? The effect of environmental variability on evolution, especially when a species faces a widening range of adaptive settings over time, is not well understood. To shed light on the problem, this chapter explores climatic, tectonic, and volcanological sources of environmental variability during the late Cenozoic, and illustrates the evolutionary paths by which terrestrial organisms have responded to novel survival conditions. This chapter considers as its main example how habitat variability may have shaped human evolution and presents the case that key human traits evolved against a background of strong inconsistencies in adaptive setting.

19.1 Sources of Cenozoic environmental change

Abiotic or physical factors have exerted a strong influence on biotic trends throughout the past 65 million years. The dominant terrestrial environmental themes of this period, the Cenozoic, were climatic cooling, drying, and oscillation. The factors underlying these changes include: change in solar heating due to cyclical variations in Earth's orbit relative to the Sun; interactions between the atmosphere and surface (e.g. heat exchange, evaporation); patterns of air and water circulation; movement of tectonic plates, which helped determine the size, shape, and location of continental masses and ocean basins; and interactions between the planet's internal properties and its surface. Moreover,

Evolution on Planet Earth
ISBN 0-12-598655-6

as in other eras, the biota of the Cenozoic has had a significant impact on atmospheric composition, surface (e.g. soil) characteristics, and other aspects of the physical planet.

The most notable developments in late Cenozoic physical environments and related changes in the terrestrial biota are shown in Figure 19.1. Since certain physical processes, such as orbital variations, operated on a global scale, the most prominent terrestrial biotic trends of the late Cenozoic were manifested in parallel on separate landmasses (Potts and Behrensmeyer, 1992). Climate, tectonic movements, and volcanism each contributed in different ways to the main themes of Cenozoic environmental history. The following sections will inspect these three physical sources of

change and assess their collective impact on the land biota.

19.1.1 Climate

Periodic oscillations have characterized the late Cenozoic climate record. These oscillations are generally attributed to variation in solar heating (insolation) related to astronomical cycles that gradually alter the shape and orientation of Earth's orbit around the Sun. The cycles (Milankovitch cycles) arise from the changing gravitational pull of the other planets and bring about periods of 19, 23, 41, and 100 thousand years in the strength and distribution of solar radiation over the globe (Hewitt, this volume, Chapter 18). Environmental oscillations of

Global Change in the Late Cenozoic

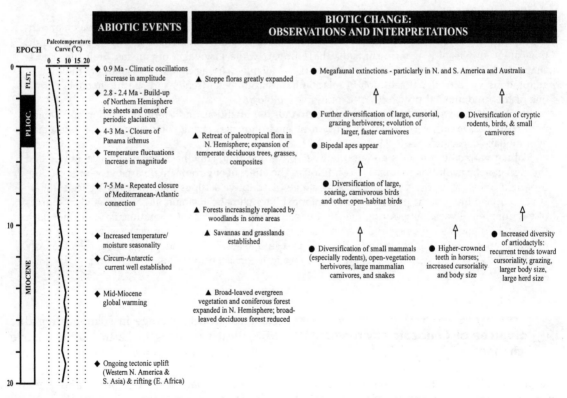

Figure 19.1 Summary of major abiotic events and biotic evolutionary change over the past 20 million years, from early Miocene through Pleistocene. Paleotemperature curve is based on composite records from Deep Sea Drilling Project sites in the Atlantic, with 0°C referring to present average temperature. Diamond symbols indicate abiotic change, while triangles and dots indicate change in flora and fauna, respectively. Modified from Potts and Behrensmeyer (1992).

these durations are very well seen in the deep-sea record of oxygen isotopes ($\delta^{18}O$), a measure of ocean temperature, evaporation, and global ice volume.

A significant change in the frequency and amplitude of $\delta^{18}O$ oscillation occurred between 3.0 and 2.5 Ma, corresponding to the onset of Northern Hemisphere glaciation and greater aridity in the tropics. Another major shift occurred between 900 000 and 600 000 years ago, associated with the development of 100 000-year-long cycles of glacial advance and interglacial warming. Progressive climatic trends, such as cooling and drying, were thus inextricably linked with changing patterns of climatic oscillation, which were largely determined by variations in Earth's orbit.

Besides the 100 000-year period of glacial growth and retreat, several other prominent cycles have been recognized in the paleoclimate record of the past 1 million years. These include the 23 000-year precessional cycle of intensified tropical African monsoons (related to periodic fluctuation in Earth's rotational axis), and millennial-scale instability related to iceberg discharge in the North Atlantic (Rossignol-Strick, 1983; Shackleton *et al.*, 1984; deMenocal *et al.*, 1993). The second of these phenomena – which involves massive ice rafting, melting, and decreased ocean salinity – has been linked to rapid (decade- to century-scale) fluctuations between interglacial warmth and near-glacial conditions (Taylor *et al.*, 1993; Bond *et al.*, 1997). Well documented during the late Pleistocene and early Holocene, such dramatic 'flickering' of climatic conditions may be traced as far back as 0.5 to 1.0 million years ago (Raymo *et al.*, 1998; McManus *et al.*, 1999). These rapid fluctuations appear to have been confined to the North Atlantic, however, as continental records from Eurasia, Africa, and Antarctica exhibit substantially longer periods between large-scale climatic shifts.

Although climatic fluctuations originate from well-defined cycles, these cycles are numerous and their interactions may amplify or buffer one another, causing a very complex temporal and geographical pattern of fluctuation (see, e.g., Clemens *et al.*, 1996; Paytan *et al.*, 1996). Because of an array of time- and place-specific physical factors (insolation, sea level, atmospheric CO_2 concentration, sea surface temperatures, ice cover, etc.), major environmental shifts occurred episodically throughout the Quaternary and did not necessarily coincide exactly with maximum changes in temperature, moisture, or any other single factor.

Climatic variations during the late Cenozoic had one other major effect, namely, the rise and fall of sea level. Fluctuations of up to 120 meters occurred over the past 600 000 years (Lister and Rawson, this volume, Chapter 16). Continental shelves were exposed, creating land bridges and opportunities for the dispersal of colonizing species in times of glacial ice growth, followed by inundation during melting. These oscillations also caused the expansion and contraction of near-shore vegetation belts. In Southeast Asia, for example, sea-level change caused large-scale fluctuation in the area devoted to lowland rainforest, a biome usually deemed to be a product of long-term environmental stability (Flenley, 1979).

19.1.2 *Tectonic movements*

Three main types of tectonic activity affected late Cenozoic terrestrial habitats: continental plate movement, uplift causing the formation of highlands, and faulting within local sedimentary basins.

Although the major landmasses reached their approximate current positions during the Miocene, the movement of continental plates has been continuous (Forster, this volume, Chapter 15). Consider the impact of the following examples. First, increasing contact between the Afro-Arabian and Eurasian continental plates between about 16 and 12 million years ago led to significant exchange of animals and plants, including the spread of early apes from Africa and of equids into Africa. Second, between 7 and 5 million years ago, as Africa drifted northward, tectonic movements caused the temporary closure of the western end of the Mediterranean from the Atlantic Ocean. Subsequent raising and lowering of sea level (due to glacial fluctuations) resulted in periodic flooding and drying of the Mediterranean basin and the build-up of huge salinity deposits. These episodes of wide-spread evaporation and salinization had a major drying influence on western Eurasia. As a final example, contact between continental plates caused uplift and the formation of the Isthmus of Panama between 4.5 and 3 million years ago. One effect was the strengthening of the 'conveyor belt' of Atlantic Ocean currents, which enhances the flow of warm Gulf Stream water into the North Atlantic and its return to the south. In the present, warmer, high-salinity North Atlantic water sinks and then spreads southward again. Thus the deep-water conveyor belt largely isolates the Arctic Ocean and prevents its warming. Its development over 3 million years ago is one of several factors that helps explain the initiation of ice ages in the Northern Hemisphere during the late Pliocene (Stanley, 1995).

Substantial uplift of large land areas has occurred during the Cenozoic. Since the early Miocene, uplift of the Tibetan Plateau and western North America (from the Sierra to the Rocky Mountains), and rift-valley formation in eastern Africa, all had cooling and drying effects on Eurasia, the Americas, and Africa. General circulation models, which describe atmospheric flow in terms of momentum, energy and heat, show that the presence of elevated plateaus would have diverted major air currents, altered the summer heating and winter cooling of air over the plateaus, and caused high and low pressure areas to form over landmasses far from the uplifted zones (Ruddiman and Kutzbach, 1989; Ruddiman et al., 1989). The ultimate effect was the creation of seasonal monsoons and greater seasonal variation (both wet/dry and warm/cool), which led to an overall drier and cooler global climate.

Due to enhanced monsoons, steepening river gradients, and faster erosional rates, uplift also promoted the geochemical weathering of rocks. It has been suggested that the weathering reactions and deposition of carbon from the highlands into the ocean would have reduced the amount of carbon dioxide available to atmospheric circulation. Since carbon dioxide is a greenhouse gas, which enables the atmosphere to retain heat, its removal from the atmosphere would have raised the

potential for global cooling (Raymo and Ruddiman, 1992). A more confined effect of tectonic uplift is the development of rainshadow drying, which results from upland precipitation and depletion of air moisture on the leeward side of the uplifted zone. This factor was partly responsible for aridification of the African rift valleys compared with terrain to the west.

A final tectonic source of environmental change, albeit even more localized, involved earthquakes that altered the shape of sedimentary basins. Throughout recent geological time, intrabasin faulting changed the size and form of lakes and fluvial systems, which governed the distribution of water and its availability to the local biota. Study of sedimentary basins in Africa and other parts of the world indicate that this factor caused substantial disruption to late Cenozoic landscapes.

19.1.3 Volcanism

Volcanic activity has had its own distinctive effect on the atmosphere, oceans, and terrestrial landscapes. While eruptions usually have a fairly localized influence, lava and tephra (ejecta such as ash and pumice) may cover hundreds to thousands of square kilometers, blanketing the landscape, temporarily eliminating certain types of habitat, and changing soil and water chemistry. Environmental change resulting from volcanism is neither continuous nor predictable with respect to an organism's lifetime. A tabulation of eruptive activity recorded in the Turkana area of northern Kenya, for example, indicates a range from 1 to 38 events per 240 000 years, the higher rate suggesting an average of one significant eruption per 6300 years (Feibel, 1999).

Volcanic eruptions can immediately change the environment in ways that may be greatly expanded by longer-term physical factors such as climate. To give one example, extensive forests of northwest North America were periodically disrupted by volcanic output during the Middle Miocene. The barren, post-disturbance landscape gradually supported patches of grasses that could survive arid, barren terrain until they were later replaced by new forest growth. Cycles of volcanic disturbance and forest regrowth continued. But over the long term, as climate became more seasonal and dry, reforestation was stalled, and the grass-dominated patches ultimately spread throughout the region to form a climax grassland community (Taggart *et al.*, 1982).

Massive eruptions may also have a wider, essentially global impact by glutting the atmosphere with particulate matter known as aerosols. Although the idea is controversial, it has even been suggested that the explosion of Mount Toba in Sumatra over 70 000 years ago, ten times bigger than any historical eruption, lofted billions of tons of matter into the stratosphere, causing sharp cooling and the shift to glacial conditions at that time (Rampino and Self, 1992).

From these examples we see that volcanic activity largely enhanced, on a local to global scale, the cooling and drying trends of the late Cenozoic and contributed to environmental instability.

19.2 Environmental change and evolution of the late Cenozoic biota

Terrestrial vegetation was modified during the late Cenozoic in two main ways: (i) from tropical to temperate latitudes, the spread of low-biomass plants, especially grasses and herbaceous species, and the development of large, continuous tracts of open vegetation (known by various terms such as prairies, grasslands, savannas, and steppes); and (ii) in the temperate to high latitudes, the spread of deciduous and coniferous trees (Figure 19.1). Between 8 and 6 million years ago, a global expansion in C_4 photosynthesis (a chemical pathway involving four carbon atoms) favored grasses and other plants found today in hot, dry conditions. Cerling *et al.* (1997) documented this global expansion and associated it with a decrease in atmospheric CO_2 concentrations below a critical level. The development of seasonal monsoons and greater seasonal variation (both wet/dry and warm/cool), moreover, favored the evolution of drought- and cold-

resistant plants. As a result, the distribution of Cenozoic forests became less widespread and continuous, and seasonally dry forest and open vegetation replaced moist rainforest in many areas.

These changes in vegetation were mirrored by the following evolutionary trends in land-based animals (see Janis, this volume, Chapter 17): an increase of large-bodied herbivores with rapid locomotion (specialized running or hopping adaptations in open terrain) and high-crowned teeth (able to process large amounts of low-quality grass and browse); expansion of small herbivores specializing in cryptic behavior (e.g. underground burrowing) in open terrain; and diversification of large predators with special locomotor and social adaptations for capturing speedy herbivores, and smaller carnivores specializing in capturing cryptic prey (Potts and Behrensmeyer, 1992).

The radiation of grasses and herbaceous species, expansion of habitats typified by such plants, and related changes in animals, directly reflected the major physical environmental trends of the late Cenozoic – namely, aridification and cooling based on tectonic uplift, episodic volcanism, and an increase in climatic seasonality and the amplitude of longer-term fluctuation.

In most groups of terrestrial organisms, examples can also be found of species whose adaptations reflect the temporal and spatial variation of environments. Omnivorous species of pigs, carnivores, and primates, which diversified since the Miocene, are generally capable of adjusting their diet seasonally. Larger habitat fluctuation, evident in the complex alternation between glacial and interglacial conditions, also left its mark on the biota. Different species manifest, for example, different rates of movement and degrees of accommodation to environmental change. As a result, significant lag effects occurred in ecological communities as plant species migrated at significantly different rates in response to climate fluctuation (Davis, 1985; Huntley and Webb, 1989), and heterogeneous assemblages of small and large mammalian species also developed as environments changed during the late Quaternary (Graham *et al.*, 1996; Potts and Deino, 1995). Thus plant and animal populations faced a shifting array of competitors, predators, symbiotes, and parasites, resulting in diverse selective environments over time and space.

Among the large-bodied apes in Africa, a moderately diverse group of bipedal species emerged during the Pliocene and then spread widely through the developing diversity of wooded, open, cold, and warm habitats of Africa, Eurasia, and eventually other landmasses. There is currently no consensus about how to explain the origin and evolution of the hominins (Hominini: the taxonomic group of humans and other bipedal apes). Their evolutionary history is, nonetheless, notable, not only because the reader and writer belong to one of its species but also because this lone surviving species has begun to significantly alter the biota and the physical factors (e.g. atmospheric gases) that have so strongly influenced Earth's environmental history over the past several million years.

19.2.1 *Escalation of environmental variability since the late Miocene*

A great deal is known from geochemical, sedimentary, and paleontological evidence about the tempo and range of Cenozoic environmental variability. Oxygen isotope and terrestrial dust records derived from deep-sea cores indicate a substantial rise in the amplitudes of temperature, ice volume, and moisture variation (Shackleton *et al.*, 1984; deMenocal, 1995). Over the past 5 million years, in particular, the amplitude of oscillation in these variables has increased by a factor of 2.

The ratio of oxygen isotopes ^{18}O and ^{16}O, called $\delta^{18}O$, can be measured in benthic (seafloor) foraminifera, which are microscopic organisms with calcium carbonate skeletons. This ratio is primarily governed by two variables – ocean temperature and global ice volume (the lighter isotope, ^{16}O, is preferentially evaporated from the ocean and captured in glacial ice). As depicted in Figure 19.2, a marked change occurred in the total amount of $\delta^{18}O$ variation per million years from the late Oligocene to the present. The range of oxygen isotope variation increased slightly during

the mid-Miocene followed by a decrease 8 to 6 million years ago. Climatic variability rose sharply, however, in the 5-to-6-million-year interval and continued to rise between 4 million years ago and the present. Escalation of environmental variability over time, with an especially marked rise around 0.7 million years ago, is also evidenced in long-term pollen records (Kukla, 1989) and the alternating loess–soil sequence of north–central China (Liu *et al.*, 1999). Variability was manifested by glacial and interglacial oscillations, and intermittent forest/grassland/desert expansions and contractions.

Thus from the seasonal time frame experienced by individual organisms to the longer scale of 10^5 to 10^6 years over which a lineage evolves, an increasing diversity of habitats and selective environments has been manifested. This increase in the variability of adap-

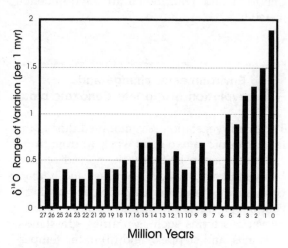

Figure 19.2 Total amount of variation in the oxygen isotope ratio ($\delta^{18}O$) for each million year interval from 27 million years to present (Potts, 1998a). The variation is expressed in parts per million, the units shown on the vertical axis. $\delta^{18}O$ of benthic (bottom-dwelling) foraminifera is a measure of ocean temperature and the amount of evaporated water tied up in glacial ice. The range of variation exceeded 0.5 parts per million only five times between 27 and 6 million years ago, and never exceeded 0.75 parts per million. A tremendous increase in total $\delta^{18}O$ variation began to occur 5 to 6 million years ago. The past 1 million years was the interval of widest recorded variation in the Cenozoic.

tive settings over time may have provided an important context for evolutionary change.

19.2.2 Environmental variability and evolutionary effects in the southern Kenyan rift

The pattern of extinction and faunal succession in the rift valley of southern Kenya provides a specific example of how Quaternary variability impinged on mammalian evolution (Potts and Deino, 1995; Potts, 1996a). Certain species of zebra, elephant, baboon, suid, and hippopotamus dominated the rift valley fauna for more than 1 million years. These lineages, however, became extinct in southern Kenya between 800 000 and 400 000 years ago. All of these species were unusually large-bodied with specialized craniodental morphologies related to grazing. These grazing specialists were replaced by the modern representatives in each of these groups, all of which were either previously rare or not even evident in the fossil record. These extant species exhibit smaller bodies and wider habitat ranges than the lineages they replaced, and can also alter their social group size and diets (e.g. proportion of grass and browse) depending on local feeding conditions. The period of replacement, between 800 000 and 400 000 years ago, coincided with large environmental change.

The sedimentary sequence in the Olorgesailie basin of southern Kenya furnishes a superb record of environmental events, such as lake expansion or disappearance, major volcanic disruption of the basin, and change between intervals of landscape erosion, deposition, and stability (e.g. soil formation). The frequency of these basinwide environmental events ranged between 1 major shift per 3600 years (the fastest rate) and 1 per 65 000 years (the slowest) in different parts of the stratigraphic record between 990 000 and 490 000 years ago. Major remodeling occurred between periods of relative landscape stability, lasting up to 150 000 years. Similar rates of change have been recorded in the Pliocene and early Pleistocene deposits of other East African localities, such as Olduvai in northern Tanzania, and the Turkana basin in northern Kenya and

southern Ethiopia (Potts, 1998b). These wide alterations in regional hydrology, vegetation, and landscape were the result of climatic variation, local basin faulting, and episodic volcanism.

It is against this backdrop of large-scale environmental change that extinction of the specialized grazers and succession of the more versatile taxa are recorded in the southern Kenya rift valley. Although, paleontologically, we observe only the coming and going of lineages and the correlated record of paleoenvironments, it is possible that the emergence of versatile adaptations in the modern lineages was a direct response to environmental variability.

19.3 Does environmental change drive biological evolution?

In a series of articles, Elisabeth Vrba (e.g. (1980, 1985, 1988, 1995a) proposed that physical environmental change has initiated evolutionary change at various points in Earth's history. Her primary example concerns the effect of global climate on species turnover (i.e. originations and extinctions) during the late Pliocene. According to Vrba's 'turnover pulse' hypothesis, the onset of global cooling and high-latitude glaciation between 2.8 and 2.4 million years ago caused change in vegetation and animal communities in lower latitudes. A well-defined period of turnover in African mammals around 2.5 million years ago coincided with the climatic shift and favored savanna/arid-adapted lineages at the expense of forest/mesic-adapted ones. This 'pulse' of change is said to be seen in the fossil records of such diverse mammalian groups as bovids, rodents, suids, cercopithecoid monkeys, and hominins.

Testing this hypothesis has rested heavily on temporal correlation – that is, synchronization of certain changes in climate, vegetation, and animal species' first and last appearances in the geological record. It is unclear, however, how precise a particular temporal correlation must be to imply a causal connection between physical environmental and evolutionary change.

In a more recent development of the idea, termed 'habitat theory', Vrba (1992, 1995a) notes that there may be a temporal lag in the response of organisms to climate change. Specialized (or stenotopic) organisms, which have a narrow ecological tolerance (e.g. exclusive browsers or grazers), are likely to be affected first by a sharp climatic shift while generalist (or eurytopic) species will be influenced later. With regard to the Pliocene cooling event, Vrba (1995a) predicts that while environmental cooling began 2.8 Ma, its effect on African mammals was registered between 2.7 and 2.5 Ma. But if the processes that link physical and biotic change may play out over hundreds of thousands of years, it becomes difficult to distinguish the potentially diverse influences of environment on evolution. Deep-sea data on oxygen isotopes and continental dust point to an increase in the amplitude of environmental oscillation during the critical period (2.8 to 2.5 Ma) (Prentice and Denton, 1988; deMenocal, 1995). Thus it is difficult to know whether the principal environmental factor affecting evolution involved the progressive shift (cooling/drying) or the rise in environmental fluctuation.

The key to the turnover pulse hypothesis is that climate is said to impose an effect on a wide variety of organisms. The impact of climate change is expected in many different lineages of organisms within a narrow time span. However, certain diverse groups of organisms, such as the bovids (Vrba, 1995b), may show a strong correspondence to the turnover pulse pattern while others groups, such as the suids, which were equally widespread and diverse, may provide no support for it (White, 1995; Bishop, 1999).

Tests between and within African basins, furthermore, show varying results. According to Behrensmeyer et al. (1997), no distinct turnover pulse occurred across mammalian taxa in the Turkana basin during the late Pliocene; rather, turnover was sustained over 700 000 years or more, and did not involve wholesale replacement of woodland-forest species by open vegetation and arid-adapted animals. Vrba believes that this finding is not representative of the whole of Africa. Although it offers

the richest and best calibrated record of African fossil mammals between 3 and 1.8 Ma, the Turkana basin, according to Vrba, may represent a refugium that was largely buffered (due to the presence of a large, persistent river system) from global climatic change to which the rest of the continent was far more sensitive. In further analysis, Behrensmeyer and Bobe (1999; Bobe, 1999) point out that a relatively abrupt shift in the abundances of bovid, suid, and primate species related to drier habitats did occur in the northern Turkana basin (i.e. the Omo valley of southern Ethiopia) about 2.8 Ma. This shift, however, was not evident throughout the entire basin and need not imply substantial species turnover or evolutionary change within lineages.

This debate underlines the potentially varied responses of organisms and assemblages of organisms in different regions to an altered physical environment. The question whether environmental change drives evolutionary events, therefore, must consider the various meanings of 'environmental change' and the diverse conditions that underlie biological evolution. Physical, biotic, and selective environments may change somewhat independently of one another. Not all shifts in the physical environment may cause a change in the conditions that affect reproductive success and survival in a given organism (selective environment). An organism may be mobile or self-regulating in a way that essentially stabilizes the selective milieu when external conditions are altered. Furthermore, alteration of the biotic environment – e.g. introduction of a novel pathogen or demographic change affecting food competition – may occur independently of change in the physical setting. While adaptive change may usually be associated with an alteration in the biotic or physical environment, modification of either one is not sufficient to cause adaptive change or speciation.

Evolutionary change may take several forms, including speciation and extinction (turnover) and adaptive modification. Alteration in the physical environment may influence these aspects of evolution. But whether change occurs at all depends on the presence of viable genetic variations, the geographic spread and subdivision of species populations, generation time (or reproductive rate) relative to the tempo of environmental change, resource distributions on which populations rely, among other variables.

Climatic and tectonic factors can either stabilize or destabilize the conditions under which organisms survive, reproduce, associate with populations of other species, and under which populations expand, contract, subdivide, and become isolated. Therefore, it is a truism of evolutionary biology that habitat change may be an important or even necessary condition for evolutionary change – but is not sufficient by itself, nor does it specify the exact type of evolutionary response.

19.4 Evolutionary responses to environmental variability

Field biologists and other patient observers of the natural world have long drawn attention to the fact that organisms are well matched to particular environments. In general, organisms live in a specific range of temperature, rainfall, or other physical settings and rely on the availability of certain types of food. By natural selection, each species becomes linked to the consistent properties of its habitat over many generations. Complex adaptations are generally considered to result from the consonant effect of directional selection – i.e. the statisti-

cally consistent interaction between organism and its local surroundings over time (e.g. Eldredge, 1989), often in the form of evolutionary 'arms races' (see Dawkins, 1987: 178–192). Thus adaptive evolution is typically linked to either a progressively changing or constant environment (Tooby and Cosmides, 1993).

But how do organisms cope with and respond to large disparities in adaptive setting over time and space? Table 19.1 summarizes three possible responses. Extinction has occurred throughout the history of life. During the Pleistocene, the background level of extinction among large mammals rose substantially as the extent of environmental variability widened. The rich species diversity of Pliocene elephants, pigs, horses, and hominins was ultimately reduced by the late Pleistocene to one or two lineages in any given geographical region (Martin and Klein, 1984; Potts, 1996a). There has been vigorous debate about the relative impact of environmental change versus human hunting, but the fact remains that the Pleistocene extinctions occurred against a backdrop of rising climatic oscillation and environmental remodeling. This suggests that many organisms were unable to adapt to the pace and degree of change – and may also have been affected by the responses of other species, including human hunters, to these highly variable settings.

A second response involves tracking a particular environmental zone or essential resource as it moves geographically. As climate

Table 19.1 Responses by organisms to large-scale variability in their adaptive settings.

1. Extinction: May result from an inability to adapt to habitat change.

2. Mobility: Allows organisms to track a favored resource or habitat.

3. Flexibility: Allows organisms to change in response to habitat change:
   (a) polymorphism: the presence of alternative alleles permits different phenotypes to be selected in different environments;
   (b) phenotypic plasticity: different phenotypes are produced by a single genotype in a specific range of settings;
   (c) adaptive tolerance: physiological or behavioral flexibility enhances fitness in an expected range of environments;
   (d) adaptive versatility: a particular anatomical structure, behavior, or complex of traits enhances fitness in novel settings.

shifts, the many species that comprise a given ecological community become displaced; but due to different rates of movement and the development of non-analogue environments, rather different biotic communities may be reassembled. Environmental variability thus results in a complexly varying sequence of species interactions and adaptive environments. It is possible, and evidently quite common, for an organism to track a fairly narrow climatic zone or essential resource and thus stabilize its adaptive setting. Indeed, most species are quite specific in their tolerances and manifest specific dietary and habitat adaptations. However, organisms that have become reliant on a broader range of resources may have a more difficult time stabilizing their adaptive surroundings in this way, since the full range of resources would seldom reassemble in exactly the same way from one environmental fluctuation to another.

A third response – the evolution of flexibility – enables a population to persist through a wide spectrum of environments. At least four types of adaptive flexibility may be defined: (i) genetic polymorphism in which alternative alleles leading to different phenotypes are selected in different environments; (ii) phenotypic plasticity in which a single genotype yields a predictable range of phenotypes in a particular range of environments; (iii) behavioral or physiological flexibility in which a single genotype and phenotype enable a certain degree of environmental tolerance, switching between alternative diets, or any similar adaptability to an expected range of environments; and (iv) adaptive versatility in which a specific structure, physiological process, or behavior is designed to enhance fitness in novel settings and survivorship in periods of unpredictable change.

Evolution of the first three types of flexibility appears to conform to the standard principles of natural selection, whereby organisms encounter a predictable range of environmental fluctuation one generation after another. How an organism may adapt to environmental novelty and unpredictable change (the fourth in the above list) is much less well understood. If such a process exists, it would seem to vio-

late the tenet that adaptive change requires long-term consistency or direction in the selective environment. The term *variability selection* has been proposed to help define the process by which inconsistent adaptive settings, created by a highly variable temporal sequence (with intervening periods of stability), may help generate adaptive versatility over many generations (Potts, 1996a, b, 1998a, b).

The basis of this hypothetical process lies not only in the physical factors that enhance environmental variability but also in the potential for a population to accumulate genes that promote fitness in novel habitats. Variability selection, as currently conceived, operates on two levels (Potts, 1998a). First is the mendelian, where genes are preserved within populations and adaptations originate. Any genetic advantage that helps an individual or kin group cope with environmental dynamics over small time frames may be retained until more major shifts arise, including radically remodeled adaptive settings. At longer timescales, environmental change becomes more episodic and largely unpredictable but is nonetheless a filter that sorts genetic variations according to how well they promote adaptive versatility. Due to the inconsistency in selective environment, therefore, certain genes and developmental patterns may be retained in a way that builds adaptations to changing circumstances rather than to specific environments.

The results are adaptive specializations that enhance the use of environmental information, favor novel responses to the surroundings, and expand the options available to the organism. These advantages may take several forms, such as cognitive functions that depend on external data or promote innovative behavior. As a consequence, an organism can become disassociated from any single specific habitat. The variability selection hypothesis does not specify how an organism *must* respond to environmental variability; rather, it suggests an alternative to the other possible outcomes noted in Table 19.1.

The second level of variability selection involves selection or sorting of lineages. Populations or species will differ in how well they survive environmental variability on the basis of their genetic and phenotypic foundation for versatility. The example of large mammalian herbivores in the southern Kenya rift illustrates a possible case of lineage-level variability selection in which the extinction or success of species is governed by a highly variable sequence of environments.

Although plausible mechanisms of variability selection are not well established or well tested, the fact remains that physical factors have caused wide variability in the adaptive settings of organisms. Survival in the planet's biota has very likely been influenced by the inconsistency of environments. The variability selection idea proposes that this inconsistency has been important in shaping the adaptations that have evolved and persisted over time.

19.5 Human evolution

While most species are tightly bound to a particular ancestral environment, others are more independent and have become decoupled from any one specific habitat. According to the variability selection hypothesis, this decoupling occurs due to an evolutionary past of large habitat disparity, which results in the evolved ability to respond effectively to instability and settings never previously encountered. The spread of the genus *Homo*, particularly *H. sapiens*, and its unusual capacity to develop novel behaviors strongly hint of an adaptive process that led to considerable independence from specific ancestral habitats (especially African savanna or northern ice-age settings) that have dominated previous scenarios of human evolutionary history (Potts, 1996a, b, 1998a, b).

The variability selection hypothesis presents two testable issues. The first considers whether past environmental variability has been extensive and, if so, whether it led to wide disparities in an organism's selective environment. The second is whether the characteristics that emerged over time furthered that organism's ability to adapt to novel environmental settings.

Regarding the period of human evolution, the past 6 million years saw a marked increase in the range of environmental oscillation (Figure 19.2). Physical and biotic settings were periodically recast in ways that substantially altered the amount and distribution of water, food, and other resources available to hominins and other organisms. These perturbations were registered in the geological record typically every several thousand to tens of thousands of years. Between these times of environmental transformation were periods of relative stability (e.g. river channel filling or continuous soil formation) that usually spanned many thousands of years in any given region. Individuals lived, died, and reproduced in a specific range of habitats, but the genetic lineages to which these individuals belonged encountered perturbations and adaptive settings of far greater diversity. The evolutionary environments of early humans, therefore, were highly variable, a property of the environment that challenged the survival of populations and helped shape the adaptive properties of long-lived lineages.

A further set of observations indicates that the evolving features of early hominins were, in fact, shaped by the inconsistency of selective environments. Figure 19.3 is a list of ten evolutionary modifications with brief explanations of how each one potentially amplified the adaptive versatility of early hominins. This list suggests that human evolution involved a sequence of heightened responses to environmental instability. The first response (at the bottom of the list) involved an alteration in how the earliest hominins moved, while later responses entailed expanded cognitive and social functions.

There has long been debate about the locomotion of the earliest australopiths, particularly *Australopithecus afarensis* (commonly represented by the skeleton known as 'Lucy'). One camp of researchers has argued that *A. afarensis*, with its broad, humanlike pelvis and angled knee joint, was a very competent biped that walked at least as efficiently as modern humans (e.g. Lovejoy *et al.*, 1973; Ohman *et al.*, 1997). The dissenting camp has focused on the long, powerful arm bones of Lucy's skele-

TIME PERIOD	EVOLUTIONARY CHANGE	POTENTIAL ADAPTIVE BENEFITS
?200 to 20 ka	Complex cultural & technological diversification	Expanded range of adaptive options at the species level
?400 to 20 ka	Enhanced symbolic behavior	Enhanced capacity to plan & communicate novel ideas
?500 to 100 ka	Complex spatial mapping, including long-distance movement & exchange of resources	Larger store of environmental information; reduced risk during episodic habitat change
?700 to 100 ka	Intensive resource reciprocity, including enhanced food-sharing & related home-base behaviors	Enhanced social memory & cognition; buffering of risk
700 to 100 ka **	Encephalization & enhanced mental functions	Plasticity of neural circuitry; better processing of external data
1.7 to 0.4 Ma	Enhanced technical & associated cognitive skills, including control of fire	Increased ability to use the environment & respond to varying habitats
2.5 to 1.5 Ma	Extensive carrying of stone & food, including parts of large animals	Accommodation to changing food distributions; harbinger of geographical spreading
3 to 2 Ma	Stone flaking; clustering of tools	Access to new foods - - e.g., meat, marrow, underground plants
5 to 2 Ma	Access to new dietary resources	A means of buffering change in food availability
5 to 2 Ma *	Locomotor skeleton enabling terrestrial bipedality & arboreal climbing	Versatile movement as settings varied between wooded & open vegetation

* Ma: millions of years ago
** ka: thousands of years ago

Figure 19.3 Ten major developments in human evolution (Evolutionary Change), the main time period when these developments took place (read from bottom to top), and a brief summary of the potential adaptive benefits in the context of environmental instability. Question marks indicate uncertainty as to the exact date when the change was first manifested.

ton and the presence of long toe and finger bones, indicative of competent tree climbing (e.g. Stern and Susman, 1983; Susman and Stern, 1991). Paleoenvironmental data from early *Australopithecus* sites (5.0 to 2.5 Ma) indicate, however, a wide variety of habitats, both heavily wooded and open settings (Potts, 1998b; Andrews and Humphrey, 1999). On this basis, the heated debate may be resolved by a different interpretation – namely, that the anatomical structure of the early australopiths was well designed to combine the advantages of both climbing and terrestrial walking. Australopith locomotion provided a means of coping with repeated shifts between relatively treeless and wooded landscapes. The variability hypothesis implies that terrestrial bipedal behavior initially emerged as part of an adapt-

able system of locomotion – a product of wide habitat fluctuation – rather than as a specialized adaptation to open terrain.

This novel strategy of moving around may have made new food resources accessible. Certainly by 2.6–2.5 Ma, certain groups of hominins, possibly later *Australopithecus* or an early member of *Homo*, had gained access to large animal carcasses by using chipped stone tools. Sharp stone flakes could penetrate hides that even the largest carnivore could not handle, while hammerstones could process hard or tough plant foods as effectively as the largest herbivore molars. By carrying stone tools many kilometers, the toolmaking hominins had developed a means of overcoming change in the spatial distribution of favored foods as environments fluctuated. Toolmaking and expansion of the diet were not necessarily adopted by all hominins. But as wider ranges of environmental oscillation were confronted, the mental, social, and manipulative skills involved in making tools leading to novel food sources would have proved advantageous.

Over the past 1 million years, cognitive and technical skills were further enhanced. Brains were enlarged relative to body size, complex spatial memory was expanded, and the cooperative exchange of resources emerged. While a variety of valid scenarios could explain the emergence of these features, each represents an important means by which human beings store information about their surroundings and buffer the risks of unpredictable change.

After several hundred thousand years of the largest environmental oscillations of the Cenozoic, symbolic behaviors began to be expressed by hominins with increasing frequency. Between 50 000 and 20 000 years ago, these behaviors took the form of cave paintings, personal adornment (e.g. beadwork), and engravings on bone and ivory. Complex symbolic systems, such as language, represented effective means of planning, social sharing of mental and temporal maps of the world, and communicating novel information in the present and beyond the immediate setting. The expression of complex symbolic functions underlies the unique aspects of human cultural behavior and its effectiveness as a system of information transfer and adjustment to the environment (social, biotic, and physical). Language and culture are the most sophisticated devices humans use to respond to their surroundings. Especially over the past 100 000 years, human behavior diversified into the distinctive units we call cultures, a development that greatly expanded the adaptive options available to the human lineage.

In light of the variability hypothesis, the unique features of human cognition, language, and culture reflect adaptations to changing conditions and novel challenges that occurred in concert with the escalation of environmental variability. This escalation has occurred since 6 million years ago (Figure 19.2), and thus the list in Figure 19.3 reflects the accretion of a package

of traits (in spite of the loss of the oldest one, locomotor versatility) that has made the surviving species, *Homo sapiens*, an extremely adaptable organism, capable of spreading globally.

19.6 Conclusion

The interplay of a number of physical factors has been responsible for environmental change during the Cenozoic. These factors include planetary phenomena, such as orbital cycles, and regional phenomena, such as tectonic uplift and volcanism. The interaction among these factors has largely determined the extent to which environmental change has been progressive (e.g. cooling), catastrophic (e.g. due to massive volcanic eruptions), and/or variable (e.g. increasing amplitude of oscillation). The last of these possibilities, environmental variability, has arisen from complex variations in solar insolation, dynamic responses by Earth's systems of heat, wind, and water circulation, and episodic tectonic and eruptive events that have altered landscapes.

By altering the settings in which organisms live, each of the three patterns of environmental change (progressive, catastrophic, and variable) may influence the diversity and adaptive properties of species. Much of the biotic change that has occurred during the late Cenozoic reflects the development of cooler, drier environments. Organisms have, however, differed widely in their evolutionary response to this directional trend. Among African animals, for example, the aardvark (*Orycteropus*) and impala (*Aepyceros*) represent very stable morphological lineages marked by low speciation and extinction rates; suids and alcelaphine bovids (wildebeests and hartebeests), by contrast, have evolved in the same overall milieu but exhibited much higher rates of morphological divergence and species turnover (Vrba, 1995a; Bishop, 1999).

Similarly, the adaptive histories of hominin lineages were quite varied. Certain species evolved adaptations largely shaped by the dominant trends toward cooling, drying, and the predominance of low-biomass vegetation. The massive craniodental anatomy of Plio-Pleistocene robust australopiths in East Africa appears to reflect the same selective pressures toward heavy chewing of tough, fibrous plants evident in most lineages of contemporaneous dry-savanna herbivores (Grine, 1988). The limb proportions and facial structure of the late Pleistocene Neanderthals, on the other hand, reflect specialized adaptations to cool temperate and cold glacial habitats (Trinkaus, 1981).

Over the same period, environmental variability also posed a decisive challenge and helped shape a different set of anatomical and behavior traits responsive to marked habitat instability. At least some hominin lineages, such as that leading to the Neanderthals, evolved a combination of habitat-specific adaptations (e.g. cold-weather physiologies) and variability-related ones (e.g. expanded neocortical processing). But certain lineages apparently had advantages over others in responding to environmental uncertainty and changing conditions. Human evolution was characterized by the relative success (i.e. origin and persistence) of versatile lineages over more habitat-specific ones. As this took place over the past several million years, the ancestors of modern humans became decoupled from specific habitats and more responsive to the dynamic pattern of environmental variability.

Bearing in mind the Pleistocene herbivores of southern Kenya, it appears that versatile adaptations evolved in other late Cenozoic mammals as environmental inconsistency was on the rise. It is virtually certain, though, that among all species, *Homo sapiens* is the most extreme case of an organism adapted to environmental novelty. A distinctive part of human adaptive behavior has been the ability to buffer environmental change by altering the immediate surroundings. These efforts – by using tools, building shelters, making fire, growing food – initially had minor effects. But as humans have spread globally, these effects have become so magnified as to impinge on large segments of Earth's biota. Through the history of life, many examples can be given of how physical processes altered our planet's environment and thus ultimately changed the

course of evolution. The latest example of this phenomenon may now be apparent in the connection between the physical causes of environmental variability and the origin of our own species.

Acknowledgements

I am grateful to Adrian Lister and Lynn Rothschild for their invitation to attend the stimulating conference at the Linnean Society of London and to contribute to this volume. Refining and testing the variability selection idea has been advanced through discussions with many colleagues. A partial list includes L.C. Aiello, P. Andrews, A.K. Behrensmeyer, J. Damuth, P. deMenocal, P. Ditchfield, J. Fleagle, L. Marino, E. Vrba, and B. Wood. My appreciation goes to J.B. Clark for support in preparing the manuscript. This is a publication of the Smithsonian's Human Origins Program.

References

Andrews, P. and Humphrey, L. (1999). African Miocene environments and the transition to early hominines. In: T.G. Bromage and F. Schrenk (eds) *African Biogeography, Climate Change, and Human Evolution*. New York: Oxford University Press, pp. 282–300.

Behrensmeyer, A.K. and Bobe, R. (1999). Change and continuity in Plio-Pleistocene faunas of the Turkana basin, Kenya and Ethiopia. In: J. Lee-Thorp and H. Clift (eds) *International Union for Quaternary Research XV International Congress: Book of Abstracts*. Rondebosch (South Africa): University of Cape Town, p. 20.

Behrensmeyer, A.K., Todd, N.E., Potts, R. and McBrinn, G.E. (1997). Late Pliocene faunal turnover in the Turkana basin, Kenya and Ethiopia. *Science*, **257**: 1589–1594.

Bishop, L.C. (1999). Suid paleoecology and habitat preferences at African Pliocene and Pleistocene hominid localities. In: T.G. Bromage and F. Schrenk (eds) *African Biogeography, Climate Change, and Human Evolution*. New York: Oxford University Press, pp. 216–225.

Bobe R. (1999). Faunal responses to climatic change in the East African Plio-Pleistocene. In: J. Lee-

Thorp and H. Clift (eds) *International Union for Quaternary Research XV International Congress: Book of Abstracts*. Rondebosch (South Africa): University of Cape Town, p. 25.

Bond, G., Showers, W., Cheseby, M. *et al.* (1997). A pervasive millennial-scale cycle in North Atlantic Holocene and glacial climates. *Science*, **278**: 1257–1266.

Cerling, T.E., Harris, J.M., MacFadden, B.J. *et al.* (1997). Global vegetation change throught the Miocene/Pliocene boundary. *Nature*, **389**: 153–158.

Clemens, S.C., Murray, D.W. and Prell, W.L. (1996). Nonstationary phase of the Plio-Pleistocene Asian monsoon. *Science*, **274**: 943–948.

Coope, G.R. (1967). The value of Quaternary insect faunas in the interpretation of ancient ecology and climate. In: E.J. Cushing and H.E. Wright (eds) *Quaternary Paleoecology*. New Haven: Yale University Press, pp. 359–380.

Davis, M.B. (1985). Climatic instability, time lags, and community disequilibrium. In: J. Diamond and T.J. Case (eds) *Community Ecology*. New York: Harper and Row, pp. 269–284.

Dawkins, R. (1987). *The Blind Watchmaker*. New York: Norton.

DeMenocal, P.B. (1995). Plio-Pleistocene African climate. *Science*, **270**: 53–59.

DeMenocal, P.B., Ruddiman, W.F. and Pokras, E.M. (1993). Influences of high- and low-latitude processes on African terrestrial climate: Pleistocene eolian records from equatorial Atlantic Ocean Drilling Program Site 663. *Paleoceanography*, **8**: 209–242.

Eldredge, N. (1989). *Macroevolutionary Dynamics*. New York: McGraw-Hill.

Feibel, C.S. (1999). Tephrostratigraphy and geological context in paleoanthropology. *Evolutionary Anthropology*, **8**: 87–100.

Flenley, J.R. (1979). *The Equatorial Rain Forest: A Geological History*. London: Butterworth.

Graham, R.W., Lundelius, Jr E.L., Graham, M.A. *et al.* (1996). Spatial response of mammals to late Quaternary environmental fluctuations. *Science*, **272**: 1601–1606.

Grine, F.E. (1988). *Evolutionary History of the 'Robust' Australopithecines*. New York: Aldine de Gruyter.

Huntley, B. and Webb, T. III (1989). Migration: species' response to climatic variations caused by changes in the earth's orbit. *Journal of Biogeography*, **16**: 5–19.

Kukla, G. (1989). Long continental records of climate – an introduction. *Palaeogeography, Palaeoclimatology, Palaeoecology*, **72**: 1–9.

Liu, T., Ding, Z. and Rutter, N. (1999). Comparison of Milankovitch periods between continental loess and deep sea records over the last 2.5 Ma. *Quaternary Science Reviews*, **18**: 1205–1212.

Lovejoy, C.O., Heiple, K.G. and Burstein, A.H. (1973). The gait of *Australopithecus*. *American Journal of Physical Anthropology*, **38**: 757–780.

Martin, P.S. and Klein, R.G. (eds) (1984). *Quaternary Extinctions*. Tucson: University of Arizona Press.

McManus, J.F., Oppo, D.W. and Cullen, J.L. (1999). A 0.5-million-year record of millennial-scale climate variability in the North Atlantic. *Science*, **283**: 971–975.

Miller, K.G., Fairbanks, R.G. and Mountain, G.S. (1987). Tertiary oxygen isotope synthesis, sea level history, and continental margin erosion. *Paleoceanography*, **2**: 1–19.

Ohman, J.C., Krochta, T.J., Lovejoy, C.O. *et al.* (1997). Cortical bone distribution in the femoral neck of hominoids: implications for the locomotion of *Australopithecus afarensis*. *American Journal of Physical Anthropology*, **104**: 117–131.

Paytan, A., Kastner, M. and Chavez, F.P. (1996). Glacial to interglacial fluctuations in productivity in the equatorial Pacific as indicated by marine barite. *Science*, **274**: 1355–1357.

Potts, R. (1996a). *Humanity's Descent: The Consequences of Ecological Instability*. New York: Avon.

Potts, R. (1996b). Evolution and climate variability. *Science*, **273**: 922–923.

Potts, R. (1998a). Variability selection and hominid evolution. *Evolutionary Anthropology*, **7**: 81–96.

Potts, R. (1998b). Environmental hypotheses of hominin evolution. *Yearbook of Physical Anthropology*, **41**: 93–136.

Potts, R. and Behrensmeyer, A.K. (1992). Late Cenozoic terrestrial ecosystems. In: A.K. Behrensmeyer, J.D. Damuth, W.A. DiMichele, R. Potts, H.-D. Sues and S.L. Wing (eds) *Terrestrial Ecosystems Through Time*. Chicago: Chicago University Press, pp. 419–541.

Potts, R. and Deino, A. (1995). Mid-Pleistocene change in large mammal faunas of the southern Kenya rift. *Quaternary Research*, **43**: 106–113.

Prentice, M.L. and Denton, G.H. (1988). The deep-sea oxygen isotope record, the global ice sheet system, and hominid evolution. In: F.E. Grine (ed.) *Evolutionary History of the 'Robust' Australopithecines*. New York: Aldine de Gruyter, pp. 383–403.

Rampino, M.R. and Self, S. (1992). Volcanic winter and accelerated glaciation following the Toba super-eruption. *Nature*, **359**: 50–52.

Raymo, M.E. and Ruddiman, W.F. (1992). Tectonic forcing of late Cenozoic climate. *Nature*, **359**: 117–122.

Raymo, M.E., Ganley, K., Carter, S. *et al.* (1998). Millennial-scale climate instability during the early Pleistocene epoch. *Nature*, **392**: 699–701.

Rossignol-Strick, M. (1983). African monsoons, and immediate climate response to orbital insolation. *Nature*, **304**: 46–49.

Ruddiman, W.F. and Kutzbach, J.E. (1989). Forcing of late Cenozoic Northern Hemisphere climate by plateau uplift in southern Asia and the American West. *Journal of Geophysical Research*, **94**: 18409–18427.

Ruddiman, W.F., Prell, W.L. and Raymo, M.E. (1989). Late Cenozoic uplift in southern Asia and the American West: rationale for general circulation modeling experiments. *Journal of Geophysical Research*, **94**: 18379–18391.

Shackleton, N.J., Backman, J., Zimmerman, H. *et al.* (1984). Oxygen isotope calibration of the onset of ice-rafting and history of glaciation in the North Atlantic region. *Nature*, **307**: 620–623.

Stanley, S.M. (1995). New horizons for paleontology, with two examples: the rise and fall of the Cretaceous Supertethys and the cause of the modern ice age. *Journal of Paleontology*, **69**: 999–1007.

Stern, J.T. Jr. and Susman, R.L. (1983). Locomotor anatomy of *Australopithecus afarensis*. *American Journal of Physical Anthropology*, **60**: 279–317.

Susman, R.L. and Stern, J.T. (1991). Locomotor behavior of early hominids. In: Y. Coppens and B. Senut (eds) *Origine(s) de la bipédie chez les hominidés*. Paris: Editions du CNRS, pp. 121–131.

Taggart, R.E., Cross, A.T. and Satchell, L. (1982). Effects of periodic volcanism on Miocene vegetation distribution in eastern Oregon and western Idaho. In: B. Mamet and M.J. Copeland (eds) *Third North American Paleontological Convention, Montreal*, pp. 535–540.

Taylor, K.C., Lamorey, G.W., Doyle, G.A. *et al.* (1993). The 'flickering switch' of late Pleistocene climate change. *Nature*, **361**: 432–436.

Tooby, J. and Cosmides, L. (1993). The psychological foundations of culture. In: J.H. Barkow, L. Cosmides and J. Tooby (eds) *The Adapted Mind*. New York: Oxford, pp. 19–136.

Trinkaus, E. (1981). Neanderthal limb proportions and cold adaptation. In C.B. Stringer (ed.) *Aspects of Human Evolution*. London: Taylor and Francis, pp. 187–224.

Vrba, E.S. (1980). Evolution, species and fossils: how does life evolve? *South African Journal of Science*, **76**: 61–84.

Vrba, E.S. (1985). Environment and evolution: alternative causes of the temporal distribution of evolutionary events. *South African Journal of Science*, **81**: 229–236.

Vrba, E.S. (1988). Late Pliocene climatic events and hominid evolution. In: F.E. Grine (ed.) *Evolutionary History of the 'Robust' Australopithecines*. New York: Aldine de Gruyter, pp. 405–426.

Vrba, E.S. (1992). Mammals as a key to evolutionary theory. *Journal of Mammalogy*, **73**: 1–28.

Vrba, E.S. (1995a). On the connection between paleoclimate and evolution. In: E.S. Vrba, G.H. Denton, T.C. Partridge and L.H. Burckle (eds) *Paleoclimate and Evolution, with Emphasis on Human Origins*. New Haven: Yale University Press, pp. 24–45.

Vrba, E.S. (1995b). The fossil record of African antelopes (Mammalia, Bovidae) in relation to human evolution and paleoclimate. In: E.S. Vrba, G.H. Denton, T.C. Partridge and L.H. Burckle (eds) *Paleoclimate and Evolution, with Emphasis on Human Origins*. New Haven: Yale University Press, pp. 385–424.

White, T.D. (1995). African omnivores: global climatic change and Plio-Pleistocene hominids and suids. In: E.S. Vrba, G.H. Denton, T.C. Partridge and L.H. Burckle (eds) *Paleoclimate and Evolution, with Emphasis on Human Origins*. New Haven: Yale University Press, pp. 369–384.

20

Climatic change and the beginnings of agriculture: the case of the Younger Dryas

David R. Harris

ABSTRACT

See Plates 31–33

The transition from hunting and gathering to agriculture initiated a profound change in human society that altered irrevocably the relationship of people to planet Earth, but why and how it occurred remains obscure. Now new palaeoclimatic and palaeoecological data, together with improved recovery and dating of plant and animal remains from archaeological sites, suggest that the transition began during the late glacial period when an abrupt, short-term change to cold, dry conditions during the Younger Dryas stadial prompted some groups of hunter-gatherers to start cultivating wild cereals and legumes. Before this evidence is reviewed, the crops and livestock that were incorporated into early agricultural systems in Southwest Asia, China, Mesoamerica and tropical Africa in the Early Holocene are briefly described. The terms *cultivation*, *domestication*, *agriculture* and *pastoralism* are defined, and the question of how they can be recognized archaeologically is considered.

Southwest Asia provides the best evidence for a causal link between the Younger Dryas and changes in the availability and exploitation of wild plant foods and the beginnings of cereal and legume cultivation. Attention focuses on the evidence from archaeological sites in the Levant, particularly Abu Hureyra. New evidence from sites in China is also discussed, and in conclusion the question is raised as to whether short-term climatic changes may have influenced the beginnings of agriculture elsewhere in the world.

20.1 Introduction

The transition from foraging to farming radically changed the relationship of people to the environment and paved the way for the development of urban civilization. Archaeological and biological evidence now suggests that the transition began in the Levantine region of Southwest Asia about 11 000 (uncalibrated radiocarbon) years ago (*c.* 13 000 calendar years before present) – see 'Radiocarbon dates' in the Glossary for an explanation of uncalibrated and calibrated dates. But why it occurred then, and there, after more than 100 000 years of hunting and gathering by anatomically modern humans, remains a controversial question. Many factors, singly or in combination, have been suggested to explain the transition – environmental change, resource availability, sedentary settlement, population growth, technological innovation, group competition and wealth accumulation – but there has been insufficient evidence to test alternative explanatory models. Now new data on climatic change and associated changes in vegetation at the end of the Pleistocene, coupled with more precise dating of plant and animal remains from archaeological sites, are combining to suggest a more conclusive answer to the question of why agriculture began when and where it did.

The earliest evidence for the transition comes from two regions of the world – Southwest Asia and China – when particular juxtapositions of environmental and cultural conditions caused some groups of hunter-gatherers to start cultivating and domesticating a limited range of plants (crops) and animals (livestock). At present we only have sufficient evidence from Southwest Asia to draw fairly firm conclusions about how and why the transition occurred, but recent archaeological investigations in China are now beginning to clarify the process there. In Mesoamerica (principally Mexico) and northern tropical Africa (the southern Sahara and the Sahel zone north of the Congo rainforest) there is evidence to suggest that hunter-gatherers independently developed agriculture in those regions too, as they may have also in southern India (Fuller *et al.*, 2001), New Guinea (Golson, 1989; Yen, 1991) and eastern North America (Smith, 1992), but all these transitions from foraging to farming appear to have occurred later, in the Early Holocene, i.e. between 10 000 and 5000 radiocarbon years ago.

In each of the four core regions of Southwest Asia, China, Mesoamerica and northern tropical Africa, shifts took place from the harvesting of wild plants, particularly the seeds of grasses and herbaceous legumes, to their cultivation and domestication, and over time their human populations became progressively more dependent for their food supply on a small selection of grain crops, and, in some areas, root and tuber crops. In each region one or more domesticated cereal became a major staple: barley and wheats in Southwest Asia, rice in China, maize in Mesoamerica and sorghum in Africa. Legumes too were domesticated in each region. These crops, known as pulses, complemented the cereals nutritionally by providing oils, and essential amino acids

such as lysine, that the cereals lacked, as well as adding to the general supply of carbohydrate and protein. Thus the cereals were complemented by lentil, pea, chickpea and other pulses in Southwest Asia, by soybean in China, by common, scarlet runner and tepary bean in Mesoamerica, and by cowpea (*Vigna unguiculata*) and groundnuts (*Macrotyloma geocarpum* and *Voandzeia subterranea*) in northern tropical Africa (Harris, 1981).

Comparison of the four core regions in terms of the domestic animal component in their Early Holocene agricultural systems reveals greater contrasts between them than in their crop complexes. Southwest Asia is distinctive in being the locale where a group of social ungulates (herd animals) were domesticated and incorporated into a unique system of agro-pastoral production that gradually, over some 3000 years from *c.* 10 500 to *c.* 7500 bp (in this chapter bp indicates uncalibrated radiocarbon dates and cal BP indicates calibrated dates), integrated cereal and pulse cultivation with the raising of goats, sheep, pigs and cattle (Harris, 1998a). This system of mixed grain—livestock farming, which spread in later prehistoric times west into Europe and North Africa and east into Central and South Asia (Bar-Yosef and Meadow, 1995: 70–75; Harris, 1996a: 554–64, 1998b; Price, 2000; Zilhão, 2001), was not paralleled in China, Mesoamerica or Africa. In China domestic pig, chicken and water buffalo became associated early with rice cultivation; domestic sheep, goats and cattle were introduced from Southwest Asia into Africa but no indigenous herd animals were domesticated there; and in Mesoamerica no domestic animals were integrated into systems of crop cultivation before the arrival of Europeans in the sixteenth century AD (although domesticated turkeys and Muscovy ducks were present in pre-European times neither was incorporated into agricultural production as were livestock in Southwest Asia and China).

This summary of some of the similarities and differences in what we can describe as the founder crop and livestock complexes of Southwest Asian, Chinese, Mesoamerican and tropical African agriculture prompts the question of whether similar processes of environmental and/or cultural change led to the transition to agriculture in all four regions. In particular, it invites consideration of the question implied by the title of this chapter: was climatic change responsible for the beginnings of agriculture? Unfortunately, we have insufficient evidence at present to attempt to answer this fundamental question in relation to all four regions, so this chapter is largely restricted to discussion of the two regions, Southwest Asia and China, where evidence is emerging for a possible causal link between the beginnings of agriculture and a late glacial interval of relatively abrupt and brief climatic change to colder and drier conditions – the Younger Dryas stadial (Roberts, 1998: 70–76), also known as the Greenland stadial 1. However, before pursuing that question further we need to clarify what is meant here by the term agriculture.

20.2 | What is 'agriculture'?

There is much confusion in archaeological and anthropological literature about the meaning of such terms as agriculture, cultivation, domestication and pastoralism, and the lack of precise definitions often hinders study of why, how, when and where agriculture began. All that needs to be established here, to facilitate comparison between the regions under discussion, is how the terms are used in this chapter. Thus, by *cultivation* I mean the sowing and planting, tending and harvesting of useful wild or domestic plants, with or without soil tillage; *domestication* means that plants and animals have been changed morphologically, physiologically and/or behaviourally as a result of cultural (inadvertent or deliberate) selection and have become dependent on humans for their long-term survival; *agriculture* is defined as the production of (domesticated) crops, normally involving systematic tillage; and *pastoralism* refers to the management of domesticated livestock in systems of mixed farming, transhumance or nomadic pastoralism (cf. Harris, 1989: 17–22, 1996b: 444–456).

The distinction between cultivation and agriculture has particular relevance to this chapter because it underlies the rationale of focusing on the regions where we have evidence for the autochthonous development of systems of agricultural production (as here defined), as opposed to the many parts of the world where 'hunter-gatherers' practised various techniques of cultivation but did not domesticate (in the above sense) any crops or livestock. The distinction also allows use of the term 'pre-domestication cultivation', which, as will become apparent below, is an essential concept in our understanding of how agriculture originated.

It is of course easier to define the terms used than to establish criteria for distinguishing archaeologically (or even historically and ethnographically) between cultivation and agriculture. But by adopting a narrow definition of domestication, which gives weight to morphological changes that occur, it becomes possible to distinguish some, although by no means all, of the main plant and animal domesticates from their wild progenitors. For example, domesticated barley, wheats and rice can be identified, if sufficiently well-preserved (usually charred) archaeological samples are available, by the presence of rough disarticulation scars on the spikelet forks which are diagnostic of the replacement, under domestication, of the brittle rachis of the wild grasses by the tough rachis of the cereals. Domesticated maize is much easier to identify because the morphology of the cob is conspicuously different, even in small primitive varieties, from the seed head of its main wild progenitor, teosinte. Distinguishing archaeologically between most pulses and their wild progenitors or other close relatives is, however, very difficult (Butler, 1989, 1992), particularly because the seeds show few changes apart from some increase in average size. Livestock too vary in the ease with which domesticates can be recognized archaeologically, with, for example, goats and sheep being more readily (although often still with difficulty) distinguished from their wild progenitors than pigs and cattle, despite the fact that the domestic forms of all four groups are usually smaller than their wild forebears.

The problems inherent in the archaeological identification of domestic plants and animals demonstrate how difficult it is to establish just when, in any given situation, agriculture can be said to have begun, and how easy it is for differences of interpretation to arise because the terms of debate are not clearly defined. These difficulties can be obviated if the terminology used is first defined and then consistently applied in interpreting the available evidence – as is attempted in the regional sections that follow. It is appropriate to take up first the case of Southwest Asia, because it is for that region that we have the most comprehensive archaeological evidence and because the impact of climatic change on the transition to agriculture can be traced most clearly there.

20.3 Southwest Asia

Over 70 years ago the archaeologist V. Gordon Childe first propounded what came to be known as his 'desiccation' or 'oasis' theory to explain the origins of agriculture in the Near East. He referred to this profound change as the Neolithic Revolution, and in successive publications proposed and elaborated the thesis that it was brought about by climatic change from humid ('pluvial') to more arid conditions at the end of the last glacial period (Childe, 1928: 42–43, 1934: 14–30, 1936: 66–104). He envisaged animals and people being drawn to the diminishing sources of water in the landscape (oases) where their proximity led to sheep, goats and cattle being domesticated, a process that, he suggested, would have been aided by the animals being attracted to cereal fields. Childe thus postulated that cereal cultivation preceded livestock raising, a view that finds confirmation today in evidence from the central and southern Levant. We now know that Childe was wrong in supposing that 'pluvial' conditions prevailed in the Near East during the last glacial period and gave way to aridity when the northern hemisphere ice caps and glaciers retreated, but we should acknowledge his prescience in suggesting that animal and plant domestication and the transi-

tion to a food-producing economy was related to a climatic change to drier conditions.

Since Childe's time, cores from deep-sea deposits, ice caps, coral reefs and lake and swamp sediments have begun to provide abundant data on climatic and vegetation change. At the same time, a range of methods of absolute dating of palaeoenvironmental events have been successfully developed, to the extent that detailed chronologies for Late Pleistocene–Early Holocene environmental change (the time span of relevance to agricultural origins) are becoming available (Roberts, 1998: 68–126). Archaeologically, the most important advance in dating has been the development of the radiocarbon method and its calibration by reference to tree-ring sequences (Baillie, 1995), in particular the advent of radiocarbon dating by accelerator mass spectrometry (^{14}C AMS) which has made possible the direct dating of very small samples, such as individual charred cereal grains (Gowlett and Hedges, 1986; Harris, 1987). As a result of these technical advances, coupled with improved recovery of plant and animal remains from archaeological sites, particularly by the flotation of bulk samples, it is now worth reopening the question – which has, with a few notable exceptions (Byrne, 1987; Whyte, 1977; Wright, 1977), been largely disregarded since Childe's desiccation hypothesis came to be generally rejected – of whether the transition to agriculture in Southwest Asia may have been caused, in part at least, by climatic change.

The period during which agriculture emerged and became established in Southwest Asia falls within a time span of some 5500 radiocarbon years from *c.* 13 000 to *c.* 7500 bp (*c.* 15 400 to *c.* 8200 cal BP) which, in the Levant, is conventionally divided into three archaeological periods: the Late Epipalaeolithic (*c.* 13 000–10 300 bp), the Pre-Pottery Neolithic A or PPNA (*c.* 10 300–9500 bp) and the Pre-Pottery Neolithic B or PPNB (*c.* 9500–7500 bp). In Table 20.1, these archaeological periods are correlated with a succession of climatic and vegetational changes inferred from stratigraphic sequences of terrestrial and marine sediments.

Table 20.1 **Simplified sequence of Late Pleistocene–Early Holocene archaeological periods and climatic and vegetational changes in the Levant.**

Years bp	Archaeological periods	Climatic changes	Vegetational changes
12 500	Late Epipalaeolithic	Lateglacial warming trend (from the LGM at 18 000 bp): rising temperature, precipitation and atmospheric CO_2	Expansion of woodland and steppe, including annual grasses, from LGM refugia
11 000	PPNA *c.* 10 300 bp	YOUNGER DRYAS: reversion to cold conditions with some reduction of precipitation	Interruption of Lateglacial vegetation expansion, die-back of trees, shrubs and grasses
10 000	PPNB *c.* 9500 bp (to *c.* 7500 bp)	Postglacial return to warmer, wetter conditions	Re-expansion of forest, woodland and steppe

The earliest of the three periods of environmental change shown in Table 20.1 saw the gradual recovery of the region from the extreme cold and aridity of the Late Glacial Maximum (LGM). During the LGM (in oxygen-isotope stage 2) sea level had fallen by *c.* 100 m, increasing the width of the coastal plains of the Levant and the Persian Gulf; mean annual temperatures across Southwest Asia were several °C below present values; and the growing season was also reduced by lowered spring and summer precipitation. Deep pollen cores that go back in time to the LGM (Baruch, 1995; Bottema and van Zeist, 1981; van Zeist and Bottema, 1982) indicate that steppe and desert widely replaced forest vegetation, which survived only in isolated mountain refugia. A fluctuating trend towards warmer and wetter conditions set in after the LGM (at the end of oxygen-isotope stage 2), associated with increases in atmospheric CO_2 (Sage, 1995). Sea level began to rise from *c.* 15 000 bp (Bard *et al.*, 1990: 405), thus reducing the formerly expanded coastal plains of the Levant and Persian Gulf (and restricting the foraging territories of hunter-gatherer groups living there). The warming trend brought increased summer solar radiation (COHMAP Members, 1988: 1048) and enhanced the seasonal differences between cool wet winters and hot dry summers. The spring–early summer growing season increased in length, and

forest, woodland and steppe vegetation expanded. Within the woodlands and steppes there were extensive stands of annual grasses with edible seeds, including wild wheats, barley and rye. Hillman (1996: 183–191) has modelled their eastward spread around the northern margins of the Fertile Crescent from refugia in the northeastern Levant between *c.* 13 000 and *c.* 11 000 bp.

It was around 11 000 bp (*c.* 13 000 cal BP) that a relatively abrupt climatic reversal to colder and drier conditions ushered in the second period of environmental change that concerns us – the Younger Dryas. This climatic oscillation takes its name from the abundance of the Alpine–Arctic species *Dryas octopetala* in European late glacial pollen diagrams where it was first recognized (as pollen zone III) by Iversen (1954). The oscillation has since been traced in ocean-core oxygen-isotope records from the South Pacific (Kudrass *et al.*, 1991) and North Atlantic (Bond *et al.*, 1993), in coral-reef cores from Barbados (Bard *et al.*, 1990; Fairbanks, 1989), in ice cores from Greenland (Alley, 2000; Alley *et al.*, 1993), in laminated lake sediments in Tibet (Gasse *et al.*, 1991) and Japan (Fukusawa, 1995), and in pollen cores and glacial deposits in northern North America (LaSalle and Shilts, 1993; Mathewes *et al.*, 1993; Mayle and Cwynar, 1995), southern South America (Heusser, 1993) and New Zealand (Denton and Hendy, 1994). It is less evident in palaeoenvironmental records from the tropics, but has been detected in northern and eastern Africa (Gasse *et al.*, 1990; Roberts *et al.*, 1993) and Mesoamerica (Islebe *et al.*, 1995; Leyden, 1995; van der Hammen and Hooghiemstra, 1995). Some specialists (e.g. Alley, 2000; Kudrass *et al.*, 1991; Peteet, 1993) regard the Younger Dryas as a worldwide phenomenon which may, however, have differed greatly in severity from region to region and may not have been synchronous throughout the Northern and Southern Hemispheres.

In Southwest Asia there is strong but not unequivocal evidence for the Younger Dryas oscillation. For example, according to Bottema (1995: 890) 'it is not easily and in some areas not at all discernible as a biozone in the pollen record'. Nevertheless, the stratigraphic record of lake and deep-sea sediments (Kuzucuoglu and Roberts, 1997: 17–18; Landmann and Reimer, 1996; Rossignol-Strick, 1995, 1997; Sanlaville, 1996) suggests that from *c.* 11 000 to *c.* 10 000 bp much of the region did experience a cold, dry climate. This reversal of the preceding trend to warmer and wetter conditions (Table 20.1) probably involved a marked reduction in spring and summer precipitation, and it appears temporarily to have halted the late glacial expansion of forest, woodland and steppe, causing extensive die-back of the trees and reducing to isolated small stands the wild grasses that were a valuable source of food for hunter-gatherer groups (see below).

In early post-glacial time, from about 10 000 bp (*c.* 11 500 cal BP) onwards, the trend towards warmer and wetter conditions resumed, reaching a maximum around 5000 bp and declining slightly thereafter (Kuzucuoglu and Roberts, 1997: 18–20; Rossignol-Strick, 1999; Sanlaville, 1996: 23–24). This trend, which would have been characterized by increased winter rainfall, greater seasonal contrasts in temperature and a lengthening of the spring and early summer growing season, is evident in the Southwest Asian pollen diagrams, which show a re-expansion of forest and woodland following the end of the Younger Dryas.

We can now shift our attention from the sequence of late glacial and post-glacial climatic and vegetational changes in Southwest

Asia to the chronologically parallel changes in the archaeological record of human subsistence in the region, and ask how the transition from foraging to farming may have been influenced by the environmental changes. The two millennia from 13 000 bp to the onset of the Younger Dryas at *c.* 11 000 bp fall within the Late Epipalaeolithic period (which in the southern and central Levant is the time of the Early and Late Natufian cultures). At this time all the human inhabitants of Southwest Asia lived by hunting and gathering, but there is evidence of significant variation among them in food-procurement strategies and patterns of settlement. Most of the region was sparsely occupied by seasonally mobile hunter-gatherers, whose prey included gazelle, ibex, fallow deer and wild boar and who gathered a wide range of plant foods. However, in the southern and central Levant the Natufian culture, which emerged about 13 000 bp at the beginning of the Late Epipalaeolithic, is distinguished by evidence for relatively large sites – such as Abu Hureyra, Ain Mallaha, Hayonim, El-Wad and Wadi Hammeh (Figure 20.1) – which are characterized by elaborate material culture and may have been occupied year-round (Bar-Yosef and Valla, 1991; Bar-Yosef and Meadow, 1995: 55–61). This evidence takes the form of semi-subterranean houses, human burials (sometimes with domestic dogs), body ornaments of shell, bone and stone, bone implements decorated with animal motifs, human and animal figurines of bone and stone, and a variety of stone tools. The larger Natufian sites are frequently interpreted as base camps or hamlets, and the presence at them of large quantities of sickle blades and ground-stone tools, including pestles and mortars, suggests that wild grasses and other seed foods were intensively exploited.

At the extensive tell (mound) site known as Abu Hureyra (Figure 20.1) in the middle Euphrates valley in Syria a large assemblage of charred plant remains was recovered from Late Epipalaeolithic (Late Natufian) levels (Hillman, 1975; Moore, 1975). Analysis and interpretation of it (Hillman, 1975; Hillman *et al.*, 1989; Hillman, 2000) indicates that the seeds

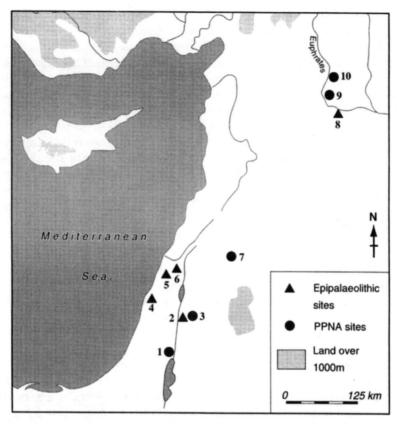

Figure 20.1   The southern and central Levant, showing the location of Late Epipalaeolithic (Natufian) and Pre-Pottery A sites mentioned in the text. 1 Jericho, 2 Wadi Hammeh, 3 Iraq ed-Dubb, 4 El-Wad, 5 Hayonim, 6 Ain Mallaha, 7 Aswad, 8 Abu Hureyra, 9 Mureybet, 10 Jerf el-Ahmar.

of wild ryes and wheats, feather grasses (*Stipa* spp.), club rush (*Scirpus* spp.), Euphrates knotgrass (*Polygonum corrigioloides*), wild lentils and other large-seeded legumes, and shrubby chenopods (species of the family Chenopodiaceae) were harvested through the growing season and provided a staple food supply rich in carbohydrates, proteins and oils (together with, probably, almonds, acorns and various root foods not identified in the plant remains). Because of the high standard of excavation and recovery at Abu Hureyra, and the extraordinary thoroughness with which the Late Epipalaeolithic plant assemblage has been analysed by Hillman and his colleagues, the site is a unique source of data on Late Epipalaeolithic plant exploitation, but unfortunately we lack equivalent sets of data from other sites of the period, including the 'classic' Natufian

sites farther south. However, the indirect evidence provided by the stone tools, and the fact that the larger Natufian sites were probably occupied year-round, strongly suggests that intensive harvesting and processing of wild seed foods was widely practised before the Younger Dryas began to have an impact on the vegetation.

To try to understand how the plant-food procurement strategies of the Late Epipalaeolithic hunter-gatherers may have been affected by the cold and dry conditions of the Younger Dryas we can turn again to the evidence from Abu Hureyra. In his analysis of changes in the relative abundance of different seed foods represented in the Late Epipalaeolithic plant remains, Hillman (2000: 376–377; Hillman *et al.*, 2001: 386) has detected a pattern of sequential decline from the least to the most drought-tolerant taxa through the Younger Dryas. Thus seed foods of oak park-woodland decline first, followed by wild lentils and other large-seeded legumes, then wild wheats and ryes, then the feather grasses and club rushes, and finally, towards the end of the period, a brief increase and subsequent decline occurs of the (most drought-tolerant) shrubby chenopods.

Such reductions in the availability of staple foods are likely to have prompted the inhabitants of Abu Hureyra (and presumably other large Late Epipalaeolithic sites) to seek new means of ensuring their supply of plant foods. They appear to have done so by initiating the (pre-domestication) cultivation of some of the staples, such as wild cereals and legumes, probably on relatively moist alluvial soils in drainage channels and on other patches of land with high water tables, as in small depressions and along breaks of slope. This supposition receives strong support from Hillman's (2000: 384–388; Hillman *et al.*, 2001: 387–378) further discovery that, when the wild cereals and feather grasses start to decline, a dramatic increase occurs in the representation of small-seeded legumes, grasses and gromwells (*Arnebia* and *Buglosoides* spp.) which are weeds typical of dryland cultivation today in areas of woodland steppe, moist steppe and dry steppe. The abundance of these three groups of weeds increases progressively through the Younger Dryas, suggesting that the area under cultivation likewise expanded. However, the most conclusive evidence that agriculture (as defined above) began during the Younger Dryas is the presence in the Abu Hureyra plant remains of several grains of fully domesticated annual rye, three of which have been directly dated by the ^{14}C AMS method to the Late Epipalaeolithic (Hillman, 2000: 379; Hillman *et al.*, 2001: 389–390). Despite the presence of numerous seeds of wild rye in the earliest (Late Epipalaeolithic but pre-Younger Dryas) levels at Abu Hureyra, no trace of domestic rye has been found among them. This supports the hypothesis that the initial cultivation of wild rye, and the retention and subsequent sowing of seed stock that led to its domestication, coincided with, and compensated for, the reduction of wild-cereal stands early in the Younger Dryas.

A further indication that grain cultivation became established as a new subsistence strategy during the Late Epipalaeolithic is the reappearance at Abu Hureyra (Hillman, 2000: 388), before the end of the Younger Dryas, of lentils and other large-seeded legumes when the still prevailing aridity would have prevented the re-establishment of wild stands of such drought-intolerant taxa. Whether these legumes were by then domesticated cannot be determined because, as has already been pointed out, the wild and domestic forms cannot be distinguished morphologically in charred seed assemblages, but it is likely that pulses were part of the mixture of wild and domesticated cereals and legumes that were being cultivated on a small scale by the end of the Late Epipalaeolithic.

The archaeological transition from the Late Epipalaeolithic to the Neolithic (PPNA) in the Levant is customarily dated to 10 300 or 10 200 bp (Bar-Yosef, 1995; Goring-Morris and Belfer-Cohen, 1998: 75), i.e. before the end of the Younger Dryas. But by *c.* 10 000 bp the trend towards warmer and wetter conditions resumed and the re-expansion of forest, woodland and steppe began (Table 20.1). These environmental changes would have facilitated the extension of grain cultivation on moisture-retentive soils near now-replenished springs, rivers and lakes. However, the archaeobotanical evidence of domesticated crops at PPNA sites is extremely meagre. Only three such sites in the Levant (Figure 20.1) have yielded definite evidence of (morphologically) domesticated cereals: wheat (probably emmer) and barley at Iraq ed-Dubb in Jordan (Colledge, 2001: 143 and Table 6.2), emmer wheat at Aswad I in Syria (van Zeist and Bakker-Heeres, 1982: 185–190), and barley, emmer and einkorn wheat at Jericho in Jordan (Hopf, 1983: 582) – although re-examination of the relationship between the cereal remains found at Jericho and the ^{14}C dates of the layers with which they were associated has now shown that the remains were chronologically contemporary not with the PPNA but with the earliest phase of the succeeding PPNB (Colledge and

Conolly, 2001/2002). Indeterminate wild/domestic remains of other Southwest Asian founder crops (wheats, lentil and field pea) have also been found at these sites, as well as wild and indeterminate cereals and legumes at other PPNA sites in the southern and central Levant such as Netiv Hagdud, Mureybet and Jerf el-Ahmar (Figure 20.1). The evidence as a whole suggests that the inhabitants were cultivating predominantly wild cereals and legumes with a minor but increasing component of domesticates during the PPNA, in a subsistence system that continued to include hunting, fishing and the gathering of wild nuts and fruits (Harris, 1998a). There is no conclusive evidence for the presence of domestic animals, other than the dog, at PPNA sites, although the remains of wild goats, sheep, pigs and cattle have been found in small numbers among the more abundant remains of gazelle, deer and equids (Bar-Yosef and Meadow, 1995: 71).

By the end of the PPNA, cereal and pulse cultivation was established in the Levant as a (probably still small-scale) activity in the vicinity of at least the larger settlements, but pastoralism had not yet developed and hunting and gathering remained important subsistence activities. It was only in the succeeding two millennia of the PPNB (*c.* 9500–7500 bp) that agriculture and pastoralism became the main system of food production supporting most of the human population of Southwest Asia (Harris, 2002).

The above outline of how agriculture originated in prehistoric Southwest Asia greatly simplifies a complex process. It ignores the probable importance in the transition from foraging to farming of many non-climatic factors, for example population growth following the establishment of sedentary settlements; the environmental impact of intensified resource use in the vicinity of the larger settlements; technological innovations such as the development of hoes for tillage, sickles for harvesting, threshing sledges for separating grain from chaff, and pestles and mortars for grinding the grain; competition for land and the emergence of the concept of the ownership of resources by kin groups; increased division of labour and more hierarchical social organization; and the elaboration of exchange networks between settlements. However, most of these factors relate more to the development of agriculture through the PPNA and PPNB than they do to its origins in the Late Epipalaeolithic. Both the palaeoenvironmental and the archaeobotanical evidence now available reinforce the hypothesis, previously advanced by Moore and Hillman (1992) and recently elaborated by Bar-Yosef and Belfer-Cohen (2002), that it was the abrupt climatic reversal to the cold, dry conditions of the Younger Dryas that initiated the transition to agriculture; and it did so by causing Late Epipalaeolithic sedentary hunter-gatherers, who were already dependent on wild-seed harvesting for much of their food supply, to begin cultivating (and domesticating) the large-seeded wild grasses and legumes that were to become the staple crops of Southwest Asian agriculture during the ensuing millennia.

20.4 China

In comparison with Southwest Asia, where archaeologists have been investigating early agriculture since the 1950s (Braidwood, 1960), there has, until recently, been relatively little archaeological research in eastern Asia focused specifically on the origins of agriculture there. However, during the last two decades there has been a dramatic increase in research on the subject by Chinese, Japanese and some Western archaeologists, most of which has been concerned with the beginnings of rice agriculture – see, for example, Cohen (1998) and the publications cited both in the special section on rice domestication published in *Antiquity*, 1998, **72**: 855–907 and others in the *Bulletin of the Indo-Pacific Prehistory Association*, 1999, **18**: 77–100.

There is now convincing evidence from east-central China, mainly from sites in the middle and lower Yangzi valley, that rice had become an important food for Neolithic populations by 8000–6000 bp (approximately equivalent in time to the Pottery Neolithic period in

Southwest Asia). The earliest grains of rice (pre-served in pottery) securely dated by the ^{14}C AMS method come from Pengtoushan in the middle Yangzi valley (Figure 20.2). They date to approximately 7000 bp (*c.* 7800 cal BP) (Crawford and Shen, 1998: 861), and by 6000 bp there is plentiful evidence of domesti-cated rice, associated with the remains of pile dwellings, spade-like implements made of bone and wood, and abundant pottery (Glover and Higham, 1996: 426–429). The oldest Neolithic sites in the Yangzi valley have also yielded the remains of domesticated dog, pig, chicken and water buffalo (Yan, 1993). In the middle Huanghe valley and on the associated loess pla-teau there is evidence by 7000 bp for villages supported by a mixed economy of hunting, fishing, the cultivation of domesticated foxtail, broomcorn and barnyard millet (*Setaria italica*, *Panicum miliaceum* and *Echinochloa utilis*) and the raising of domestic dogs, pigs and probably chicken (Chang, 1986: 87–95; Crawford, 1992: 13–14). The broad conclusion to be drawn

from all this evidence is that by 7000 bp, in the Chinese Early Neolithic, substantial settle-ments, supported by grain agriculture based on rice and millets, associated with the raising of domesticated pigs, chicken and water buf-falo, were well established in east-central China at such sites as Hemudu and Luojiajiao southeast of the lower Yangzi, Jiahu north of the Yangzi, and Cishan and Peiligang in the region of the middle Huanghe valley (Figure 20.2).

It is when we turn to the earlier archaeolo-gical record in search of the antecedents of the Chinese Neolithic that we encounter major dif-ficulties. Very few Late Pleistocene–Early Holocene occupation sites are known and none has yielded a sequence of well-identified and dated plant or animal remains that throw light on the beginnings of cultivation and domestication, as Abu Hureyra so impressively does in Southwest Asia. In an attempt to over-come this problem, a joint Chinese–American project, focused on the origins of rice cultiva-tion, investigated cave sites in the Dayuan

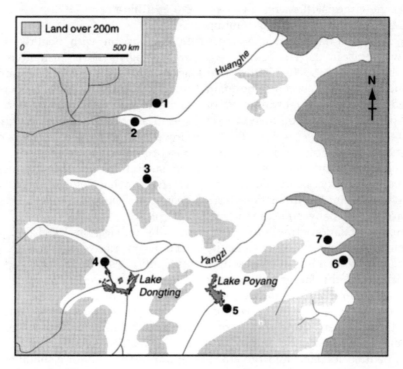

Figure 20.2 East-central China, showing the location of Early Neolithic agricultural sites mentioned in the text. 1 Cishan, 2 Peiligang, 3 Jiahu, 4 Pengtoushan, 5 Diaotonghuan, 6 Hemudu, 7 Luojiajiao.

basin south of the middle Yangzi (MacNeish *et al.*, 1997). One site, Diaotonghuan (Figure 20.2), was extensively excavated and yielded abundant animal bones, pottery, and stone, bone and shell artefacts, but plant remains are poorly preserved and the ^{14}C dates obtained, which are subject to contamination by groundwater, appear too early in relation to the cultural assemblage. The period of occupation of the cave therefore remains uncertain, and interpretation depends on establishing a relative chronology, the tentative nature of which Zhao (1998: 887–889) emphasizes. He has recovered rice phytoliths from the stratigraphic sequence of deposits and argues that the lowest level in which they occur (labelled Zone G) probably dates to between 12 000 and 11 000 bp, in the Late Palaeolithic. He interprets this as signalling the beginning of wild rice exploitation at the site during a 'warm peak in the late Pleistocene', and postulates that a marked reduction in Zone F (above Zone G) in the abundance of rice phytoliths and a return to high counts in Zones E and D reflect the impact of, and the subsequent recovery after, the cold, dry climate of the Younger Dryas (Zhao, 1998: 894). He further argues (1998: 895) that the phytoliths in Zones E and D, which he equates with the beginning of the Neolithic, derive from domesticated rice, and that by Zone B, later in the Early Neolithic at *c.* 7500–7000 bp, the transition to rice agriculture had occurred.

Zhao's interpretation remains speculative, particularly because the dating of the site is problematic, but it corresponds remarkably closely to the model for the origins of cereal cultivation in Southwest Asia derived from Hillman's analysis of the Abu Hureyra plant remains (see above). Bellwood (1996: 478) and Higham (1995: 146–147) have also suggested that the Younger Dryas may have been a factor in the transition in China from the gathering of wild rice in the Late Palaeolithic to its cultivation and domestication in the Early Neolithic. However, the occurrence and nature of a Younger Dryas effect in eastern Asia is not well established. An *et al.* (1993) suggest that between 11 000 and 10 000 bp the climate did change abruptly as the summer

monsoon strengthened, but that the main effect was not an overall colder climate but increased seasonal contrast between warmer summers and colder winters. There is some pollen evidence (Sun and Chen, 1991) from east-central China that implies colder and drier conditions between *c.* 11 000 and *c.* 10 000 bp, but until more conclusive palaeoenvironmental and archaeological evidence can be obtained the hypothesis that the Younger Dryas initiated the transition in China to (rice and perhaps millet) agriculture must remain tentative.

20.5 Conclusion

At the beginning of this chapter I highlighted the particular importance of cereal crops in the early agricultural systems of four regions of the world in which transitions from foraging to farming appear to have occurred independently at different times during the Holocene: Southwest Asia, China, Mesoamerica and northern tropical Africa. In each case the transition took place gradually over several millennia and was a prelude to the eventual establishment of urban civilization (Wheatley, 1971: 225–240). The recent acquisition of much new palaeoenvironmental and archaeological data on Late Pleistocene–Early Holocene climatic change and human subsistence has stimulated renewed interest in the proposition – first formulated by Gordon Childe in the late 1920s in his desiccation hypothesis – that a Lateglacial period of climatic deterioration could have initiated shifts to the cultivation and domestication of wild food plants, particularly annual grasses and legumes. Evidence drawn from independent sources of palaeoclimatic data has now demonstrated that the cold, dry interval known as the Younger Dryas stadial (*c.* 11 000–10 000 bp, *c.* 13 000–11 500 cal BP) had widespread, but not uniformly severe, effects in both the Northern and the Southern Hemispheres; and the archaeobotanical record from a (regrettably) small number of well-researched sites in Southwest Asia and China points to a close correlation between the effects of the Younger Dryas on the late glacial vegetation of the two regions and the beginnings of agriculture there.

In Mesoamerica there is very little local evidence of the Younger Dryas and a link between it and the beginnings of agriculture cannot at present be demonstrated. The evidence appears to be limited to records from the highlands of Colombia (van der Hammen and Hooghiemstra, 1995) and Costa Rica (Islebe *et al.*, 1995) together with a lowland record from Guatemala (Leyden, 1995). Farther south in Central and northern South America the Younger Dryas has not been detected; and in the central highlands of Mexico precipitation and temperature appear to have increased between 11 000 and 10 000 bp (Markgraf, 1993). In summarizing this meagre evidence, Piperno and Pearsall (1998: 104–105) conclude that the Younger Dryas had little effect on vegetation in the American tropical lowlands. However, they do suggest (1998: 105–107) that large-seeded grasses such as teosinte (the main progenitor of maize) and the wild ancestors of the indigenous root and tuber crops were increasingly incorporated into human diets, and eventually taken into cultivation around 10–000 bp, in response to the replacement of Lateglacial megafauna and open-land plants by expanding tropical forests. Increasing seasonality of rainfall associated with changes in the extent and intensity of tropical monsoonal circulation in the Early Holocene may also have influenced the transition to agriculture in some parts of Mesoamerica north of the equator.

Investigation of the interplay between climatic change and the origins of agriculture has advanced rapidly in recent years as the volume of palaeoenvironmental data available to archaeologists has escalated. The apparent correlation between the Younger Dryas and the beginnings of cereal cultivation in western and eastern Asia provides two persuasive examples of how an abrupt, short-term change in climate may have led to the inception of agriculture by reducing the availability of staple wild plant foods, such as the seeds of grasses and legumes, and prompting groups of hunter-gatherers (particularly those that were already occupying some sites year-round) to start cultivating these dietary staples. In other regions of agricultural origins, for example northern tropical Africa and eastern North America, where agriculture emerged later in the Early Holocene, the process may have been linked to other short-term climatic changes (Hassan, 1997; Smith, 1995: 210–213), including, in the case of Africa, changes in monsoonal circulation.

Progress towards a fuller understanding of the relation between climate and the origins of agriculture will depend critically on the recovery, accurate identification and direct dating of plant and animal remains from appropriately selected archaeological sites. But it is already apparent that climatic change was a major factor in the transformation of society from dependence on wild foods to dependence on domesticated crops and livestock, which made possible the long-term growth of the human population and led ultimately to humanity's dominion over planet Earth.

Acknowledgements

I wish particularly to thank my colleague Gordon Hillman for sharing with me over many years the results of his analyses of the plant remains from the site of Abu Hureyra and his interpretations of the beginnings of cultivation there. I also wish to thank Sue Colledge for our discussions of her re-assessment of the archaeobotany of Pre-Pottery Neolithic sites in the Levant, and especially Adrian Lister for inviting me to participate in the symposium 'Evolution on Planet Earth: the Impact of the Physical Environment' that led to publication of this book.

References

Alley, R.B. (2000). The Younger Dryas cold interval as viewed from central Greenland. *Quaternary Science Reviews*, **19**: 213–226.

Alley, R.B., Meese, D.A., Shuman, C.A. *et al.* (1993). Abrupt increase in Greenland snow accumulation at the end of the Younger Dryas event. *Nature*, **362**: 527–529.

An, Z., Porter, S.C., Zhou, W. *et al.* (1993). Episode of strengthened summer monsoon climate of Younger Dryas age on the loess plateau of central China. *Quaternary Research*, **39**: 45–54.

Baillie, M.G.L. (1995). *A Slice Through Time: Dendrochronology and Precision Dating.* London: Batsford.

Bard, E., Hamelin, B., Fairbanks, R.G. and Zindler, A. (1990). Calibration of the ^{14}C timescale over the past 30 000 years using mass spectrometric U-Th ages from Barbados corals. *Nature*, **345**: 405–409.

Baruch, U. (1994). The Late Quaternary pollen record of the Near East. In: O. Bar-Yosef and R.S. Kra (eds) *Late Quaternary Chronology and Palaeoclimates of the Eastern Mediterranean.* Tucson, Arizona: Radiocarbon, Department of Geosciences, University of Arizona, pp. 103–119.

Bar-Yosef, O. (1995). Earliest food producers – Pre-Pottery Neolithic (8000–5500). In: T.E. Levy (ed.) *The Archaeology of Society in the Holy Land.* London: Leicester University Press, pp. 190–204.

Bar-Yosef, O. and Belfer-Cohen, A. (2002). Facing environmental crisis: societal and cultural changes at the transition from the Younger Dryas to the Holocene in the Levant. In: R.T.J. Cappers and S. Bottema (eds) *The Dawn of Farming in the Near East.* Berlin: ex oriente, Studies in Early Near Eastern Production, Subsistence, and Environment 6, pp. 55–66.

Bar-Yosef, O. and Meadow, R.H. (1995). The origins of agriculture in the Near East. In: T.D. Price and A.B. Gebauer (eds) *Last Hunters–First Farmers: New Perspectives on the Prehistoric Transition to Agriculture.* Santa Fe, New Mexico: School of American Research Press, pp. 39–94.

Bar-Yosef, O. and Valla, F.R. (eds) (1991) *The Natufian Culture in the Levant.* Ann Arbor, Michigan: International Monographs in Prehistory.

Bellwood, P.S. (1996). The origins and spread of agriculture in the Indo-Pacific region: gradualism and diffusion or revolution and colonization? In: D.R. Harris (ed.) *The Origins and Spread of Agriculture and Pastoralism in Eurasia.* London: UCL Press and Washington DC: Smithsonian Institution Press pp. 465–498.

Bond, G., Broeker, W., Johnsen, S. *et al.* (1993). Correlations between climate records from North Atlantic sediments and Greenland ice. *Nature*, **365**: 143–147.

Bottema, S. (1995). The Younger Dryas in the eastern Mediterranean. *Quaternary Science Reviews*, **14**: 883–891.

Bottema, S. and van Zeist, W. (1981). Palynological evidence for the climatic history of the Near East, 50,000–6,000 BP. In: *Préhistoire du Levant: chronologie et organisation de l'espace depuis les origines jusqu'au VIe Millenaire.* Paris: Editions du CRNS, pp. 111–132.

Braidwood, R.J. (1960). The agricultural revolution. *Scientific American*, **203**: 130–148.

Butler, A. (1989). Cryptic anatomical characters as evidence of early cultivation in the grain legumes. In: D.R. Harris and G.C. Hillman (eds) *Foraging and Farming: The Evolution of Plant Exploitation.* London: Unwin Hyman, pp. 390–405.

Butler, A. (1992). The Vicieae: problems in identification. In: J.M. Renfrew (ed.) *New Light on Early Farming: Recent Developments in Palaeoethnobotany.* Edinburgh: Edinburgh University Press, pp. 61–73.

Byrne, R. (1987). Climatic change and the origins of agriculture. In: L. Manzanilla (ed.) *Studies in the Neolithic and Urban Revolutions.* Oxford: British Archaeological Reports International Series, **349**: 21–34.

Chang, K.C. (1986). *The Archaeology of Ancient China*, 4th edition. New Haven, Connecticut: Yale University Press.

Childe, V.G. (1928). *The Most Ancient East: The Oriental Prelude to European Prehistory.* London: Kegan Paul, Trench, Trubner.

Childe, V.G. (1934). *New Light on the Most Ancient East.* London: Routledge & Kegan Paul.

Childe, V.G. (1936). *Man Makes Himself.* London: Watts.

Cohen, D.J. (1998). The origins of domesticated cereals and the Pleistocene–Holocene transition in East Asia. *The Review of Archaeology*, **19**: 22–29.

COHMAP Members (1988). Climatic changes of the last 18,000 years: observations and model simulations. *Science*, **241**: 1043–1052.

Colledge, S.M. (2001). *Plant Exploitation on Epipalaeolithic and Early Neolithic Sites in the Levant.* Oxford: British Archaeological Reports International Series, **986**.

Colledge, S. and Conolly, J. (2001/2002). Early Neolithic agriculture in Southwest Asia and Europe: re-assessing the archaeobotanical evidence. In: *Archaeology International 2001/2002.* Institute of Archaeology, University College London, pp. 44–46.

Crawford, G.W. (1992). Prehistoric plant domestication in East Asia. In: C.W. Cowan and P.J. Watson (eds) *The Origins of Agriculture: An International*

Perspective. Washington DC: Smithsonian Institution Press, pp. 7–38.

Crawford, G.W. and Shen C. (1998). The origins of rice agriculture: recent progress in East Asia. *Antiquity*, **72**: 858–866.

Denton, G.H. and Hendy, C.H. (1994). Younger Dryas age advance of the Franz-Josef glacier in the southern alps of New Zealand. *Science*, **264**: 1434–1437.

Fairbanks, R.G. (1989). A 17,000-year glacio-eustatic sea level record: influence of glacial melting rates on the Younger Dryas 'event' and deep-ocean circulation. *Nature*, **342**: 637–642.

Fukusawa, H. (1995). Non-glacial carved lake sediments as natural timekeeper and detector of environmental changes. *Quaternary Research Japan*, **34**: 135–149.

Fuller, D.Q., Korisettar, R. and Venkatasubbiah, P.C. (2001). Southern Neolithic cultivation systems: a reconstruction based on archaeobotanical evidence. *South Asian Studies*, **17**: 171–187.

Gasse, F., Téhet, R., Durand, A. *et al.* (1990). The arid–humid transition in the Sahara and Sahel during the last deglaciation. *Nature*, **346**: 141–146.

Gasse, F., Arnold, M., Fontes, J.C. *et al.* (1991). A 13,000-year climate record from western Tibet. *Nature*, **353**: 742–745.

Glover, I.C. and Higham, C.F.W. (1996). New evidence for early rice cultivation in South, Southeast and East Asia. In: D.R. Harris (ed.) *The Origins and Spread of Agriculture and Pastoralism in Eurasia*. London: UCL Press and Washington DC: Smithsonian Institution Press, pp. 413–441.

Golson, J. (1989). The origins and development of New Guinea agriculture. In: D.R. Harris and G.C. Hillman (eds) *Foraging and Farming: The Evolution of Plant Exploitation*. London: Unwin Hyman, pp. 678–687.

Goring-Morris, N., Belfer-Cohen, A. (1998). The articulation of cultural processes and Late Quaternary environmental changes in Cisjordan. *Paléorient*, **23**: 71–93.

Gowlett, J.A.J. and Hedges, R.E.M. (1986). *Archaeological Results from Accelerator Dating*. Oxford: Oxford University Committee for Archaeology, Monograph 11.

Harris, D.R. (1981). The prehistory of human subsistence: a speculative outline. In: D.N. Walcher and N. Kretchmer (eds) *Food, Nutrition and Evolution*. New York: Masson, pp. 15–35.

Harris, D.R. (1987). The impact on archaeology of radiocarbon dating by accelerator mass spectrometry. *Philosophical Transactions of the Royal Society, London, Series A*, **323**: 23–43.

Harris, D.R. (1989). An evolutionary continuum of people–plant interaction. In: D.R. Harris and G.C. Hillman (eds) *Foraging and Farming: The Evolution of Plant Exploitation*. London: Unwin Hyman, pp. 11–26.

Harris, D.R. (1996a). The origins and spread of agriculture and pastoralism in Eurasia: an overview. In: D.R. Harris (ed.) *The Origins and Spread of Agriculture and Pastoralism in Eurasia*. London: UCL Press and Washington DC: Smithsonian Institution Press, pp. 552–573.

Harris, D.R. (1996b). Domesticatory relationships of people, plants and animals. In: R. Ellen and K. Fukui (eds) *Redefining Nature: Ecology, Culture and Domestication*. Oxford: Berg, pp. 437–463.

Harris, D.R. (1998a). The origins of agriculture in Southwest Asia. *The Review of Archaeology*, **19**: 5–11.

Harris, D.R. (1998b). The spread of Neolithic agriculture from the Levant to western Central Asia. In: A.B. Damania, J. Valkoun, G. Willcox and C.O. Qualset (eds) *The Origins of Agriculture and Crop Domestication: The Harlan Symposium*. Aleppo, Syria: ICARDA, pp. 65–82.

Harris, D.R (2002). Development of the agro-pastoral economy in the Fertile Crescent during the Pre-Pottery Neolithic period. In: R.T.J. Cappers and S. Bottema (eds) *The Dawn of Farming in the Near East*. Berlin: ex oriente, Studies in Early Near Eastern Production, Subsistence, and Environment 6, pp. 67–83.

Hassan, F.A. (1997). Holocene palaeoclimates of Africa. *African Archaeological Review*, **14**: 213–229.

Heusser, C.J. (1993). Late-Glacial of southern South America. *Quaternary Science Reviews*, **12**: 345–350.

Higham, C.F.W. (1995). The transition to rice cultivation in Southeast Asia. In: T.D. Price and B. Gebauer (eds) *Last Hunters–First Farmers: New Perspectives on the Transition to Agriculture*. Santa Fe, New Mexico: School of American Research Press, pp. 127–155.

Hillman, G.C. (1975). The plant remains from Abu Hureyra: a preliminary report, pp. 70–73. In: A.M.T. Moore (ed.) The excavation of Tell Abu Hureyra in Syria: a preliminary report. *Proceedings of the Prehistoric Society*, **41**: 50–77.

Hillman, G.C. (1996). Late Pleistocene changes in wild plant-foods available to hunter-gatherers of the northern Fertile Crescent: possible preludes to cereal cultivation. In: D.R. Harris (ed.) *The Origins and Spread of Agriculture and Pastoralism in Eurasia*.

London: UCL Press and Washington DC: Smithsonian Institution Press, pp. 159–203.

Hillman, G.C. (2000). The plant food economy of Abu Hureyra 1 and 2, Abu Hureyra 1: the Epipalaeolithic. In: A.M.T. Moore, G.C. Hillman and A.J. Legge, *Village on the Euphrates: From Foraging to Farming at Abu Hureyra*. New York: Oxford University Press, pp. 327–399.

Hillman, G.C., Colledge, S.M. and Harris, D.R. (1989). Plant-food economy during the Epipalaeolithic period at Tell Abu Hureyra, Syria: dietary diversity, seasonality, and modes of exploitation. In: D.R. Harris and G.C. Hillman (eds) *Foraging and Farming: The Evolution of Plant Exploitation*. London: Unwin Hyman, pp. 240–268.

Hillman, G.C., Hedges, R., Moore, A. *et al.* (2001). New evidence of Lateglacial cereal cultivation at Abu Hureyra on the Euphrates *The Holocene*, **11**: 383–393.

Hopf, M. (1983). The plants found at Jericho. In: K.M. Kenyon and T.A. Holland (eds) *Excavations at Jericho V: The Pottery Phases of the Tell and other Finds*. London: British School of Archaeology in Jerusalem, pp. 580–621.

Islebe, G.A., Hooghiemstra, H. and van der Borg, K. (1995). A cooling event during the Younger Dryas chron in Costa Rica. *Palaeogeography, Palaeoclimatology, Palaeoecology*, **117**: 73–80.

Iversen, J. (1954). The Late-Glacial flora of Denmark and its relation to climate and soil. *Danmarks Geologiske Undersøgelse* II, **80**: 87–119.

Kudrass, H.R., Erlenkeuser, H., Vollbrect, R. and Weiss, W. (1991). Global nature of the Younger Dryas cooling event inferred from oxygen isotope data from Sulu Sea cores. *Nature*, **349**: 406–409.

Kuzucuoglu, C. and Roberts, N. (1997). Evolution de l'environnement en Anatolie de 20 000 à 6 000 BP. *Paléorient*, **23**: 7–24.

Landmann, G. and Reimer, A. (1996). Climatically induced lake level changes at Lake Van, Turkey, during the Pleistocene/Holocene transition *Global Biogeochemical Cycles*, **10**: 797–808.

LaSalle, P. and Shilts, M.W. (1993). Younger Dryas-age readvance of Laurentide ice into the Champlain Sea. *Boreas*, **22**: 25–37.

Leyden, B. (1995). Evidence of the Younger Dryas in Central America. *Quaternary Science Reviews*, **14**: 833–839.

Lin, H. (1992). The origin of rice horticulture in China and its spread to Japan. *Agricultural Archaeology (Nongye Kaogu)*, **1**: 52–73.

MacNeish, R.S., Cunnar, G., Zhao, Z. *et al.* (1997). *Second Annual Report of the Sino-American Jiangxi*

(PRC) Origin of Rice Project (SAJOR). Andover, Massachusetts.

Markgraf, V. (1993). Climatic history of Central and South America since 18,000 yr B.P.: comparison of pollen records and model simulations. In: H.E. Wright, Jr, J.E. Kutzbach, T. Webb, III, W.F. Ruddiman, F.A. Street-Perrott and P.J. Bartlein (eds) *Global Climates Since the Last Glacial Maximum*. Minneapolis: University of Minnesota Press, pp. 375–385.

Mathewes, R.W., Heusser, L.E. and Patterson, R.T. (1993). Evidence for a Younger Dryas-like cooling event on the British Columbia coast. *Geology*, **21**: 101–104.

Mayle, F.E. and Cwynar, L.C. (1995). A review of multi-proxy data for the Younger Dryas in Atlantic Canada. *Quaternary Science Reviews*, **14**: 813–821.

Moore, A.M.T. (1975). The excavation of Tell Abu Hureyra in Syria: a preliminary report. *Proceedings of the Prehistoric Society*, **41**: 50–77.

Moore, A.M.T. and Hillman, G.C. (1992). The Pleistocene to Holocene transition and human economy in Southwest Asia: the impact of the Younger Dryas. *American Antiquity*, **57**: 482–494.

Peteet, D.M. (ed.) (1993). Global Younger Dryas? *Quaternary Science Reviews*, **12**: 277–355.

Piperno, D.R. and Pearsall, D.M. (1998). *The Origins of Agriculture in the Lowland Neotropics*. San Diego: Academic Press.

Price, T.D. (ed.) (2000). *Europe's First Farmers*. Cambridge: Cambridge University Press.

Price, T.D., Taieb, M., Barker, P. *et al.* (1993). Timing of Younger Dryas climatic event in East Africa from lake-level changes *Nature*, **366**: 146–148.

Roberts, N. (1998). *The Holocene: An Environmental History*, 2nd edition Oxford: Blackwell.

Rossignol-Strick, M. (1995). Sea–land correlation of pollen records in the eastern Mediterranean for the Glacial–Interglacial transition: biostratigraphy versus radiometric time-scale. *Quaternary Science Reviews*, **14**: 893–915.

Rossignol-Strick, M. (1997). Paléoclimat de la Mediterranée orientale et de l'Asie du sud-ouest de 15 000 à 6 000 BP. *Paléorient*, **23**: 175–186.

Rossignol-Strick, M. (1999). The Holocene climatic optimum and pollen records of sapropel 1 in the eastern Mediterranean, 9000–6000 BP. *Quaternary Science Reviews*, **18**: 515–530.

Sage, R.F. (1995). Was low atmospheric CO_2 during the Pleistocene a limiting factor for the origin of agriculture? *Global Change Biology*, **1**: 93–106.

Sanlaville, P. (1996). Changements climatiques dans le région Levantine à la fin du Pléistocène supérieur et au début de l'Holocène. Leurs relations avec l'évolution des sociétés humaines. *Paléorient*, **22**: 7–30.

Smith, B.D. (ed.) 1992). *Rivers of Change: Essays on Early Agriculture in Eastern North America*. Washington DC: Smithsonian Institution Press.

Smith, B.D. (1995). *The Emergence of Agriculture*. New York: Scientific American Library.

Sun, X. and Chen, Y. (1991). Palynological records of the last 11,000 years in China. *Quaternary Science Reviews*, **10**: 537–544.

Van der Hammen, T.J.F. and Hooghiemstra, H. (1995). The El Abra Stadial, a Younger Dryas equivalent in Colombia. *Quaternary Science Reviews*, **14**: 841–851.

Van Zeist, W. and Bakker-Heeres, J.A.H. (1982). Archaeobotanical studies in the Levant. 1. Neolithic sites in the Damscus Basin: Aswad, Ghoraifé, Ramad. *Palaeohistoria*, **24**: 165–256.

Van Zeist, W. and Bottema, S. (1982). Vegetational history of the eastern Mediterranean and the Near East during the last 20,000 years. In: J.L. Bintliffe and W. van Zeist (eds) *Palaeoclimates, Palaeoenvironments and Human Communities in the Eastern Mediterranean Region in Later Prehistory*. Oxford: British Archaeological Reports International Series 133, pp. 277–321.

Wheatley, P. (1971). *The Pivot of the Four Quarters: A Preliminary Enquiry into the Origins and Character of the Ancient Chinese City*. Edinburgh: Edinburgh University Press.

Whyte, R.O. (1977). The botanical Neolithic Revolution. *Human Ecology*, **5**: 209–222.

Wright, H.E., Jr. (1977). Environmental change and the origin of agriculture in the Old and New Worlds. In: C.A. Reed (ed.) *Origins of Agriculture*. The Hague: Mouton, pp. 281–318.

Yan, W. (1993). Origins of agriculture and animal husbandry in China. In: C.M. Aikens and N.R. Song (eds) *Pacific Northeast Asia in Prehistory*. Pullman: Washington State University Press, pp. 113–123.

Yen, D.E. (1991). Domestication: the lessons from New Guinea. In: A. Pawley (ed.) *Man and a Half: Essays in Honour of Ralph Bulmer*. Auckland: Polynesian Society, pp. 558–569.

Zhao, Z. (1998). The Middle Yangtze region in China is one place where rice was domesticated: phytolith evidence from the Diaotonghuan Cave, northern Jiangxi. *Antiquity*, **72**: 885–897.

Zilhão, J. (2001). Radiocarbon evidence for maritime pioneer colonization at the origins of farming in west Mediterranean Europe. *Proceedings of the National Academy of Sciences*, **98**: 14180–14185.

PART **8**

Implications for life elsewhere

21

The physical constraints on extraterrestrial life

Monica M. Grady

ABSTRACT

See Plate 34

The likelihood of life beyond the Earth has been a subject of interest to humanity for many years, but it is only recently that it has become possible to consider the prospect of extraterrestrial life on a serious, scientific basis. Owing to the range of conditions over which microbes exist and flourish on Earth, it seems fairly clear that there are many niches throughout the solar system that could provide a suitable habitat in which life might emerge. Terrestrial fossil record indicates that life on Earth started up the evolutionary pathway only a matter of a few hundreds of millions of years following Earth's differentiation, although it has been suggested that primitive life might have started more than once during this first epoch, only to be eliminated during periods of bombardment. When conditions stabilized, after the late heavy bombardment at around 4 Gyr, the evolutionary process got underway.

The evolution of life on Earth records variations that result from environmental changes, or stresses, combined with chance events, for example catastrophic impacts. Although there has been a solar system-wide delivery mechanism via comets and asteroids for the building blocks of life, and there are many environments suitable on other planets and satellites for life to emerge, different stresses and chances acting on the same starting materials would presumably result in the evolution of different types of life from that on Earth. It is far from clear that life exists on other bodies, which is why there is currently so much interest in the study of exobiology and the planning of space missions to investigate and explore other worlds.

In this chapter, I will explore the niches on Earth in which life forms survive, and then consider other places in the solar system, and beyond, where such habitable niches might exist.

21.1 Introduction

The question of whether we are alone in the solar system, galaxy or universe is one which has fascinated humanity since the earliest of times. Stories of mysterious beings from the sky permeate the mythology of many cultures, and make a regular appearance in fiction. The number of 'UFOs' recorded over the past few years also continues to rise, even though fairly mundane explanations account for practically all observations. Despite the more sensational aspects that any serious science purporting to investigate extraterrestrial life has to suffer, both the search for extraterrestrial intelligence (SETI) and the study of exobiology (or astrobiology) have become widely accepted among the academic community as valid and important areas of research. The purpose of this chapter is not to discuss SETI, but to take a very broad view of the origin and evolution of life on Earth, and the possibilities for analogous developments in other planetary environments within the solar system and beyond. These are subjects that come together under the umbrella

Evolution on Planet Earth
ISBN 0-12-598655-6

theme of exobiology, or astrobiology, the study of the origin, evolution, distribution, and future of life in the universe. As a starting point to the search for life elsewhere in the solar system, we can examine the environments in which life exists on Earth, then look for where similar environments might exist elsewhere, both now and in the past.

21.2 Life on Earth

Any attempt to study the origin of life on bodies other than the Earth is necessarily based on comparisons with the Earth. There are two sets of parameters that must be fulfilled if life is to emerge successfully: the correct ingredients, and favourable conditions. The assumptions used to draw this picture are that the building blocks for life are the same wherever we choose to look for life, and the processes that operate on these building blocks are governed by the same rules of physics and chemistry as on the Earth – rules that arise from the fundamental atomic structure and bonding properties of the different elements. The building blocks of life on Earth are the elements C, N, O and H, four of the five most abundant elements in the cosmos (the fifth being helium); phosphorus and sulphur are also essential.

All life is carbon-based: carbon has an atomic structure that imparts unique properties to the element. It can form chains and rings of atoms, giving rise to an infinite host of organic compounds capable of solution in either polar or non-polar solvents, depending on the composition of attached functional groups. No other element has the ability to produce stable chains and rings in this way. The only element that approaches carbon in its versatility is silicon, which forms extended chains, rings and polyhedra when combined with oxygen. The $Si-O$ bond is the basis of the vast inorganic chemistry of the rock-forming minerals; clay minerals have been proposed as possible templates for a mineral-based origin of life (Cairns-Smith, 1982), but the hypothesis is not widely accepted. If, for the reason of its

atomic structure, carbon is the basis of life on Earth, then it is perhaps a logical assumption to make that it is also the basis of life elsewhere in the solar system.

How did life emerge on Earth? This is a question that has occupied humanity for generations. Charles Darwin, in a letter to J.D. Hooker in 1871, wrote: 'But if (and oh! what a big if!) we could conceive in some warm little pond, with all sorts of ammonia and phosphoric salts, light, heat, electricity, etc. present, that a protein compound was chemically formed ready to undergo still more complex changes ... ' This prescient description encompasses most of the conditions now generally believed necessary for the emergence of life.

We do not know with absolute certainty the precise details or mechanisms, but first Oparin (1924, translated in 1936), and independently Haldane (1929) outlined a sequence of events that started with the abiotic synthesis of simple molecules, possibly brought to the early Earth from space. Miller (1953) carried out a series of experiments that showed very successfully that a mixture of simple molecules such as H_2, H_2O, NH_3 and CH_4, when exposed to a spark discharge, could form more complex molecules, including amino acids. These experiments were among the first indicators that biologically significant molecules could be produced from inorganic starting materials by abiogenic processes.

The evolutionary sequence of simple molecules reacting to form more complex molecules was presumably followed by development of a template for self-replication (e.g. Joyce, 1989). Subsequently, an outer membrane separated the molecules from the surrounding medium (e.g. Deamer, 1998). However, as Miller (1953) showed, simple molecules by themselves will not react to form more complex systems: the building blocks must be brought together with sufficient energy to allow reactions to occur. A suitable substrate is also required: liquid water (a polar solvent stable over a very wide temperature range) is the most effective medium for solution and transport of organic molecules, and for that reason its presence is regarded as a prime requirement for the development of life.

There are many steps between the potpourri of carbon-bearing molecules in solution and a self-replicating organism, steps that could not occur on the Earth until a stable system had emerged from the chaos of the planetary formation process.

The geological history of the early Earth is inferred from the study of meteorites and asteroids: the Sun, the Earth and other planets, along with asteroids, comets and other minor solar system bodies formed from the cloud of gas and dust that was the pre-solar nebula (Figure 21.1). Gradual accretion of dust into proto-planetary bodies was accompanied by aggregation, collision, break-up, re-aggregation, melting and differentiation of material in a complex history of planetary build-up. The final turbulent stages of solar system formation were traced out by intense cratering of the planets by asteroids and comets. The surface of the Earth was inimical to life, as it was heated and melted during bombardment (e.g. Maher and Stevenson, 1988; Sleep *et al.*, 1989). It is probable that during this period of the Earth's history, an atmosphere built up and was stripped away more than once, in a cycle of increasing stability punctured by episodes of bombardment. Gradually, however, the inner solar system became more quiescent; the Earth's surface cooled; its atmosphere was retained; oceans formed and conditions were

set to allow life to emerge (e.g. Chyba, 1993). Of this stage, the interface between the geological and biological history of the Earth, little is known with certainty.

Where on Earth did life begin? We have already considered the importance of water as a solvent or transport medium in the steps leading up to the formation of biotic molecules. It is therefore probable that life originally emerged from a watery environment – but whether from surface waters or the deep ocean floor is not known. Until fairly recently, the likely energy source for pre-biotic reactions was thought to be the Sun, implying that surface waters were where organisms first originated. However, the recognition of hydrothermal vents on the deep ocean floor (Francheteau *et al.*, 1979), where enormous amounts of geothermal energy are available for driving chemical reactions (Edmond *et al.*, 1982), has opened up the possibility that the emergence of life might have occurred at depth (e.g. Corliss *et al.*, 1981; Gold, 1992), although there are problems with this hypothesis (Miller and Bada, 1988).

Wherever it first took place, life appears to have arisen on Earth almost as soon as it was possible to do so – traces in the fossil record date back to ~3.85 Gyr (Mojzsis *et al.*, 1996), a gap of only ~700 Myr from accretion and differentiation of the planet. In the time intervening between the emergence of life and the

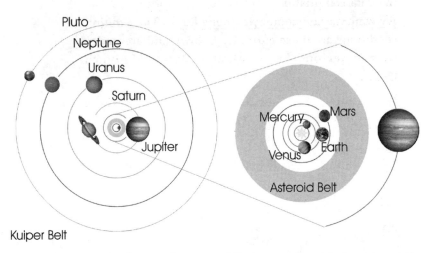

Figure 21.1 The solar system. The distance between orbits is approximately to scale; relative planet sizes are not to scale.

present day, there has been colonization by microbes of every environment in which it is possible for life to survive. The conditions prevailing in these 'habitable niches' range from extremes of temperature, salinity, etc., and are taken to define a 'biological envelope', within which life can survive, grow and evolve (e.g. Cowan, 1999). The currently recognized limits of the biological envelope are summarized in Table 21.1; the search for life elsewhere within the solar system is guided by these limits.

21.3 Life elsewhere within the solar system

A voyage through the solar system to search for traces of past or present life is, in effect, a search for liquid water. The concept of a 'habitable zone' (HZ) around a star was taken by Hart (1978) as a fairly narrow region in which the surface temperature of a planet was appropriate for liquid water to be stable. Within our solar system, only the Earth orbits at the requisite distance from the Sun for this parameter to be met. The 'Goldilocks hypoth-

esis', where conditions are not too hot or too cold, but are just right (e.g. Rampino and Caldeira, 1994) is an extension of the HZ concept, taking account of changes in solar luminosity, atmospheric composition, albedo, cloud cover, etc.

Even though more recent models predict a slightly wider HZ around the Sun and similar stars (Kasting et al., 1993), the HZ does not reach out as far as Mars or Jupiter. Other effects must also be considered to account for the possibility of viable habitats on these planets and their satellites, such as the presence of thicker atmospheres, or tidal heating.

But why should water, or indeed any of the ingredients necessary for life, be present on other bodies? The planets and their satellites all formed from the same primordial material as the Earth, and so presumably originally contained similar relative amounts of volatiles (water, organic compounds, etc.). But the thermal and evolutionary histories of the planets and their satellites differ considerably, leading to variations in processes such as degassing, atmosphere formation and loss. Indeed, it is conceivable that strongly heated bodies might become almost volatile-free.

Table 21.1 Environmental constraints for extremophilic microorganisms. Adapted from D.A. Cowan (1999) using the data compilation of D.M. Roberts (http://www.nhm.ac.uk/zoology/extreme.html). The limiting conditions are those currently known, and are under constant revision. See also Rothschild and Mancinelli (2001).

Parameter	Limiting conditions	Type of organism
Water	Liquid water required	
Temperature	−2° C (minimum) 50–80° C 80–115° C	Psychrophiles Thermophiles Hyperthermophiles
Salinity	15%–37.5% NaCl	Halophiles
pH	0–4 8–12.5	Acidophiles Alkalophiles
Atmospheric pressure	Up to 110 MPa	Barophiles

One process that reversed volatile loss became significant during the final stages of solar system formation, and still occurs today – the addition of volatiles from cometary bombardment. Relative age-dating of the distribution of craters on the Moon shows that at around 4 Gyr ago, the Earth/Moon system suffered intense bombardment, presumably by asteroids and comets. Although the Earth's surface would initially be sterilized by the bombardment (Sleep *et al.*, 1989), as the impacts decreased in frequency it is possible that incoming impactors delivered water and organic compounds to the surface of the Earth, allowing the atmosphere to stabilize, oceans to form and providing the ingredients for life to start (e.g. Oró, 1961; Anders, 1989; Chyba *et al.*, 1990). Comets, dominantly icy bodies rich in organic compounds, could deliver such materials to the Earth and also deliver them by the same mechanism to the other planets and their satellites. The volatile inventory of solar system bodies is, therefore, a combination of original material that has undergone fractionation during the heating that accompanied the planetary accretion process, plus relatively unprocessed primordial material from cometary bombardment.

There is thus the potential for life to have started in other habitable niches throughout the solar system (although Ward and Brownlee (2000) conclude that specific properties of the Earth, such as its large satellite, are also necessary for the evolution and viability of higher species). The planet Mars has long been favoured as a possible host for extraterrestrial life; more recently, the Galilean satellites of Jupiter (particularly Europa) have been proposed as appropriate bodies on which to search for traces of life, and Titan, the giant satellite of Saturn, exhibits many properties that relate to conditions at the surface of the early Earth prior to the emergence of life. Each of these bodies will be considered in the following sections.

21.3.1 *Mars*

Ever since the *canali* of Schiaparelli's nineteenth century map of Mars were mistranslated into English as canals (rather than grooves or channels), there has been speculation that life might have existed on Mars. Although the *canali* were found to be mainly illusions resulting from the poor optics used at the time to observe Mars, images returned by the *Mariner* and *Viking* spacecraft clearly showed that the Martian surface was marked by features similar to those carved out by rivers and streams on the Earth's surface. Like the Earth, Mars is a rocky planet. It has a radius approximately half, and a mass around one tenth that of the Earth; in consequence, gravity on Mars is only about 40% that of the Earth. Mars' atmosphere is also different from that of the Earth: it is much thinner, around 6 mbar (600 Pa), compared with 1000 mbar (10^5 Pa), and is dominantly carbon dioxide ($\sim 95\%$) rather than nitrogen. The thin atmosphere provides the Martian surface with little protection from heat loss, thus the average daily temperature is around $-60°C$. Temperatures may reach $+30°C$ at the equator in summer, and fall to $-130°C$ at the poles in winter (Kieffer *et al.*, 1992). Mars has a core–mantle structure similar to the Earth, but seems to have a rigid crust rather than the more flexible plate structure of the Earth, although recent results from the magnetometer on NASA's *Mars Global Surveyor* indicate that there might be evidence for limited tectonic spreading (Connerney *et al.*, 1999).

Mars exhibits extreme examples of the features shown by the Earth. For example, the biggest volcano in the solar system occurs on Mars: Olympus Mons is a shield volcano almost three times as high as Mount Everest. Shield volcanoes on Earth are formed as piles of magma that build up when the crust is above a hot spot in the mantle. Plate movement over the mantle hot spot prevents continuous accumulation of a magma pile in a single location. In contrast, on Mars where there is little or no plate motion, the volcanoes simply increase in size.

The first detailed topographic maps of Mars were produced with data from NASA's *Mariner 9* orbiter mission of 1971–1972, in which channel and valley networks, volcanoes, canyons and craters were imaged at reasonable resolution for the first time (e.g. Masursky,

1973). *Mariner 9* was also the mission that returned pictures of layered (possibly sedimentary) terrain in Mars' polar regions (e.g. Murray *et al.*, 1972; Blasius *et al.*, 1982). Five years after *Mariner 9*, in 1976, NASA's two *Viking* landers sent back many images of Mars' landscape, showing panoramic scenes of broken boulders distributed over flat dusty plains. *Viking* also measured the elemental composition of both Mars' atmosphere (e.g. Nier *et al.*, 1976; Owen *et al.*, 1977) and surface soils (e.g. Clark *et al.*, 1976, 1982).

In combination with the *Mariner* data, the *Viking* results have allowed a picture of Mars as a rocky planet with a significant geological history to be built up (e.g. Tanaka *et al.*, 1992). The two hemispheres of Mars exhibit very different geological histories. Most of the northern hemisphere consists of almost flat, low-lying plains, showing little cratering. In contrast, the terrain of the southern hemisphere appears more ancient, with cratered highland regions cross-cut by canyons, channels and valley networks.

But many questions about Mars remained unanswered, particularly about the fluvial and seismic history of the planet. The *Pathfinder* mission of 1997 landed on a rocky plain at the mouth of the Ares Vallis in Chryse Planitia (19.33°N, 33.55°W). During its month of operation, it recorded spectacular images of a rock-strewn plain, with tantalizing glimpses of rounded pebbles and possible layered structures and hollows within some of the rocks (Golombek *et al.*, 1997). Chemical and image data for rocks and soil were acquired by the *Sojourner* rover, a mobile robotic probe. Interpretations of the *Sojourner* and *Pathfinder* data were published in a series of papers by several authors in a special issue of the journal *Science* (1997, **278**: 1734–1774). Inferences drawn from these data are that some of the rocks might be sedimentary, even conglomerates, thus implying a fluvial history. Paler, almost cream-coloured patches in the soil might be areas leached by fluid action, hard-grounds or evaporite deposits. The poorly sorted landscape of part-rounded pebbles and boulders has been interpreted as the type of landscape remaining after catastrophic

flooding, further evidence of the stability of liquid water at some time in Mars' past.

Some of the latest images from the orbital camera on *Mars Global Surveyor* suggest that liquid water might have percolated to the Martian surface in much more modern times than was previously thought (Figure 21.2), perhaps as recently as 1 million years ago, and indeed might even still be present in subsurface locations today (Malin and Edgett, 2000). Water is the chief agent of weathering and erosion on Earth. Mars is a much drier, colder planet on which liquid water cannot exist very long at the surface because it will immediately begin to boil, evaporate, and freeze – all at the same time. However, new pictures from the Mars Orbiter Camera (MOC) onboard the *Mars Global Surveyor* (MGS) have provided observations which suggest that liquid water may very recently have played a role in shaping some gully-like features found on the slopes of various craters, troughs, and other depressions on the red planet.

The pictures in Figure 21.2 compare a Martian gully (left) with one on Earth (right). The following information is taken from the figure caption originally released with publication of the image (NASA image PIA 01031). The terrestrial gully was formed by rainwater flowing under and seeping along the base of a recently deposited volcanic ash layer. For Mars, water is not actually seen, but is inferred from the similarity of the landforms to examples on Earth.

The MOC image covers an area 1.3 km (0.8 mile) wide by 2 km (1.2 miles) long. The pictures from the flank of the Mount St Helens volcano in Washington (right; large image and inset) were taken by MGS MOC Principal Investigator Michael C. Malin in the 1980s after the eruptions of May 1980. They are illuminated from the left; note footprints on left side of the picture for scale, also note the bar, which is 30 cm (11.8 in) long.

Data from the laser altimeter aboard the *Mars Global Surveyor* have been interpreted as evidence for a very large ocean across Mars' northern hemisphere (Head *et al.*, 1999). For water to have been present at Mars' surface, the Martian atmosphere must have been

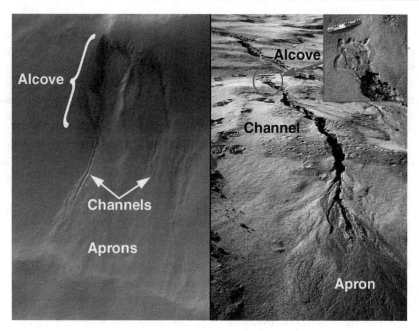

Figure 21.2   Water on Mars – see text for details (NASA/JPL/Malin Space Science Systems).

much thicker, and surface temperatures much warmer than they are today (e.g. Pollack *et al.*, 1987; Squyres and Kasting, 1994). A thicker atmosphere engenders greater protection from solar radiation; in a previous epoch, Mars would have had a warmer and wetter climate and provided all the conditions suitable for the emergence of life (e.g. McKay *et al.*, 1992). Given this framework, it is not surprising that so much interest has been focused on a potential Martian biosphere.

The observational evidence that the surface environment of Mars has been more amenable for life in the past inspired experiments on the *Viking* landers. Both craft had instrument payloads that included a gas chromatograph-mass spectrometer (GC-MS) and three separate biological experiments (e.g. Klein *et al.*, 1976). The GC-MS was designed to test surface soils for the presence of organic compounds, and to identify them if present. The biological experiments were designed to test for metabolic action. Unfortunately, the results obtained were, on balance, negative (e.g. Klein, 1978, 1979). Although one of the experiments did give a positive signal that might have implied the presence of a metabolizing agent, the over-

all conclusion from the mission was that there was no detectable trace of organic matter in the surface soils at either of the two landing sites, and thus no likelihood of the presence of extant life (Klein *et al.*, 1992).

Despite these conclusions, Mars has remained a prime target in the search for extraterrestrial life. The recognition of extremophile microorganisms that can survive (and thrive) in conditions previously thought inimical to life, but that might exist at or below the surface of Mars has given impetus to remote Martian exploration. Since the *Viking* mission, there have been two very successful missions to Mars (NASA's *Pathfinder* and *Global Surveyor*). Although neither of these missions included experiments specifically intended to test for traces of past or present life at the Martian surface, they are part of NASA's programme of missions leading up to the return of samples from Mars to Earth.

Before the sample return mission takes place, several other spacecraft will visit Mars. NASA's *Mars Odyssey* orbiter arrived at the planet in October 2001; one of its mission objectives is to search for signs of water in the Martian surface layers. The European Space

Agency's (ESA) *Mars Express* mission will be launched in June 2003, arriving at Mars in late December 2003. The payload comprises an orbiter and the *Beagle2* lander. The orbiter will pass around Mars in an elliptical polar orbit for approximately 1 Martian year (2 terrestrial years), with a closest approach of around 250 km, probing the atmosphere and mapping the Martian surface. The *Beagle2* lander will have on board a fully integrated package designed to search for the chemical traces of past (and present?) life on Mars, by examining surface and subsurface soil and rock samples and the atmosphere (Pillinger *et al.*, 1999). Part of the landing package will be a gas analysis package (GAP; Wright *et al.*, 2000) which, like the GC-MS on the *Viking* landers, will determine the amount and type of organic carbon in the Martian soil. The *Beagle2* instrument, however, also will measure the abundance of inorganic carbon (e.g. as carbonate minerals) in the soil, and the carbon, nitrogen, hydrogen and noble gas isotopic compositions of components within soils, rocks and the atmosphere (Wright *et al.*, 2000).

Vital to interpretation of much of the chemical data returned by *Viking* and *Pathfinder* have been data obtained from Martian meteorites (e.g. McSween, 1994). Analyses of Martian meteorites have also added impetus to the resurgence of interest in exobiology. The SNCs (named for the type specimens of the three original subgroups: Shergotty, Nakhla and Chassigny) are all igneous rocks and comprise a group of (currently) 20 meteorites (25 fragments in total) that can be distinguished from 'regular' meteorites (that originate in the asteroid belt) on the basis of their younger (and therefore planetary) crystallization ages. The observation that at least one of the SNCs, EETA79001 (a specimen found in the Elephant Moraine region of Antarctica in 1979), contained indigenous (i.e. Martian) organic material associated with carbonate minerals (Wright *et al.*, 1989) sparked a debate on the possibility of Martian meteorites containing evidence for extraterrestrial life.

This work was followed, in 1994, by the discovery of similarly enhanced levels of organic carbon in close association with the abundant rosettes of bright orange-coloured carbonate that are so abundant in a second Antarctic Martian meteorite, ALH 84001 (Grady *et al.*, 1994, 1999) (Figure 21.3). ALH 84001 became a further focus of interest in 1996 when a team of scientists led by David McKay of

Figure 21.3 Rosettes of orange carbonate on the surface of Martian meteorite ALH 84001. Image size approx. 3 mm.

NASA's Johnson Space Center in Houston reported the description of nanometre-sized features within carbonate patches in ALH 84001, and claimed to have found evidence for a primitive 'fossilized martian biota' (McKay *et al.*, 1996). Identification of the 'nanofossils' remains controversial, since much of the evidence is circumstantial and relies on the coincidence between a number of otherwise unrelated characteristics of the meteorite (carbonate minerals in ALH 84001; organic compounds and magnetite associated with the carbonates, and the external morphology of the carbonates). Indeed, the formation conditions of both the carbonate and magnetite, the relevance of the organic compounds and interpretation of the morphology of the 'nanofossils' have all been subject to detailed investigation by several groups of scientists. A brief outline of the main arguments for and against the interpretation of McKay *et al.* (1996) follows; an expanded version of the discussion is contained in Grady *et al.* (1999).

The abundant carbonates within ALH 84001 are the host of the features portrayed as 'fossils', and so understanding the environment and mode of formation of the carbonates is crucial to correct interpretation of the features. Although carbonates are a common Antarctic weathering product, it is certain that these particular minerals formed on Mars: they have a carbon isotopic composition well outside the range of terrestrial carbonates (Romanek *et al.*, 1994). Based on mineralogy alone, the carbonates were originally believed to have formed at temperatures $\sim 700°C$ (Mittlefehldt, 1994), possibly a consequence of alteration by fluids during impact heating (Harvey and McSween, 1996); supporting evidence came from other petrographic studies (e.g. Scott *et al.*, 1997). Such conditions are clearly incompatible with primitive life. In contrast, stable oxygen isotope compositions of the carbonates indicate that low-temperature (<200°C, and closer to 0–80°C) hydrothermal processes were more likely to be responsible for production of the minerals (Romanek *et al.*, 1994; Valley *et al.*, 1997; Leshin *et al.*, 1998; Saxton *et al.*, 1998), an observation corroborated by detailed mineralogical studies (Treiman, 1995). Since few

hydrated minerals have been identified among the alteration products in ALH 84001, McSween and Harvey (1998) and Warren (1998) independently proposed that the carbonates in ALH 84001 were produced at the surface of Mars in a region of restricted water flow, such as an evaporating pool of brine. The new hypothesis satisfactorily accounts for the chemical and isotopic characteristics of the carbonates, and is also a mechanism that is compatible with an environment in which microorganisms might survive.

The elevated abundance of organic carbon in ALH 84001 was first noted by Grady *et al.* (1994), although its identification as Martian or terrestrial was not settled. McKay *et al.* (1996) claimed that a proportion of the organic carbon occurred as polycyclic aromatic hydrocarbons (PAHs) concentrated in the interior of the sample. PAHs are common on Earth, produced during diagenetic alteration of biological precursors in sedimentary rocks. The mass abundance distribution of the PAHs in ALH 84001 is different from that in terrestrial sediments, and thus might possibly be indigenous to the meteorite. PAHs, however, are present in many meteorites and probably also occur in interplanetary space as molecules evaporated from comets, where they have a purely nonbiogenic origin, so their presence in ALH 84001 cannot be taken as firm evidence for past biological activity on Mars. Further analysis of the organic carbon in ALH 84001 has shown that while much of it is terrestrial contamination (Bada *et al.*, 1998), at least some of it is Martian (Jull *et al.*, 1998; Wright *et al.*, 1998; Grady *et al.*, 1999), still leaving open the possibility that trace evidence of past life occurs in ALH 84001.

One of the main observations that McKay *et al.* (1996) used to argue for the presence of fossilized bacteria in ALH 84001 was of the morphology of various structures found within the carbonate patches, particularly of the tiny (10–20 nm) magnetite grains concentrated within the carbonates. McKay *et al.* (1996) suggested that the grains were morphologically similar to biologically produced magnetite. Features in other types of magnetite have been interpreted as indicative of production by direct condensation from a vapour at high temperatures (Bradley *et al.*, 1996, 1998), conditions under which bacteria are unlikely to survive. Unfortunately, since the magnetite grains are small and widely distributed, different investigators cannot be certain they are looking at the same populations of magnetite grains, resulting in a lack of consensus as to the origin and significance of the magnetite.

The freshly broken surfaces of the carbonate grains exhibit parallel ridge-like features which were interpreted as a fossilized biota (McKay *et al.*, 1996). The features, however, are smaller by about two orders of magnitude than most common bacteria known on Earth (e.g. Schopf, 1983). Although this is indeed the case, subsequent investigation by Thomas-Keprta *et al.* (1998a) has revealed that nanobacteria might be more prevalent than previously anticipated and nanometre-sized organisms have been isolated from terrestrial sedimentary rocks (Uwins *et al.*, 1998). Thomas-Keprta *et al.* (1998b) have also suggested that the smallest features might be biofilms. Even so, many doubts still arise as to the validity of interpreting morphological data as fossil nanobacteria, and it has been suggested that the features are simply irregularities in the surface of the carbonates (Bradley *et al.*, 1998).

The same Houston-based team announced, in March 1999, that they had identified features similar to those in ALH 84001 in a second Martian meteorite, Nakhla (McKay *et al.*, 1999). Despite this announcement, and the vast amount of research that has been undertaken on ALH 84001, there is no consensus as to whether or not the features described by McKay *et al.* (1996) are indeed the remnants of Martian fossils. In the absence of suitable meteoritic material, the answer to the question of the existence (either now or in the past) of life on Mars must await results from future space missions to the red planet.

21.3.2 *The Jovian satellites*

Jupiter has 16 satellites, the four brightest of which (Io, Europa, Ganymede and Callisto) were discovered in 1610 by Galileo Galilei (hence Galilean satellites). Astronomical observations over the past 360 years allowed basic information about the satellites to be built up: diameter, mass, density, orbital period, albedo, spectral signature, etc. From these parameters the bulk composition of the satellites could be deduced, from which it became apparent that the four satellites were very different from each other. Both the *Voyager 1* and *2* missions of 1979, and the *Galileo* mission that started in 1995 (and is still in progress) have provided spectacular images and a wealth of data for the Galilean satellites. It is clear that the four bodies exhibit a range of properties and environments that maintain the potential to harbour life.

21.3.3 *Io*

Io is the innermost of the Galilean satellites (but not the innermost satellite of all: there are four additional minor moons of Jupiter that orbit the giant planet closer than Io), orbiting at a distance of $\sim 400\,000$ km. Io has a bulk density ($\sim 3.5\,\mathrm{g\,cm^{-3}}$) that implies it is a silicate body; recent gravitational and magnetic data from the *Galileo* mission suggest that Io has an iron-rich core surrounded by a molten silicate interior overlain by a thin silicate crust (Anderson *et al.*, 1996a).

Io is strongly heated by the tidal forces of Jupiter's gravitational attraction, a frictional force so great that Io exhibits substantial thermal activity. In fact, Io has one of the most energetic surfaces in the solar system, with silicate lava flows from active volcanoes frosted by sulphur deposits emanating from geysers and vents (McEwan *et al.*, 1998; LopesGautier *et al.*, 1999). The surface temperature range is between $-150°C$ and $+300°C$, depending on locality and proximity to thermal vents.

Even though there is no indication that water is present on Io, either at its surface as ice, or as water vapour, hydrogen or oxygen in its extremely tenuous atmosphere, Io is included in the list of potential habitable bodies on the basis of its thermal properties. There are several sulphur-loving species on Earth that survive more than adequately in hot springs, around geysers and in volcanic craters (e.g. Stetter, 1998). It is therefore possible that similarly extremophilic organisms might have carved out a niche on Io.

21.3.4 Europa

Europa is the smallest of the Galilean satellites, with a radius a little under that of the Earth's Moon. Plate 35 (in the plate section) shows two views of the trailing hemisphere of Europa.

Europa orbits Jupiter at a distance of ~ 600 000 km, a distance sufficiently close for the satellite to be heated by the tidal forces of Jupiter's gravitational attraction. Most of the information and images that we have of Europa have been acquired by NASA's recent *Galileo* mission, and so we now have a fairly accurate picture of Europa's bulk density, surface composition, magnetic properties and appearance.

Europa's density ($\sim 2.97\,\mathrm{g\,cm^{-3}}$) implies that it is a silicate body, but with a significant water ice content. Recent models based on gravity and magnetic data have suggested a differentiated structure for Europa, comprising a metallic core, overlain by a silicate mantle, over which resides a crust of water ice (Anderson *et al.*, 1997a, 1998a). The icy crust appears to be ~ 150 km thick, and might also have a layered structure, with a substantial subsurface salt-rich ocean beneath a thinner shell of water ice (Carr *et al.*, 1998; Pappalardo *et al.*, 1999).

If there is indeed a subsurface liquid ocean on Europa, then the heat source that keeps it liquid is likely to come from the complex interplay of the orbiting Galilean satellites with Jupiter: both Io and Ganymede exert minor tidal influences on Europa, in addition to the much larger tidal effect emanating from the parent planet itself (Greenberg *et al.*, 1998). There has been much speculation that Europa's ocean might be heated from the bottom upwards, by hydrothermal vents analogous to those found on the deep ocean floor of the Earth.

It has been suggested recently that sufficient chemical energy could be produced from radiation to maintain microbial life at Europa's surface (Chyba, 2000). Study of the subsurface inland Lake Vostok beneath the ice of the Antarctic plateau (Kapitsa *et al.*, 1996) is being employed as a pathfinder case study prior to the future exploration of Europa and its ocean (e.g. Bell and Karl, 1998).

21.3.5 Ganymede and Callisto

Ganymede and Callisto are the two largest of Jupiter's satellites, with similar radii (~ 2500 km) and densities ($\sim 2\,\mathrm{g\,cm^{-3}}$). They are, however, very different in appearance, testament to their differing evolutionary histories. Even so, both satellites have properties that imply they might provide habitable niches for extraterrestrial organisms. Magnetic and gravitational data from the *Galileo* mission have allowed models to be proposed for the internal structures of Ganymede and Callisto. The former appears to be differentiated: it has a small iron core overlain first by a silicate and then by an icy mantle, encased in a crust of ice (Anderson *et al.*, 1996b), whereas Callisto is an undifferentiated mixture of rock and ice with no core, and is again topped by a crust of ice (Anderson *et al.*, 1997b, 1998b). Like Europa and Io, Ganymede is influenced by Jupiter's gravitational heating.

Ganymede also appears to have an intrinsic magnetic field, implying that its core is still partially molten (Schubert *et al.*, 1996). By analogy with Europa, it is possible that at least part of Ganymede's icy mantle might yet be liquid. In contrast, Callisto has no intrinsic magnetic field, and thus no molten core (Khurana *et al.*, 1997). However, the Galileo mission did record magnetic anomalies when it flew by Callisto, suggesting interaction between Jupiter's magnetosphere and its satellite (Khurana *et al.*, 1998). One interpretation of the measurements is that the magnetic field was induced by a salt-rich ocean; models suggest that such a global ocean on Callisto would be between 10 and 100 km deep. The heat source that has kept such a layer liquid is unknown: Callisto is unaffected by tidal disruption from Jupiter (Khurana *et al.*, 1998).

Table 21.2   Web-site addresses for planetary missions

Body	Mission	Web site address
Mars	Mariner	http://nssdc.gsfc.nasa.gov/planetary/planets/
	Viking	http://nssdc.gsfc.nasa.gov/planetary/Viking.html
	Pathfinder	http://mars.jpl.nasa.gov/
	Global Surveyor and Odyssey	
	Mars Express	http://sci.esa.int/marsexpress
	Beagle2	http://www.beagle2.com
Jupiter	Voyager	http://voyager.jpl.nasa.gov
	Galileo	http://galileo.jpl.nasa.gov
Saturn (Titan)	Huygens	http://sci.esa.int/huygens
Earth-like planets	Darwin	http://sci.esa.int/darwin
	Terrestrial Planet Finder	http://tpf.jpl.nasa.gov/

21.3.6   *Titan*

The gas giant Saturn orbits the Sun at a mean distance of $\sim 9.5\,AU$. With its spectacular ring system composed of a myriad of blocks of ice and rock, Saturn is probably the most readily recognized of all planets. The complexity of the ring system, and the variety of properties shown by its numerous satellites were recorded by the *Voyager* fly-by missions of 1980 and 1981.

Titan is by far and away Saturn's largest satellite, and is the second largest satellite in the solar system (after Jupiter's Ganymede). Its density of $\sim 1.88\,\mathrm{g\,cm^{-3}}$ implies that it is a mix of rock and ice. Titan was a prime target for *Voyager 1*, since data from ground-based observations and also from the 1979 *Pioneer 11* encounter had shown that Titan had an atmosphere of nitrogen, methane and ammonia. There were great hopes that *Voyager 1's* encounter with Titan would provide images of the surface of this enigmatic body. Unfortunately, complete coverage of the satellite by thick clouds prevented successful acqui-

sition of surface pictures. Even so, the instruments on board the *Voyager* craft were able to acquire sufficient data to build up a much better picture of the enigmatic satellite.

Titan is enveloped by a thick atmosphere, the main component of which, like the Earth, is nitrogen (e.g. Raulin, 1998). However, unlike the Earth, the remaining constituents are methane and argon, with trace amounts of higher hydrocarbons (ethane, propane, ethyne, etc.), carbon monoxide, hydrogen cyanide and cyanogen (Smith *et al.*, 1982). Radar measurements from Earth (Muhleman *et al.*, 1990), coupled with observations with the Hubble Space Telescope (Smith *et al.*, 1996), have built up a picture of Titan as having a rough rocky and icy surface with a temperature $\sim -200°C$. Condensation of ethane from the atmosphere has resulted in the build-up of lakes and ponds of ethane in which ammonia and methane are dissolved (Lunine *et al.*, 1983).

Although there is no suggestion that the environment of Titan is in any way suitable to harbour life, the atmospheric composition and surface conditions are sufficiently primitive that they have been taken as possible analogues to the conditions extant on the early Earth, prior to development of the biosphere. Hence understanding the atmospheric and surface processes on Titan is a key to understanding the processes that led to the evolution of life on Earth (e.g. Clarke and Ferris, 1997; Raulin, 1998). As a consequence of this significance, Titan is the target of *Huygens,* an ESA-led probe (part of the joint NASA-ESA *Cassini-Huygens* mission to the Saturnian system, scheduled to arrive in January 2005), which will descend for several hours through Titan's atmosphere. The probe will acquire images, measure the elemental and isotopic composition of the atmosphere and record its environmental conditions right down to its crash- (or splash-) landing on the surface.

21.4 Life beyond the solar system

The study of exobiology is not confined to the solar system, but extends to the search for planets orbiting other stars. Astronomical observations over the past few years led first to the detection of three approximately Earth-sized planets orbiting pulsar PSR 1257+12 (Wolszczan and Frail, 1992; Wolszczan, 1994), and then to the detection of a Jupiter-sized planet orbiting a Sun-type star (51 Pegasi; Mayor and Queloz, 1995). Approximately 100 giant planets have now been detected around nearby solar-type stars, using variations in stellar motion as an indirect indicator of their presence (Marcy and Butler, 1998).

None of the planets discovered thus far has had the characteristics that would designate it as a likely candidate for further study as the host for life: the planets have been Jupiter-like gas giants orbiting close in to the central stars, frequently much closer than Mercury is to the Sun in our own planetary system. The next stage in searching for extra-solar planets can only come with improvements in techniques that will allow the detection of Earth-like planets (ELPs) orbiting Sun-type stars. Both ESA and NASA have missions in the planning stage (*Darwin* and *Terrestrial Planet Finder* (TPF),

respectively), which utilize space-based interferometer arrays, and will have the capability both to detect ELPs and to determine their spectral characteristics.

21.5 Signatures of life

The signature of life on Earth is indisputable – any spacecraft passing in close proximity to our planet would detect evidence for life (abundant water and organic material), and even of intelligence (regular thermal and radio emissions). Life was successfully detected on Earth by the *Galileo* spacecraft as it flew by on its journey to Jupiter (Sagan *et al.,* 1993). Unfortunately such clear indication of life on other solar system bodies is very unlikely. What should any mission to detect extraterrestrial life be looking for? Signatures might remain as traces in an atmosphere, in surface layers or in subsurface regions; all possible niches must be examined. For life on Earth to succeed, the importance of water and organic compounds has been stressed; any search for extraterrestrial life is therefore likely to include, at the very least, the search for water and carbon.

The relevant signatures fall into two groups: those that can be measured directly, and those that are observed remotely. In the former category, measurements of organic compounds are likely to be of the greatest significance. The simple presence of organic molecules is, of course, insufficient evidence to infer a biota, since organic compounds can have a purely abiogenic origin: they are found in comets and meteorites, on Titan, etc. There are, however, two analytical techniques that can be used to distinguish the by-products of biological processes if organic molecules are detected *in situ.* The first method measures the optical chirality of the molecules. On Earth, stereoisomers of an organic compound (e.g. the amino acid alanine) produced abiologically generally form a racemic mixture (i.e. contain an equal mixture of the D- and L-enantiomers), whereas the same compound produced biologically is present as only the L-enantiomer. It is possible that a similar effect might occur with extraterrestrial biology, although there are some astrophysical environments in which enantiomeric excesses can be generated, for example irradiation of presolar material by circularly polarized ultraviolet light (e.g. Cronin and Pizzarello, 1997).

The second method for determining the biogenic, or otherwise, nature of extraterrestrial organic compounds uses the isotopic composition of the material as a marker. Non-biological systems tend towards equilibrium. Biology introduces disequilibrium, for example during carbon fixation the isotopic composition of carbon becomes fractionated, such that the ratio of ^{13}C to ^{12}C is lower in the product relative to the starting material. As a result, biologically produced organic matter is depleted in ^{13}C relative to the source carbon. Assuming an optimistic scenario, when searching for life, say on Mars, the final products might be waste products, or fossilized remains that have different carbon isotopic signatures from the inorganic carbon also present in the soil, and from carbon dioxide in Mars' atmosphere. The GAP experiment (Wright *et al.*, 2000), designed to search for just such a disequilibrium isotopic signature in soils excavated from below the Martian surface, is part of the *Beagle2* lander due for launch in 2003 on board *Mars Express*. However, other factors can alter isotopic discrimination, thus making its use in searching for life on Mars difficult to interpret (Rothschild and DesMarais, 1989).

Fortunately, since it is not always possible to analyse planetary materials directly, there are indicators of life for which searches can be made using remote observation techniques. The utility of these techniques was shown when the *Galileo* spacecraft, *en route* to Jupiter, flew by the Earth and found evidence for life (Sagan *et al.*, 1993). The most common tool is the spectral signature of a planetary body, most usually of its atmosphere. As outlined in the preceding paragraph, abiological systems tend towards equilibrium, and disequilibrium products, if present, can be taken to infer the presence of life. Specifically, the presence of ozone (which implies oxygen) or methane would be most diagnostic – both molecules are rapidly removed from the atmosphere,

and so their signature would indicate a source of continual renewal from a biological source. Future searches for Earth-like planets (ELPs) will concentrate on the detection of conditions that would indicate environments conducive to the survival of life. Thus infrared spectrometers will be employed to measure the thermal emission characteristics of ELPs, to determine surface temperatures, and the presence or otherwise of water.

21.6 Summary

Based on the range of conditions over which microbes exist and flourish on Earth, it seems fairly clear that there are many niches throughout the solar system that could provide a suitable habitat in which life might emerge. There is, however, a distinction that must be made between the *emergence* of life, and its successful *evolution*. The terrestrial fossil record indicates that life on Earth started up the evolutionary pathway only a matter of a few hundreds of millions of years following Earth's differentiation (see Nisbet and Sleep, this volume, Chapter 1). Nonetheless, it has been suggested that primitive life might have started more than once during this first epoch, only to be eliminated during periods of bombardment. It was not until conditions stabilized, after the late heavy bombardment at around 4 Gyr, that the evolutionary process got underway. The evolution of life on Earth records variations that result from environmental changes, or stresses, combined with chance events (such as catastrophic impacts, etc.). Even though there has been a solar system-wide delivery mechanism (via comets and asteroids) for the building blocks of life, and there are many environments suitable on other planets and satellites for life to emerge, different stresses and chances acting on the same starting materials would presumably result in the evolution of different types of life from that on Earth. It is far from clear that there has been any emergence or evolution of life on other bodies, which is why there is currently so much interest in the study of exobiology and the planning of appropriate space missions to investigate and explore other worlds.

So, are we alone? Given the abundance of organic materials and water within the solar system and the range of conditions under which microbial life seems to be able to survive and even grow, if results from missions such as ESA's *Mars Express* to Mars are as inconclusive as those of the *Viking* biology experiments, then perhaps a more pertinent question should be 'Why are we alone?'

Acknowledgements

The manuscript benefited from comments by Ian Wright and Lynn Rothschild, to whom gratitude is expressed. The work was carried out under the auspices of the *Earth Materials, Histories and Processes* Theme at the Natural History Museum, with additional financial support from the Particle Physics and Astronomy Research Council (Grant No. GR/L21518). Thanks are extended to the Linnaean Society for their invitation to address the Society and undertake this chapter.

References

Anders, E. (1989). Pre-biotic organic matter from comets and asteroids. *Nature*, **342**: 255–257.

Anderson, J.D., Sjogren, W.L. and Schubert, G. (1996a). Galileo gravity results and the internal structure of Io. *Science*, **272**: 709–712.

Anderson, J.D., Lau, E.L., Sjogren, W.L. *et al.* (1996b). Gravitational constraints on the internal structure of Ganymede. *Nature*, **384**: 541–543.

Anderson, J.D., Lau, E.L., Sjogren, W.L. *et al.* (1997a). Europa's differentiated internal structure: Inferences from two Galileo encounters. *Science*, **276**: 1236–1239.

Anderson, J.D., Lau, E.L., Sjogren, W.L. *et al.* (1997b). Gravitational evidence for an undifferentiated Callisto. *Nature*, **387**: 264–266.

Anderson, J.D., Schubert, G., Jacobson, R.A. *et al.* (1998a). Europa's differentiated internal structure: inferences from four Galileo encounters. *Science*, **281**: 2019–2202.

Anderson, J.D., Schubert, G., Jacobsen, R.A. *et al.* (1998b). Distribution of rock, metals and ices in Callisto. *Science*, **280**: 1573–1576.

Bada, J.L., Glavin, D.P., McDonald, G.D. and Becke, L. (1998). A search for endogenous amino acids in martian meteorite ALH 84001. *Science*, **279**: 362–365.

Bell, R.E. and Karl, D.M. (1998). Lake Vostok: a curiosity or a focus for interdisciplinary study? *Final Report of NSF Workshop*, 83pp.

Blasius, K.R., Cutts, J.A. and Howard, A.D. (1982). Topography and stratigraphy of martian polar layered deposits. *Icarus*, **50**: 140–160.

Bradley, J.P., Harvey, R.P. and McSween, H.Y. (1996). Magnetite whiskers and platelets in the ALH 84001 martian meteorite: evidence of vapor phase growth. *Geochim. Cosmochim. Acta*, **60**: 5149–5155.

Bradley, J.P., McSween, H.Y. and Harvey, R.P. (1998). Epitaxial growth of nanophase magnetite in martian meteorite Allan Hills 84001: implications for biogenic mineralization. *Meteoritics Planet. Sci.*, **33**: 765–773.

Cairns-Smith, A.G. (1982). *Genetic Takeover and the Mineral Origins of Life*. Cambridge: Cambridge University Press.

Carr, M.H., Belton, M.J.S., Chapman, C.R. *et al.* (1998). Evidence for a subsurface ocean on Europa. *Nature*, **391**: 363–365.

Chyba, C.F. (1993). The violent environment of the origin of life: progress and uncertainties. *Geochim. Cosmochim. Acta*, **57**: 3351–3358.

Chyba, C.F. (2000). Energy for microbial life on Europa. *Nature*, **403**: 381–382.

Chyba, C.F, Thomas, P.J., Brookshaw, L. and Sagan, C. (1990). Cometary delivery of organic molecules to the early Earth. *Science*, **249**: 366–373.

Clark, B.C., Baird, A.K., Rose, H.J. *et al.* (1976). Inorganic analysis of martian surface samples at the Viking landing sites. *Science*, **194**: 1283–1288.

Clark, B.C., Baird, A.K., Weldon, R.J. *et al.* (1982). Chemical composition of martian fines. *J. Geophys. Res.*, **87**: 10059–10067.

Clarke, D.W. and Ferris, J.P. (1997). Chemical evolution on Titan: comparisons to the prebiotic Earth. *Origins Life Evol. Biosphere*, **27**: 225–248.

Connerney, J.E.P., Acuña, M.H., Wasilewski, P.J. *et al.* (1999). Magnetic lineations in the ancient crust of Mars. *Science*, **284**: 794–798.

Corliss, J.B., Baross, J.A. and Hoffman, S.E. (1981). An hypothesis concerning the relationship between submarine hot springs and the origin of life on Earth. *Proc. 26th IGC, Geology of Oceans Symp. Oceanol. Acta*, **4**: 59–69.

Cowan, D.A. (1999). Life in extreme thermal environments: implications for exobiology. In: J.A. Hiscox (ed.) *The Search for Life on Mars*. London: British Interplanetary Society, pp. 37–48.

Cronin, J.R. and Pizzarello, S. (1997). Enantiomeric excesses in meteoritic amino acids. *Science*, **275**: 951–955.

Deamer, D.W. (1998). Membrane compartments in prebiotic evolution. In: A. Brack (ed.) *The Molecular Origins of Life*. Cambridge: Cambridge University Press, pp. 189–205.

Edmond, J.M., Von Damm, K.L., McDuff, R.E. and Measures, C.I. (1982). Chemistry of hot springs on the East Pacific Rise and their effluent dispersal. *Nature*, **297**: 187–191.

Francheteau, J., Needham, H.D., Choukroune, P. *et al.* (1979). Massive deep-sea sulphide ore deposits discovered on the East Pacific Rise. *Nature*, **277**: 523–528.

Gold, T. (1992). The deep, hot biosphere. *Proc. Natl. Acad. Sci. USA*, **89**: 6045–6049.

Golombek, M.P., Cook, R.A., Economou, T. *et al.* (1997). Overview of the Mars Pathfinder mission and assessment of landing site predictions. *Science*, **278**: 1743–1748.

Grady, M.M., Wright, I.P. and Pillinger, C.T. (1999). The search for life in Allan Hills 84001. In: J.A. Hiscox (ed.) *The Search for Life on Mars*. London: British Interplanetary Society, pp. 78–82.

Grady, M.M., Wright, I.P., Douglas, C. and Pillinger, C.T. (1994). Carbon and nitrogen in ALH 84001. *Meteoritics*, **29**: 469.

Greenberg, R., Geissler, P., Hoppa, G. *et al.* (1998). Tectonic processes on Europa: tidal stresses, mechanical response, and visible features. *Icarus*, **135**: 64–78.

Haldane, J.B.S. (1929). The origin of life. *Ration. Ann.*, **148**: 3–10.

Hart, M.H. (1978). The evolution of the atmosphere of the Earth. *Icarus*, **33**: 23–39.

Harvey, R.P. and McSween, H.Y. (1996). A possible high-temperature origin for the carbonates in the martian meteorite ALH84001. *Nature*, **382**: 49–51.

Head, J.W. III, Hiesinger, H., Ivanov, M.A. *et al.* (1999). Possible ancient oceans on Mars: evidence from Mars Orbiter Laser Altimeter data. *Science*, **286**: 2134–2137.

Joyce, G.F. (1989). RNA evolution and the origins of life. *Nature*, **338**: 217–224.

Jull, A.J.T., Courtney, C., Jeffrey, D.A. and Beck, J.W. (1998). Isotopic evidence for a terrestrial source of organic compounds found in martian meteorites Allan Hills 84001 and Elephant Moraine 79001. *Science*, **279**: 366–369.

Kapitsa, A.P., Ridley, J.K., Robin, G. de Q. *et al.* (1996). A large deep freshwater lake beneath the ice of central East Antarctica. *Nature*, **381**: 684–686.

Kasting, J.F., Whitmire, D.P. and Reynolds, R.T. (1993). Habitable zones around main-sequence stars. *Icarus*, **101**: 108–128.

Khurana, K.K., Kivelson, M.G., Russell, C.T. *et al.* (1997). Absence of an internal magnetic field at Callisto. *Nature*, **387**: 262–264.

Khurana, K.K., Kivelson, M.G., Stevenson, D.J. *et al.* (1998). Induced magnetic fields as evidence for subsurface oceans in Europa and Callisto. *Nature*, **395**: 777–780.

Kieffer, H.H., Jakosky, B.M. and Snyder, C.W. (1992). The planet Mars: from antiquity to the present. In: H.H. Kieffer, B.M. Jakosky, C.W. Snyder and M.S. Matthews (eds) *Mars*. Tucson: University of Arizona Press, pp. 1–33.

Klein, H.P. (1978). The Viking biological experiments on Mars. *Icarus*, **34**: 666–674.

Klein, H.P. (1979). The Viking mission and the search for life on Mars. *Rev. Geophys. Space Physics.*, **17**: 1655–1662.

Klein, H.P., Horowitz, N.H. and Biemann, K. (1992). The search for extant life on Mars. In: H.H. Kieffer, B.M. Jakosky, C.W. Snyder and M.S. Matthews (eds) *Mars*. Tucson: University of Arizona Press, pp. 1221–1233.

Klein, H.P., Horowitz, N.H., Levin, G.V. *et al.* (1976). The Viking biological investigation: preliminary results. *Science*, **194**: 99–105.

Leshin, L.A., McKeegan, K.D. and Harvey, R.P. (1998). Oxygen isotopic constraints on the genesis of carbonates from martian meteorite ALH84001. *Geochim. Cosmochim. Acta*, **62**: 3–13.

LopesGautier, R., McEwan, A.S., Smythe, W.B. *et al.* (1999). Active volcanism on Io: global distribution and variations in activity. *Icarus*, **140**: 243–254.

Lunine, J.I., Stevenson, D.J. and Yung, J.L. (1983). Ethane ocean on Titan. *Science*, **222**: 1229–1230.

Maher, K.A. and Stevenson, D.J. (1988). Impact frustration of the origin of life. *Nature*, **331**: 612–614.

Malin, M.C. and Edgett, K.S. (2000). Evidence for recent groundwater seepage and surface runoff on Mars. *Science*, **288**: 2330–2335.

Marcy, G.W. and Butler, R.P. (1998). Detection of extrasolar giant planets. *Ann. Rev. Astron. Astrophys.*, **36**: 57–97.

Masursky, H. (1973). An overview of geologic results from Mariner 9. *J. Geophys. Res.*, **78**: 4009–4030.

Mayor, M. and Queloz, D. (1995). A Jupiter-mass companion to a solar-type star. *Nature*, **378**: 355–359.

McEwan, A.S., Keszthelyi, L., Spencer, J.R. *et al.* (1998). High temperature silicate volcanism on Jupiter's moon Io. *Science*, **281**: 87–90.

McKay, C.P., Mancinelli, R.L., Stoker, C.R. and Wharton, R.A. Jr. (1992). The possibility of life on Mars during a water-rich past. In: H.H. Kieffer, B.M. Jakosky, C.W. Snyder and M.S. Matthews (eds) *Mars*. Tucson: University of Arizona Press, pp. 1234–1245.

McKay, D.S., Gibson, E.K. Jr, Thomas-Keprta, K.L. *et al.* (1996). Search for past life on Mars: possible relic biogenic activity in martian meteorite ALH 84001. *Science*, **273**: 924–930.

McKay, D.S., Wentworth, S.W., Thomas-Keprta, K.L. *et al.* (1999). Possible bacteria in Nakhla. *Lunar Planet Sci.*, **XXX**: 1816 (CD-ROM).

McSween, H.Y. Jr. (1994). What we have learned about Mars from SNC meteorites. *Meteoritics*, **29**: 757–779.

McSween, H.Y. Jr. and Harvey, R.P. (1998). Brine evaporation: an alternative model for the formation of carbonates in the ALH 84001 martian meteorite. *Int. Geol. Rev.*, **40**: 774–783.

Miller, S.L. (1953). Production of amino acids under possible primitive Earth conditions. *Science*, **117**: 528–529.

Miller, S.L. and Bada, J.L. (1988). Submarine hot springs and the origin of life. *Nature*, **334**: 609–611.

Mittlefehldt, D.W. (1994). ALH 84001, a cumulate orthopyroxenite member of the martian meteorite clan. *Meteoritics*, **29**: 214–221.

Mojszis, S.J., Arrhenius, G., McKeegan, K.D. *et al.* (1996). Evidence for life on Earth before 3,800 million years ago. *Nature*, **385**: 55–59.

Muhleman, D.O., Grossman, A.W., Butler, B.J. and Slade, M.A. (1990). Radar reflectivity of Titan. *Science*, **248**: 975–980.

Murray, B.C., Soderblom, L.A., Cutts, J.A. *et al.* (1972). Geological framework of the south polar region of Mars. *Icarus*, **17**: 328–345.

Nier, A.O., McElroy, M.B. and Yung, Y.L. (1976). Isotopic composition of the martian atmosphere. *Science*, **194**: 68–70.

Oparin, A.I. (1936). *The Origin of Life*. New York: McMillan.

Oró, J. (1961). Comets and the formation of biochemical compounds on the primitive Earth. *Nature*, **190**: 442–443.

Owen, T., Biemann, K., Rushneck, D.R. *et al.* (1977). The composition of the atmosphere at the surface of Mars. *J. Geophys. Res.*, **82**: 4635–4639.

Pappalardo, R.T., Belton, M.J.S., Brenemann, H.H. *et al.* (1999). Does Europa have a subsurface ocean? Evaluation of the geological evidence. *J. Geophys. Res.*, **104**: 24015–24055.

Pillinger, C.T., Wright, I.P. and Sims, M.R. (1999). Science activities with Beagle 2, the Mars Express lander. *Lunar Planet Sci.*, **XXX**: 1560 (CD-ROM).

Pollack, J.B., Kasting, J.F., Richardson, S.M. and Poliakoff, K. (1987). The case for a wet, warm climate on early Mars. *Icarus*, **71**: 203–224.

Rampino, M.R. and Caldeira, K. (1994). The Goldilocks problem: climatic evolution and long-term habitability of terrestrial planets. *Ann. Rev. Astron. Astrophys.*, **32**: 83–114.

Raulin, F. (1998). Titan. In: A. Brack (ed.) *The Molecular Origins of Life*. Cambridge: Cambridge University Press, pp. 365–385.

Romanek, C.S., Grady, M.M., Wright, I.P. *et al.* (1994). Record of fluid-rock interactions on Mars from the meteorite ALH 84001. *Nature*, **372**: 655–657.

Rothschild, L.J. and DesMarais, D. (1989). Carbon isotope fractionation and the search for life on Mars. *Adv. Space Res.*, **9**: 159–165.

Rothschild, L.J. and Mancinelli, R.L. (2001). Life in extreme environments. *Nature*, **409**: 1092–1101.

Sagan, C., Thompson, W.R., Carlson, R. *et al.* (1993). A search for life on Earth from the Galileo spacecraft. *Nature*, **365**: 715–721.

Saxton, J.M., Lyon, I.C. and Turner, G. (1998). Correlated chemical and isotopic zoning in carbonates in the Martian meteorite ALH84001. *Earth Planet. Sci. Lett.*, **160**: 811–822.

Schopf, J.W. (1983). (ed.) *Earth's Earliest Biosphere: Its Origin and Evolution*. Princeton: Princeton University Press.

Schubert, G., Zhang, K.K., Kivelson, M.G. and Anderson, J.D. (1996). The magnetic field and internal structure of Ganymede. *Nature*, **384**: 544–545.

Scott, E.R.D., Yamaguchi, A. and Krot, A.N. (1997). Petrological evidence for shock melting of carbonates in the martian meteorite ALH 84001. *Nature*, **387**: 377–379.

Sleep, N.H., Zahnle, K.J., Kasting, J.F. and Morowitz, H.J. (1989). Annihilation of ecosystems by large asteroid impacts on the early Earth. *Nature*, **342**: 139–142.

Smith, G.R., Strobel, D.F., Broadfoot, A.L. *et al.* (1982). Titan's upper atmosphere: composition and temperature from the EUV solar occultation results. *J. Geophys. Res.*, **87**: 1351–1359.

Smith, P.H., Lemmon, M.T., Lorenz, R.D. *et al.* (1996). Titan's surface revealed by HST imaging. *Icarus*, **119**: 336–349.

Squyres, S.W. and Kasting, J.F. (1994). Early Mars: how warm and how wet? *Science*, **265**: 747–749.

Stetter, K.O. (1998). Hyperthermophiles and their possible role as ancestors of modern life. In: A. Brack (ed.) *The Molecular Origins of Life.* Cambridge: Cambridge University Press, pp. 315–335.

Tanaka, K.L., Scott, D.H. and Greeley, R. (1992). Global stratigraphy. In: H.H. Kieffer, B.M. Jakosky, C.W. Snyder and M.S. Matthews (eds) *Mars.* Tucson: University of Arizona Press, pp. 345–382.

Thomas-Keprta, K.L., McKay, D.S., Wentworth, S.J. *et al.* (1998a). Bacterial mineralization patterns in basaltic aquifers: implications for possible life in martian meteorite ALH84001. *Geology*, **26**: 1031–1034.

Thomas-Keprta, K.L., McKay, D.S., Wentworth, S.J. *et al.* (1998b). Mineralization of bacteria in terrestrial basaltic environments; comparison with possible life forms in martian meteorite ALH 84001. *Lunar Planet Sci.*, **XXIX**: 1489 (CD-ROM).

Treiman, A.H. (1995). A petrographic history of martian meteorite ALH 84001: two shocks and an ancient age. *Meteoritics*, **30**: 294–302.

Uwins, P.J.R., Webb, R.I. and Taylor, A.P. (1998). Novel nano-organisms from Australian sandstones. *Amer. Min.*, **83**: 1541–1550.

Valley, J.W., Eiler, J.M., Graham, C.M. *et al.* (1997). Low-temperature carbonate concretions in the martian meteorite, ALH84001: evidence from stable isotopes and mineralogy. *Science*, **275**: 1633–1637.

Ward, P.D. and Brownlee, D. (2000). *Rare Earth: Why Complex Life is Uncommon in the Universe.* New York: Copernicus.

Warren, P.H. (1998). Petrologic evidence for low-temperature, possible flood-evaporitic origin of carbonates in the Allan Hills 84001 meteorite. *J. Geophys. Res.*, **103**: 16759–16773.

Wolszczan, A. (1994). Confirmation of Earth-mass planets orbiting the millisecond pulsar PSR1257+12. *Science*, **264**: 538–540.

Wolszczan, A. and Frail, D. (1992). A planetary system around the millisecond pulsar PSR1257+12. *Nature*, **255**: 145–147.

Wright, I.P., Grady, M.M. and Pillinger, C.T. (1989). Organic materials in a martian meteorite. *Nature*, **340**: 220–222.

Wright, I.P., Grady, M.M. and Pillinger, C.T. (1998). On the ^{14}C and amino acids in martian meteorites. *Lunar Planet Sci.*, **XXIX**: 1594 (CD-ROM).

Wright, I.P., Morgan, G.H., Praine, I.J. *et al.* (2000). Beagle 2 and the search for organic compounds on Mars using GAP. *Lunar Planet Sci.*, **XXXI**: 1573 (CD-ROM).

Epilogue

Adrian Lister and Lynn Rothschild

This volume will, we trust, stimulate the imagination, at the very least. Where else might one have noticed, for example, that a cooling of the Earth's surface was a prerequisite both for the origin of life, and for sowing the seeds of the first human civilizations, albeit in fundamentally different ways? While the kaleidoscopic diversity of the present volume defies easy summary, certain common themes emerge. We briefly suggest three broad areas of synthesis.

Physical factors influence many aspects of evolution

- *The fundamental characteristics of all life on this planet*, both at its inception and in its maintenance throughout its history, have been influenced and constrained by basic parameters of physics, chemistry, the make-up of the Earth and its position in the solar system. The properties of key elements such as oxygen and nitrogen, and the nature of the Sun's radiation, are obvious examples. Such factors continue to operate at higher levels of adaptation too, for example in the role of gravity in constraining developmental growth and animal locomotion.
- *The waxing and waning of major groups of organisms* through geological time have been strongly influenced by changing conditions on the Earth. Key evolutionary innovations have been favoured at certain times (e.g. the origin of insect flight when oxygen tension was high), and the success of groups

bearing them has depended on shifts in ambient conditions. 'Mass extinction' is the extreme end of this spectrum.
- *Taxonomic separation, the prerequisite for diversification*, has depended on contingent features of the Earth's geography. This has operated at levels ranging from continental break-up leading to Class-level divergence, to mountain-building leading to diverse topography and the potential for divergence of populations into subspecies or species. Climatic and sea-level change can also shift and divide populations, fuelling taxic divergence.
- *Adaptations of individual species or groups* are a response to local or regional physical conditions. Hot springs harbour hyperthermophilic bacteria, animals and plants in fast-moving water have holdfasts. In some cases, the temporal pattern of environmental change (e.g. cyclicity) may be important in its own right, for example in the evolution of human adaptive flexibility.
- *Genomic evolution* may be directly influenced by the physical environment; for example, gene transfer between species of microorganisms may be facilitated or prevented by the chemical composition of their medium.

There are complex chains of causality among physical factors

The contributors to this volume, individually and collectively, not only describe a remarkable array of physical factors affecting evolution, but make abundantly clear that these

Evolution on Planet Earth
ISBN 0-12-598655-6

interact in complex chains, cascades and networks of causality. From the outset, the properties of oxygen or nitrogen would be of little use if gravity did not bind them to the Earth's surface. Every chapter of this book indicates how the physical factor(s) which are its main subject-matter can be fully understood only within a mesh of other, interacting factors, often in positive or negative feedback systems. We had considered drawing up, on the basis of the links enunciated by the various contributors, a grand network diagram indicating all the postulated effects; a preliminary survey showed that such a diagram would be so dense with arrows as to be virtually unreadable.

A few examples must therefore serve to illustrate the theme. A massive impact early in Earth's history tilted the axis of rotation and gave us pole-to-equator gradients and seasonality of climate, forcing tremendous biotic diversity and adaptation. The Earth is protected from ultraviolet and other radiation both by its magnetosphere, and by the ozone layer (itself a product of biotically produced oxygen). Regular variations in the Earth's orbit combine with continental positioning resulting from continental drift, to produce ice ages which in turn alter both global climate, and the relative level of land and sea, which results in shifting barriers, passageways and habitats. Increased carbon dioxide levels create a 'greenhouse Earth' that raises global temperatures, in turn (among other things) promoting chemical weathering, and hence increasing the nutrient supply to the biosphere.

If one factor appears more often than any other in these discussions, it is the tectonic activity of the Earth's crust. Driven ultimately by geothermal energy, the various manifestations of tectonic movement are implicated in numerous chains of physical causality that, in their multifarious ways, impact evolution. Vulcanism affects the composition of both the atmosphere and lithosphere. Sea-floor spreading and submarine vulcanism affect the size and shape of ocean basins, and hence their depth and relative encroachment onto land. Mass lava flows have been implicated in heightened extinction. The movement of continental plates (itself requiring also liquid water for its operation) affects ocean circulation, climate and glaciation, crafts marine and terrestrial habitat areas, and opens gateways for dispersal or splits for divergence. Burial or subduction of fixed carbon (e.g. as coal) reduces atmospheric carbon dioxide, lowering the greenhouse effect and reducing global temperature (this is opposed by vulcanism effecting the inverse). Mountain-building influences climate, and provides a varied and dissected topography encouraging floral and faunal diversification. Hydrothermal activity is a source of nutrients into the ocean and may have provided the energy for the earliest life.

Many of these links are deduced from a knowledge of Earth processes, but their history of operation in the past is known only from geological data, and the links are essentially correlative. As emphasized in these pages, this does not preclude the rigorous testing of alternative models, especially where nature has provided repeated patterns through the immensity of geological time.

Many physical influences on evolution are mediated biotically

To do justice to this fascinating subject would require a treatment at least as long again as the present work. Nonetheless, some of the key features of the abiotic/biotic interaction have been touched upon in this volume. First, if we consider individual species, their response to environmental perturbation will clearly vary depending on their intrinsic attributes of adaptation, geographical distribution, flexibility and so on. Second, more complex biotic interactions among species in a community can be modified by physical effects. Thus, if one species is modified by physical environmental change, others that interact with it will be affected indirectly: the influence of the physical environment has been biotically mediated. Extinction, again, is the extreme end of the spectrum if it vacates niches that other species then evolve to fill.

Examples from the present work touch on a wide range of likely interactions. Enhanced inorganic nutrient input at various times in Earth history (itself triggered by a variety of more distal physical forcing factors) likely expanded biological diversity and intensified biotic interactions. Plate tectonics resulted in the forced cohabitation of previously separate floras and faunas, with resulting new biotic interactions, as in the Great American Interchange when North and South America collided in the late Cenozoic. Climate change (again with a range of distal triggers) has many times affected the nature and distribution of vegetation types around the globe, secondarily affecting mammalian and other faunal taxic and adaptational diversity. This whole area, like the others mentioned above, is clearly ripe for further work.

We hope that the unprecedented scope of this volume will have as inspirational an effect on its readers as it has had on us, and on the many who crowded the Linnean Society's rooms at the original meeting. For rising to the spirit of the enterprise and collectively providing such a breathtaking vision, we offer our heartfelt thanks to all our contributors.

Glossary

Adaptive scope Range of adaptive possibilities.

Allopatric speciation The origin of a new species from a population of the ancestral species which has become geographically isolated, usually at the periphery of the ancestral species' range.

Allopatry When two or more species or populations occupy separate geographical areas.

AMS (Accelerator Mass Spectrometric) dating A method of radiocarbon dating (q.v.) that depends on the direct measurement of the radioactive isotope ^{14}C with a mass spectrometer. The method allows much smaller samples (weighing as little as 5 mg) to be dated than conventional radiocarbon dating, e.g. individual charred seeds.

Anagenesis Evolutionary change through time in a single lineage, i.e. without 'splitting' into two or more lineages.

Anhydrobiotic organisms Organisms that are capable of surviving complete dehydration.

Anoxygenic photosynthesis The use of light energy to synthesize ATP by cyclic photophosphorylation without O_2 production.

Anthracotheres An early group of suine mammals, primarily of the Old World, persisting from the Late Eocene to the Pleistocene. Early members were terrier-sized, some later ones as big as a hippopotamus.

Antifreeze Any compound which inhibits the process of freezing, thereby providing a degree of protection for the organism. Antifreezes work dynamically, so they affect the freezing point, but not the melting point once the solution has frozen.

Antilocaprids Members of the pronghorn (=American 'antelope') family of artiodactyl mammals.

Aptation Any organic structure or process modified by natural selection including those structures or processes specifically modified to perform their current functions (adaptations) and those structures or processes originally modified to perform a function other than their current function or function of interest (exaptations).

Archaea Phylogenetic domain of prokaryotes consisting of the methanogens, extreme halophiles, and hyperthermophiles.

Archaean The second aeon of Earth history, from about 4.0 to 2.5 billion years ago.

Artiodactyls Even-toed ungulates, such as cattle, antelopes, deer, pigs, camels and hippos.

Astrobiology A science that combines the disciplines of astronomy, biology, chemistry, geology and physics to study the origin, evolution, distribution, and future of life in the universe.

Astronomical unit (AU) The mean distance between the Earth and the Sun, $149\,597\,890 \pm 500$ km.

Atmosphere The mass of air enveloping the Earth.

Australopith An informal taxonomic category that includes the oldest known members of the early human clade (see **Hominini**). Australopiths include the genera *Australopithecus*, *Paranthropus*, and *Ardipithecus*. Due to taxonomic revisions that place African apes and humans together in the subfamily Homininae (or 'hominine'), the term 'australopith' now replaces the more familiar subfamily term 'australopithecine'.

Bacterial spores Resistant form of bacteria, able to survive hostile conditions. Bacterial endospores are especially resistant against environmental extremes, such as desiccation, heat, ionizing and UV radiation.

Banded iron formations (BIFs) Layered deposits of ferric oxide (Fe_2O_3) in marine sedimentary rocks. These indicate an anoxic early ocean rich in ferrous iron (Fe^{2+}). Ferrous iron can be oxidized by O_2 from oxygenic photosynthesis, or by a non-oxygenic photosynthetic pathway.

Bauplan The archetypal body plan of a clade of animals. Used here mainly in the context of the morphological and functional design of those organisms, but often also interpreted in a developmental context.

Biogeography The study of the distributions and the causes of the distributions of organisms through time.

Biosphere The portion of the Earth in which life exists.

Biostack concept Method to investigate the effects of individual HZE particles (q.v.) of cosmic radiation in individual biological systems or cells.

Bolide Generic term for a relatively small astronomical object, usually a comet or asteroid.

Boreal Pertaining to the northern coniferous forest zone of the Northern Hemisphere.

Brachydont Cheek teeth (molars and premolars) with a low crown.

Brontotheres Large Eocene rhino-like browsing perissodactyl mammals with a frontal horn.

Browsers Herbivorous mammals with a diet dominated by the leaves of trees and shrubs.

Bunodont Cheek teeth with rounded, low cusps.

C_3 carbon cycle The process by which plants make a three-carbon compound as the first stable product of carbon fixation. In conditions of high moisture and low temperature (i.e. in temperate climate zones, or at high latitudes), and in conditions of high CO_2, C_3 plants have the advantage over those using the C_4 cycle.

C_4 carbon cycle A biochemical pathway used by certain plants to obtain carbon during photosynthesis, resulting initially in a four-carbon compound. The net rate of photosynthesis (and consequently also the net production of biomass) of C_4 plants is greater under conditions of low moisture, high temperatures and low levels of CO_2 than that of C_3 plants. C_4 plants, nearly always herbs or grasses, are more successful in the open country of warmer zones.

Camelids Members of the camel and llama family of artiodactyl mammals.

Carbonaceous chondrite A meteoritic stone containing carbon compounds and rounded granules of cosmic origin.

Carbonate A class of inorganic salts of carbon, where a central carbon atom is surrounded by three oxygen atoms. The resulting molecule has an overall negative charge, and so bonds readily with positive ions such as Ca^{2+}, Mg^{2+}, etc. Calcium carbonate is the mineral that forms the rock types limestone and chalk.

Centrifugal force or centrifugal acceleration The apparent force in a rotating system, deflecting masses radially outward from the axis of rotation, with magnitude per unit mass $\omega^2 R$, where ω is the angular speed of rotation and R is the radius of curvature of the path.

Centripetal acceleration The acceleration on a particle moving in a curved path, directed toward the instantaneous centre of curvature of the path, with magnitude v^2/R, where v is the speed of the particle and R is the radius of curvature of the path.

Chemolithotrophy Process by which organisms grow using inorganic chemical reactions as their energy source and inorganic chemicals as their source of carbon.

Chemoorganotrophy Process by which organisms grow using organic chemicals as their source of both energy and carbon.

Chronospecies In the fossil record, a species occupying a certain interval of geological time. Generally used in the sense of a series of chronospecies that replace each other through time and are assumed to form a continuously evolving lineage.

Clade Generic term denotes an evolutionary lineage that contains the lineage's ancestor and all descendant species. See monophyletic group.

Cladogenesis The process of phylogenetic branching into two derivative evolutionary lineages or clades.

Conjugation In prokaryotes the transfer of genetic material from one cell to another by a mechanism involving cell-to-cell contact.

Convection Movement within a fluid resulting from differential heating and cooling of the fluid. Convection produces mass transport or mixing of the fluid

Craniodental Any characteristic of the bony cranium and teeth (dentition) of an animal.

Dansgaard–Oeschger (D–O) events. Short-term warming events, each lasting *c.* 500–2000 years, occurring in the Northern Hemisphere roughly every 1500 years during the last glacial period (*c.* 80 000–20 000 years ago).

Decomposition Transformation of dead organic matter by organisms to organic compounds available for uptake by plants and other organisms capable of fixing carbon by photosynthesis.

Denitrification The reduction of NO_3^- to N_2O or N_2.

Density (of a fluid) The mass of unit volume of a fluid. Important in swimming and flying locomotion as density determines the mechanical properties of fins or wings as force-generating surface. Also important to all air-breathing animals as it can determine the physiology of oxygen uptake.

Diagenesis The chemical, physical or biological changes that occur to a sediment as it is buried then lithified.

Diazotroph Organism capable of fixing nitrogen (i.e. $N_2 \rightarrow NH_3$).

Dinocerata The largest mammals of the Late Paleocene and Early Eocene, also known as uintatheres. Occurring in North America and Asia, they were herbivorous but with relatively small teeth (and brains). *Uintatherium* was rhino-sized with bony protuberances on its head. Their affinities are debated.

Dispersal The movement or migration of a taxon across a barrier.

Double-wedge A pattern in the fossil record where the gradual disappearance of one taxon is coincident with the gradual increase of another. It is thought to indicate competitive replacement.

Drag The hydro- or aerodynamic resistance to the movement of a body through a fluid, and (along with weight) is the main force determining energy expenditure in swimming or flying animal locomotion. Some animals use the drag force on their appendages to produce propulsive forces (by rowing), but this is generally regarded as less efficient than lift- (q.v.) based propulsion.

Drosophila A small fly of the genus *Drosophila*, especially the fruit fly *D. melanogaster* used extensively in genetic studies.

Earth system Open system comprising atmosphere, oceans, life, soils and crust, bounded by outer space and the molten inner Earth.

Endemic species A species which evolved in a particular limited geographical area, and which remains restricted to that area.

Endemism The distribution of a taxon in a single, restricted area.

Endolithic microbial communities Microbial communities that grow in a layer inside sandstone, a few mm below the surface, thereby forming a microclimate in an otherwise hostile environment, e.g. in cold and hot deserts.

Entelodonts Large pig-like mammals of the Eocene to Miocene, common in Europe and North America, ranging up to the size of a bison.

Eocytes A group of predominately hyperthermophilic prokaryotes, including *Sulfolobus*, *Theroproteus*, and other organisms. Many of their genes closely resemble those found in eukaryotes, and they have been proposed to be the sister taxon of eukaryotes.

Epipalaeolithic An archaeological period in Southwest Asia that extends from *c.* 20 000 to *c.* 10 300 radiocarbon years ago, equivalent to *c.* 20 000 to 12 000 calendar years ago.

Erosion (physical) The breakage of rock into smaller particles.

ESA The European Space Agency, an inter-governmental organization of 15 member states, with a mission to provide and promote – for exclusively peaceful purposes – the exploitation of space science, research and technology.

Escalation Adaptation to enemies, as in the evolutionary elaboration of a prey's defences against its predators.

Escherichia coli A common Gram negative rod-shaped enteric organism belonging to the Bacteria. Commonly found in the intestines of mammals. Most strains are non-pathogenic.

Essential element A chemical element which is essential for completion of the life cycle of all organisms (e.g. carbon) or some organisms (e.g. nickel).

Eubacteria A speciose group containing many prokaryotes, including the cyanobacteria, the gram-positive prokaryotes, the purple bacteria, and many other groups.

Eukaryote A cell or organism usually having a unit membrane enclosed by true nucleus and usually other organelles.

Eurytopic Pertaining to a broad diet, habitat, or other type of ecological tolerance. An animal that is eurytopic in its habitat is able to occupy a wider range of environments compared to other species. See **Stenotopic**.

Eustatic Global; an effect that is globally distributed, such as sea level.

Eutrophic Applied to nutrient-rich waters with high primary productivity.

Evaporites Salt crusts crystallized from brine. They may provide an oasis for osmophilic microbial communities (q.v.).

Exobauplan The Bauplan (q.v.) of organisms of other planets.

Exozoology The biology of organisms inhabiting, but not necessarily evolving on, other planets and which are capable of independent locomotion.

Extraterrestrial solar UV radiation Full spectrum of solar radiation which is a continuum from vacuum UV to UV-A. The different UV regions are: vacuum UV (≤ 190 nm), UC-C (190–280 nm) UV-B (280–315 nm), UV-A (315–400 nm).

Extrinsic biotic factors Aspects of the surrounding biological environment (predation, competition, parasites, pathogens, etc.) that influence the probability of death/extinction or reproduction/speciation of an organism, population, species or lineage.

Feedback occurs when a change in a variable triggers a response that also affects that variable, said to be 'negative' when it tends to damp the initial change and 'positive' when it tends to amplify it.

Fitness The relative reproductive success of a genetic variation (or of the individual carrying that variation), which results in the increase or decrease of that variation in the gene pool.

Flood basalt Large accumulation of volcanic rock (basalt) erupted on a large landmass in one or more very extensive episodes of volcanic activity over a geologically brief interval (often less than 1 million years).

Flotation A technique for separating lighter materials in archaeological samples, such as charred plant remains and very small bones, from heavier materials by agitating the samples in water and collecting the lighter materials that float to the surface.

Frugivore A fruit-eating animal.

Ga Giga-annum, 10^9 years; billion years (in US English).

Gaia theory Theory of the Earth as a system in which the evolution of organisms and environment are tightly coupled, and self-regulation of climate and chemical composition are emergent properties.

Gait The movement pattern of the limbs or appendages of an animal in locomotion. Normally the gaits used by an animal are distinguished by discontinuities in limb kinematics which correlate with mechanical, morphological and physiological factors in movement, and animals often change gait as they move at different speeds. Examples are the walk, trot and canter/gallop adopted by most horses.

Galactic *cosmic rays* (GCRs) are the high-energy particles that flow into our solar system from far away in the Galaxy. GCRs are about 98% atomic nuclei which have their electrons stripped off during passage through the Galaxy, and 2% electrons and positrons. The magnetic fields of the Galaxy, the solar system, and the Earth have scrambled the flight paths of these particles so much that they no longer point back to their sources in the Galaxy.

Gene orthologs Orthologous genes are descended from a common ancestral gene, and share the most recent common gene ancestor. For example, haemoglobin contains two types of subunits, called alpha and beta. All alpha and beta genes are descended from an ancestral haemoglobin gene. Hence, the alpha haemoglobin subunit in humans is the ortholog of the mouse alpha subunit, since these two genes diverged from their common ancestor most recently. Similarly, human and mouse beta genes are orthologs. In contrast, human alpha and human beta genes are not orthologs because they do not share a most recent common gene ancestor (they share an earlier ancestor).

Geopotential The work required to raise a unit mass from mean sea level to a specified height.

Geotherm Increase in temperature with depth in the Earth. Geothermal gradients are around 10°C per kilometre in the most ancient shields today, but 30–40°C/km in young active areas. In the Archaean they may typically have been 20–40°C/km in stable areas.

Gigantism The tendency for organisms to increase in size substantially compared to relatives or to other representatives of their clade.

Gomphotheres A diverse group of Miocene to Pliocene mammals related to elephants, but with shorter legs and more bunodont teeth indicative of a more omnivorous/browsing type of diet.

Gravity The force imparted by Earth (or any heavenly body), to a mass, relative to its position on Earth. The gravity force is the resultant of the force of gravitation and the centrifugal force arising from Earth's rotation. On Earth, gravity is directed perpendicular to sea level.

Grazers Herbivorous mammals with a diet dominated by grass.

Greenhouse Earth A long geological interval when a predominately warm global climate prevailed.

Greenhouse gas A component of the atmosphere which contributes to the retention of energy received from solar radiation by absorbing re-radiated energy from the atmosphere and Earth's surface. Maintains Earth's temperature at 20°C or so higher than it would be in the absence of such gases.

Hadean The first great division, or aeon, of Earth history, from about 4.55 to 4.0 billion years ago.

Haplotype The genetic composition of any haploid part of a genome (e.g. mitochondrial DNA).

Hominini The taxonomic category (Tribe level) of humans and related bipedal species. The associated term 'hominin' replaces 'hominid' (Hominidae), which now refers to the evolutionary group of the great apes (humans, chimpanzees, gorillas, orangutans, and related fossil species).

Horizontal gene transfer (HGT) The transfer of genes from one organism (prokaryotic or eukaryotic) to another. The transfer may be between organisms closely, or distantly, phylogenetically related. In prokaryotes, the principal methods of transfer are conjugation, transformation, and transduction (q.v.). In eukaryotes, mechanisms of HGT are not as completely understood, but include transformation, viral transfer, and endosymbiotic events.

Hydrosphere The aqueous portion of the Earth.

Hydrothermal activity Release of hot, chemically reactive fluids and gases from the mantle to the ocean at deep-sea vents.

Hydrothermal system Circulation system of water in fractures and pores around a hot body of rock. Major hydrothermal systems occur around large volcanoes, especially at mid-ocean ridges and around volcanoes above subduction zones, but hydrothermal circulation occurs anywhere that cool water encounters hot rock.

Hyperthermophile An organism having an optimal growth temperature of 80°C or higher.

Hypertragulids A group of primitive ruminant artiodactyls from the Late Eocene of North America, of similar size and proportions to modern mouse-deer or chevrotains.

Hypsodont Cheek teeth (molars and premolars) having a high crown: part or most of this crown is embedded in the jaw bone on the initial eruption of the tooth, and erupts further over time as the exposed crown is worn down.

Hysteresis The difference between the temperature at which a solution freezes and melts. This difference is brought about by the presence of an antifreeze (sometimes called a thermal hysteresis protein) which inhibits the process of freezing, thereby providing protection, but does not affect the melting point.

HZE particles Heavy particles of galactic cosmic rays of high charge Z (>2) and high energy E.

Icehouse Earth A long geological interval when a predominately cool global climate prevailed and continental ice sheets developed.

Impactor Large body hitting the Earth during accretion.

Incumbency Competitive advantage enjoyed by organisms, populations, species or lineages, owing to established presence in an ecological community.

Infrared radiation Radiation with wavelengths between about 750 nm and 2.5 μm. Infrared radiation is divided into two parts: near infrared (750 nm to 2.5 μm) which reaches the surface of the Earth, and infrared (2.5 to 25 μm) which is attenuated by the atmosphere.

Interferometer A device that combines radiation (light, radio waves, etc.) detected from a source by several small receivers to mimic a large receiver.

Interstadial A relatively short interval of mild climate within a cold stage of the Pleistocene.

Intrinsic biotic factors Aspects of an organism, population, species or lineage that influence the probability of death/extinction or reproduction/speciation.

Lift Lift is the hydrodynamic or aerodynamic force generated by a wing moving through a fluid. It is associated with the generation of wake vortices by aerofoil action on the wing. It is the locomotion mechanism of all flying vertebrates and most insects, and of many swimming vertebrates. See also **Drag**.

Lithosphere The solid part of the Earth.

Loph A ridge-like structure on the occlusal (chewing) surface of a cheek tooth.

Macroevolution Evolution above the species level.

Mammutids A group of Miocene to Pleistocene mammals related to elephants, including the 'American mastodon'.

Martian meteorites Meteorites that originate from Mars. The Martian origin is identified by comparing their gas inclusions and/or isotopic ratios with data from the Martian surface obtained by the Viking missions.

Mesodont Cheek teeth with a medium-height crown.

Metazoa A subkingdom comprising all multicellular animals.

Methanotroph An organism capable of oxidizing methane.

Microfossils Fossilized microorganisms.

Microsatellite DNA Stretches of DNA consisting of numerous repeats (arrays) of short (1–5 nucleotide) sequences. Distributed throughought the genome, variations in the length of the arrays are used as genetic markers for families or populations.

Milankovitch cycles Long-term cycles of climatic change, triggered by variations in the Earth's spin and orbit.

Mitochondrial DNA A small circular DNA molecule, found in mitochondria of animal and plant cells, which specifies tRNAs, rRNAs, and some mitochondrial proteins.

Molecular clock A way of measuring the time since the divergence of two present-day lineages, based on the assumption that the rate of nucleotide substitution in certain unselected ('neutral') genes is proportional to the time elapsed.

Monophyletic group A group of taxa derived from a common ancestor and including all its descendants.

Natural selection The adaptive process of biased reproductive and survival in an environmental context. The process requires the presence of genetic variation and inherited characteristics (e.g. morphological, physiological, and behavioural traits) that enhance an individual's reproduction success and survival relative to other individuals that have different genetic variations.

Nearest living equivalent An extant species with ecological or structural similarity to an extinct species with which it is being compared.

Nematodes Worms of the phylum Nematoda that have unsegmented threadlike bodies. The best known nematode for research purposes is the roundworm *Caenorhabditis elegans*.

Nemoral Living at the edges of woodlands, or in open woodland.

Neolithic An archaeological period defined by the presence of ground stone tools and (usually) pottery. It is also often associated with the earliest (local) evidence of agriculture. Its chronology varies from region to region. In the Levant it spans the period from *c.* 10 300 to *c.* 6200 radiocarbon years ago, equivalent to *c.* 12 000 to 7000 calendar years ago, and is divided into the Pre-Pottery Neolithic and the Pottery Neolithic.

Niche The functional position of an organism in a community together with the environmental position (with respect to resource utilization, temperature, etc.) and the spatial position it occupies.

Nitrification The oxidation of NH_4^+ to NO_3^-.

Nitrogen fixation The conversion of N_2 into ammonia (NH_3), ammonium (NH_4^+), nitrogen oxides, or N that is chemically bound to either inorganic or organic molecules and releasable by hydrolysis to NH_3 or NH_4^+.

Nitrogen pools consist of N_2, N_2O, NH_4^+/NH_3, NO_3^-, NO_2^- or organic N.

Nomadic pastoralism A system of livestock management in which people and their herd animals move together in seasonal migrations from pasture to pasture. The people rely heavily on their animals for food, in the form of meat and milk products, which they supplement with grain obtained from farmers.

Non-analogue (1) A past community of species whose members, while still alive today, have geographical distributions which no longer overlap. (2) A community type that no longer exists, e.g. diversities and combinations of ecomorphological types that can't be found in any modern community.

Normal (Gaussian) dispersal Dispersal whereby the distance of propagules from the parent follows a normal distribution.

Nuclear DNA DNA found in the cell nucleus, i.e. the majority of the DNA of a eukaryotic cell, localized in the chromosomes.

Occlusal Referring to the chewing surface of a tooth.

Oligotrophic Applied to nutrient-poor waters with low primary productivity.

Oreodonts Primitive artiodactyls of the group Merycoidodontoidea, common and diverse mammals ranging from the Eocene to the Late Miocene, and entirely restricted to North America. They had the size and proportions of pigs or sheep.

Osmophilic microorganisms Microorganisms that grow at high osmotic pressures, usually due to high salt concentrations.

Oxidant (Oxidizing agent). A substance that causes oxidation (loss of electrons) in other substances by being reduced itself (gaining electrons). Oxidizing agents contain atoms with high oxidation numbers that have lost electrons, which they gain in the process of oxidation.

Oxygen isotope ratio ($\delta^{18}O$). An index of the proportion of the rare ^{18}O isotope in comparison to ^{16}O, providing a relative measure of global temperature or ice volume in the past. The isotope ratios are generally determined in fossil shells or ice sheets.

Oxygenic photosynthesis The chemical process in which organic compounds are synthesized from carbon dioxide and water using the free energy from sunlight and liberating oxygen. The process originated in cyanobacteria and is present in all algae and plants.

Palaeogeophysics Properties of the Earth's environment and atmosphere over geological time.

Paleosols Remnants of ancient soils preserved within geological strata.

Palynology The study of pollen, especially the reconstruction of past vegetation based on pollen recovered from sediments.

Pantodonts A group of Late Cretaceous herbivorous placental mammals, which ranged up to the size of small rhinos.

Paramecium Ciliated protozoans of the genus *Paramecium*.

Parapatry When the ranges of two or more species or populations abut.

Paraphyletic group A group of taxa descended from a common ancestor but not including all the descendants of that common ancestor.

Perissodactyls Odd-toed ungulates, such as horses, rhinos and tapirs.

Photic zone Region in the biosphere where light is available for photosynthesis.

Photochemistry Process by which a chemical reaction is energized by light.

Photolithotrophy Process by which organisms grow using light as their energy source and inorganic chemicals as their source of carbon.

Photosynthesis Biological process in which light energy is converted into chemical energy, which is then used with a reductant for the production of organic compounds from carbon dioxide. In oxygenic photosynthesis (plants, algae, cyanobacteria), water is the reductant. Anoxygenic photosynthesis uses other reductants such as hydrogen sulphide.

Phylogeny The genealogical relationships among organisms.

Phylogeography The origins of, and relationships among, subspecies and species in relation to their geographic distributions.

Phytoplankton Single-celled organisms floating or moving in water and capable of fixing carbon by photosynthesis

pO$_2$ Partial pressure of oxygen, usually expressed in atmospheres (1 atm = 101 325 pascals = 760 mm Hg).

Precambrian Hadean, Archaean, and Proterozoic aeons (q.v.).

Productivity Rate of formation of biomass per unit time.

Proglacial lake A lake formed just beyond the frontal margin of an advancing or retreating glacier, generally in direct contact with the ice.

Prokaryote An organism lacking a nucleus and other membrane enclosed organelles, and usually having its DNA as a single stranded circular molecule.

Prosimians A paraphyletic group of primitive primates, including lemurs, lorises and tarsiers.

Proterozoic The third aeon, from about 2.5 to 0.55 billion years ago.

Protozoan Single-celled, usually microscopic, organisms including the most primitive forms of animal life.

Pulses Crops such as beans, peas and lentils that are domesticated forms of wild herbaceous legumes (plants of the legume subfamily Fabaceae).

Radiation Energy that comes from a source and travels through some material or through space. Light, heat and sound are types of radiation.

Radiocarbon dates Dates determined on organic materials such as charcoal, wood and bones up to *c.* 40 000 years old by measuring the decay of the radioactive isotope ^{14}C since the death of the organism. Uncalibrated ^{14}C dates give ages in radiocarbon years, which diverge from calendar years as a result of variations through time in atmospheric ^{14}C. Calibrated ^{14}C dates give ages in calendar years. See also **AMS dating**.

Reaction centre. Site in photosynthesis at which photochemistry occurs.

Red beds Terrestrial deposits of ferric oxide (Fe_2O_3) indicative of oxidative weathering on land.

Reductant (Reducing agent). A substance that causes reduction (gain of electrons) in other substances by being oxidized itself (losing electrons). Reducing agents contain atoms with low oxidation numbers that have gained electrons, which they lose in the process of oxidation.

Refugium An isolated area of habitat that retains the environmental conditions that were once widespread.

Respiration Process by which organisms make energy available for a range of processes by oxidizing an organic substrate.

Reverse Tricarboxylic Acid Cycle A CO_2-assimilation pathway in some chemolithotrophic and photolithotrophic organisms (q.v.). Formally the reverse of the Tricarboxylic Acid Cycle of many aerobic organisms.

Rubisco Ribulose 1,5 bisphosphate carboxylase/oxygenase is the enzyme that catalyses the fixation of CO_2 during photosynthesis. It is probably the most abundant protein in nature.

Scaling Analysis of the functional and morphological consequences of changes in size in otherwise similar animals (see also **Similarity**).

Selective environment The set of environmental characteristics that influence the survival and reproductive success of a particular species or population of organisms. Selective environment represents a subset of all the physical and biotic environmental variables in an organism's surroundings.

Self-regulation An emergent property of a system: the ability to counteract perturbations and gradual changes in external forcing, and hence to maintain a dynamic stable state.

SETI The Search for Extraterrestrial Intelligence, an effort to detect extraterrestrial civilizations by listening for radio signals that are either being deliberately beamed to Earth, or are inadvertently transmitted from another planet.

Sibling species Two very closely related species, generally with the implication that they have separated relatively recently.

Sierran Pertaining to a mountain range with jagged peaks.

Similarity Analysis of quantities which remain independent or weakly influenced by scaling (q.v.) changes in size or equivalent factors.

Speciation The process of origin of a species. Often taken to imply lineage splitting (cladogenesis), although this is not necessarily entailed by the term itself.

Sputnik The first artificial Earth satellite which was launched by the former Soviet Union on 4 October 1957.

Stadial An interval of cold climate within a generally cold stage of the Pleistocene, e.g. the Younger Dryas (cf. **Interstadial**).

Stenotopic Pertaining to a narrow diet, habitat, or other type of ecological tolerance. An animal with a stenotopic diet is one whose digestive physiology may require it to eat a smaller range of foods than another species. The ant-eating aardvark, for example, has a stenotopic diet (ants and termites) but is eurytopic in its habitat, as its food occurs in a wide variety of environments. See **Eurytopic**.

Stepping-stone dispersal A series of comparatively short dispersal events that allow a species to reach a distant area.

Stoma (pl., stomata) Pores on the surface of plants, involved in gas exchange (inward movement of carbon dioxide, outward diffusion of water vapour and oxygen); aperture controlled by two guard cells.

Stomatal density The number of stomatal pores (stomata) per square mm.

Stomatal index The ratio of the number of stomata to the number of epidermal cells plus stomata, within a unit area of the plant (leaf) surface.

Stomatal ratio The ratio of the stomatal density of the nearest living equivalent divided by that of a given fossil.

Subduction Process in which a tectonic plate falls back into the Earth's interior. Subduction zones are typically characterized by volcanism above the sinking slab of old plate.

Suines Members of the Suina, a group of artiodactyl mammals including pigs, peccaries and hippos.

Suture zone A geographical area (usually long and narrow) where hybrid zones of several species or subspecies coincide, as a result of colonization from similar refugial areas.

Sweepstakes dispersal The colonization of an area by a rare, chance event. Frequently used for the crossing of a sea barrier between areas of land. The implication is that there is a chance element to which species cross and which do not.

Sympatry When the ranges of two or more species or populations overlap.

Taeniodonts A small group of North American Palaeocene to Eocene herbivorous mammals, of uncertain affinity, that ranged up to pig-sized.

Tectonic Pertaining to the movement of Earth's crust, including continental and ocean plate drift (plate tectonics), mountain-building and uplift, and the development of faults due to earthquakes.

Tectonics The large-scale processes that collectively deform the Earth's crust.

Temperature Average kinetic energy of molecules of a substance.

Tethys Sea A vast seaway filling the east–west depression which developed in the Mesozoic between Laurasia and Gondwanaland, and covering southern Europe, North Africa, the Near East and Himalayan region. Its western part was the precursor of the Mediterranean Sea.

Thermohaline circulation Global ocean circulation, driven by differences in the density of the seawater which is controlled by temperature (thermal) and salinity (haline).

Tillodonts A group of Palaeocene to Eocene herbivorous mammals, of uncertain affinity, that ranged up to pig-sized.

Transduction DNA transfer to host prokaryotes by bacterial viruses.

Transformation Transfer of DNA from the environment into the host cell performed by the host cell.

Transhumance The seasonal pasturing, often at summer grazing grounds in the mountains, of herds of domestic sheep, goats, cattle and other livestock by herders who leave their agricultural villages for weeks or months at a time to guard and care for the animals.

Ultraviolet radiation Radiation less than visible radiation, but longer than X-rays (10–400 nm).

Ungulates An informal term covering various groups of hoofed mammals.

Upwelling Replacement of surface waters by nutrient-rich water from greater depth in a body of water.

Uraninite Ore deposits containing uranium oxide (UO_2). Uranium is readily oxidized to higher oxidation states, indicating that the minerals formed when atmospheric oxygen was scarce.

Variability selection The adaptive process of biased reproduction and survival of genes and genetic lineages through a protracted period of highly variable selective environments. This process is a variant of natural selection in which certain genetic variations and gene combinations are favoured to the extent that they enhance the adaptive versatility of a lineage of organisms through an interval of novel and increasingly diverse adaptive settings.

Vicariance Division of a species range into two separate populations by the erection of a barrier.

Viscosity (of a fluid) A measure of the response of the fluid to a shear stress. Viscosity is equivalent to the 'stickiness' of the fluid; treacle is more viscous than water. Viscosity determines the friction drag force acting on a body in the fluid, and is also essential for the generation of aerodynamic lift.

Visible radiation Radiation with wavelengths between 400 and 750 nm, roughly the range for human vision.

Weathering (chemical) The transformation of minerals in rocks to soluble ions and clay, mainly through the activity of organisms.

Weight An object's mass times the local value of gravitational acceleration (1 G on Earth).

Index

(Page numbers in italics refer to figures and tables)